How to Be a Quantum Mechanic

How to Be a Quantum Mechanic is an introduction to quantum mechanics at the upper-division level. It begins with wave-particle duality and ends with a brief introduction to the Dirac equation. Two attitudes went into its writing: examples are the best way to get into a subject, and numbers and equations do not always sum to understanding—results need exploration.

Charles Wohl taught for 40 years at the University of California, Berkeley. He earned his PhD at Berkeley in experimental elementary-particle physics in the group led by Luis Alvarez.

How to Be a Quantum Mechanic

Charles G. Wohl

CRC Press
Taylor & Francis Group
Boca Raton London New York

CRC Press is an imprint of the
Taylor & Francis Group, an **informa** business

First edition published 2023
by CRC Press
6000 Broken Sound Parkway NW, Suite 300, Boca Raton, FL 33487-2742

and by CRC Press
4 Park Square, Milton Park, Abingdon, Oxon, OX14 4RN

CRC Press is an imprint of Taylor & Francis Group, LLC

Library of Congress Cataloging-in-Publication Data

Names: Wohl, C. G., author.
Title: How to be a quantum mechanic / Charles G. Wohl.
Description: First edition. | Boca Raton : CRC Press, 2022. | Series:
Frontiers in physics | Includes bibliographical references and index. |
Summary: "These lecture notes comprise an advanced undergraduate course
in quantum mechanics as taught by Charles G. Wohl for over 30 years at
the University of California, Berkeley. Each chapter covers a major
subject in quantum mechanics, beginning with an accessible introduction
and unfolding in subsections to signpost the reader's progression
through the topic. And, because examples are the best way to get into a
subject, every chapter ends with a series of problems-over 175 total in
the book-to provide plenty of hands-on practice in calculating. Targeted
to upper-division physics students and lecturers, this textbook and its
worked examples will teach students how to think like a quantum
mechanic"-- Provided by publisher.
Identifiers: LCCN 2021061434 (print) | LCCN 2021061435 (ebook) | ISBN
9781032256030 (hardback) | ISBN 9781032256023 (paperback) | ISBN
9781003284185 (ebook)
Subjects: LCSH: Quantum theory.
Classification: LCC QC174.12 .W633 2022 (print) | LCC QC174.12 (ebook) |
DDC 530.12--dc23/eng20220412
LC record available at https://lccn.loc.gov/2021061434
LC ebook record available at https://lccn.loc.gov/2021061435

ISBN: 978-1-032-25603-0 (hbk)
ISBN: 978-1-032-25602-3 (pbk)
ISBN: 978-1-003-28418-5 (ebk)

DOI: 10.1201/9781003284185

Typeset in CMR10
by KnowledgeWorks Global Ltd.

Arwen—not for, but because of.

CONTENTS

1. Strangest Things   1
1.1. Planck, Einstein, Compton, de Broglie   2
1.2. Neutron Interference   4
1.3. Photon Interference   7
1.4. Bohr and Hydrogen   9
    Problems   14

2. The Schrödinger Equation. Bound States   17
2.1. The Time-Dependent Schrödinger Equation   18
2.2. The Wave Function   21
2.3. The Time-Independent Equation. Energy Eigenstates   23
2.4. The Infinite Square Well   26
2.5. The Finite Square Well   28
2.6. The Delta-Function Well   34
2.7. Schrödinger in Three Dimensions   37
2.8. Two Important Energy Eigenstates   41
2.9. Qualitative Properties of Bound Energy Eigenstates   44
    Problems   45

3. Simple Approximations for Bound States   54
3.1. Dimensions and Scaling   55
3.2. Fitting Wavelengths in a Well   58
3.3. Guessing the Ground-State Wave Function   62
3.4. Useful Integrals   67
    Problems   68

4. Scattering in One Dimension   70
4.1. Particle Densities and Currents   71
4.2. Scattering by a Step   74
4.3. A General Rectangular Barrier   76
4.4. A Simple Rectangular Barrier   80
4.5. Designing with Rectangular Barriers   82
4.6. Thin Films and Light   86
4.7. Weak Tunneling   87
    Problems   90

5. Mathematical Formalism   93
5.1. Vector Spaces. Dirac Notation   94
5.2. States as Vectors   98
5.3. Operators   100
5.4. Successive Operations. Commutators   102
5.5. Operators as Matrices   104

5.6. Expectation Values   105

5.7. More Theorems   106

5.8. Revised Rules   108

Problems   109

6. The Harmonic Oscillator   111

6.1. The Classical Oscillator   112

6.2. The Quantum Oscillator: Series Solution   113

6.3. The Operator Solution   119

6.4. States as Vectors, Operators as Matrices   123

Problems   125

7. Uncertainty Relations. Simultaneous Eigenstates   130

7.1. Heisenberg Uncertainty Relations   131

7.2. The Schwarz Inequality   132

7.3. Proof of Uncertainty Relations   133

7.4. Fourier Transforms. Momentum Space   135

7.5. Time and Energy   138

7.6. When Operators Commute   142

Problems   144

8. Angular Momentum   147

8.1. Central Forces. Separation of Variables   148

8.2. Angular Momentum Commutation Relations   150

8.3. The Operator Solution   152

8.4. Certainty and Uncertainty   156

8.5. States as Vectors, Operators as Matrices (Again)   157

8.6. Differential Operators for Orbital Angular Momentum   159

8.7. Spherical Harmonics   161

8.8. Angular Momentum and the Oscillator   163

Problems   163

9. Hydrogen. The Isotropic Oscillator   165

9.1. The Effective Potential Energy   166

9.2. The Hydrogen Bound-State Energies   167

9.3. The Hydrogen Eigenfunctions   170

9.4. The Isotropic-Oscillator Energies   174

9.5. The Oscillator Eigenfunctions   177

9.6. Hydrogen and the Oscillator   177

Problems   180

10. Spin-1/2 Particles   183

10.1. Spinors. Eigenvalues and Eigenstates   184

10.2. The Polarization Vector   187

10.3. Magnetic Interactions and Zeeman Splitting   189

10.4. Time Dependence and Larmor Precession   191

10.5. Time Dependence and Magnetic Resonance   192

10.6. Stern-Gerlach Experiments   196

10.7. Polarization and Light   199

    Problems   201

11. Hyperfine Splitting. Two Angular Momenta. Isospin   204

11.1. Hyperfine Structure of the Hydrogen Ground State   205

11.2. The 21-cm Line and Astronomy   208

11.3. Coupling Two Spin-1/2 Particles   209

11.4. Coupling Any Two Angular Momenta   212

11.5. Clebsch-Gordan Coefficients   214

11.6. Particle Multiplets and Isospin   216

    Problems   218

12. Cryptography. The EPR Argument. Bell's Inequality   223

12.1. Quantum Cryptography   224

12.2. The EPR Argument   226

12.3. Bell's Inequality   227

    Problems   231

13. Time-Independent Perturbation Theory   233

13.1. The Nondegenerate Recipes   234

13.2. Examples of Nondegenerate Theory   236

13.3. The Nondegenerate Derivations   239

13.4. The Degenerate Recipe   241

13.5. Two Selection Rules. A Useful Relation   243

13.6. The Stark Effect in Hydrogen (Strong-Field Case)   245

13.7. Hydrogen Fine Structure: Experiment   248

13.8. Hydrogen Fine Structure: Theory   250

13.9. Atomic Magnetic Moments   255

13.10. The Zeeman Effect in Hydrogen (Weak-Field Case)   257

    Problems   259

14. Identical Particles   263

14.1. Electrons in a Box   264

14.2. Electrons in an Atom: the Periodic Table   267

14.3. Electrons in an Atom: More Pauli   273

14.4. Two-Electron Symmetries   275

14.5. The Helium Ground State   276

14.6. The Electron-Electron Repulsion Integral   281

14.7. Helium Excited States. Exchange Degeneracy   283

14.8. Fermions and Bosons. How to Count. Symmetries 288

Problems 291

15. Time-Dependent Perturbations. Planck and Einstein 296

15.1. Sudden Changes 297

15.2. Time-Dependent Perturbation Theory 298

15.3. Magnetic Resonance (Again) 300

15.4. Hydrogen in an Electromagnetic Wave 302

15.5. Averaging over Polarizations and Frequencies 305

15.6. The Boltzmann Factor 307

15.7. Planck's Oscillators 308

15.8. Black-Body Radiation 311

15.9. Einstein's A and B Coefficients 313

15.10. Decay Lifetimes 314

Problems 316

16. Scattering 320

16.1. Solid Angle 321

16.2. Classical Particle Scattering. Rutherford 322

16.3. The Scattering Amplitude 328

16.4. The Born Approximation 334

16.5. Kinematics 337

16.6. Partial-Wave Theory 339

16.7. Partial-Wave Examples 344

16.8. Scattering of Identical Particles 348

Problems 350

17. The Dirac Equation 356

17.1. Dirac at Play 357

17.2. Spin! 360

Problems 362

Sources of Quotes 365

Index 367

PREFACE

This text is an introduction to quantum mechanics at the upper-division level. The usual courses in physics and mathematics a physics student takes in the first two years of college should be sufficient preparation. Two quotations express the principal attitudes that informed the writing of the text: "One must know concrete instances first, for ... one can see no farther into a generalization than just so far as one's acquaintance with particulars enables one to take it in" (William James), and "The purpose of computing is insight, not numbers" (Richard Hamming). Examples are the best way into a subject, and numbers and equations by themselves often do not sum to understanding.

The rest of this preface consists of some generalities, a run-through of the chapters, a word of advice for students, an example for lecturers, and acknowledgements.

Generalities

Separate major subjects get separate chapters. Each chapter begins with a page of introduction. Most sections of chapters are further divided into titled subsections, as signposts to where we are going. Important equations are boxed. There are about 130 figures; unlike a Dirac, most of us do not think entirely algebraically. There are about 175 problems, each with a title to tell what it is about; in this, I follow French and Taylor, *An Introduction to Quantum Physics*. There are no equation numbers; in this, I follow the 600 pages of Bracewell, *The Fourier Transform and Its Applications*. Instead, at some slight cost in length, I usually repeat what is needed where it is needed. Hats go on operators that represent observables ($\hat{H}$, $\hat{p}$, etc.), to distinguish them from the observables themselves. A short list of textbooks I am familiar with, from a large literature, follows this preface. A few more specialized books are cited where relevant. A scattering of quotations, mostly in the early chapters, from Planck, Einstein, Heisenberg, Dirac, and others, aims to add some color and insight.

Chapters (Skimming is recommended here.)

• A text on quantum mechanics ought to begin with some of the mysteries. Chapter 1 describes a double-slit experiment with neutrons, an interferometer experiment with photons (with "quantum seeing in the dark"), and the Bohr model of hydrogen.

• Chapter 2 introduces the Schrödinger equations and provisional interpretative rules of the theory; detailed formalism can await some experience with what the formalism addresses. So examples, examples, examples! The time-independent Schrödinger equation is solved for the energy eigenstates of the one-dimensional infinite and finite square wells, the delta-function well, the two- and three-dimensional infinite square wells, the infinite spherical well when the angular-momentum is zero, and the ground states of the harmonic oscillator and the hydrogen atom. The problems introduce other wells. Most of these "models" are put to use in later chapters.

• Following the exact solutions of Chapter 2, Chapter 3 develops three methods to get approximate energy eigenvalues when getting exact solutions is difficult or impossible. One method gets eigenvalue estimates and scaling rules using just the parameters in the Schrödinger equation, not the equation itself. Another method fits position-dependent de Broglie wavelengths into a well—exactly first-order WKB, but with less mathematics. The third method starts with a guess of the form of the ground-state wave function, one with parameters to be varied for a best result—the variational method. As usual, perturbation methods are left until much later.

• Chapter 4 is about finding the fractions of particles reflected and transmitted when a plane wave of them is incident normally upon an arbitrary series of "rectangular" potential-energy barriers and wells. The results, expressed in terms of wave numbers, are identical to those for light incident normally on the designer thin films on lenses, screens, mirrors, and other optical devises. Finding common solutions across fields multiplies understanding. Waves are waves!

• Chapter 5 introduces some formal theory: Dirac notation, states as vectors, properties of operators for physical observables, operators as matrices, commutators of operators, expectation values, theorems on conservation laws and the time dependence of expectation values, and correspondences with classical relations with the Ehrenfest relations and the virial theorem.

• Chapter 6 is about the harmonic oscillator. The eigenvalue problem is solved using a series solution of the Schrödinger equation, and then is solved all over again using an operator algebra based on the formalism of Chapter 5. States appear as vectors, operators as matrices.

• Chapter 7 starts with a derivation of the uncertainty relations associated with non-commuting operators, then introduces position-momentum and time-energy Fourier transforms, and finally considers sets of operators that *do* commute with one another, a requirement for them to have simultaneous eigenstates.

• Chapter 8 is about angular momentum. The eigenvalue problem is solved along the same lines as the operator solution of the oscillator problem. Half of the solutions—the spherical harmonics—are the angular solutions of the Schrödinger equation for central forces. The other half are not solutions of that equation, but Nature likes them anyway.

• In Chapter 9, the radial equations for hydrogen and the isotropic oscillator are solved.

• Chapter 10 is about spin-1/2 particles and their interactions with magnetic fields—the Zeeman effect, spin precession, magnetic resonance, and Stern-Gerlach experiments. Magnetic resonance is the first example with an interaction that is time dependent. Spin 1/2 is a lovely little two-state quantum system, one we see much of in the rest of the text. The polarization states of spin-1/2 particles are compared with those of light.

• Chapter 11 is about how the angular-momentum states of a system are related to the angular-momentum states of its constituents. It begins with the hyperfine splitting of the ground state of hydrogen and the 21-cm line used in radio astronomy, and ends with calculations of branching fractions of elementary particles.

• Chapter 12 is about entangled states of two spin-1/2 particles. The subjects are quantum cryptography, the Einstein-Podolsky-Rosen argument, which insists that physical properties are real whether or not they are measured, and Bell's test for deciding experimentally between local-realistic theories and quantum mechanics.

• Chapter 13 is about time-independent perturbations. The ever finer shifts and splittings of hydrogen energy levels and spectra, as revealed by experiment, were a testing ground for quantum theory—and nearly the whole chapter is about calculating perturbations due to relativity, spin-orbit interactions, and external electric and magnetic fields in hydrogen.

• Chapter 14 is about identical particles. The Pauli exclusion principle is used to explain the stacking up of the conduction electrons in a block of metal, the structure of the periodic table of the elements, the angular-momentum states of the ground states of the elements, and the structure of the energy levels of the two-electron atom, helium. Then the larger classes of particles, the fermions and bosons, are introduced, and the symmetries they obey and how to count states in three different ways are explained.

• Chapter 15 is about perturbations with time dependence. First-order theory for magnetic resonance between spin-up and -down states is compared with the exact result from Chapter 10. Then the theory is applied to the analogous electromagnetic-wave resonance between atomic states, and the results are averaged over random polarizations and frequencies. This sets up sections on Planck's quantized oscillators, black-body radiation, Einstein's A and B coefficients, and a calculation of the lifetime $\tau = 1/A$ of a hydrogen excited state.

• Chapter 16 is about the angular distributions of particle-particle elastic scattering in three dimensions. One approach starts by deriving an integral equation for the scattering amplitude, but the equation is only useful if we know the interaction potential energy. A more open-ended approach is to formally break down the full scattering amplitude into "partial-wave amplitudes," one for each spherical harmonic in an expansion of the full amplitude. Then the partial amplitudes can be deduced from experimental angular distributions. The chapter concludes by discussing the remarkable differences between scattering angular distributions of identical particles and of non-identical particles.

• Chapter 17 is a very short introduction to Dirac's relativistic theory for spin-1/2 particles. It only goes far enough to show where spin-1/2 comes from.

For students

There are various ways you can do physics: You can build experimental apparatus, run experiments, analyze data, write programs, ... But to learn how to calculate you have to calculate. You can't learn to play a piano by simply listening to music and reading sheet music. You have to put hands to keyboard—you learn up your arms in some alchemy of brain, nerve, and muscle that slowly changes you. And you can't learn to calculate by simply listening to lectures or reading a text. You have to put pencil to paper (lots of it), do problems, try to recreate something from the text while away from the text, try to calculate something new. You learn up your arm here too.

For lecturers

Following is an abbreviated example, simple but not in any of the introductory texts I know of. Familiarity with the usual notation for orbital, spin, and total angular-momentum quantum numbers is assumed. The example is not about iron as such, or even about atoms; rather it is about seeing that sometimes complicated problems have special cases where the solutions come almost for free. The example understood, one could find the $^{2S+1}L_J$ ground states of the p through p^6 elements, 30 elements in all, in three or four minutes. Of course, you may already know all this.

The $^{2S+1}L_J$ ground state of iron—Iron is element 26 in the periodic table. The subshell structure of its ground state is $1s^2\, 2s^2\, 2p^6\, 3s^2\, 3p^6\, 4s^2\, 3d^6$. The incompletely filled $n\ell = 3d$ subshell has six electrons (10 would fill it). Following rules of Pauli and Hund, we put six arrows in a table, with $\uparrow$ and $\downarrow$ for spin components $m_s = +\frac{1}{2}$ and $-\frac{1}{2}$. There are 210 ways to put six arrows in this table consistent with Pauli, but this one does the job:

$m_\ell =$	+2	+1	0	−1	−2	$M_L = \sum m_\ell$	$M_S = \sum m_s$	$^{2S+1}L_J$	μ/μ_B
	$\uparrow\downarrow$	$\uparrow$	$\uparrow$	$\uparrow$	$\uparrow$	+2	+2	$^5D_4$	−6

Putting two up or two down arrows in the same orbital m_ℓ bin would violate Pauli's exclusion principle: No two of the $3d$ electrons can have both of m_ℓ and m_s the same. Hund's first rule (as used here) is: Make $M_S = \sum m_s$ as large as possible—so use as many up-arrows as you can. Hund's second rule (if there is still a choice) is: Make $M_L = \sum m_\ell$ as large as possible—so put the arrows as far to the left as you can. Then $S = M_S = 2$ and $L = M_L = 2$. (Why?) Hund's third rule is: The total J is $L + S$ or $|L - S|$ as the subshell is more or less than half filled. So $J = L + S = 4$ for iron, and the ground state is $^{2S+1}L_J = {}^5D_4$. See Sec. 14·2.

The magnetic moment μ of iron—You do this one. The magnitude of the magnetic moment of a $^{2S+1}L_J$ state in units of the Bohr magneton μ_B is the expectation value of $M_L + 2M_S$ in the state $|L, S; J, M_J\rangle$ when $M_J = +J$. So, without any recourse to the Landé g factor or to any pencil-pushing at all, explain why the magnetic moment of iron in its ground state is $\mu = -6\mu_B$. And why for the infinite half of hydrogen states with $J = L + \frac{1}{2}$ (the $^2S_{1/2}$, $^2P_{3/2}$, $^2D_{5/2}$, ... states), $\mu = -(L+1)\mu_B = -\mu_B, -2\mu_B, -3\mu_B, \ldots$ See Sec. 13·9.

Again, this is not about iron, or even just about getting easy answers. It is about looking to see if there *are* easy answers—for the insight, and for a good way to start at getting all the answers.

Acknowledgements

I am indebted to the Physics Department of the University of California, Berkeley for allowing me to teach for 40 years. I should first of course thank the students, and especially those who asked questions in class and enlivened office hours. By the time I retired, perhaps 80% of this text was, in at least rough draft, available to students on-line. I had many fine teaching assistants, but shall name three who taught with me (and occasionally to me) for from two to five semesters—Jason Zimba, Damon English, and Erik Aldape. I must also thank Donna Sakima, Claudia Trujillo, and Kathy Lee, the advisors for students taking physics; I too over many years sought their advice on all kinds of administrative matters. One cannot write a text dense with equations without feeling enormous gratitude to the creator of TeX, Donald E. Knuth. Finally, two friends, Drs. Jerry Anderson and Piotr Zyla, read large parts of the text at various stages of the writing and made many valuable comments. Any remaining errors are mine. Please let me know of any you find.

Charles Wohl
January 2022
cgwohl@berkeley.edu

REFERENCES

Most of these books listed here are quite old, but there is not much that is out of date in any of them. Some of them I have little more than skimmed, but I have learned at least a little from every one of them. As you learn, you should from time to time browse a few books for a wider range of approaches, examples, and problems. A few more specialized books are cited in the text where relevant.

J.-L. Basdevant, J. Dalibard, *The Quantum Mechanics Solver*, Springer, 2000.

P.A.M. Dirac, *The Principles of Quantum Mechanics*, 4th ed., Oxford University Press, 1958.

Richard F. Feynman, Robert B. Leighton, Matthew Sands, *The Feynman Lectures on Physics*, Vol. III, Addison-Wesley, 1965.

A.P. French, Edwin F. Taylor, *An Introduction to Quantum Physics*, W.W. Norton & Company, 1978.

George Greenstein, Arthur G. Zajonc, *The Quantum Challenge*, 2nd ed., Jones and Bartlett Publishers, 2006.

Robert Gilmore, *Elementary Quantum Mechanics in One Dimension*, Johns Hopkins University Press, 2004.

David J. Griffiths, Darrell F. Schroeter, *Introduction to Quantum Mechanics*, 3rd ed., Cambridge University Press, 2018.

Gerhard Herzberg, *Atomic Spectra and Atomic Structure*, Dover Publications, 1945.

L.D. Landau, E.M. Lifshitz, *Quantum Mechanics,, Non-Relativistic Theory*, 2nd ed., Pergamon Press, Oxford, 1965.

The first of Mandl's texts was adapted from lectures given to experimenters, the second is a standard textbook.

F. Mandl, *Quantum Mechanics*, 2nd ed., Butterworths, 1957.

—— *Quantum Mechanics*, John Wiley & Sons, 1992.

David S. Saxon, *Elementary Quantum Mechanics*, Holden-Day, 1968.

Ramamurti Shankar, *Principles of Quantum Mechanics*, 2nd ed., Plenum Press, 1980.

C.W. Sherwin, *Introduction to Quantum Mechanics*, Holt-Dryden, 1959.

John S. Townsend, *A Modern Approach to Quantum Mechanics*, 2nd ed., University Science Books, 2000.

G.K. Woodgate, *Elementary Atomic Structure*, 2nd ed., Clarendon Press, 1986.

Secondary

Ronald N. Bracewell, *The Fourier Transform and Its Applications*, 3rd ed., McGraw Hill, 2000.

Max Jammer, *The Conceptual Development of Quantum Mechanics*, McGraw-Hill Book Company, 1966.

Abraham Pais, *Subtle is the Lord*, Oxford University Press, Oxford, 1982;

—— *Inward Bound*, Clarendon Press, 1986.

SOME CONSTANTS AND RELATIONS

speed of light in vacuum $c = 2.9979 \times 10^8$ m/s

Planck constant $h = 6.6261 \times 10^{-34}$ J s

reduced Planck constant $\hbar = 1.0546 \times 10^{-34}$ J s $= 6.5821 \times 10^{-16}$ eV s

electron charge magnitude $e = 1.6022 \times 10^{-19}$ C

$\qquad$ 1 eV $= 1.6022 \times 10^{-19}$ J

electron mass $m = 9.1094 \times 10^{-31}$ kg$= 0.51100$ MeV$/c^2$

proton mass $m_p = 1.6726 \times 10^{-27}$ kg$= 938.27$ MeV$/c^2$

permittivity of free space $\epsilon_0 = 8.8542 \times 10^{-12}$ C^2/N m^2

$\qquad$ useful constant $k \equiv 1/4\pi\epsilon_0 = 8.9876 \times 10^9$ N m^2/C^2

fine-structure constant $\alpha \equiv ke^2/\hbar c = 7.2974 \times 10^{-3} = 1/137.04$

Bohr radius $a_0 \equiv \hbar^2/mke^2 = 0.52918 \times 10^{-10}$ m

Bohr magneton $\mu_B \equiv e\hbar/2m = 5.7884 \times 10^{-5}$ eV/T

hydrogen ground-state $E_1 = -mk^2e^4/2\hbar^2 = -13.606$ eV $(-13.599$ eV with reduced mass)

$$i\hbar\frac{\partial\Psi}{\partial t} = \hat{H}\Psi, \quad \Psi(\text{space}, t) = \psi(\text{space})T(t), \quad \hat{H}\psi_E = E\psi_E, \quad T_E(t) = e^{-iEt/\hbar}$$

$$\hat{H}\psi = -\frac{\hbar^2}{2m}\frac{d^2\psi}{dx^2} + V(x)\psi$$

$$\hat{H}\psi = -\frac{\hbar^2}{2m}\left[\frac{1}{r}\frac{\partial^2(r\psi)}{\partial r^2} + \frac{1}{r^2\sin\theta}\frac{\partial}{\partial\theta}\left(\sin\theta\frac{\partial\psi}{\partial\theta}\right) + \frac{1}{r^2\sin^2\theta}\frac{\partial^2\psi}{\partial\phi^2}\right] + V(r)\psi$$

$$\infty \text{ square well, } L \text{ wide}: E_n = n^2\frac{\pi^2\hbar^2}{2mL^2}, \ n = 1, 2, 3, \ldots, \ \psi_n(x) = \sqrt{\frac{2}{L}}\sin\left(\frac{n\pi x}{L}\right)$$

$$\text{1D harmonic oscillator}: E_n = (n + \tfrac{1}{2})\hbar\omega, \ n = 0, 1, 2, \ldots, \ \psi_0(x) = \left(\frac{m\omega}{\pi\hbar}\right)^{1/4}e^{-m\omega x^2/2\hbar}$$

$$\text{hydrogen bound states}: E_n = -\frac{1}{n^2}\frac{mk^2e^4}{2\hbar^2}, \ n = 1, 2, 3, \ldots, \ \psi_{100}(r) = \frac{1}{\sqrt{\pi a_0^3}}e^{-r/a_0}$$

hydrogen wave functions $\psi_{n\ell m}(r, \theta, \phi)$ for $n = 1, 2, 3$ 173

periodic table of the elements 268

subshell structures, $^{2S+1}L_J$ states, and ionization energies of the elements 271

1. STRANGEST THINGS

1.1. *Planck, Einstein, Compton, de Broglie*

1.2. *Neutron Interference*

1.3. *Photon Interference*

1.4. *Bohr and Hydrogen Problems*

We trace two themes—wave-particle duality, and atomic spectra and structure—that ran through the years of "modern physics" (roughly 1900 to 1925), an era that culminated in the discovery of quantum mechanics. We assume that the reader already knows something of this physics, and is also familiar with the wave optics of interference and diffraction.

The first theme: At the heart of quantum mechanics is the bizarre matter of wave-particle duality. (Webster's: "bizarre ... involving sensational contrasts or incongruities.") After a very brief sketch of early developments, two modern examples are given: the interference of neutrons, one at a time, by a double slit; and the interference of photons, one at a time, in an interferometer. The discussion includes "interaction-free measurements," a quantum-mechanical way to see in the dark.

The second theme: Ernest Rutherford gave us the picture of the atom as roughly like a tiny solar system, with negative electrons surrounding a positive, massive nuclear sun. However, on classical principles Rutherford's atom seemed to be completely unstable—yet the world is made of extremely stable atoms. Niels Bohr made a model of the simplest atom, hydrogen, that involved quantized planetary orbits. The stability of these orbits violated classical laws, but the model gave the known spectral lines and predicted an infinity of new lines. Although Bohr's orbits have been replaced by the nebulous states of quantum mechanics, his model was a giant step forward.

In the following, there are occasional historical notes and quotes. In any physics text (this one included), such notes scarcely begin to convey the confusion and controversy that often accompany major discoveries. And they often also fail to mention the developers of scientific instruments, without which experimenters could not have made their discoveries and theorists would have had little to ponder. For a proper history, one might start with

Max Jammer, *The Conceptual Development of Quantum Mechanics*

Abraham Pais, *Subtle is the Lord*, and *Inward Bound*

These often require some understanding of the physics of the times and so can be difficult reading, but they are full of interesting and accessible passages. Pais, a prominent physicist himself, knew many of the founders of quantum mechanics.

1

1·1. PLANCK, EINSTEIN, COMPTON, DE BROGLIE

This first section is a very brief *review* of one theme running through modern physics ($\approx$ 1900-1925)—the particle nature of waves and the wave nature of particles. Here History will be greatly simplified. The next two sections then give detailed examples.

(a) Planck—We see by light of wavelengths from about 400 to 700 nm (many birds and insects see over somewhat different ranges). The infrared warms us, the ultraviolet can burn us. The Sun's radiations of course result from its high surface temperature. All objects continuously absorb and emit electromagnetic radiation. At room temperatures, the emissions peak in the far infrared. The intensity of the radiation is a function of the wavelength λ, the absolute temperature T, and the surface properties of the object. A good absorber of radiation, such as a black surface, is a good emitter. A small hole into a cavity in a block of material is a near-perfect absorber and emitter—a universal "black." The walls of the cavity and the "black-body radiation" that fills the cavity are in thermal equilibrium. If the block is heated to incandescence, the emissions from the hole to the outside make it the brightest spot on the body. The Sun is scarcely black, but it radiates nearly as a black body with a surface temperature of about 5800 K.

The measurements of black-body radiation required the development of instruments that see at wavelengths at which our eyes cannot. By 1900, the spectral distribution of emissions from a hole as a function of λ and T was measured with considerable accuracy. However, no one could derive the distribution from theory.

In 1900, Max Planck made a guess at the mathematical function that would fit the spectral distribution, and then tried to derive that function from physical principles. He considered the thermal equilibrium of the cavity radiation with the cavity walls, and he modeled the matter of the walls as harmonic oscillators. His success required making *two* remarkable assumptions: (1) The energies of the oscillators are *quantized* in elements of magnitude $\epsilon = h\nu$, where ν is the oscillation frequency of the oscillators and h is a new constant, its value found by fitting the spectral distribution. (The energies of classical oscillators are of course continuous.) (2) In equilibrium, the energies are statistically distributed over the oscillator states in a new and unheard of way. "... the only justification for [his assumptions] was that they gave him what he wanted. His reasoning was mad, but his madness has that divine quality that only the greatest transitional figures can bring to science" (Pais, *Subtle*, p. 371). Thus was born the constant h that rules over quantum mechanics. We come to this calculation in a late chapter.

And a very small constant h is indeed. Beginning 20 May 2019, h is *defined* by the Committee on Data for Science and Technology (CODATA) to be *exactly*

$$h = 6.626\,070\,15 \times 10^{-34} \text{ joule seconds (J s \quad or \quad J/Hz)}$$

(see the inside front cover). We shall usually use $\hbar \equiv h/2\pi$ ("h-bar") instead of h,

$$\boxed{\hbar = 1.054\,571\,817\cdots \times 10^{-34} \text{ J s} = 6.582\,119\,569\cdots \times 10^{-16} \text{ eV s} .}$$

The first value here is in SI units, joule seconds. However, energies in atomic, nuclear, and particle physics are usually given in electron volts (eV) or multiples of 10^3 thereof (keV, MeV, GeV, TeV). A particle with charge q coulombs accelerated through a potential difference of V volts acquires a kinetic energy of qV joules. A particle with the electronic charge e accelerated through one volt acquires a kinetic energy of 1 eV $= 1.602\,176\,634 \times 10^{-19}$ J (e too now has an exact value); this then is the conversion factor between joules and electron volts. Examples: the ionization energy of hydrogen is 13.6 eV; the rest-mass energy $E = mc^2$ of the proton is 938.3 MeV; in 2016, the Large Hadron Collider (near Geneva) was colliding protons on protons at a center-of-mass energy of 13 TeV.

In terms of the fundamental dimensions of mass M, length L, and time T, the dimensions of $\hbar$, energy times time, are $(ML^2/T^2) \times T$. This may be rearranged as $L \times ML/T$, so that, dimensionally, $\hbar$ is

$$\boxed{\text{energy} \times \text{time} = \text{length} \times \text{momentum} = \text{angular momentum}.}$$

This dimensional equivalence will be apparent in expressions (to come) such as $e^{i(px-Et)/\hbar}$, because the arguments of transcendental functions have to be dimensionless.

(We have already used T for both temperature and the dimension of time. This presents no problem if you know the context. Example: A couple goes into a cafe late in the evening and asks at the counter, "Too late to buy two brews?" Say it aloud. Why is this relevant?)

(b) Einstein—"... at least until 1905, nobody in fact seems to have realized that Planck's was indeed 'a discovery comparable perhaps only to the discoveries of Newton'" (Jammer, *Development*, p. 23). Enter Einstein. In 1906 he wrote, "We must consider the following theorem to be the basis of Planck's radiation theory: the energy of a [Planck oscillator] can only take on those values that are integral multiples of $h\nu$; in emission and absorption the energy of a [Planck oscillator] changes by jumps which are multiples of $h\nu$."

In 1906, Planck himself might not have quite fully agreed with that statement (see below). Already in 1905, Einstein had taken Planck's theory a step farther. It concerned another obscure phenomenon, the photoelectric effect, in which light incident upon a clean metallic surface can eject electrons. Measurements were poor, but two things were known: (1) Weak blue light might eject electrons from a given metal, whereas intense red light might not. This was puzzling because the rate that energy is delivered by an electromagnetic *wave* is proportional to the intensity, independent of the frequency. (2) If electrons are ejected, the ejection begins as soon as light strikes the surface, even with weak light. This was puzzling because the energy of an incident *wave* is spread over the whole wave front, and it was hard to see how weak radiation could be concentrated instantly on a particular electron.

Suppose, Einstein said, that Planck's *exchanges* of energy are quantized because the energy in the electromagnetic radiation is itself quantized—in packets with energies $\epsilon = h\nu$. In this view, light is a hail of "photons" (as we now call the light quanta). Einstein's equation could not be simpler: The maximum kinetic energy of an ejected electron will be

$$K_{max} = h\nu - \phi = h(\nu - \nu_0) .$$

Here ϕ is the minimum energy needed to eject an electron from the metal (ϕ, called the work function, varies from metal to metal but is of order 3 eV); and $\nu_0 = \phi/h$ is the corresponding threshold frequency: If $\nu < \nu_0$, no electrons are ejected. And if ν_0 fell between ν_{blue} and ν_{red}, blue light would eject electrons when red light would not, because $\nu_{blue} > \nu_{red}$.

A century of experiments on interference, diffraction, and polarization had shown that light is a wave, not a hail of particles. In 1913, Planck himself, together with three other eminent physicists, in a letter of recommendation for Einstein, wrote: "That he may sometimes have missed the target in his speculations, as, for example, in his hypothesis of light quanta, cannot really be held too much against him ..." Nor should their disbelief be held too much against *them*—the first experiment to accurately confirm the relation $K_{max} = h\nu - \phi$ was not done until 1915. And not until 1923, when Arthur Holly Compton explained the scattering of x-rays by electrons by treating the radiation as photons, was the dual nature of light—wave *and* particle—finally uneasily accepted by (nearly) everyone.

(c) **Compton**—Compton's kinematical calculation treated a photon as a particle having energy E and momentum p, where

$$\boxed{E = h\nu \quad \text{and} \quad p = h/\lambda .}$$

The equation for p comes from the relativistic relation $E^2 = p^2c^2 + m^2c^4$: The photon is massless, so $p = E/c = h\nu/c = h/\lambda$, where $\lambda = c/\nu$ is the wavelength of the light. Notice how particle aspects E and p on the left sides of the equations are related to wave aspects ν and λ on the right. In the scattering of a photon by an initially stationary electron ($\gamma e \to \gamma e$), the photon loses some momentum, and so its wavelength λ' is then larger. An elementary exercise in conserving (relativistic) energy and momentum shows that

$$\lambda' - \lambda = \Delta\lambda = \frac{h}{mc}(1 - \cos\theta) .$$

Here m is the mass of the electron, and θ is the angle through which the photon is scattered. And how are those wavelengths before and after the scattering determined? By *wave diffraction* in crystals. In experimental sequence, the radiation is wave, particle, wave.

(d) **de Broglie**—In 1923, Louis de Broglie advanced another absurd idea. If waves sometimes exhibit particle-like behavior, perhaps particles sometimes exhibit wave-like behavior. And he proposed that the relation between particle and wave aspects was again

$$\boxed{p = h/\lambda .}$$

This gave a new way to look at the Bohr quantization rule for angular momentum L (1913; see Sec. 4). The allowed orbits are those for which an integral number of wavelengths fit round a circle. If $2\pi r = n\lambda$, where $n = 1, 2, 3, \ldots$, then $2\pi r = nh/p$, and the angular momentum is $L = rp = nh/2\pi = n\hbar$, which is Bohr's rule. Furthermore, de Broglie's hypothesis provided an explanation for patterns already beginning to be observed for electron beams scattered by crystals. The electrons are *diffracted* by the regular array of atomic sites in the crystal. Experiments soon showed that the scattering of both x rays and electrons by crystals made similar diffraction patterns.

1·2. NEUTRON INTERFERENCE

(a) **A double-slit experiment**—Neutrons are particles. A neutron is slightly more massive than a proton plus an electron: A free neutron decays to a proton, an electron, and an antineutrino ($n \to pe^-\bar{\nu}$) with an average lifetime of 880 s. Aside from ordinary hydrogen and the rare isotope ^{3}He, all stable nuclei have at least as many neutrons as protons. Therefore, aside from any hydrogen content, at least half the mass of all the stable matter we know of comes from neutrons. Neutrons are particles!

4

Figure 1 shows an experimental set-up used to investigate the *wave* nature of neutrons. A flux of neutrons from a nuclear reactor is collimated by slits S_1 and S_2 and directed upon a quartz prism. The prism fans out neutrons according to wavelength, just as an ordinary prism fans out light (but light would be bent upward, not downward—the neutrons *speed up* in the quartz). Slit S_3 then selects neutrons in a fairly narrow band of wavelength and directs them toward the double slit S_5.

The slits in S_5, shown magnified in Fig. 1(bottom), are 22 μm wide; the center-to-center distance is 126 μm. A tiny arrangement, but just visible with the naked eye. The slit widths and separation can be varied; the wire can be removed to make a single slit of variable width; and a jaw can be removed to leave a single sharp edge. The pattern produced by S_5 is measured at the far right by "stepping" the scanning slit S_4 across the pattern and counting neutrons behind the slit at each step for some length of time.

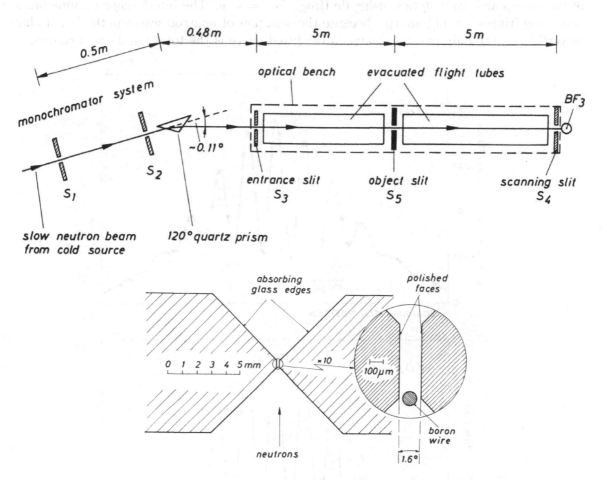

Figure 1. Top: Neutrons from a reactor are collimated by slits S_1 and S_2, refracted by a quartz prism, selected on momentum by S_3, and directed upon the object slit S_5. The patterns produced are measured at S_4. Note the changes of scale in the figure. Bottom: Looking down on the double-slit arrangement at S_5. The figures here and Fig. 2(a) are reproduced from R. Gähler and A. Zeilinger, "Wave-optical experiments with very cold neutrons," Am. J. Phys. **59** (1991) 316, with the permission of the American Association of Physics Teachers.

With $p = mv$ (the neutrons here are nonrelativistic), the de Broglie wavelength is

$$\lambda = \frac{h}{m}\frac{1}{v} = \frac{396}{v} \ ,$$

where λ is in nm and v is in m/s. The neutrons are made slow ("cold") by passing them through a bath of liquid deuterium at 25 K before they enter the set-up. Lower speed means lower momentum and larger wavelength. Otherwise λ would be too small to observe interference effects in a set-up of (barely) macroscopic dimensions. As it is, the wavelength, about 2 nm, is much smaller than the 400-to-700 nm of visible light.

Figure 2(a) shows the pattern at S_4 produced by the double slit at S_5. The data points are built up neutron by neutron. (The paper also shows the diffraction patterns of a single slit and a single edge.) The curve through the data points is calculated from the geometry of the set-up and *wave optics*, using de Broglie's $\lambda = h/p$. The interference minima are not zero—the fringes are not sharp—because the selection of neutron wavelengths is not sharp. Nevertheless, the only free parameter in the fitted curve is the total numbers of counts.

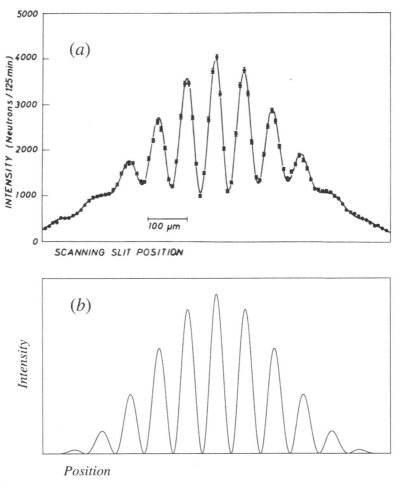

Figure 2. (a) The neutron interference pattern from the experiment, R. Gähler and A. Zeilinger, Am. J. Phys. **59** (1991) 316. (b) The double-slit pattern for light of a single sharp wavelength and the same ratio of slit separation to slit width as in (a).

6

Thomas Young using a double-slit arrangement to show that light was a wave phenomenon in 1800. Figure 2(b) shows the double-slit pattern produced by *light* of a single wavelength when the center-to-center distance between the slits is $126/22=5.73$ times the width of each slit, the same geometry as in (a). The interference fringes produced by two very narrow slits are modulated by the diffraction pattern of a single slit of finite width. This pattern is derived in elementary textbooks using Huygens wavelets, and in Sec. 7·4 using a Fourier integral. Aside from not having a sharp wavelength, the neutron interference pattern is exactly as it is for light in the same geometry. Neutrons are waves!

(b) What are neutrons?—This discussion—this explanation—has been in terms of waves, but neutrons are particles! The neutrons are produced in the fission of uranium atoms in the reactor. The rate at which neutrons pass through the set-up is such that nearly always one neutron has been counted at S_4 before the next one to reach the counter is even born in the decay of another uranium atom. Thus the patterns are produced one neutron at a time. If the scanning slit and counter were replaced by a pixelated detector covering the whole plane at S_4, each neutron would arrive at a single pixel, not spread out. Surely a particle that arrives at some one place must on its way to contributing to the double-slit pattern pass through only one slit or the other. But if that is so, how can the neutron then even "know" that the other slit exists? But just as surely, the pattern built up neutron by neutron requires the interference of *something* that involves the presence of both slits—the two-slit interference pattern is not remotely like the sum of two one-slit patterns of particles. *Neutrons are waves! Neutrons are particles!*

We are driven to conclude that neutrons (and all other "particles") have a dual nature, being neither purely particle nor wave. We have a theory—*Quantum Mechanics*—that predicts the results of all such experiments. But our thoughts cannot fully get round the duality of waves and particles. Richard Feynman said, "Nobody understands how it can be like that." Chapter 1 of Vol. III of *The Feynman Lectures* is all about this duality: "[It] is impossible, *absolutely* impossible, to explain [the duality] in any classical way ..."

Double-slit experiments with electrons and other particles—even with buckyballs (soccer-ball-like molecules made of 60 carbon atoms!)—have produced better fringe patterns. However, the experiment with neutrons described here is conceptually the simplest of them all, and the relatively primitive nature of the apparatus makes it the best one to use here.

1·3. PHOTON INTERFERENCE

(a) An interferometer experiment—A double slit selects two pieces out of a wave front (*wave-front division*) and spreads all degrees of interference—constructive to destructive—into fringes over a plane. An interferometer splits one beam into two beams (*amplitude division*), brings them back together, and there asks an "either/or" question.

Figure 3 shows a Mach-Zehnder interferometer. At the lower left (ignore the "downconverter" for a moment), a parallel beam of light of wavelength λ is incident upon a 50-50 beam splitter—half of the intensity of the light is reflected and half is transmitted. Each half beam is then reflected by a mirror toward a second beam splitter, where the beams interfere. In general, some of the light will go to detector $C1$ and some will go to $C2$. We might simply observe the illumination on screens placed at positions 1 and 2; or we might hook detectors up to loudspeakers and compare the loudness of two tones. To see circular fringes on a screen at $C1$ or $C2$, we might introduce a diverging lens in front of the first beam splitter; to see linear fringes, we might tilt one of the mirrors. But here we do neither.

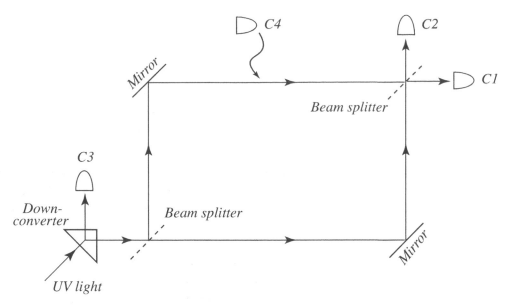

Figure 3. A Mach-Zehnder interferometer.

Suppose that the distance to $C1$ is the same along the two paths, or that it differs by an integral number of wavelengths. Then all the light goes to $C1$. This is because the lower path to $C1$ involves a transmission through the first beam splitter, a reflection from a mirror, and a reflection from the second beam splitter; and exactly the same things happen to the upper beam, although in opposite order. Therefore, the two beams arrive at $C1$ in phase and interfere constructively. The happenings along the two paths to $C2$ are *not* the same: The upper path reflects from, the lower path passes through, both beam splitters. Evidently, the difference between two reflections and two transmissions introduces what amounts to a path difference of half a wavelength between the two paths to $C2$ (see an optics text).

We can reduce the intensity of the incoming beam to the point where only one photon is in the interferometer at a time. We can know when a photon (particle language) enters the interferometer by placing an "ultraviolet down-converter," shown in Fig. 3, at the entrance. This device "splits" an ultraviolet photon into two visible photons, one of which goes to $C3$, while the other enters the interferometer. Let the two path lengths to $C1$ be equal or differ by an integral number of wavelengths; and set the detectors to "click" when a photon arrives. Then when $C3$ clicks so does $C1$.

We can ask *which* path to $C1$ a particular photon takes by placing a counter $C4$ in, say, the upper path. Then about half the time a photon enters the system, we get a click at $C4$, but no click at either $C1$ or $C2$; the photon took the upper path and was absorbed at $C4$. (There is no way to both record the arrival of the photon at $C4$ and have it continue *undisturbed* on its way.) The other half of the time, we get a click, with equal probabilities, at either $C1$ or $C2$, but no click at $C4$. The photon took the lower path to the second 50-50 beam splitter, but now it is equally likely to go to $C1$ and $C2$, because the upper path is now blocked by $C4$. The interference—getting all the counts at $C1$—has been destroyed because we looked to see which path was taken.

Adding $C4$ to the setup changed a "both-paths-at-once" to an "either/or." To get interference, we must not interfere. If we look, the photon takes either the upper path or the lower path. If we don't look, the photon, like the neutron in the double-slit experiment,

8

somehow knows about both paths. It follows the advice of the philosopher Yogi Berra, "When you come to a fork in the road, take it." But now we have the same kind of question we asked about a neutron in the double-slit experiment: Since the photon only ever *appears* at one place or another, how can it somehow take—or even know about—two paths, as the interference (when $C4$ is absent) seems to require? Is light particles or is it waves?

(b) Seeing in the dark—Interest in the interferometer arrangement just discussed was enlivened by A.C. Elitzur and L. Vaidman, in "Quantum Mechanical Interaction-Free Measurements," Found. Phys. **24** (1993) 987. Imagine a stock of bombs, each one having a trigger sensor so sensitive that the slightest touch or even a single stray photon will explode the bomb. Thus the bombs are stored in total darkness. However, some of the sensors have fallen out, and if a photon were to reach where a sensor should be, it would pass through undisturbed. Can the "good" bombs—those with sensors in place—be sorted out and saved? With the present arrangement, the answer is—about one-third of them can.

In total darkness, we place a bomb in our Mach-Zehnder interferometer, with the sensor—or where the sensor should be—in the upper path. The sensor of a good bomb blocks the upper path; the sensor of a bad bomb is absent and the upper path is open. We send in a photon, and three things can happen:

(1) The bomb explodes. As Elitzur and Vaidman put it, "the bomb *was* good." The photon took the upper path. An explosion occurs with a probability of 50% when the bomb is good. We hire another undergraduate research assistant.

(2) $C2$ clicks. This means that the bomb *is* good! The loss of constructive interference at $C1$ means that the upper path is blocked by a sensor. $C2$ clicks with a probability of 25% when the bomb is good. We quickly stop any more photons from entering the interferometer, because the next one might explode the bomb.

(3) $C1$ clicks. We can't tell whether the bomb is good or not. For a good bomb this happens with a probability of 25%; for a bad bomb (no sensor) it always happens. We send in another photon. If the bomb is good, eventually we either blow up the apparatus or we get a click at $C2$. But if $C1$ clicks for, say, 20 photons in a row, we become convinced that the sensor is missing and go on to test the next bomb.

We can identify all the bad bombs: Each one gives a long series of clicks at $C1$. And we can save one-third of the good bombs, because the ratio of clicks at $C2$ to explosions is 25-to-50 when the bomb is good. Getting a click at $C2$ tells us that the upper path is blocked, even though the photon did not take that path. We know that a sensor is there *even though no photon reaches a bomb we save*. With a more complicated arrangement, it is in theory even possible to save *all* of the good bombs, without touching any of them. See P. Kwiat, H. Weinfurter, and A. Zeilinger, "Quantum Seeing in the Dark," Sci. Amer. **275** (November 1996) 72.

1·4. BOHR AND HYDROGEN

(a) Instabilities—Another theme of modern physics involved atomic structure, particularly as revealed by atomic spectra. In 1911 Ernest Rutherford showed that an atom must be roughly like a little solar system, with negative electrons surrounding a tiny positive nuclear sun that carries nearly all the mass. However, it seemed to be impossible to construct a model in which the electrons are held in orbit around the nucleus by electromagnetic forces. But since the world around us is made of extremely stable atoms, either other unknown forces were in play, or classical physics was inadequate for the task.

Consider a *static* model with only electrostatic forces. Obviously, a single electron set down at rest near a nucleus will fall straight into it. And two electrons set down on opposite sides of a nucleus with charge $2e$ will suffer the same fate. In fact, the only stable, static arrangement of any number of point charges interacting only through Coulomb forces is one in which the charges are sitting on top of one another or have flown off to infinity. To see this, suppose there exists a static arrangement of Z electrons occupying distinct positions in space around a nucleus with charge Ze. If an electron at position P is nudged in any direction, it must, for stability, be pushed back toward P by the electrostatic forces due to all those other charges. This requires that the electric field due to those other charges point outward in every direction from P. Now draw a little closed (Gaussian) surface around P. There is a net flux of field out of the surface, and therefore, by Gauss's law, there is a positive charge at P. This is contrary to the assumption that the electron at P sits at an otherwise empty point in space. Field lines due to all those other charges flow *through* P.

What about a *dynamic* model? Systems of planets held in orbit about stars by that other inverse-square-law force, gravity, are certainly stable. But an electron circling a nucleus acts as a little antenna and radiates its energy away, spiraling into the nucleus in a tiny fraction of a second. To show this, we need a result from classical electromagnetism. The rate at which an accelerating charge e radiates away energy E is

$$\frac{dE}{dt} = -\frac{2}{3}\frac{ke^2}{c^3}a^2 \ .$$

Here $k \equiv 1/4\pi\epsilon_0$, ϵ_0 is the permittivity constant, c is the speed of light, and a is the acceleration of the charge. See, for example, E.M. Purcell and D.J. Morin, *Electricity and Magnetism*, 3rd ed., App. H. The radiating charge *loses* energy at this rate.

We begin with an electron in a circular orbit of radius r around a proton, and assume (as you could show after the following) that as the electron radiates it follows a nearly circular, spiral path inward. From $F = ma$, where $F = ke^2/r^2$ is the inward Coulomb force and $a = v^2/r$ is the centripetal acceleration, we get

$$\frac{ke^2}{r^2} = m\frac{v^2}{r}, \qquad \text{or} \qquad mv^2 = \frac{ke^2}{r} \ .$$

The potential energy is $V(r) = -ke^2/r$, where the zero of $V(r)$ is the electron at rest at infinity. The total energy, kinetic plus potential, of the orbiting electron is

$$E = K + V = \frac{1}{2}mv^2 - \frac{ke^2}{r} = -\frac{ke^2}{2r} \ .$$

Note that $K = -\frac{1}{2}V$, so that $E = \frac{1}{2}V = -K$. The same holds for planets in circular orbits.

To find how r varies with t, we take the time derivative of E, put that into the above equation for dE/dt, and use $a = F/m = ke^2/mr^2$. We get

$$\frac{dE}{dt} = +\frac{ke^2}{2r^2}\frac{dr}{dt} = -\frac{2}{3}\frac{ke^2}{c^3}a^2 = -\frac{2}{3}\frac{ke^2}{c^3}\left(\frac{ke^2}{mr^2}\right)^2 \ .$$

Rearranging the equality between the second and last expressions, we get

$$dt = -\frac{3}{4}\frac{m^2c^3}{k^2e^4}r^2\,dr \ .$$

Integrating to get the time τ it takes for the radius to fall from some initial value r_0 to any much smaller value, we get

$$\tau \simeq \frac{1}{4} \frac{m^2 c^3}{k^2 e^4} r_0^3 \; .$$

For $r_0 = 10^{-10}$ m, the approximate size of an atom, we get $\tau \simeq 10^{-10}$ s. Check the numbers.

(b) Bohr's model—We are not here so much interested in the Bohr model (1913) as we are in its results, because the energy levels derived from the model are the same we eventually get from quantum mechanics. We shall want to use those levels in examples before we derive them properly from quantum mechanics.

Niels Bohr had two things to explain—the stability of atoms, and the frequencies of light they emit and absorb. He succeeded with hydrogen (and with other single-electron ions such as singly ionized helium) by breaking some of the rules of classical physics. At the time, only a few spectral lines of hydrogen were known, and Johann Balmer, a Swiss school teacher, had found an empirical formula for their frequencies. Bohr was guided by this experimental data, and by the idea that in the limit of large orbits the electron ought to radiate like a classical antenna.

As did Planck, Bohr replaced the continuous with the discrete. And as with Planck, the rules of his model were justified only by success:

(1) An electron in an atom can only exist in a denumerable set of circular orbits. The energies E_n, where $n = 1, 2, 3, \cdots$, of the orbits are discrete, with $E_1 < E_2 < E_3 < \cdots < 0$.

(2) A transition between orbits having energies E_n and $E_{n'}$ involves the emission or absorption of radiation of frequency ν given by $h\nu = |E_{n'} - E_n|$, thus conserving energy.

Bohr's original derivation involved the energies directly, but the results are most easily obtained using his famous rule for the quantization of orbital angular momentum, $L_n = n\hbar$, where $\hbar \equiv h/2\pi$ and $n = 1, 2, 3, \cdots$ This is sometimes just pulled out of a hat, but here we can call upon de Broglie's argument (10 years later) that the allowed orbits are those for which an integral number n of wavelengths $\lambda = h/p$ fit round a circle. Then (as was already shown in Sec. 1) $2\pi r = n\lambda = nh/p$, and $L_n = rp = nh/2\pi = n\hbar$, Bohr's rule.

The quantization of angular momentum leads to quantization of the speeds, radii, and energies of the orbits. Consider a single electron in a circular orbit around a proton. From $F = ma$ again, we get

$$\frac{ke^2}{r^2} = m\frac{v^2}{r}, \quad \text{or} \quad ke^2 = mv^2 r = (mvr)v = Lv \; ,$$

where $L = mvr$ is the angular momentum. With $L_n = n\hbar$, we get

$$v_n = \frac{1}{n}\frac{ke^2}{\hbar}, \quad \text{or} \quad \frac{v_n}{c} = \frac{1}{n}\frac{ke^2}{\hbar c} = \frac{1}{n}\alpha \; ,$$

where c is the speed of light, and

$$\boxed{\alpha \equiv ke^2/\hbar c \simeq 7.2974 \times 10^{-3} \simeq 1/137 \; .}$$

This is the *fine-structure constant* (its value is usually given in this inverse form). We shall eventually find that α pervades atomic physics. It is a *dimensionless* constant, and as such

11

its value is independent of the units we use. It is the same *number* (in decimal), just as is π, in advanced civilizations throughout the Universe. In hydrogen, the largest speed of the electron is $c/137$, when $n = 1$. Thus non-relativistic dynamics is a good approximation. Eventually, we shall worry about a small relativistic correction.

Next, r. From $L_n = mvr = n\hbar$, we get

$$r_n = \frac{n\hbar}{mv_n} = n^2 \frac{\hbar^2}{mke^2} = n^2 a_0 \ ,$$

where

$$\boxed{a_0 \equiv \hbar^2/mke^2 \simeq 0.52918 \times 10^{-10} \text{ m} \ .}$$

This is the *Bohr radius*. The radii of the Bohr orbits are $n^2 a_0 = a_0, 4a_0, 9a_0, \cdots$

Finally, and most importantly, the total energy E. This is, again,

$$E = K + V = \frac{1}{2}mv^2 - \frac{ke^2}{r} = -\frac{ke^2}{2r} \ .$$

Using the quantized r_n, we get

$$\boxed{E_n = -\frac{1}{n^2} \frac{mk^2e^4}{2\hbar^2} = \frac{1}{n^2} E_1 \ ,}$$

where $|E_1| = mk^2e^4/2\hbar^2 \simeq 13.606$ eV; $|E_1|$ is the *binding energy* or *ionization energy* of hydrogen, the minimum energy that would be needed to remove the electron from its most tightly bound state to infinity. (But see below for a small correction.) Two other useful ways to write $|E_1|$ are in terms of the Bohr radius a_0 and the fine-structure constant α:

$$|E_1| = \frac{\hbar^2}{2m}\left(\frac{mke^2}{\hbar^2}\right)^2 = \frac{\hbar^2}{2m}\frac{1}{a_0^2}, \qquad \text{and} \qquad |E_1| = \frac{1}{2}\left(\frac{ke^2}{\hbar c}\right)^2 mc^2 = \frac{1}{2}\alpha^2 mc^2 \ .$$

Here $mc^2 \simeq 0.511$ MeV is the rest-mass energy of the electron. Since $|E_1| = 13.6$ eV is $(1/137)^2$ times smaller than mc^2, we see again that using the non-relativistic $K = p^2/2m$ in these calculations should be a good approximation.

(c) Energy levels and the radiation spectrum—Figure 4(a) shows as horizontal lines the lowest few bound-state energy levels. An infinite number of levels crowd in just below $E = 0$. Any positive energy is allowed (the electron is then unbound); this is indicated by the continuous shading.

Experiments usually observe the frequencies of electromagnetic radiation emitted or absorbed, not the energy levels directly. The *second* part of Bohr's model is that the radiation spectrum comes from transitions between the levels. Radiation is emitted when an electron drops down to a lower energy level and is absorbed when it is raised up to a higher level; and energy is conserved. The frequencies ν are given by $|\Delta E| = h\nu$,

$$\boxed{h\nu = E_{n'} - E_n = |E_1|\left(\frac{1}{n^2} - \frac{1}{n'^2}\right) \ .}$$

Here n' and n are the quantum numbers of the upper and lower states. Figure 4(a) shows as vertical lines some of the transitions; their lengths are proportional to the frequencies of the radiations. Transitions to or from the $n = 1$ level are in the *Lyman series*; those with the $n = 2$ level as the lower state are in the *Balmer series*; and so on.

12

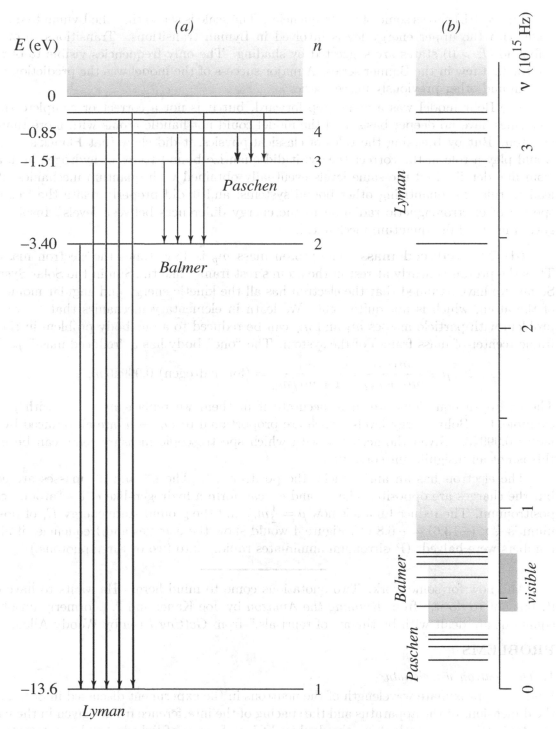

Figure 4. (a) The lowest few energy levels of hydrogen, with some transitions indicated. (b) The radiation spectrum. The energy differences and frequencies are related by $\Delta E = h\nu$. Shading in (b) indicates transitions to or from the continuum of unbound ($E > 0$) states. The only discrete lines in the visible are from the Balmer series.

Figure 4(b) shows some of the frequencies. The scale is set so that the Lyman frequencies align with the upper energy levels involved in Lyman transitions. Transitions to or from unbound ($E > 0$) states are suggested by shading. The only frequencies visible to our eyes are the first few in the Balmer series. A major success of the model was the prediction of the Lyman and other previously unseen series.

The Bohr model was a giant step forward, but it is not a correct or complete theory. The rules have no deeper basis, and the model could not handle atoms with more than one electron. But by breaking the rules of classical physics, it did show that Planck's constant would play a role in the correct theory; it did quantize bound states of hydrogen and, aside from fine details, find the same levels eventually obtained with quantum mechanics; it did lead to rules for quantizing other bound systems; and it did properly relate the frequency spectra of electromagnetic radiation to the energy differences between levels. Insofar as it goes, Fig. 4 is fully quantum mechanical.

(d) The reduced mass—The proton mass m_p is 1836 times the electron mass m_e. Thus the proton is nearly at rest in the atom's rest frame (like the Sun in the Solar System). So far, we have assumed that the electron has all the kinetic energy and angular momentum of the atom, which is not quite right. We learn in elementary mechanics that a two-body problem with particle masses m_1 and m_2 can be reduced to a one-body problem in the rest frame (center-of-mass frame) of the system. The "one" body has a "reduced mass" μ,

$$\mu \equiv \frac{m_1 m_2}{m_1 + m_2} = \frac{m_1}{1 + m_1/m_2} = \text{(for hydrogen) } 0.99946 \, m_e \; .$$

The hydrogen equations are more accurate if in them we replace $m = m_e$ with μ. For example, the Bohr energy levels, which are proportional to $m_e \to \mu$, are all reduced by that factor 0.99946. Given the accuracy with which spectroscopic measurements can be made, this is not an insignificant correction.

The electron has an antiparticle, the positron e^+. The e^+ and e^- masses are equal, but the charges are opposite. The e^+ and e^- can form a hydrogen-like $e^+ e^-$ "atom," called positronium. The reduced mass is now $\mu = \frac{1}{2} m_e$, and the ground-state energy E_1 of positronium is $\frac{1}{2} \times (-13.6) = -6.8$ eV. Figure 4 would show the energies and frequencies if all the numbers were halved. (Positronium annihilates rapidly into two or three photons.)

———————

And now for some work. Two quotations come to mind here: "He wants to have done it, but not to do it," from *Running the Amazon* by Joe Kane; and "...formerly unsolvable equations are dealt with by threats of reprisals," from *Getting Even* by Woody Allen.

PROBLEMS

1. *The neutron wavelength.*
Find the approximate wavelength of the neutrons in the experiment discussed in Sec. 2 using the dimensions of the apparatus and the spacing of the interference fringes given in the figures and text. (If necessary, look up the double-slit interference of light in an elementary text.)

A duck landing on a still pond and bobbing up and down makes circular waves. Two ducks landing together a few feet apart make an interference pattern—fingers of outgoing wavelets between fingers of calm (the fingers lie along hyperbolas). The contrast of skylight reflected from the surfaces tilted toward and away from you makes the pattern easy to see. A couple of large pine cones tossed in, or anything else that will bob up and down, will do.

2. *Energies in electron volts.*

(a) The range of visible wavelengths of light is about 400 to 700 nm. What range of photon energies is this, in eV? (This range is worth remembering.)

(b) Show that the de Broglie wavelength of a nonrelativistic electron with a kinetic energy K is $\lambda = 1.23/\sqrt{K}$, where λ is in nm and K is in eV. Find the corresponding relation for a proton.

3. *Maximum energy transfer in Compton scattering.*

Show that the maximum energy transferred to the electron in Compton scattering is $T_{max} = E/(1 + mc^2/2E)$, where E is the initial energy of the photon.

4. *Standing waves in a box.*

(a) Show that the quantized energies of a nonrelativistic particle of mass m in a one-dimensional box of length L are

$$E_n = n^2 \frac{\pi^2 \hbar^2}{2mL^2}.$$

Do this by requiring that the round-trip distance $2L$ be an integral number of de Broglie wavelengths. (Quantum mechanics gets the same energies.)

(b) Do the same for a photon, which of course is *not* nonrelativistic.

5. *Quantizing a harmonic oscillator.*

A particle with a mass m is bound to the origin by the Hooke's-law force $\mathbf{F}(r) = -k\mathbf{r}$, where k is the spring constant. Use Bohr's rule, $L_n = n\hbar$, where $n = 1, 2, 3, \ldots$, to show that the quantized energies for circular orbits are $E_n = n\hbar\omega$, where $\omega = \sqrt{k/m}$ is the angular frequency of the motion. (Quantum mechanics in two dimensions gets $E_{n_x n_y} = (n_x + n_y + 1)\hbar\omega$, where $n_x, n_y = 0, 1, 2, 3, \ldots$ Note the value 0 here; and n_x and n_y are independent.)

6. *A classical limit.*

(a) Show that the *orbital* frequency $\nu(\text{orbit}) = v/2\pi r$ of the electron in the nth Bohr orbit is related to the energy E_n of the orbit by

$$\nu_n(\text{orbit}) = \frac{2|E_n|}{n\hbar}.$$

Classically, a charged particle moving in a circular orbit radiates at the frequency of its motion, and if not constrained will spiral in as it radiates.

(b) In Bohr's model, the frequency of radiation for transitions between *adjacent* orbits is $\nu_n(\text{Bohr}) = (E_n - E_{n-1})/h$. Find the ratio $\nu_n(\text{Bohr})/\nu_n(\text{orbit})$ for $n = 1, 2, 5, 10, 100$, and 1000, and show that it approaches one as $n \to \infty$. This is an example of Bohr's *correspondence principle*: In the limit of large quantum numbers, results of quantum mechanics approach those of classical mechanics.

7. *Derivation of the reduced mass.*

Let $\mathbf{r}_1$ and $\mathbf{r}_2$ be the position vectors of masses m_1 and m_2 with respect to an origin O in an arbitrary inertial frame. Then the vector from m_2 to m_1 is $\mathbf{r} = \mathbf{r}_1 - \mathbf{r}_2$, and the vector to the center of mass is $\mathbf{R} = (m_1\mathbf{r}_1 + m_2\mathbf{r}_2)/M$, where $M = m_1 + m_2$.

(a) Show that

$$\mathbf{r}_1 = \mathbf{R} + \frac{m_2}{M}\mathbf{r} \qquad \text{and} \qquad \mathbf{r}_2 = \mathbf{R} - \frac{m_1}{M}\mathbf{r} \ .$$

(b) Show that the total kinetic energy can be written in terms of either v_1 and v_2 or V and v,

$$K = \tfrac{1}{2}m_1 v_1^2 + \tfrac{1}{2}m_2 v_2^2 = \tfrac{1}{2}MV^2 + \tfrac{1}{2}\mu v^2 \ ,$$

where $\mathbf{v}_1 = d\mathbf{r}_1/dt$, $\mathbf{v}_2 = d\mathbf{r}_2/dt$, $\mathbf{v} = d\mathbf{r}/dt$, $\mathbf{V} = d\mathbf{R}/dt$, and $\mu = mM/(m+M)$ is the reduced mass. Thus in an inertial frame in which the center of mass is at rest, the kinetic energy is just $\tfrac{1}{2}\mu v^2$.

8. *Oxygen with only one electron.*

A recent experiment claims to have discovered that the Milky Way is surrounded by an enormous halo of highly ionized gas. One of the spectral lines detected is the $n=2$ to $n=1$ line for an oxygen atom so highly ionized that only one electron remains in orbit around its nucleus. As oxygen is the eighth element in the Periodic Table, the nuclear charge is $8e$ (the number of protons, Z, is eight). What is the energy, in eV, of a photon emitted in this transition?

9. *Taking apart a figure.*

(a) With a ruler, check that the energy levels in Fig. 4(a) are indeed spaced according to $1/n^2$ (measured from where?). With a calculator, check that the energies, given in eV, are correct.

(b) On a copy of Fig. 4, label the vertical lines in Fig. 4(a), from left to right, 1, 2, 3, ... With a ruler, match the length of each line with a frequency in Fig. 4(b), and label the frequency with the number on its transition line. [There are a few more frequencies in (b) than there are transitions in (a).] Check that the continuum lower limits are properly given.

(c) Check that the frequency scale in Fig. 4(b), in Hz, agrees with the energy scale in Fig. 4(a), in eV. Check that the range visible to the human eye, roughly 400 to 700 nm, is properly given.

Take every figure of any complexity apart like this.

10. *A mystery element.*

The analysis of a meteorite found in Antarctica discovered a trace of a mysterious element. Its spectral lines were all multiples of a frequency ν_0. The multiples were

$$1, 2, \mathbf{3}, \mathbf{5}, 6, \mathbf{8}, 10, 11, \mathbf{13}, 16, 18, 19, \mathbf{21}, 26, 29, 31, 32, 34, 42, 47, 50, 52, 53 \ .$$

The frequencies in bold had twice the intensity expected from the pattern of intensities of the other lines. The investigating spectroscopists assumed that these were cases in which two distinct transitions had the same frequency.

(a) Make a figure like Fig. 4, scaling the frequency spectrum to the simplest energy-level diagram that gives rise to those frequencies (and to *no others*). Assume that transitions between all pairs of levels occur. Make the height of the energy diagram $0.4 \times 53 = 21.2$ cm. *Use a ruler!* It is only energy *differences* that matter here, but take the zero of the energy scale to be at the top (as in Fig. 4), and let the levels get closer together near the top (as in Fig. 4).

(b) The symbol given to the new element was Fb. What was its name? (You won't find it in the Periodic Table.)

2. THE SCHRÖDINGER EQUATION. BOUND STATES

2.1. *The Time-Dependent Schrödinger Equation*
2.2. *The Wave Function*
2.3. *The Time-Independent Equation. Energy Eigenstates*
2.4. *The Infinite Square Well*
2.5. *The Finite Square Well*
2.6. *The Delta-Function Well*
2.7. *Schrödinger in Three Dimensions*
2.8. *Two Important Energy Eigenstates*
2.9. *Qualitative Properties of Bound Energy Eigenstates*
    Problems

Quantum mechanics is new physics—it cannot be derived from what went before. "There is no logical way to the discovery of these elemental laws. There is only the way of intuition ..." (Albert Einstein). The central object in quantum mechanics is the wave or state function Ψ (psi), which is a function of the space coordinates of the system and time (spin will come later). Anything knowable about the system can be obtained from Ψ, but what is knowable is less complete than what is assumed in classical mechanics: From Ψ we get probability distributions for where a particle is at, and for what its momentum and energy are. The evolution of Ψ in time is given by a partial differential equation, the time-dependent Schrödinger equation. "The motion of particles is subject to probabilistic laws, but the probability itself evolves in accord with causal laws." (Max Born)

For a while, our system will be a single spinless particle, which we introduce first in one space dimension. Separation of space and time variables for this equation leads to the time-independent Schrödinger equation, an eigenvalue equation whose eigenvalues are the possible results of measuring the energy of the system. In quantum mechanics, the possible energies of a particle that cannot escape to infinity are discrete (i.e., quantized), not continuous. For example, in Figure 1·4 the $E < 0$ states are quantized, the $E > 0$ states are not.

After setting provisional "rules" of quantum mechanics—detailed formalism is better left until some experience is gained with what the formalism addresses—we solve some problems. We first find the energy eigenstates of a particle bound in these one-dimensional potential-energy wells: an infinite square well, a finite square well, and a delta-function well. Mathematically, the problems are similar to finding the normal frequencies and modes of classical vibrating systems, such as guitar strings and sound waves in an enclosed space.

We then step (in this chapter, only briefly) into three space dimensions, and find the energy states of a particle in a three-dimensional rectangular box and (for the case of no angular momentum) a spherical box. We also solve for the ground states of the harmonic oscillator and the hydrogen atom, two systems each with a chapter of its own later on. Two or three other one-dimensional wells are left for the problems.

17

2·1. THE TIME-DEPENDENT SCHRÖDINGER EQUATION

New physics can only be guessed at using intelligence and imagination, guided by attention to results of experiments. Quantum mechanics was first discovered, in a matrix form, by Werner Heisenberg in 1925. A few months later Erwin Schrödinger discovered a differential-equation form, and after a brief period of confusion Schrödinger showed that the two forms were equivalent. In 1925, physicists were more familiar with differential equations than with matrices (Heisenberg at first did not know that arrays he constructed were matrices), and Schrödinger's form is, at least at this level, the one more used. We follow Schrödinger, but later will see something of Heisenberg and much of matrices.

Schrödinger's search for a fundamental equation began with a talk he gave on de Broglie's ideas about associating a wave with a particle. After the talk, someone remarked that where there was a wave there ought to be a wave equation. So we begin by reviewing very briefly the simplest *classical* wave equations.

(a) Classical wave equations—A string of mass per unit length μ is stretched along the x axis under tension T. The string undergoes small transverse (y) oscillations, in which case the tension remains, to a very good approximation, constant. Application of $F = ma$ to the infinitesimal mass $\mu\,dx$ of an infinitesimal length dx of the string leads to

$$\text{tension} \times \text{curvature} = \text{mass per unit length} \times \text{transverse acceleration} \ ,$$

or

$$\frac{\partial^2 y}{\partial x^2} = \frac{\mu}{T}\frac{\partial^2 y}{\partial t^2} = \frac{1}{v^2}\frac{\partial^2 y}{\partial t^2} \ ,$$

where $y = y(x,t)$ is the transverse displacement. (You might try to derive this.) Here $v \equiv \sqrt{T/\mu}$. If you shake one end of a long stretched string transversely, a wave will sweep down the string with speed v. Or if you pluck a string that is fixed at its ends, you excite various of a *discrete* set of frequencies. The spectrum of possible frequencies for *traveling* waves in an *open* medium (such as the ocean or space) is continuous; the spectrum of frequencies for *standing* waves in a *closed* system (such as a guitar string) is discrete.

The equation for plane electromagnetic waves in vacuum, derived from Maxwell's equations, has the same form as the string equation:

$$\frac{\partial^2 \Psi}{\partial x^2} = \frac{1}{c^2}\frac{\partial^2 \Psi}{\partial t^2} \ .$$

Here x is the direction in which the wave is advancing, $\Psi = \Psi(x,t)$ is any transverse component of the electric or the magnetic field ($\mathcal{E}_y$, $\mathcal{E}_z$, B_y, B_z), and c is the speed of light. And again there are both traveling waves, as when light radiates from the Sun or radio waves radiate from an antenna, and standing waves, as in a laser cavity. There are many other physical systems for which an equation of the same or a similar form applies, and there are more complicated equations for waves in, for example, inhomogeneous or absorbing media.

When a *sinusoidal* wave sweeps down a long string, or when one of the single-frequency (normal) modes of a string fixed at its ends is excited, each bit of the string undergoes simple harmonic motion. In either case, the *energy density* at a point along the string is proportional to the square of the amplitude of the motion at that point. Similarly, electric and magnetic energy densities $u_{\mathcal{E}}$ and u_B are proportional to the squares of the electric and magnetic field amplitudes $\mathcal{E}$ and B: In free space, $u_\mathcal{E} = \epsilon_0 \mathcal{E}^2/2$ and $u_B = B^2/2\mu_0$, where ϵ_0 and μ_0 are the permittivity and permeability constants of free space.

The above string and electromagnetic wave equations (one nonrelativisitic, the other completely relativistic) have solutions of the form

$$\Psi(x,t) = A \sin 2\pi \left(\frac{x}{\lambda} - \frac{t}{\tau} \right) = A \sin(kx - \omega t) \ .$$

Here λ and τ, the wavelength and the period, are the repetition intervals in space and in time. A cosine would do as well here as a sine. It simplifies writing to define a wave number $k \equiv 2\pi/\lambda$ and an angular frequency $\omega \equiv 2\pi/\tau = 2\pi\nu$, where ν is the cycle frequency. Putting the sine wave into the wave equations shows that the speed of the wave is v (or c) $= \lambda/\tau = \omega/k$. (Show this.)

In almost any advanced treatment of waves (for light, sound, circuits, etc.), it simplifies the mathematics to use a complex-exponential form for the wave function. An exponential includes both sines and cosines:

$$\boxed{e^{i\theta} = \cos\theta + i\sin\theta \ .}$$

Feynman has called this "the most remarkable formula in mathematics." For example, $e^{i\pi} = -1$ relates the four fundamental numbers e, i, π, and -1. One advantage of the exponential form is that it factors, even when an arbitrary phase factor is thrown in:

$$e^{i(kx - \omega t + \phi)} = e^{ikx} e^{-i\omega t} e^{i\phi} \ .$$

Sines and cosines do not so factor. Furthermore, exponentials are eigenfunctions of derivatives of any order:

$$\frac{d^n}{dx^n}(e^{ikx}) = (ik)^n e^{ikx} \ .$$

Sines and cosines are only eigenfunctions of derivatives of even order.

In classical calculations, the physical answer (after exponentials have simplified the math) is usually the real part of the mathematical answer.

(b) Pieces of a puzzle—Three pieces go into guessing a wave equation for a particle:

(1) The non-relativistic relation between the kinetic energy K, the mass m, and the momentum p of a particle is $K = p^2/2m$. Then the total energy is

$$E = \frac{p^2}{2m} + V(x) \ ,$$

where $V(x)$ is the potential energy. (For a while, we stay in one space dimension and consider only time-independent potential energies.) The total energy, written in terms of p (instead of the speed v) and x is called the Hamiltonian H.

(2) For a photon, the completely relativistic relations between the particle properties E and p and the wave properties ν and λ are $E = h\nu = \hbar\omega$ and $p = h/\lambda = \hbar k$, where $\hbar \equiv h/2\pi$ (Sec. 1·1(c)). De Broglie borrowed the relation $p = h/\lambda$ to apply it to particles, and thereby was able to "explain" (in 1923) Bohr's angular-momentum quantization rule (of 1913), $L_n = n\hbar$. See Sec. 1·1(d).

19

(3) The complex expression for a wave of well-defined wavelength and frequency in a uniform medium, whether a wave on a string or a sound wave in air or a light wave in vacuum, is

$$\Psi(x,t) = Ae^{2\pi i(x/\lambda - t/\tau)} = Ae^{i(kx-\omega t)} .$$

For a photon, the relations $E = \hbar\omega$ and $p = \hbar k$ let us write $\Psi(x,t)$ as

$$\Psi(x,t) = Ae^{i(px-Et)/\hbar} .$$

(c) Inventing an equation—For now, the physical system will be a single particle with mass m. We *invent* a partial differential equation—a wave equation—in the variables x and t that embodies the physical relation $E = p^2/2m + V$ in (1) above. We assume that *both* of the relations $E = \hbar\omega$ and $p = \hbar k$ in (2) apply to particles as well as to photons—even for non-relativistic particles, where the relation between E and p is very different from $E = pc$. We also assume that the *solution* of the (unknown) equation for a *free* particle having well-defined energy and momentum has the form of $\Psi(x,t)$ in (3) above,

$$\Psi(x,t) = Ae^{i(px-Et)/\hbar} .$$

Taking partial derivatives of this wave function, we get

$$\frac{\partial\Psi}{\partial x} = \frac{ip}{\hbar}\Psi, \qquad \text{or} \qquad -i\hbar\frac{\partial\Psi}{\partial x} = p\Psi ,$$

and

$$\frac{\partial\Psi}{\partial t} = -\frac{iE}{\hbar}\Psi, \qquad \text{or} \qquad i\hbar\frac{\partial\Psi}{\partial t} = E\Psi .$$

Since the kinetic energy involves p^2, we take a second derivative with respect to x:

$$\frac{\partial^2\Psi}{\partial x^2} = -\frac{p^2}{\hbar^2}\Psi, \qquad \text{or} \qquad -\hbar^2\frac{\partial^2\Psi}{\partial x^2} = p^2\Psi .$$

Thus we may write

$$i\hbar\frac{\partial\Psi}{\partial t} = E\Psi \qquad \text{and} \qquad -\frac{\hbar^2}{2m}\frac{\partial^2\Psi}{\partial x^2} = \frac{p^2}{2m}\Psi .$$

Finally, since for a free particle $E = p^2/2m$, we guess that we can equate the left sides of these two equations,

$$i\hbar\frac{\partial\Psi}{\partial t} = -\frac{\hbar^2}{2m}\frac{\partial^2\Psi}{\partial x^2} .$$

And that is the *time-dependent Schrödinger equation for a free particle* having energy E and momentum p. It is first order in the time derivative and second order in the space derivative because that is how E and p occur in $E = p^2/2m$. The solution of the equation is, by (reverse) construction,

$$\Psi(x,t) = Ae^{i(px-Et)/\hbar} .$$

A Schrödinger equation based on $E = p^2/2m + V(x)$ is non-relativistic because $K = p^2/2m$ is non-relativistic. We come briefly to a relativistic Schrödinger equation in Chap. 17. But since the exponential *form* of Ψ applies to both non-relativistic and relativistic wave equations, the differential operations $p = -i\hbar\partial/\partial x$ and $E = i\hbar\partial/\partial t$ are the same in both realms.

When a force acts on the particle, and the associated potential energy is $V(x)$, we add a term $V(x)\Psi(x,t)$ to get the full *time-dependent Schrödinger equation*,

$$i\hbar\frac{\partial\Psi}{\partial t} = -\frac{\hbar^2}{2m}\frac{\partial^2\Psi}{\partial x^2} + V(x)\Psi \ .$$

The terms on the right here came from the Hamiltonian, $H = p^2/2m + V$. In classical mechanics, $H(x,p)$ is a stand-alone function of the position and momentum of a particle (and perhaps of the time). In simple cases, it gives the energy. In quantum mechanics, H is an *operator*,

$$H = -\frac{\hbar^2}{2m}\frac{\partial^2}{\partial x^2} + V(x) \ ,$$

and it *operates* on $\Psi(x,t)$. When $V(x)$ varies with x, the form of the solution will no longer be as simple as $\Psi(x,t) = Ae^{i(px-Et)/\hbar}$.

We shall mark the quantum-mechanical H, and later on other operators, with a hat. Then a short form of the Schrödinger equation is

$$i\hbar\frac{\partial\Psi}{\partial t} = \hat{H}\Psi \ .$$

There will be other Hamiltonians. For example, for the interaction of a magnetic moment $-\boldsymbol{\mu}$ with a magnetic field $\mathbf{B}$, $\hat{H}$ is $-\boldsymbol{\mu}\cdot\mathbf{B}$.

Of course it took Schrödinger rather longer to discover his equation than the above might suggest—and then he had to show that it explained results of experiments. This he did in a remarkable series of papers that laid the foundations for much of non-relativistic quantum mechanics. *The equation worked!* For the story of the brilliant work of Heisenberg, Schrödinger, and the other founders of quantum mechanics—an amazing community—see the books listed in the introduction to Chap. 1. They are well worth a browse.

2·2. THE WAVE FUNCTION

(a) **The probabilistic interpretation**—What does the wave function $\Psi(x,t)$ for a particle represent? Max Born gave the answer. The infinitesimal *probability* dP that the particle would, on making a measurement, be found between x and $x+dx$ at time t is

$$dP(x,t) = |\Psi(x,t)|^2\,dx = \Psi^*(x,t)\Psi(x,t)\,dx \ .$$

Thus $|\Psi(x,t)|^2$ is a probability *density*, and $\Psi(x,t)$ is a probability *amplitude*. As with classical waves, the square of an amplitude gives a density, but here it is a density of probability, not of energy. And now the amplitude is, as we shall see, unavoidably complex. Thus the square of $|\Psi(x,t)|$, not of $\Psi(x,t)$, is needed to make $dP(x,t)$ real. Since the particle must be somewhere, $\Psi(x,t)$ is *normalized* when

$$\int_{-\infty}^{+\infty} dP = \int_{-\infty}^{+\infty} |\Psi(x,t)|^2\,dx = 1 \ .$$

Probabilities are dimensionless, so $|\Psi(x,t)|^2$ has (in one space dimension) the dimension of inverse length.

(b) Probabilities and mean values—We throw an unbiased die. Then all six outcomes, $n = 1, 2, \ldots, 6$, are equally likely, and some outcome is certain. Thus the probabilities are *normalized* if

$$\sum_1^6 P(n) = 1 .$$

And so $P(n) = 1/6$ for each n—the probability of getting a 3 on a throw is $P(3) = 1/6$. *Before* a throw we have a probability distribution. *After* a throw we have an experimental result.

Given the $P(n)$, we can calculate $\bar{n}$, the expected mean or average of the value of the showing face were we to throw the die, say, 1000 times (or throw 1000 dice once). We simply weight the various possible results with their probabilities and sum over the results:

$$\bar{n} = \sum_{\text{possible results}} \text{result} \times \text{probability of result} = \sum_1^6 n P(n) = 3\tfrac{1}{2} .$$

Of course, an actual experiment with 1000 throws is unlikely to get exactly $\bar{n} = 3\tfrac{1}{2}$. In the same way, we can calculate the mean of n^2, or of any other function $f(n)$ of n:

$$\overline{n^2} = \sum_1^6 n^2 P(n) = \frac{91}{6}, \qquad \text{and} \qquad \overline{f(n)} = \sum_1^6 f(n) P(n) .$$

When the possible results of a measurement are continuous, such as for the position x of a particle, the sum over a discrete set of results is replaced by an integral over the continuous range. The normalization is

$$\int_{\text{range of } x} dP(x, t) = 1 ,$$

where $dP(x, t)$ is the probability of getting a result between x and $x + dx$ on making a measurement at time t. In terms of the quantum-mechanical $dP(x, t) = |\Psi(x, t)|^2 \, dx$, the mean value of x at t is

$$\boxed{\langle x \rangle = \int_{\text{results}} \text{result} \times \text{probability of result} = \int_{-\infty}^{+\infty} x \, |\Psi(x, t)|^2 \, dx .}$$

Here, anticipated notation to be made formal in Chap. 5, we denote quantum-mechanical mean values with angular brackets, not overbars. The integral is over space, but not over time. If $|\Psi(x, t)|^2$ is time-dependent, then $\langle x \rangle$ will likely be time dependent too. The mean value of, say, a potential energy $V(x)$ at time t would be

$$\langle V(x) \rangle = \int_{-\infty}^{+\infty} V(x) \, |\Psi(x, t)|^2 \, dx .$$

The rule for calculating mean values of quantities such as momentum that involve derivatives requires more discussion and will come later.

In quantum mechanics, a mean or average value is often called an *expectation value*. It is the mean value we *expect* to get on making repeated measurements on a system. A crucial point is that each measurement must start with a system in the *original state*, *not* the state left *after* a measurement. This is because in general the states before and after a measurement will not be the same. This will need some discussion.

(c) Normalizability—Physical wave functions, whether classical or quantum mechanical, are single-valued functions of position x and time t. They are also finite in magnitude and fall off rapidly enough as $x \to \pm\infty$ that the integral $\int_{-\infty}^{+\infty} |\Psi(x,t)|^2 \, dx$ is finite (or "square integrable"), so that $\Psi(x,t)$ can be normalized.

Now in fact the only wave function we have seen so far, $\Psi(x,t) = A\, e^{i(px-Et)/\hbar}$, is not normalizable. It was assumed to have an exact momentum p (or wavelength $\lambda = h/p$), and Fourier analysis tells us that no wave finite in extent does so. Thus $\int_{-\infty}^{+\infty} |\Psi(x,t)|^2 \, dx = |A|^2 \int_{-\infty}^{+\infty} dx$ is infinite. Nevertheless, as a mathematical simplification and a physical idealization, $\Psi(x,t) = A\, e^{i(px-Et)/\hbar}$ is highly useful. The idealization of a single wavelength is made throughout much of every textbook on light. It will sometimes be useful in this text too, where it will represent not just one particle but a stream of them, all having the same momentum. We can defer discussion of such amplitudes until Chap. 4. All other wave functions in the this chapter will be clearly normalizable.

(d) First rules—Here and in the next section, we state five "rules" about the meaning and interpretation of quantum mechanics. The full meaning of the rules will not be immediately apparent, and in fact they are here provisional. They will be revised and extended in Chap. 5.

Rule 1: The state of a particle is completely specified by a function $\Psi(x,t)$. Anything knowable about the particle can be obtained from this function, but what is knowable is less complete than what is assumed in classical mechanics. For example, a particle that has a definite energy has only a probability distribution for its position.

Rule 2: The Schrödinger equation $i\hbar\, \partial\Psi/\partial t = \hat{H}\Psi$ tells how $\Psi(x,t)$ evolves in time—until a measurement is made. Measurement can cause an abrupt change in $\Psi(x,t)$.

2·3. THE TIME-INDEPENDENT EQUATION. ENERGY EIGENSTATES

(a) Product solutions—We stay in one space dimension until Sec. 7, and consider only time-independent potential energies until Chap. 10. (A time-varying electric field, say, acting on a charged particle, would make V time-dependent.) We look for solutions of the time-dependent Schrödinger equation that have the product form

$$\Psi(x,t) = \psi(x)\, T(t) \,.$$

The aim is to get separate ordinary differential equations in x and t, one for $\psi(x)$, one for $T(t)$. Divide and conquer: Ordinary differential equations are easier to solve than partial differential equations. Afterwards, we can put the pieces, $\psi(x)$ and $T(t)$, back together again.

Putting the product $\psi(x)\, T(t)$ into the time-dependent Schrödinger equation,

$$i\hbar \frac{\partial \Psi}{\partial t} = -\frac{\hbar^2}{2m} \frac{\partial^2 \Psi}{\partial x^2} + V(x)\Psi(x,t) \,,$$

we get

$$i\hbar\psi \frac{dT}{dt} = T\left(-\frac{\hbar^2}{2m} \frac{d^2\psi}{dx^2} + V(x)\,\psi \right).$$

We write ordinary derivatives here because T depends only of t, ψ only on x. Dividing by ψT, we get

$$\frac{i\hbar}{T}\frac{dT}{dt} = \frac{1}{\psi}\left(-\frac{\hbar^2}{2m}\frac{d^2\psi}{dx^2} + V(x)\,\psi\right).$$

Now all the t dependence is on the left side of the equation, and all the x dependence is on the right. This separation of x from t would not work if V were a function of both x and t.

Now comes the crux of the argument. The variables x and t are independent—either can be varied at will without regard for the other. Suppose we fix x and vary t. Then only the left side of the equation would change; the right side depends only on x, which is fixed. However, if the right side doesn't change, then neither can the left side, because the two sides are to be equal. The same result follows if we fix t and vary x. Therefore, *for a product solution to exist*, each side of the separated equation must stay constant, no matter how x and t vary—and the two sides must be equal. The constant to which they are both equal is a *separation constant*, E. Thus we get two ordinary differential equations,

$$\boxed{\;i\hbar\frac{dT}{dt} = ET \qquad \text{and} \qquad -\frac{\hbar^2}{2m}\frac{d^2\psi}{dx^2} + V(x)\,\psi = \hat{H}\psi = E\psi\;.}$$

We call the separation constant E because $\hat{H}$ is the energy operator. The equation in x is the time-*independent* Schrödinger equation, and it will play the lead role in this text.

(b) The other Schrödinger equation—The time-dependent equation tells how the wave function $\Psi(x,t)$ evolves in time. The time-independent equation is a different kind of thing; it is an *eigenvalue* equation. An operation on ψ—do the various things to ψ that are specified by the Hamiltonian $\hat{H}$—is to give ψ back again, multiplied by a constant.

Each different $V(x)$ presents a different eigenvalue problem. Suppose, for some $V(x)$, we have found the eigenvalues E_n and eigenfunctions $\psi_n(x)$ of $\hat{H}\psi_n = E_n\psi_n$. (This chapter will be about bound states, which can be labeled by an index n.) The *physical* significance is:

Rule 3: The only possible result of an accurate measurement of the energy of the system is one of the eigenvalues. For a particle bound to a limited region of space, the set of energy eigenvalues is discrete, not continuous.

Rule 4: If, say, the result of a measurement is E_6, then immediately afterwards the system is in the corresponding eigenstate, $\psi_6(x)$. Conversely, if the particle is in the state $\psi_6(x)$, then E_6 will be obtained on measuring the energy.

These two rules will need to be modified and generalized later on.

(c) Time dependence—To find the time dependence of an eigenstate having eigenvalue E_n, we go back to the separated equation for $T(t)$:

$$\frac{i\hbar}{T}\frac{dT}{dt} = E_n, \qquad \text{or} \qquad \int_{T(0)}^{T(t)}\frac{dT}{T} = -\frac{iE_n}{\hbar}\int_0^t dt\;.$$

The solution is $T_n(t) = e^{-iE_n t/\hbar}$. (We will absorb a multiplicative constant $T_n(0)$ into $\psi_n(x)$.) Whenever V is independent of time, $T_n(t)$ will have the above form $e^{-iE_n t/\hbar}$. Of course, the possible *values* of E_n will depend on the Hamiltonian, and in particular on $V(x)$.

The *product solutions* of the time-*dependent* Schrödinger equation are

$$\boxed{\Psi_n(x,t) = \psi_n(x)\, e^{-iE_n t/\hbar}\, .}$$

An example is the free-particle wave function of Sec. 1(c), $\Psi(x,t) = Ae^{ipx/\hbar}e^{-iEt/\hbar}$. For bound states, we normalize the eigenstates so that

$$\int_{-\infty}^{+\infty} |\Psi_n(x,t)|^2\, dx = \int_{-\infty}^{+\infty} |\psi_n(x)|^2\, dx = 1\, .$$

Note that $|\Psi_n(x,t)|^2$ is independent of time. For this reason, energy eigenstates are sometimes called *stationary* states, even though $\Psi_n(x,t)$ itself is time dependent.

(d) **Superposition states**—The product solutions $\Psi_n(x,t) = \psi_n(x)\, e^{-iE_n t/\hbar}$, taken separately, are not the only solutions of the time-*dependent* Schrödinger equation. Consider the linear sum—the *superposition*—of any two product solutions, $\Psi_1(x,t)$ and $\Psi_2(x,t)$:

$$\Psi(x,t) = c_1\Psi_1(x,t) + c_2\Psi_2(x,t) = c_1\psi_1(x)e^{-iE_1 t/\hbar} + c_2\psi_2(x)e^{-iE_2 t/\hbar}\, ,$$

where c_1 and c_2 are arbitrary constants, subject only to the normalization constraint. Now $i\hbar\, \partial/\partial t$ operating on each term multiplies it by its energy, E_1 or E_2. And $\hat{H}$ operating on each term does exactly the same thing. So, putting $\Psi(x,t)$ into the time-dependent equation $i\hbar\, \partial\Psi/\partial t = \hat{H}\Psi$, we get the same thing on both sides of the equation, namely

$$c_1 E_1 \Psi_1 + c_2 E_2 \Psi_2\, .$$

Thus the superposition of two product solutions is another solution of $i\hbar\, \partial\Psi/\partial t = \hat{H}\Psi$. By extension, a linear superposition $\Psi(x,t) = \sum c_n \Psi_n(x,t)$ of any number of product solutions would be a solution. The last rule is

> *Rule 5*: Any (renormalized) linear superposition of possible states $\Psi(x,t)$ of a system is a possible state of the system.

Any time dependence of a probability density $|\Psi(x,t)|^2$ will come from the cross terms in $|\Psi(x,t)|^2$ when $\Psi(x,t)$ is a superposition of two or more states with different energies. A classical particle has a definite energy. A quantum particle can exist in a state involving several energies.

Note that $c_1\psi_1(x) + c_2\psi_2(x)$ is *not* a solution of the time-*independent* Schrödinger equation. What would we use for E in $\hat{H}\psi = E\psi$?

Each normalized product solution $\Psi_n(x,t)$ is a kind of "unit vector," and the full set of product solutions will provide a "basis" for constructing general solutions, much as the unit vectors $\hat{\mathbf{x}}$, $\hat{\mathbf{y}}$, and $\hat{\mathbf{z}}$ in ordinary space provide a basis for constructing ordinary three-dimensional vectors. The quantum-mechanical "components" then are the constants c_1, c_2, c_3, ..., in the expansion $\Psi(x,t) = \sum c_n \Psi_n(x,t)$. We develop this geometrical analogy in Chap. 5; it makes matters more intuitive. We need all such help we can get.

Some of the deepest consequences of quantum mechanics arise from the superposition of states. For the rest of this chapter, however, we simply solve the time-independent Schrödinger equation, $\hat{H}\psi = E\psi$, for several different forms of $V(x)$. The examples are important in their own right; and we need some familiarity with the methods and results of quantum mechanics before we address the general formalism.

2·4. THE INFINITE SQUARE WELL

A particle with mass m is confined to a one-dimensional box of length L; see Fig. 1. The potential energy is

$$V(x) = \begin{cases} 0 & \text{for } 0 < x < L \\ \infty & \text{outside the box .} \end{cases}$$

The classical analogue is a ball bouncing elastically between two rigid walls, in which case any kinetic energy would be possible. However, since we are making a *wave* mechanics, the solutions will resemble those for a guitar string tied down at $x = 0$ and L.

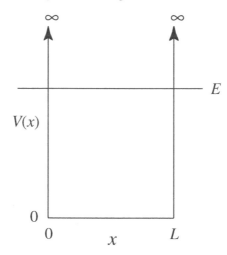

Figure 1. The infinite square well (*up* in the figure is in *energy*, not in *space*).

(a) Solution—Inside the box, the time-independent Schrödinger equation is

$$-\frac{\hbar^2}{2m}\frac{d^2\psi}{dx^2} = E\psi .$$

We may write this as $\psi'' + k^2\psi = 0$, where primes indicate differentiation and $k^2 \equiv 2mE/\hbar^2$. The general solution of this ordinary, linear, homogeneous, second-order, constant-coefficient differential equation is

$$\psi(x) = A\cos kx + B\sin kx ,$$

where A and B are arbitrary constants. Since the particle is confined to the box, the wave function outside the box is zero. Continuity of $\psi(x)$, like continuity of a guitar string tied down at $x = 0$ and L, requires that $\psi(0) = 0$ and $\psi(L) = 0$. From $\psi(0) = 0$, it follows that $A = 0$. Then from $\psi(L) = 0$, it follows either that $B = 0$, in which case $\psi(x)$ is zero everywhere and we have no physics, or that $kL = n\pi$, where $n = 1, 2, 3, \ldots$ Thus $k = n\pi/L$ and the eigenfunctions are

$$\psi_n(x) = B\sin\left(\frac{n\pi x}{L}\right)$$

for $0 \leq x \leq L$.

The probability of finding the particle in dx at x is $|\psi(x)|^2\, dx$. Since the probability of finding it somewhere in the box is 100%, we require that

$$\int_0^L |\psi_n(x)|^2\, dx = |B|^2 \int_0^L \sin^2\left(\frac{n\pi x}{L}\right) dx = 1 .$$

26

The average height of $\sin^2(n\pi x/L)$ over $0 \leq x \leq L$ is $1/2$, and so the integral itself is $L/2$. Therefore $|B|^2 = 2/L$, and we choose B to be real and positive. So finally,

$$\psi_n(x) = \sqrt{\frac{2}{L}} \sin\left(\frac{n\pi x}{L}\right).$$

Since $kL = n\pi$ and $k^2 = 2mE/\hbar^2$, the corresponding energy eigenvalues are

$$E_n = \frac{\hbar^2 k_n^2}{2m} = n^2 \frac{\pi^2 \hbar^2}{2mL^2} = n^2 E_1 \, ,$$

where $E_1 \equiv \pi^2 \hbar^2 / 2mL^2$ and $n = 1, 2, 3, \ldots$

This method, to be used over and over in what follows, is:

(1) Find the general solution of the time-independent Schrödinger equation.

(2) Fix the undetermined coefficients using boundary conditions and normalization.

Figure 2(a) shows the lowest energy levels and their eigenfunctions. Although the physical system here is of course very different, the patterns themselves are the same as the fundamental and overtone oscillation patterns (normal modes) of a guitar string. In the ground state, one half wavelength fits between 0 and L; in the $n = 2$ state, two half wavelengths fit in L; for the nth level, n half-wavelengths fit in L.

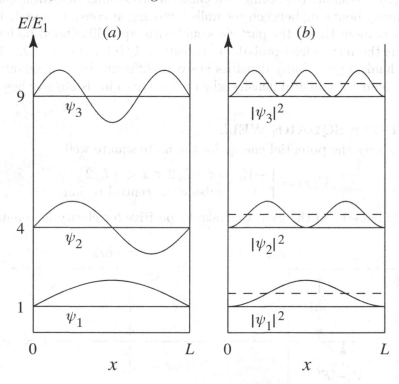

Figure 2. (a) The lowest energy levels $E_n = n^2 E_1$ and eigenfunctions $\psi_n(x)$ of the infinite square well. The energy lines are used as the x axes for drawing the corresponding eigenfunctions. (b) The quantum-mechanical probability densities $|\psi_n(x)|^2$ and the classical densities (dashed).

(b) Properties—*Quantization*: The possible energies of a particle in a one-dimensional box are discrete, or *quantized*. No states with in-between energies exist! For each E_n, there is only one $\psi_n(x)$; and there is only a probability density, $|\psi_n(x)|^2$, not a definite value, for the position of the particle. The energy of a particle bound to a limited region of space is always quantized, but in two or three dimensions there can be more than one wave function for each energy (see Sec. 7).

Nodes: In Fig. 2(a), the ground-state wave function $\psi_1(x)$ never crosses the x axis; that is, $\psi_1(x)$ has no nodes. (The tie-downs at $x = 0$ and L due to the infinities in $V(x)$ don't count here as nodes.) And $\psi_2(x)$ crosses the x axis once, $\psi_3(x)$ crosses it twice, and so on. Higher energy states have more nodes because higher energies mean higher momenta and thus, from $\lambda = h/p$, shorter wavelengths. Count nodes and you know which eigenstate it is.

Symmetries: The potential energy here is symmetric about the midpoint $x = L/2$,

$$V(L/2 - x) = V(L/2 + x) \ .$$

If (in one dimension) $V(x)$ is symmetric about some point, then the first, third, fifth, ..., eigenfunctions are symmetric about this point (have *even* or *positive parity*); but the second, fourth, ..., eigenfunctions are antisymmetric (have *odd* or *negative parity*). Thus in Fig. 2(a), $\psi_1(x)$ is symmetric about $L/2$, $\psi_2(x)$ is antisymmetric, and so on. It follows that the probability *densities* $|\psi_n(x)|^2$ of energy eigenstates in symmetric one-dimensional potential wells are symmetric about the point of symmetry; see Fig. 2(b).

Figure 2(b) also shows the normalized *classical* probability densities. Classically, a particle of given energy bouncing between the walls is moving at constant speed. If 1000 snapshots were taken at random times, the particle would with equal likelihood be found anywhere in the well. Thus the normalized probability density is $1/L$ for $0 < x < L$. The classical and quantum-mechanical probability densities are very different, but the quantum density oscillates rapidly about the classical density when n is large. This is already beginning to happen in Fig. 2(b).

2·5. THE FINITE SQUARE WELL

Figure 3 shows the potential energy for the finite square well:

$$V(x) = \begin{cases} -V_0 & \text{for } -L/2 < x < +L/2 \\ 0 & \text{outside the central region} \ . \end{cases}$$

By setting $V(x) = -V_0$ in the well, we make V_0 positive for clarity in equations to come.

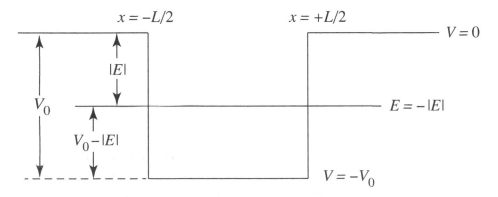

Figure 3. The finite square well.

28

The origin of x is now at the midpoint of the well—the symmetry point—so that $V(-x) = V(+x)$ for all x. This will make the mathematics simpler and the symmetries of the solutions more apparent. We are after the *bound-state* energy eigenvalues and eigenfunctions. Bound states will have negative energies in the range $-V_0 < E < 0$. Taking $V = 0$ at the top of the well (as for hydrogen) instead of at the bottom (as for the infinite square well) makes $E = 0$ the dividing line between bound and free. And we need to take $V = 0$ at the top in order to apply the results of this section to the problem of the next section.

Chapter 4 is about what happens when a beam of particles with $E > 0$ approaches this or other abrupt changes of potential energy from far away. It is very like what happens when light meets abrupt changes of index of refraction.

(a) Piecewise solutions—The time-independent Schrödinger equation is

$$\hat{H}\psi = -\frac{\hbar^2}{2m}\frac{d^2\psi}{dx^2} + V(x)\psi = E\psi, \qquad \text{or} \qquad \frac{d^2\psi}{dx^2} + \frac{2m}{\hbar^2}\big[E - V(x)\big]\psi = 0 \ .$$

Let ψ_-, ψ_c, and ψ_+ be $\psi(x)$ in the left-hand, central, and right-hand regions of x. In the central region, where $V(x) = -V_0$, we write the Schrödinger equation (with primes indicating differentiation) as

$$\psi_c'' + k^2\psi_c = 0, \qquad \text{where} \qquad k^2 \equiv \frac{2m}{\hbar^2}(E - V) = \frac{2m}{\hbar^2}(V_0 - |E|) > 0 \ ,$$

so that k is real and positive. The general solution in the central region then is

$$\psi_c(x) = B\cos kx + C\sin kx \ ,$$

where B and C are arbitrary constants.

Classically, a particle with $E < 0$ would bounce between the walls at $x = \pm L/2$, the *classical turning points*. However, the quantum wave functions penetrate into the classically forbidden regions, but with a change of character. Between the turning points, the wave functions are *concave* (oscillatory) with respect to the x axis; outside the turning points, they are are *convex* (exponential) with respect to the axis. To the left of the well, the Schrödinger equation is

$$\psi_-'' - q^2\psi_- = 0, \qquad \text{where} \qquad q^2 \equiv -\frac{2mE}{\hbar^2} = \frac{2m|E|}{\hbar^2} > 0 \ ,$$

so that q is real and positive. The general solution is

$$\psi_-(x) = A\,e^{+qx} + A'e^{-qx} \ ,$$

where A and A' are arbitrary constants. However, e^{-qx} increases without limit as x goes toward $-\infty$. Since this would place the particle off at $-\infty$, we discard this part of the mathematical solution on physical grounds, leaving $\psi_-(x) = Ae^{+qx}$. For like reasons, the physical solution to the right of the well must be $\psi_+(x) = De^{-qx}$.

We shall need both the wave functions and their first derivatives. All together,

$$\psi_-(x) = Ae^{+qx} \qquad\qquad d\psi_-/dx = +qAe^{+qx}$$
$$\psi_c(x) = B\cos kx + C\sin kx \qquad\qquad d\psi_c/dx = -kB\sin kx + kC\cos kx$$
$$\psi_+(x) = De^{-qx} \qquad\qquad d\psi_+/dx = -qDe^{-qx} \ .$$

(b) Boundary conditions—Let there be a *finite* discontinuity in a $V(x)$ at $x = x_0$ (as there is in the finite well at $x = \pm L/2$). The Schrödinger equation gives $\psi''(x)$ in terms of E, $\psi(x)$, and $V(x)$, all of which are everywhere finite, even where $V(x)$ jumps. Therefore $\psi''(x)$ is everywhere finite too. Now integrate $\psi''(x)$ across the discontinuity in $V(x)$:

$$\int_{x_0-\epsilon}^{x_0+\epsilon} \psi''(x)\, dx = \psi'(x_0 + \epsilon) - \psi'(x_0 - \epsilon) \,,$$

where ϵ is infinitesimal. Since $\psi''(x)$ is finite, the integral approaches zero as $\epsilon \to 0$. Therefore $\psi'(x)$ is continuous across x_0. (For the *in*-finite square well, $\psi'(x)$ was discontinuous at the *in*-finities in V.) By the same argument, if $\psi'(x)$ if finite everywhere, as the derivatives listed in the preceding paragraph certainly are, the wave function $\psi(x)$ itself must be continuous everywhere, including across a jump in $V(x)$. (What about $\psi''(x)$? Is it continuous or is it discontinuous at a jump in V?)

There are altogether five unknowns in ψ_-, ψ_c, and ψ_+: the four multiplicative constants A, B, C, and D, and the energy E, which is buried in the definitions of k and q. There are four constraints: the continuity of $\psi(x)$ and its derivative at each wall. These constraints lead to the energy eigenvalues and to elimination of three of the four multiplicative constants. The last constant would come from normalizing $\psi(x)$.

At $x = -L/2$, we set $\psi_- = \psi_c$ and $d\psi_-/dx = d\psi_c/dx$ and use $\sin(-\theta) = -\sin\theta$ and $\cos(-\theta) = \cos\theta$:

$$Ae^{-qL/2} = B\cos(\tfrac{1}{2}kL) - C\sin(\tfrac{1}{2}kL)$$

$$qAe^{-qL/2} = kB\sin(\tfrac{1}{2}kL) + kC\cos(\tfrac{1}{2}kL) \,.$$

Letting $\theta \equiv \tfrac{1}{2}kL$ and dividing the second equation by the first, we get

$$q = k\left(\frac{B\sin\theta + C\cos\theta}{B\cos\theta - C\sin\theta}\right).$$

In the same way, over at $x = +L/2$, we get

$$q = k\left(\frac{B\sin\theta - C\cos\theta}{B\cos\theta + C\sin\theta}\right).$$

Equating the two expressions for q/k and cross multiplying, we get

$$(B^2 + C^2)\sin\theta\cos\theta + BC(\sin^2\theta + \cos^2\theta) = (B^2 + C^2)\sin\theta\cos\theta - BC(\sin^2\theta + \cos^2\theta) \,.$$

This reduces at once to $BC = -BC$, or to $BC = 0$. Thus either $C = 0$ or $B = 0$. (Why is the case $B = C = 0$ of no interest?) We consider the two cases in turn.

(c) Symmetry and antisymmetry—When $C = 0$, the solution in the central region is $\psi_c(x) = B\cos kx$, an even function of x. Thus $\psi_c(-L/2) = \psi_c(+L/2)$, and the continuity of $\psi(x)$ then requires that $A = D$. The solutions are even functions of x, with oscillations within the well and mirror-image decaying exponentials to either side. Also, when $C = 0$, we have $q = k\tan\theta$.

When $B = 0$, the solution in the central region is $\psi_c(x) = C\sin kx$, an odd function of x. Thus $\psi_c(-L/2) = -\psi_c(+L/2)$, and the continuity of $\psi(x)$ requires that $A = -D$. Now the solutions are odd functions of x, with oscillations within the well and upside-down mirror-image decaying exponentials to either side. And now $q = -k \cot \theta$. But $\cot \theta = -\tan(\theta + \pi/2)$ (show this), so we can write $q = -k \cot \theta$ as $q = k \tan(\theta + \pi/2)$.

The eigenfunctions of the infinite square well were either symmetric or antisymmetric about the symmetry point of $V(x)$, and we have now found the same to be true of the bound eigenfunctions of the finite well. In fact, this is *always* the case for bound eigenstates in symmetric one-dimensional potential wells.

The energy eigenvalues for the even- and odd-parity solutions will come from the equations $q = k \tan \theta$ and $q = k \tan(\theta + \pi/2)$, which we rewrite. The definitions of k^2 and q^2 were

$$k^2 \equiv \frac{2m}{\hbar^2}(V_0 - |E|) \quad \text{and} \quad q^2 \equiv \frac{2m|E|}{\hbar^2} .$$

The ratio of q^2 and k^2 is

$$\frac{q^2}{k^2} = \frac{|E|}{V_0 - |E|} = \frac{V_0}{V_0 - |E|} - 1 .$$

We recall the definition $\theta \equiv kL/2$ and define a new parameter θ_0:

$$\theta^2 \equiv k^2 \frac{L^2}{4} = \frac{2m}{\hbar^2}(V_0 - |E|)\frac{L^2}{4} \quad \text{and} \quad \theta_0^2 \equiv \frac{2m}{\hbar^2} V_0 \frac{L^2}{4} .$$

And then

$$\frac{q^2}{k^2} = \frac{V_0}{V_0 - |E|} - 1 = \frac{\theta_0^2}{\theta^2} - 1 .$$

Thus, finally, the equations $q = k \tan \theta$ and $q = k \tan(\theta + \pi/2)$ may be written like this:

$$\tan \theta = \left(\frac{\theta_0^2}{\theta^2} - 1\right)^{1/2} \quad \text{and} \quad \tan\left(\theta + \frac{\pi}{2}\right) = \left(\frac{\theta_0^2}{\theta^2} - 1\right)^{1/2} ,$$

where $\theta_0^2 = mV_0L^2/2\hbar^2$. All the constants and parameters in the problem, m, $\hbar$, L, and V_0, are in θ_0, and θ_0 is dimensionless (show this directly). And only in the combination θ_0 do the parameters determine the energy eigenvalues.

(d) **Roots and eigenvalues**—The last equations have the form $f(\theta) = g(\theta)$, and we want the roots, the values of θ for which the equations are true. Because the equations are transcendental, we can only find the roots graphically or numerically, not algebraically. Figure 4 shows a graphical solution of a (non-transcendental!) equation. Figure 5 shows $\tan \theta$ and $\tan(\theta + \pi/2)$ as alternating solid and dashed curves, repeating forever. The down-sloping curves are

$$g(\theta) \equiv \left(\frac{\theta_0^2}{\theta^2} - 1\right)^{1/2}$$

for three *different* values of θ_0, and thus for three *different* sets of wells. For each value of θ_0, $g(\theta)$ is large when $\theta \ll \theta_0$, swoops down as θ increases, and strikes the θ axis vertically at $\theta = \theta_0$. Since $\theta_0^2 = mV_0L^2/2\hbar^2$, a wider and/or deeper well means a larger value of θ_0.

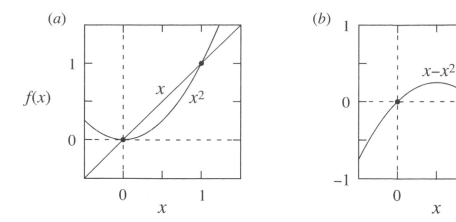

Figure 4. (a) The solutions of $x = x^2$ are the values of x at the crossings of the curves x and x^2. (b) They are also the values of x at the crossings of $x - x^2$ with the x axis.

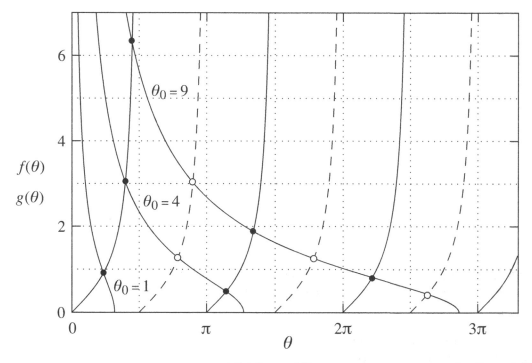

Figure 5. The solutions of $\tan\theta = (\theta_0^2/\theta^2 - 1)^{1/2}$ and $\tan(\theta + \pi/2) = (\theta_0^2/\theta^2 - 1)^{1/2}$ are the values of θ at the crossings of the curves. Here $(\theta_0^2/\theta^2 - 1)^{1/2}$ is drawn for three different values, $\theta_0 = 1$, 4, and 9 rad.

The positions of the intersections, and thus the values of the roots, are entirely determined by the single parameter θ_0. For $\theta_0 < \frac{1}{2}\pi$, there is only one root; for $\frac{1}{2}\pi \leq \theta_0 < \pi$, there are two roots; and so on. The three values of θ_0 shown in Fig. 5 lead to one, three, and six roots, respectively. Let θ_1, θ_2, θ_3, ..., be the roots in order of increasing magnitude. Each root gives an energy eigenvalue. The symmetries of the corresponding eigenfunctions will alternate between even and odd.

32

In (c), θ_n^2 was defined to be proportional to $(V_0 - |E_n|)$ and θ_0^2 to V_0, so the ratio is

$$\boxed{\frac{\theta_n^2}{\theta_0^2} = \frac{V_0 - |E_n|}{V_0}} \,.$$

As Fig. 3 showed, V_0 is the depth of the well and $|E_n|$ is how far down the energy level is from the top of the well. Thus the ratio in the box is the *fraction of the way the level is up from the bottom of the well.* Or solving for E_n, we get

$$E_n = -\left(1 - \frac{\theta_n^2}{\theta_0^2}\right) V_0 \,.$$

Larger roots give higher (less negative) values of E_n.

Figure 6 shows the energies for $\theta_0 = 4$. It also shows where the first two energies of the infinite square well of the same width would be. If, for a given L, V_0 is made very large, then so is θ_0, and in Fig. 5 the left-most intersections would be high up on the tangent curves, where $\theta_n \approx \frac{1}{2} n\pi$. The corresponding energies, measured from the bottom of the well, are

$$V_0 - |E_n| = V_0 \frac{\theta_n^2}{\theta_0^2} = \frac{2\hbar^2}{mL^2} \theta_n^2 \approx n^2 \frac{\pi^2 \hbar^2}{2mL^2} \,,$$

approximately those of the infinite square well.

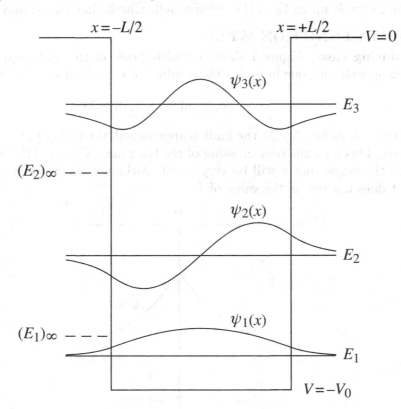

Figure 6. The bound states of the finite square well when $\theta_0 = 4$. The dashed lines show where the first two levels of the infinite square well of the same width would be.

(e) Some numbers—Here for the values of θ_0 shown in Fig. 5 are the energies as *fractions* of V_0 when measured from the *bottom* of the well:

θ_0 (rad) $n =$	1	2	3	4	5	6
1	0.5462					
4	0.0980	0.3827	0.8079			
	(1.0)	(3.905)	(8.244)			
9	0.0247	0.0984	0.2203	0.3886	0.5995	0.8427
	(1.0)	(3.984)	(8.919)	(15.73)	(24.27)	(34.12)

For example, when $\theta_0 = 9$, E_4 is 38.86% up from the bottom of the well. The numbers in parentheses are the ratios of energies (as measured from the bottom) to E_1, the ground-state energy. As θ_0 gets larger, the ratios approach the 1:4:9:... ratios of the infinite square well. For example, when $\theta_0 = 9$, $E_4/E_1 = 15.73 \approx 4^2$.

Note from Fig. 5 that the roots, and therefore the energy levels, of a finite square well would alternate with those of the infinite well of the same width. In the finite well, the eigenfunctions spread out a little past the walls, and the finite E_n lies lower than the infinite E_n. Thus if, say, there are 20 levels in a finite square well of width L and depth V_0, there will be either 19 or 20 levels up to V_0 in the infinite well. Check that this squares with Fig. 6.

2·6. THE DELTA-FUNCTION WELL

(a) A limiting case—Figure 7 shows a limiting case of the finite square well—a very narrow, very deep well, but one in which the product of the width and depth is finite:

$$L \to 0, \quad V_0 \to \infty, \quad \text{and} \quad g \equiv LV_0 \text{ is finite .}$$

(This g is not the $g(\theta)$ of Sec. 5.) As the limit is approached, the ruling parameter $\theta_0 \propto \sqrt{V_0}\, L$ approaches zero. This puts the only crossing of the $\tan\theta$ and $(\theta_0^2/\theta^2 - 1)^{1/2}$ curves very near the origin, and the single root θ will be very small. Although this tells us there is only one bound state, it does not tell us the value of E.

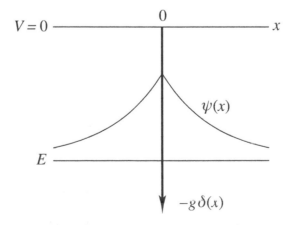

Figure 7. A very narrow, very deep square well, and its only bound state; $\psi(x)$ is measured from the E line.

However, we can back up a bit, sometimes a useful tactic. We got $\tan\theta = (\theta_0^2/\theta^2 - 1)^{1/2}$ from $q = k\tan\theta$, and in Sec. 5(b) we defined $\theta \equiv kL/2$. Since the solution θ is here very small, $q = k\tan\theta$ reduces to $q \simeq k\theta = k^2L/2$. Then, since $q^2 \equiv 2m|E|/\hbar^2$ and $k^2 \equiv 2m(V_0 - |E|)/\hbar^2$, $q = k^2L/2$ becomes

$$\left(\frac{2m|E|}{\hbar^2}\right)^{1/2} = \frac{2m}{\hbar^2}(V_0 - |E|)\frac{L}{2} = \frac{mg}{\hbar^2} \ .$$

The last step follows because $V_0L \to g$, and $|E|L \to 0$ (E is finite). Solving for E, we get

$$\boxed{E = -|E| = -\frac{mg^2}{2\hbar^2} \ .}$$

Figure 7 shows the wave function, which has the usual decaying exponentials outside the well and a kink at $x = 0$.

(b) The Dirac delta function—Figure 8 shows two sequences of functions that grow ever taller and narrower while the areas under the curves remain constant at unity. The common limit of these (and of similar) sequences is the Dirac function $\delta(x)$, which has these defining properties:

$$\boxed{\delta(x) = 0 \text{ for } x \neq 0, \quad \text{and} \quad \int \delta(x)\,dx = 1 \ ,}$$

where the range of integration includes $x = 0$. The integral is (by definition) dimensionless, so $\delta(x)$ has the dimension of inverse length. To put the spike at $x = a$ instead of at $x = 0$, we write $\delta(x - a)$. In engineering circles, the δ function is often called the impulse function. It predates Dirac (whose college training was in engineering).

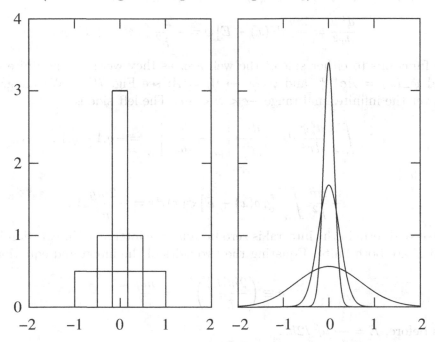

Figure 8. Two sequences of functions that grow ever taller and narrower while the areas under the curves remain constant at unity.

35

In applications, $\delta(x-a)$ is most often found multiplying an ordinary function $f(x)$ in an integrand, as in $\int f(x)\,\delta(x-a)\,dx$. Since $f(x)$ is then multiplied by zero everywhere except at $x = a$, only its value at $x = a$ matters:

$$\int f(x)\,\delta(x-a)\,dx = \int f(a)\,\delta(x-a)\,dx = f(a)\int \delta(x-a)\,dx = f(a)\;,$$

when the range of integration includes $x = a$. Thus, for example,

$$\int_0^\pi e^{i\theta}\,\delta(\theta - \pi/2)\,d\theta = i\;.$$

We can do no better here than quote from Chap. 5, "The Impulse [or δ] Symbol," of the fine book, *The Fourier Integral and Its Applications*, R.N. Bracewell (3rd ed.), p. 74:

> The important attribute of an impulse is the integral; the precise details of form [rectangular, gaussian, ...] are unimportant ... Point masses, point charges, point sources, concentrated forces ... and the like are familiar and accepted entities in physics. Of course, these things do not exist. Their conceptual value stems from the fact that the impulse response ... may be indistinguishable ... from the response due to a physically realizable pulse. It is then a convenience to have a name for pulses which are so brief and intense that making them any briefer and more intense does not matter.

(c) The delta-function well—We solve the deep, narrow well again, writing the potential energy in terms of $\delta(x)$ and integrating the Schrödinger equation directly. The integral $\int V(x)\,dx$ across a finite square well L wide and V_0 deep is $-V_0 L$, so in our limiting case we may write $V(x) = -V_0 L\,\delta(x) = -g\,\delta(x)$. The Schrödinger equation is

$$\frac{d^2\psi}{dx^2} = \frac{2m}{\hbar^2}\big[V(x) - E\big]\psi = -\frac{2m}{\hbar^2}\big[g\,\delta(x) + E\big]\,\psi\;.$$

The wave functions to either side of the well are, as they were for the finite well, $\psi_+(x) = Ae^{-qx}$ and $\psi_-(x) = Ae^{+qx}$, and $\psi_\pm(x \to 0) = A$; see Fig. 7(b). We integrate the above equation over the infinitesimal range $-\epsilon < x < +\epsilon$. The left side is

$$\int_{-\epsilon}^{+\epsilon} \frac{d^2\psi}{dx^2}\,dx = \frac{d\psi_+}{dx}\bigg|_{+\epsilon} - \frac{d\psi_-}{dx}\bigg|_{-\epsilon} = -qA - qA\;.$$

The right side is

$$-\frac{2m}{\hbar^2}\int_{-\epsilon}^{+\epsilon}\big[g\,\delta(x) + E\big]\,\psi(x)\,dx = -\frac{2mg}{\hbar^2}A\;,$$

where the second term in the integral is zero because the interval of integration is infinitesimal and ψ and E are both finite. Equating the two sides of the integrated equation, we get

$$q \equiv \left(\frac{2m|E|}{\hbar^2}\right)^{1/2} = \frac{mg}{\hbar^2}\;,$$

or, just as before, $E = -mg^2/2\hbar^2$.

The results would be the same for a very deep, very narrow gaussian or triangular well: "... the precise details of form are unimportant ..."

36

2·7. SCHRÖDINGER IN THREE DIMENSIONS

(a) Rectangular coordinates—In three dimensions and rectangular coordinates, the classical Hamiltonian is

$$H = \frac{1}{2m}(p_x^2 + p_y^2 + p_z^2) + V(x, y, z) .$$

By symmetry, the quantum-mechanical operators for p_x, p_y, and p_z all have the same form, so $\hat{p}_y = -i\hbar\,\partial/\partial y$ and $\hat{p}_z = -i\hbar\,\partial/\partial z$ (hats on operators, like $\hat{H}$). Thus the time-dependent Schrödinger equation in three dimensions is

$$i\hbar\frac{\partial\Psi}{\partial t} = \hat{H}\Psi = -\frac{\hbar^2}{2m}\left(\frac{\partial^2\Psi}{\partial x^2} + \frac{\partial^2\Psi}{\partial y^2} + \frac{\partial^2\Psi}{\partial z^2}\right) + V(x, y, z)\,\Psi ,$$

where now $\Psi = \Psi(x, y, z, t)$. The quantity in parentheses is the Laplacian operator ∇^2 operating on Ψ, so we may write the equation more compactly as

$$i\hbar\frac{\partial\Psi}{\partial t} = \hat{H}\Psi = -\frac{\hbar^2}{2m}\,\nabla^2\Psi + V\Psi .$$

When $V = V(x, y, z)$, with no time dependence, we again look for product solutions that separate space and time: $\Psi(x, y, z, t) = \psi(x, y, z)\,T(t)$. In the same way as before, we get

$$\frac{i\hbar}{T}\frac{dT}{dt} = \frac{1}{\psi}\left(-\frac{\hbar^2}{2m}\nabla^2\psi + V(x, y, z)\,\psi\right) = E ,$$

where E is the separation constant. The equation for $T(t)$ is the same as it was for one dimension. However, now the time-independent equation is still a partial differential equation,

$$\boxed{\hat{H}\psi = -\frac{\hbar^2}{2m}\nabla^2\psi + V(x, y, z)\,\psi = E\psi .}$$

Suppose that, for some V, we find an eigenvalue E of this equation. Then the time dependence of the corresponding wave function $\Psi(x, y, z, t)$ will be $e^{-iEt/\hbar}$, just as it was in one dimension.

Sometimes $V(x, y, z)$ is a *sum* of three functions, one for each variable,

$$V(x, y, z) = V_x(x) + V_y(y) + V_z(z) .$$

Then the eigenvalue equation may be rearranged like this:

$$-\frac{\hbar^2}{2m}\frac{\partial^2\psi}{\partial x^2} + V_x(x)\psi - \frac{\hbar^2}{2m}\frac{\partial^2\psi}{\partial y^2} + V_y(y)\psi - \frac{\hbar^2}{2m}\frac{\partial^2\psi}{\partial z^2} + V_z(z)\psi = E\psi .$$

The form suggests that we try to break this into separate ordinary differential equations. Trying a product form $\psi(x, y, z) = X(x)Y(y)Z(z)$ and dividing by XYZ, we get

$$\frac{1}{X}\left(-\frac{\hbar^2}{2m}\frac{d^2X}{dx^2} + V_x(x)X\right) + \frac{1}{Y}\left(-\frac{\hbar^2}{2m}\frac{d^2Y}{dy^2} + V_y(y)Y\right) + \frac{1}{Z}\left(-\frac{\hbar^2}{2m}\frac{d^2Z}{dz^2} + V_z(z)Z\right) = E .$$

37

The first term depends only on x, the second only on y, etc. And x, y, and z are independent variables. Suppose we fix y and z and vary x. Then only the first term in the above equation could vary; but if it did the sum on the left would not remain constant at E. We conclude that, for a product solution to exist, each of the three terms on the left of the equation must be constant. Let the constant values be E_x, E_y, and E_z. Then $E_x + E_y + E_z = E$, and

$$-\frac{\hbar^2}{2m}\frac{d^2X}{dx^2} + V_x(x)X = E_xX \ ,$$

with equations of the same form for Y and Z. These are just one-dimensional Schrödinger equations. If we can solve them, getting eigenvalues and eigenfunctions E_{n_x} and $X_{n_x}(x)$, etc., then the three-dimensional product solutions are

$$\psi_{n_xn_yn_z}(x,y,z) = X_{n_x}(x)Y_{n_y}(y)Z_{n_z}(z) \qquad \text{with} \qquad E_{n_xn_yn_z} = E_{n_x} + E_{n_y} + E_{n_z} \ .$$

The eigenvalues are sums of one-dimensional eigenvalues and the eigenfunctions are products of one-dimensional eigenfunctions.

As an example, consider a particle confined to a three-dimensional rectangular box of sides a, b, and c. The potential energy is the *sum* of

$$V_x(x) = 0 \text{ for } 0 \le x \le a, \ \infty \text{ otherwise}$$
$$V_y(y) = 0 \text{ for } 0 \le y \le b, \ \infty \text{ otherwise}$$
$$V_z(z) = 0 \text{ for } 0 \le z \le c, \ \infty \text{ otherwise} \ .$$

The sum confines the particle to the box. The separation of variables leads to three one-dimensional infinite-square-well problems, solved in Sec. 4. The boundary conditions are that $\psi(x,y,z) = 0$ on the six faces of the box. The eigenvalues are

$$E_{n_xn_yn_z} = \frac{\pi^2\hbar^2}{2m}\left(\frac{n_x^2}{a^2} + \frac{n_y^2}{b^2} + \frac{n_z^2}{c^2}\right),$$

where each of n_x, n_y, and n_z can (independently) be any of 1, 2, 3, ... And for points inside the box, the normalized eigenfunctions are

$$\psi_{n_xn_yn_z}(x,y,z) = \sqrt{\frac{8}{abc}}\sin\left(\frac{n_x\pi x}{a}\right)\sin\left(\frac{n_y\pi y}{b}\right)\sin\left(\frac{n_z\pi z}{c}\right) \ .$$

The x-dependent factor in $\psi(x,y,z)$ makes $\psi = 0$ on the $x = 0$ and $x = a$ faces of the box, etc.

The energy levels in one-, two-, and three-dimensional boxes, where all sides are the same length L, are given by the first one, two, or three terms of

$$\boxed{E_{n_xn_yn_z} = (n_x^2 + n_y^2 + n_z^2)\frac{\pi^2\hbar^2}{2mL^2} \ .}$$

Figure 9 shows the lowest energy levels. There is a level for each value of n_x, and for each set of (n_x, n_y) and of (n_x, n_y, n_z) quantum numbers. The states with, say, $(n_x, n_y) = (2,3)$ and $(3,2)$ have the same energy (are *two-fold degenerate*) but have linearly independent wave functions. The states $(2,2,3)$ and $(1,3,4)$ belong to three- and six-fold degenerate levels. (What are the *four* sets of quantum numbers that give $n_x^2 + n_y^2 + n_z^2 = 27$? The *nine* sets that give $n_x^2 + n_y^2 + n_z^2 = 38$?)

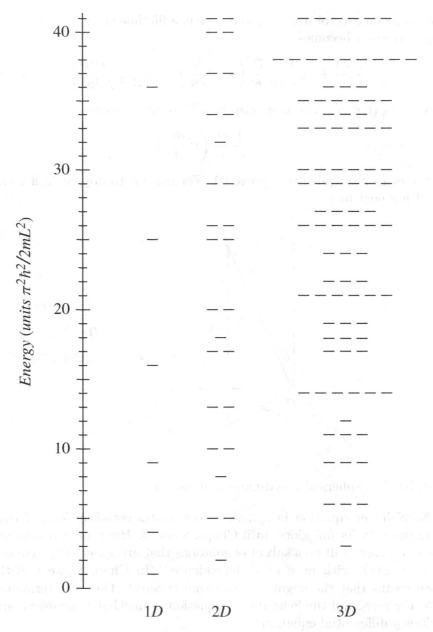

Figure 9. The lowest energy levels for one-, two-, and three-dimensional infinite square wells of side L.

In the figure, the densities of levels in the two and three-dimensional cases are rather ragged. However, on a larger scale, the average density of levels is constant in two dimensions and increases with energy in three (see Chap. 14).

(b) **Spherical coordinates**—Figure 10 shows the spherical coordinates—the radial distance r, the polar angle θ, and the azimuthal angle ϕ. Their ranges are $0 \leq r \leq \infty$, $0 \leq \theta \leq \pi$, and $0 \leq \phi \leq 2\pi$. (*Warning*: θ and ϕ are interchanged in many mathematical writings.) When the potential energy is central—that is, when $V = V(r)$—it is far easier to work in spherical than rectangular coordinates. The Laplacian operator in spherical coordinates is derived in courses on vector analysis, and once in a lifetime is enough. The time-independent Schrödinger equation becomes

$$-\frac{\hbar^2}{2m}\left[\frac{1}{r}\frac{\partial^2(r\psi)}{\partial r^2}+\frac{1}{r^2\sin\theta}\frac{\partial}{\partial\theta}\left(\sin\theta\frac{\partial\psi}{\partial\theta}\right)+\frac{1}{r^2\sin^2\theta}\frac{\partial^2\psi}{\partial\phi^2}\right]+V(r)\,\psi=E\psi\ ,$$

where now $\psi=\psi(r,\theta,\phi)$. The first term in $\nabla^2\psi$ is often written

$$\frac{1}{r^2}\frac{\partial}{\partial r}\left(r^2\frac{\partial\psi}{\partial r}\right),$$

but the two forms are equivalent (prove it). For reasons to appear in a moment, we prefer the form with r next to ψ.

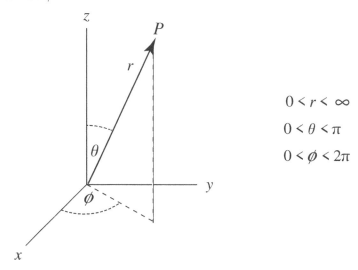

$$0<r<\infty$$
$$0<\theta<\pi$$
$$0<\phi<2\pi$$

Figure 10. The spherical coordinates r, θ, and ϕ.

The Schrödinger equation in spherical coordinates certainly looks formidable, and we shall not tackle it in its full glory until Chaps. 8 and 9. However, when as we are assuming here, $V=V(r)$, there will be a subset of solutions that are *spherically symmetric*—solutions in which $\psi=\psi(r)$, with no θ or ϕ dependence. (In Chap. 8, we find that no angular dependence means that the angular momentum is zero.) Then the terms involving $\partial\psi/\partial\theta$ and $\partial\psi/\partial\phi$ are zero, and the Schrödinger equation, somewhat rearranged, greatly simplifies to the *ordinary* differential equation

$$\frac{d^2(r\psi)}{dr^2}+\frac{2m}{\hbar^2}\big[E-V(r)\big]\,r\psi=0\ .$$

With $u(r)\equiv r\psi(r)$, this becomes

$$\frac{d^2u}{dr^2}+\frac{2m}{\hbar^2}\big[E-V(r)\big]\,u=0\ .$$

The equation is for $u(r)$, not $\psi(r)$, but it has exactly the same form as the *one*-dimensional equation for $\psi(x)$,

$$\frac{d^2\psi}{dx^2}+\frac{2m}{\hbar^2}\big[E-V(x)\big]\,\psi=0\ .$$

40

However, whereas the range of x is $-\infty$ to $+\infty$, the range of r, again, is 0 to $+\infty$. Also, because $u(r) = r\psi(r)$ and $\psi(r)$ is everywhere finite, $u(0)$ is zero—a boundary condition.

In spherical coordinates, the infinitesimal volume element is

$$\boxed{d^3\mathbf{r} = r^2 \sin\theta \, dr \, d\theta \, d\phi \, .}$$

The normalization integral is

$$\int_{r=0}^{\infty} \int_{\theta=0}^{\pi} \int_{\phi=0}^{2\pi} |\psi(r,\theta,\phi)|^2 \, r^2 \sin\theta \, dr \, d\theta \, d\phi = 1 \, .$$

When ψ depends only on r, this becomes

$$4\pi \int_0^\infty |\psi(r)|^2 \, r^2 \, dr = 4\pi \int_0^\infty |u(r)|^2 \, dr = 1 \, .$$

For the integral to be finite, it is necessary that $u(r)$, not just $\psi(r)$, vanish as $r \to \infty$.

As an example, consider a particle of mass m confined to the interior of a sphere of radius R:

$$V(r) = \begin{cases} 0 & \text{for } 0 \le r \le R \\ \infty & \text{for } r > R \, . \end{cases}$$

For the subset of solutions that are spherically symmetric, where $\psi = \psi(r)$, the eigenvalue problem is

$$\frac{d^2 u}{dr^2} + \frac{2mE}{\hbar^2} u = 0 \, ,$$

with $u(0) = u(R) = 0$. This is mathematically identical to the one-dimensional infinite-square-well problem for $\psi(x)$,

$$\frac{d^2 \psi}{dx^2} + \frac{2mE}{\hbar^2} \psi = 0 \, ,$$

with $\psi(0) = \psi(L) = 0$. Equations of the same form have solutions of the same form. Thus the *spherically symmetric* solutions here are

$$E_n = n^2 \frac{\pi^2 \hbar^2}{2mR^2} \quad \text{and} \quad \psi_n(r) = \frac{u_n(r)}{r} = \frac{1}{\sqrt{2\pi R}} \frac{1}{r} \sin\left(\frac{n\pi r}{R}\right)$$

for $0 \le r \le R$. (Why is the normalization different from that for the one-dimensional well?)

2·8. TWO IMPORTANT ENERGY EIGENSTATES

The Hooke's-law and Coulomb forces are the most important forces giving bound states. It will take a full chapter to solve the Schrödinger equation for each of them. In the meantime it will be useful to have as examples the important *ground-state* solutions.

(a) Forces and potential energies—The Hooke's-law force is linear in displacement:

$$F(x) = -kx \text{ in one dimension}, \qquad \mathbf{F}(\mathbf{r}) = -kr\hat{\mathbf{r}} \text{ in three},$$

where k is the force constant and $\hat{\mathbf{r}}$ is the unit vector in the radial direction. The angular frequency of the classical oscillator is $\omega = \sqrt{k/m}$, and we usually use $m\omega^2$ in place of k. The corresponding potential energies, from $F(x) = -dV/dx$ or $F(r) = -dV/dr$, are

$$V(x) = -\int_0^x (-m\omega^2 x) \, dx = \frac{1}{2} m\omega^2 x^2 \quad \text{and} \quad V(r) = -\int_0^r (-m\omega^2 r) \, dr = \frac{1}{2} m\omega^2 r^2 \, ,$$

where $V(0) = 0$. Figure 11(a) shows $V(x)$. The oscillator energies are non-negative.

We have now used the symbol k for the wave number $2\pi/\lambda$, and for the oscillator spring constant, and we are about to use it for still something else. But if you know what it going on, it should cause no confusion. Context is all.

The Coulomb force is

$$\mathbf{F}(\mathbf{r}) = \frac{q_1 q_2}{4\pi\epsilon_0 r^2}\,\hat{\mathbf{r}} = -\frac{ke^2}{r^2}\,\hat{\mathbf{r}}\ ,$$

where on the right we take the charges to be $+e$ and $-e$, as for the proton and electron in hydrogen; and for brevity we let $k \equiv 1/4\pi\epsilon_0$ (*not* a force constant). The potential energy is

$$V(r) = -\int_\infty^r \left(-\frac{ke^2}{r^2}\right) dr = -\frac{ke^2}{r}\ ,$$

where $V(\infty) = 0$. Figure 11(b) shows $V(r)$. *Bound* states of hydrogen have negative energies.

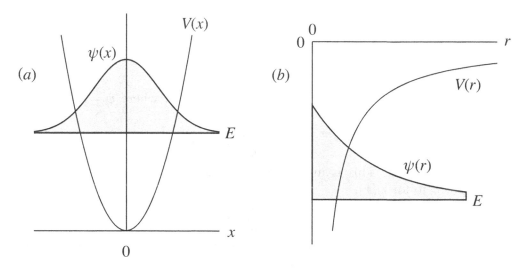

Figure 11. Potential energies (a) $V(x)$ for the harmonic oscillator, and (b) $V(r)$ for the hydrogen atom. Also shown are the ground-state energies and wave functions. The E lines are used as the axes for the ψ functions.

We took the oscillator potential energy to be zero at the origin, where the force is zero. We took the Coulomb potential energy to be zero at infinity, where the force is zero. Provided that we do so consistently, we can always add a constant to potential energies, but we cannot calculate the oscillator potential energy integrating from $x = \infty$ or the Coulomb energy integrating from $r = 0$. (Why not?)

(b) Oscillator ground state—The time-independent Schrödinger equation for the one-dimensional oscillator is

$$-\frac{\hbar^2}{2m}\frac{d^2\psi}{dx^2} + \frac{1}{2}m\omega^2 x^2\psi = E\psi\ .$$

The ground-state wave function will be symmetric about $x = 0$, have no nodes, vanish as $x \to \pm\infty$ fast enough for $\psi(x)$ to be square integrable, and be finite, continuous, and differentiable over the entire range $-\infty < x < +\infty$. Since there are no discontinuities in $V(x)$, there are no *piecewise* solutions. One function has to do the job for all x.

42

The Schrödinger equation tells us that we need a $\psi(x)$ whose second derivative has the form $(cx^2 + d)\psi$. A moment's scribbling shows that

$$\psi(x) = e^{-ax^2}$$

is such a function—and it also satisfies all the other requirements. This is the gaussian function, the bell-shaped curve that occurs in statistics and other fields. Exponentials with negative exponents die away nicely, and the exponent $-ax^2$ makes the function smooth and symmetric. Putting this $\psi(x)$ into the Schrödinger equation shows that it is indeed a solution, provided that $a = m\omega/2\hbar$, and that $E = \hbar\omega/2$. The details are left for a problem. The lowest energy is not at the bottom of the well; nor was it at the bottom of the square wells.

It remains to normalize $\psi(x)$. The integral is important and the method is clever. We calculate the *area* integral

$$\left(\int_{-\infty}^{+\infty} e^{-2ax^2}\,dx\right)\left(\int_{-\infty}^{+\infty} e^{-2ay^2}\,dy\right) = \int_{-\infty}^{+\infty}\int_{-\infty}^{+\infty} e^{-2a(x^2+y^2)}\,dx\,dy \ .$$

This is the *square* of the integral we are after. (Why are there 2's in the exponents?) First, we transform to polar coordinates, where $x^2 + y^2 = r^2$ and the area element is $r\,dr\,d\theta$. The integral becomes

$$\int_0^\infty \int_0^{2\pi} e^{-2ar^2} r\,dr\,d\theta = 2\pi \int_0^\infty e^{-2ar^2} r\,dr \ .$$

Then we substitute s for r^2, in which case $ds = 2r\,dr$, and the integral is trivial:

$$\pi \int_0^\infty e^{-2as}\,ds = \pi \left(\frac{e^{-2as}}{-2a}\right)_0^\infty = \frac{\pi}{2a} \ .$$

Thus, since for the oscillator $a = m\omega/2\hbar$,

$$\int_{-\infty}^{+\infty} e^{-2ax^2}\,dx = \sqrt{\frac{\pi}{2a}} = \sqrt{\frac{\pi\hbar}{m\omega}} \ .$$

Then the ground-state energy and normalized wave function are

$$\boxed{E_0 = \frac{\hbar\omega}{2} \quad \text{and} \quad \psi_0(x) = \left(\frac{m\omega}{\pi\hbar}\right)^{1/4} e^{-m\omega x^2/2\hbar} \ .}$$

Figure 11(a) shows E_0 and $\psi_0(x)$. For the oscillator (as opposed to square wells), the numbering of states starts with zero, not one. This is because of the way polynomials that occur in the full solution of the oscillator problem are numbered (see Chap. 6).

(c) **Hydrogen ground state**—In Sec. 7, we saw that the Schrödinger equation with a central potential energy greatly simplifies when ψ has no θ or ϕ dependence. It is then

$$\frac{d^2(r\psi)}{dr^2} + \frac{2m}{\hbar^2}\big[E - V(r)\big]r\psi = 0 \ .$$

For the Coulomb $V(r) = -ke^2/r$, this is

$$\frac{d^2(r\psi)}{dr^2} + \frac{2m}{\hbar^2}\left(E + \frac{ke^2}{r}\right)r\psi = 0 .$$

Even in this case of no angular dependence, there are an infinite number of bound-state ($E < 0$) solutions, but here we are only after the ground state. The equation tells us that we need a $u(r) = r\psi(r)$ whose second derivative has the form $(c + d/r)u$. The function $u(r) = re^{-br}$, or $\psi(r) = e^{-br}$, satisfies this and all other requirements, provided that $b = mke^2/\hbar^2 = 1/a_0$, where $a_0 = 0.529 \times 10^{-10}$ m is the Bohr radius, and that $E = -mk^2e^4/2\hbar^2 = -13.6$ eV. The energy and normalized wave function are

$$E_1 = -\frac{mk^2e^4}{2\hbar^2} \quad \text{and} \quad \psi_1(r) = \frac{1}{\sqrt{\pi a_0^3}}e^{-r/a_0} .$$

We again leave details for a problem. Figure 11(b) shows the wave function.

It will sometimes be useful to consider the electron to be smeared out over space with a volume charge density of $\rho(r) = -e\,|\psi(r)|^2$. The volume integral of this $\rho(r)$ is of course $-e$.

2·9. QUALITATIVE PROPERTIES OF BOUND ENERGY EIGENSTATES

Here, without proofs, are some properties of bound energy eigenstates. Most of the properties are illustrated by the examples of this chapter and the problems that follow. You might try to prove (3) from how the Schrödinger equation relates $d^2\psi(x)/dx^2$ to $\psi(x)$ in one dimension, or $d^2u(r)/dr^2$ to $u(r)$, where $u(r) = r\psi(r)$, in three.

(1) The energies of bound states are quantized. See Secs. 4 through 8.

(2) All the energies lie above the minimum of the potential energy $V(x)$. See Secs. 4 through 8.

(3) In one dimension, the eigenfunctions $\psi_n(x)$ are concave toward the x axis (oscillatory) in regions of x where $E > V(x)$, but convex toward the axis (exponential) where $E < V(x)$. See Figs 6 and 11(a); in Fig. 11(b), it would be true for $u(r) \equiv r\psi(r)$.

(4) The number of nodes of $\psi_n(x)$ increases with energy. See Figs. 2 and 6.

(5) If $V(-x) = V(+x)$, the eigenfunctions are alternately even and odd about $V(0)$. See Figs. 2 and 6, and Fig. 6·3.

(6) In one dimension, there is only one eigenstate per bound energy eigenvalue (the levels are non-degenerate).

(7) In one dimension, the eigenfunctions may be taken to be real. See Figs. 2 and 6.

(8) For large n, the quantum probability densities $|\psi_n(x)|^2$ on average track classical densities. See Fig. 6·3.

(9) For large n, in $E > V(x)$ regions, the wave-function amplitude is larger where $V(x)$ is larger. See Fig. 6·3.

PROBLEMS

Any assignment should include the first four problems, which are are simple but important.

1. *Even and odd functions.*

Illustrate your answers here with sketches, using $\cos\theta$ and $\sin\theta$ as examples of even and odd functions.

    (a) If $f(x)$ is an even function about $x = 0$ (has *even parity*), and $g(x)$ is an odd function (has *odd parity*), what is the parity of $f^2(x)$? Of $g^2(x)$? Of $f(x)g(x)$?

    (b) If $f(x)$ is an odd function about $x = 0$, what is the value of $\int f(x)\,dx$ over the interval from $-a$ to $+a$?

    (c) What is the parity of df/dx if $f(x)$ is even? If $f(x)$ is odd? What is the value of $\int f(x)\,(df/dx)\,dx$ over the interval $-a$ to $+a$ in either case?

    It saves a lot of work to be able to see at a glance that an integral is zero.

2. *A sloshing particle.*

The probability density of a single energy eigenstate, $|\Psi_n(x,t)|^2 = |\psi_n(x)|^2$, has no time dependence. Time dependence in a probability density comes from the cross terms in a superposition of states with different energies.

    At $t = 0$, the wave function for a particle in an infinite square well L wide is

$$\Psi(x,0) = \frac{1}{\sqrt{2}}\left[\psi_1(x) + \psi_2(x)\right] ,$$

where ψ_1 and ψ_2 are the two lowest normalized energy eigenfunctions.

    (a) Show that $\Psi(x,0)$ is normalized.

    (b) Write down $\Psi(x,t)$ for times $t > 0$ and prove that it too is normalized.

    (c) Calculate the mean value $\langle x \rangle = \int_0^L x\,|\Psi(x,t)|^2\,dx$. The integral is over x but not t, so $\langle x \rangle$ is still a function of t. An identity $\sin\theta_1\sin\theta_2 = \frac{1}{2}[\cos(\theta_1 - \theta_2) - \cos(\theta_1 + \theta_2)]$ might make finding an integral easier. The answer is $\langle x \rangle = \frac{1}{2}L[1 - 0.3603\cos(\omega t)]$, where the oscillation angular frequency is $\omega = (E_2 - E_1)/\hbar$.

    (d) Sketch $|\Psi(x,t)|^2$ versus x at $t = 0$ and after half a period.

3. *Half wells.*

Let the energy eigenvalues and normalized eigenfunctions for a potential energy $V(x)$ that is symmetric about $x = 0$ be $E_1 < E_2 < E_3 \ldots$ and $\psi_1(x)$, $\psi_2(x)$, $\psi_3(x)$, ... What are the eigenvalues and normalized eigenfunctions for

$$V_{1/2}(x) = \begin{cases} \infty & \text{for } x < 0 \\ V(x) & \text{for } x > 0 \end{cases} ?$$

This requires *no* calculation. Once seen, the answer ought to be obvious.

4. *Reading a wave function.*

Part (a) of the figure below shows an energy eigenfunction for the potential well shown in (b).

    (a) Which energy level is it? Count the ground state as level 1.

    (b) Is the energy of this state above or below the top of the central plateau in the well?

(c) Sketch the ground-state and first-excited-state wave functions. In particular, get the curvatures of the wave functions in the various regions right.

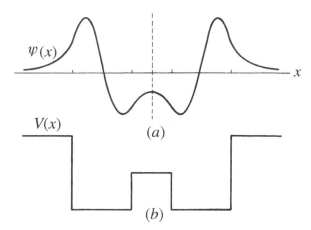

See Chap. 3 of A.P. French and E.F. Taylor, *Introduction to Quantum Physics*, from which the figure is taken, for a number of examples of the relations between $\psi(x)$ and $V(x)$.

5. *The finite square well, I.*

Below is the hunt-and-peck way I used a calculator to find that the $n = 4$ value in the $\theta_0 = 9$ rad row of the table in Sec. 5(e) is 0.3886. This is an *odd*-parity solution, so it involves $\tan(\theta + 90°)$, not $\tan\theta$. One can of course work with radians, but degrees gave me a better sense of where I was at: $\theta_0 = 9$ rad $= 515.66°$.

My starting guess, $\theta = 320°$, came from looking at Fig. 5. The fourth crossing of the $\theta_0 = 9$ curve is a little past halfway between $3\pi/2$ rad $= 270°$ and 2π rad $= 360°$. Then because $\theta = 320°$ makes $\tan(\theta + 90°)$ a little smaller than $[(\theta_0/\theta)^2 - 1]^{1/2}$, I increased θ a little, which *increases* $\tan(\theta + 90°)$ and *decreases* $[(\theta_0/\theta)^2 - 1]^{1/2}$. After that, it is just a matter of homing in:

$\theta(°)$	$\tan(\theta + 90°)$	$[(\theta_0/\theta)^2 - 1]^{1/2}$
320	1.1918	1.2636
322	1.2799	1.2508
321.5	1.2572	1.2540
321.45	1.2549	1.2543
321.44	1.2545	1.2544

The fourth energy level, as a fraction of the way up from the bottom of the well, is

$$V_0 - |E_4| = \frac{\theta_4^2}{\theta_0^2} = \left(\frac{321.44}{515.66}\right)^2 = 0.3886 \; ,$$

as given in the table in Sec. 5(e).

Your problem: Use a calculator to calculate the $n = 1$ and 6 entries in the $\theta_0 = 9$ row of that table. Draw to scale the energy-level diagram, and also show where the first five energies of the infinite square well of the same width would be. Of course, you could use a spreadsheet or a root-finding routine, but there is more to be learned from hunting and pecking, so do it that way.

46

6. *The finite square well, II.*

A particle of mass m in a finite square well of width L and depth V_0 has just barely three bound states ($E_3 = 0$). Find the energy levels (in terms of V_0) to 3-place accuracy. Use a calculator to hunt and peck, as in the previous problem. Draw to scale the energy-level diagram.

7. *The Heaviside step function.*

The Heaviside step function is

$$H(x) = \begin{cases} 0 & \text{for } x < 0 \\ \frac{1}{2} & \text{for } x = 0 \\ 1 & \text{for } x > 0 \,. \end{cases}$$

(a) Show that $\int_{-\infty}^{x} \delta(x)\, dx = H(x)$.

(b) Show that $dH(x)/dx = \delta(x)$. One way is to consider the derivative of an approximation to $H(x)$ that ramps up in a straight line from 0 at $x = -\epsilon$ to 1 at $x = +\epsilon$. Then let $\epsilon \to 0$.

(c) The symbol $H(x - a)$ places the step at $x = a$. Sketch $H(x) - H(x - a)$ and its derivative.

8. *A one-dimensional crystal.*

We might model the potential energy of a particle in a one-dimensional crystal using a very long line of identical δ functions at the sites of the atoms:

$$V(x) = -g\big[\cdots + \delta(x + 3b/2) + \delta(x + b/2) + \delta(x - b/2) + \delta(x - 3b/2) + \cdots \big]\,.$$

(a) Sketch the ground-state wave function. Write down the form of $\psi(x)$ for $-b/2 \leq x \leq +b/2$, using symmetry to simplify it. Let $q^2 \equiv 2m|E|/\hbar^2$.

(b) Use the discontinuity of $d\psi/dx$ at $x = b/2$ to show that

$$\coth z = z/z_0 \,,$$

where $z \equiv qb/2$ and $z_0 \equiv mgb/2\hbar^2$. The root z of this equation, which depends the parameters m, g, b, and $\hbar$ only in the combination z_0, would give E.

(c) Snip off a string of N of these "atomic sites" and join the cut ends to make a ring molecule of length Nb. What is the lowest energy of a particle in this ring?

9. *The double-delta well.*

The double-δ potential energy,

$$V(x) = -g\left[\delta(x - a) + \delta(x + a) \right]\,,$$

has, depending on the parameters a, g, and the mass m, either one or two bound states.

(a) Sketch the even- and odd-parity bound-state wave functions, assuming both exist.

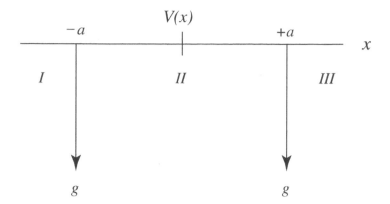

(b) For the even-parity state, write down the form of the wave functions in regions I and III (use a coefficient A) and region II (use B). Let $q^2 \equiv 2m|E|/\hbar^2$.

(c) Use the (dis)continuity conditions on $\psi(x)$ and $d\psi/dx$ at $x = +a$ to get

$$e^{-2qa} = 2qa/\Lambda - 1$$

as the equation that determines E_1. Here $\Lambda \equiv 2mga/\hbar^2$.

(d) Sketch $\exp(-2qa)$ and $(2qa/\Lambda - 1)$ as functions of qa. Show that the only crossing occurs for qa somewhere between $\Lambda/2$ and Λ. Thus find bounds on E_1 in terms of m, g, a, and $\hbar$.

(e) Repeat parts (b-d) for the odd-parity state and thus show that the equation that determines E_2 is

$$-e^{-2qa} = 2qa/\Lambda - 1 .$$

Sketch the two sides and show that there is no physical ($qa > 0$) solution unless $\Lambda > 1$.

10. *Integrating by differentiating.*
Let

$$I_n \equiv \int_{-\infty}^{+\infty} x^n e^{-bx^2} \, dx .$$

Then $I_n = 0$ when n is odd; and we showed in Sec. 8 that $I_0 = \sqrt{\pi/b}$. By taking derivatives of I_0 with respect to b, show that

$$I_2 = \frac{1}{2b} I_0 , \quad I_4 = \frac{3}{4b^2} I_0 ,$$

and get a general expression for I_n when n is even.

From *Surely You're Joking, Mr. Feynman:*

...when guys at MIT or Princeton had trouble doing a certain integral, it was because they couldn't do it with the standard methods they had learned in school ... Then I'd come along and try differentiating under the integral sign, and often it worked. So I got a great reputation for doing integrals, only because my box of tricks was different from everybody else's.

11. *Oscillator and hydrogen ground states.*

(a) Show that $\psi(x) = e^{-ax^2}$ is a solution of the Schrödinger equation for the oscillator, provided that $a = m\omega/2\hbar$, and that $E = \hbar\omega/2$.

(b) Show that $\psi(r) = e^{-br}$ is a solution of the Schrödinger equation for the hydrogen atom, provided that $b = mke^2/\hbar^2 = 1/a_0$, where a_0 is the Bohr radius, and that $E = -mk^2e^4/2\hbar^2$. Normalize the wave function.

12. *The hydrogen ground state.*

A hydrogen atom is in its ground state.

(a) If space is divided into identical infinitesimal cubes, in which cube is the electron most likely to be found?

(b) If instead space is divided into shells of infinitesimal thickness, like the layers of an onion, centered on the proton, what is the radius of the shell in which the electron is most likely to be found?

(c) Show that the mean radius $\langle r \rangle = 4\pi \int_0^\infty r\, |\psi(r)|^2\, r^2\, dr$ is 3/2 times the most probable radius.

(d) Calculate the mean value of the potential energy $V(r)$. What then, without further calculation, is the mean value of the kinetic energy K?

13. *Polarizability of hydrogen.*

When a hydrogen atom is placed in an electric field $\mathcal{E}$, the proton and electron are pushed in opposite directions. Thus an electric dipole moment $p = ed$ is induced, where d is the offset of the center of charge of the electron "cloud" from the point proton. For fields attainable in the laboratory, d will be far smaller than the Bohr radius a_0, and p is proportional to $\mathcal{E}$: $p = \alpha\mathcal{E}$, where α is the *polarizability*.

Show that $\alpha = 3\pi\epsilon_0\, a_0^3$ by modeling the electron cloud as a spherical charge density given by $\rho(r) = -e\, |\psi(r)|^2$, where $\psi(r)$ is the ground-state wave function. Assume that for attainable fields, ρ is not appreciably distorted.

This semi-classical model is too simple. The proper quantum-mechanical model gives an answer six times larger, in agreement with experiment.

14. *An infinite system of wells.*

(a) Calculate the second derivative of the (unnormalized) wave function $\psi(x) = \text{sech}(ax)$. Sort out what you get into the form of the Schrödinger equation, $-(\hbar^2/2m)\psi'' + V(x)\psi = E_1\psi$, choosing $V(\pm\infty)$ to equal 0. Let $E_0 \equiv \hbar^2a^2/2m$. You should find that the energy associated with $\psi(x)$ is $E_1 = -E_0$, and that $V(x) = -2E_0\, \text{sech}^2(ax)$. See (a) in the figure below.

(b) Let $\psi(x) = \text{sech}^N(ax)$, where N is any positive integer. Show that now $E_1 = -N^2E_0$ and $V_N(x) = -N(N+1)E_0\, \text{sech}^2(ax)$. For $N = 2$, see (b) in the figure.

(c) Let $\psi(x) = \text{sech}^M(ax)\tanh(ax)$, where M is any positive integer. Show that $E_2 = -M^2E_0$ and $V_M(x) = -(M+1)(M+2)E_0\, \text{sech}^2(ax)$. Why are these the first excited states?

(d) Show that the potential energies $V_M(x)$ for $M = 1, 2, 3, \ldots$ are the same as the potential energies $V_N(x)$ for $N = 2, 3, 4, \ldots$ Thus E_2 for $M = 1$ belongs to the same well as E_1 does for $N = 2$, as in (b) in the figure.

(e) Part (c) of the figure shows the energy levels for $N = 1, 2, 3,$ and 4 (and for $M = 1, 2,$ and 3). We haven't proved all the equivalences of energies shown there, but given the pattern, write down the energy levels for $N = 5$.

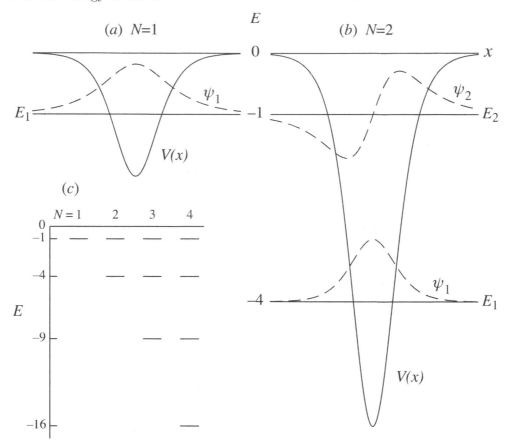

15. *Simplest solutions of eigenvalue equations.*

Finding the full set of solutions of an eigenvalue equation can be a formidable task, but finding the simplest one or two solutions, which are the often those of most interest, is sometimes easy. Always take a minute to look. The simplest eigenfunctions are sometimes as simple as 1 or x. And 0 can be an eigenvalue.

(a) Find *by inspection* (that is, in your head) two linearly independent eigenfunctions and the corresponding eigenvalues ϵ of this equation for $h(x)$, $-\infty < x < +\infty$:

$$-h'' + 2xh' + h = \epsilon h .$$

Normalizability is not a requirement here or in (b). This equation appears in Chap. 6.

(b) This one, for $g(r)$, $0 \le r < \infty$, may take a bit of scribbling:

$$g'' - 2\sqrt{\epsilon}\, g' + \left(\frac{2}{r} - \frac{\ell(\ell + 1)}{r^2} \right) g = 0 .$$

It contains *two* eigenvalues, ϵ and ℓ, both of which must be ≥ 0. Is there a choice of ϵ and ℓ that lets $g = 1$ be a solution? How about $g = r$? How about $g = r^2$? Try for an infinite set of eigenfunctions, with the corresponding eigenvalues (ϵ, ℓ). This equation appears in Chap. 9.

50

(c) Find by inspection the simplest eigenfunction and the corresponding eigenvalue λ of this equation for $P(\theta)$, $0 \le \theta \le \pi$:

$$-\frac{1}{\sin\theta}\frac{d}{d\theta}\left(\sin\theta\frac{dP}{d\theta}\right) = \lambda P \; .$$

One of $\sin\theta$ and $\cos\theta$ is another eigenfunction. Which one, and what is its λ? This equation appears in Chap. 8.

> *The last two problems invoke the Pauli exclusion principle (the subject of Chap. 14), which is probably familiar from a course on chemistry or modern physics. There it is used to explain the periodicity of the chemical properties of the elements. The principle states that no more than two electrons can go in a given spatial quantum state in a well. Thus in one-, two-, or three-dimensional wells, no more than two electrons can have the same quantum number(s) n, or (n_x, n_y), or (n_x, n_y, n_z).*

16. *Absorption frequencies of certain molecular ions.*

Certain simple hydrocarbon ions consist of a linear chain of an odd number, $N = 3, 5, 7,$ ..., of carbon atoms, with hydrogen atoms attached as shown schematically for $N = 3$. And each such chain has $(N + 1)$ "π electrons"—electrons that have free run of the length of the chain. (Ask a chemist.)

Model the chain as a one-dimensional infinite square well of width $L = Nd$, where $d = 1.4 \times 10^{-10}$ m is the distance between adjacent carbon atoms. The chain effectively extends a distance $d/2$ beyond each of the end carbon atoms, so when $N = 3$ the width is $3d$, not $2d$ as the figure might suggest. The $N = 3$ ion has got four π electrons. Then, according to the Pauli principle, in its ground state, two of them are in the $n = 1$ level of the well and two are in the $n = 2$ level. The $N = 5$ ground state will also have two electrons in the $n = 3$ level.

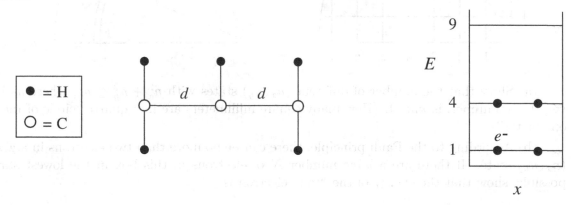

Show that the wavelength of light that will excite an electron to the nearest open level is $\lambda_N = 64.6N^2/(N + 2)$ nm. Find λ_N for $N = 3, 5, 7,$ and 9. In what part of the electromagnetic spectrum—infrared, visible, ultraviolet—are each of these wavelengths?

In evaluating equations that involve $\hbar^2/m$ numerically, it is often simplest to multiply the top and bottom by c^2, and then use the mass energy mc^2 in eV units together with $\hbar c = 197.33$ eV nm.

51

17. *Electrons in a box.*

(See the note in front of Prob. 16.) In a block of metal such as copper, there is about one free electron per atom. These *conduction electrons* move about almost freely, like the molecules of a gas in a box, through the fixed neutralizing background of the positive ions; the walls of this box are the faces of the block. The conduction electrons belong to no particular atom and, given their wave nature, each electron is better thought of as belonging to the whole block. If a voltage difference is applied to opposite sides of the block—or to the ends of a wire—the conduction electrons drift in response. At normal temperatures, the electrons nearly all lie in the lowest states possible consistent with the Pauli principle.

It is easier to start with a two-dimensional square box of side L. The energy levels are

$$E_{n_x n_y} = (n_x^2 + n_y^2) \frac{\pi^2 \hbar^2}{2mL^2} \ ,$$

where n_x and $n_y = 1, 2, \ \ldots$ We may associate each pair of quantum numbers (n_x, n_y) with a point and with a square on a grid, as shown in the figure.

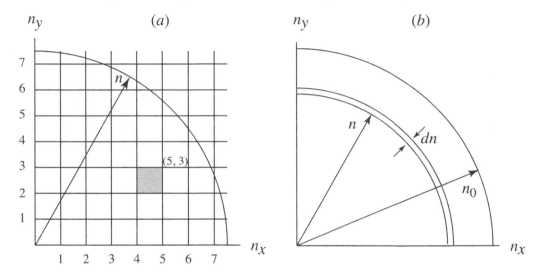

(a) Show that the number of different (n_x, n_y) states with $n_x^2 + n_y^2 \leq n_0^2$, where n_0 is a very large number, is $\pi n_0^2 / 4$. (How many square millimeters are in a quarter circle of radius one meter?)

(b) According to the Pauli principle, there can be no more than two electrons in a given (n_x, n_y) state. If there are a large number N of electrons in this box in the lowest states possible, show that the energy of the "top" electron is

$$E_F = 2\pi \frac{N}{A} \frac{\hbar^2}{2m} \ ,$$

where $A = L^2$ is the area of the box. Here E_F is the *Fermi energy*.

(c) Show that the average energy of an electron in this lowest state is $\overline{E} = E_F / 2$.

(d) Three dimensions should now be easy. For a cube of side L and volume V, show that the number of different (n_x, n_y, n_z) states with $n_x^2 + n_y^2 + n_z^2 \leq n_0^2$, where n_0 is a large

52

number, is $\pi n_0^3/6$. Then show that

$$E_F = \left(3\pi^2 \frac{N}{V}\right)^{2/3} \frac{\hbar^2}{2m} \quad \text{and} \quad \overline{E} = \frac{3}{5}E_F \ .$$

Note that E_F (and so also $\overline{E}$) increases as the number density N/V increases.

18. *The free-electron gas in copper.*

The atomic number of copper is 29, the molar mass is 63.5 g/mol, and the density is 8.96 g/cm^3. There is one free electron per atom. Show that the Fermi energy, E_F, is 7.0 eV, and that the corresponding "Fermi temperature," T_F, is 8.2×10^4 K, where $E_F = kT_F$ and $k = 8.617 \times 10^{-5}$ eV/K is the Boltzmann constant.

The high Fermi temperature indicates that, on a scale of kT for normal temperatures ($kT = 1/40$ eV at $T = 290$ K), the "Fermi sea" is deep. Thus nearly all the free electrons indeed lie far below the surface and have no way to absorb thermal energies—all nearby states are already occupied.

The most loosely bound electrons in a block of metal are those at the top of the Fermi sea. The figure shows the relation between the depth of the Fermi sea and the minimum energy, ϕ, needed to remove a free electron from a metal; ϕ is called the work function and appears in Einstein's photoelectric equation (see Sec. 1·1).

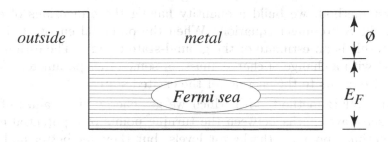

3. SIMPLE APPROXIMATIONS FOR BOUND STATES

3.1. *Dimensions and Scaling*

3.2. *Fitting Wavelengths in a Well*

3.3. *Guessing the Ground-State Wave Function*

3.4. *Useful Integrals*

    Problems

A short chapter, in which we develop three ways to get partial or approximate information about the energy eigenvalues of bound states. We need methods of approximation because many problems are too hard to solve (at least at this level) in closed form, or they don't have closed-form solutions. Approximation methods are usually left until later in a text, but those introduced here follow naturally upon Chapter 2, where getting exact solutions was within our abilities. The methods are also used in other fields besides quantum mechanics.

There is another class of approximation methods, called perturbation methods, which are used to calculate previously neglected small corrections to already known eigenvalues. Some such corrections are those due to relativity, to interactions of magnetic moments of constituent particles, and to external electric and magnetic fields. Perturbation methods are nearly always relegated to the latter part of a text, and they are in this one too.

In the first method, we build a quantity having the dimensions of energy out of the parameters in the Schrödinger equation. When the potential energy V is a simple enough function, the result is an estimate of the ground-state energy. The estimate shows how the energy "scales" with a change of charge, mass, or some other parameter. And from this first result we can also estimate the energies of the excited states.

Next is the WKB method, with which we get energy eigenvalues of bound states by fitting de Broglie wavelengths between the turning points of a potential energy $V(x)$. The values are sometimes poor for the lowest levels, but they get better and better for higher levels. We develop the method using simple examples rather than the more complicated mathematics that usually accompanies the subject.

Third is the variational method, in which we guess at the ground-state wave function and calculate the expectation value of the energy for that wave function. The trial wave function can have built-in parameters, and then we minimize the expectation value with respect to those parameters. The value we get is always an upper limit on the true ground-state energy. Thus we can try different trial functions, and whichever one gives the lowest energy is the winner.

We have omitted the most powerful means for solving difficult equations—with numerical calculations on a computer.

3·1. DIMENSIONS AND SCALING

Sometimes we can learn a lot from an equation, without actually solving it, simply by considering the quantities upon which the solutions must depend. One result, sometimes useful as a check, comes directly from the Schrödinger equation,

$$-\frac{\hbar^2}{2m}\frac{d^2\psi}{dx^2} + V(x)\psi = E\psi .$$

As with any physics equation, all three terms here have to be dimensionally the same. In the first term, $d^2\psi/dx^2$ has (with L for length) the dimensions of ψ/L^2. Therefore that first term has the dimensions of $\psi \times \hbar^2/mL^2$. And since ψ is common to all three terms, the dimensions of $\hbar^2/mL^2$ have to be those of $V(x)$ and E: energy. We have already seen this with the hydrogen atom (Sec. 1·3(b)) and the infinite square well (Sec. 2·4):

$$E_n = -\frac{1}{n^2}\frac{\hbar^2}{2ma_0^2} \qquad \text{and} \qquad E_n = n^2\frac{\pi^2\hbar^2}{2mL^2} ,$$

where $a_0 = \hbar^2/mke^2$ is the Bohr radius and L is the width of the square well. It is useful to have a check that a solution is at least dimensionally correct—or not!

(a) Example: the oscillator—The Schrödinger equation for the harmonic oscillator, where $V(x) = \frac{1}{2}m\omega^2 x^2$, is

$$-\frac{\hbar^2}{2m}\frac{d^2\psi}{dx^2} + \frac{1}{2}m\omega^2 x^2\psi = E\psi .$$

The energy eigenvalues can only depend upon these quantities: a quantum number n; perhaps some other pure numbers such as 2 or π; and the parameters that appear in the equation, $\hbar$, m, and ω. Parameters not in the equation are of course irrelevant—if you are not part of the problem, you are not part of the solution.

What can we learn? We construct a quantity out of $\hbar$, m, and ω that is dimensionally an energy. Let M, L, and T represent the physical dimensions of mass, length, and time. Then the dimensions of energy are

$$\dim(E) = ML^2T^{-2} \qquad\qquad\qquad (\text{think } E = mc^2) .$$

The dimensions of m, ω, and $\hbar$ are

$$\dim(m) = M$$
$$\dim(\omega) = T^{-1} \qquad\qquad\qquad (\text{think } \omega = 2\pi/\tau)$$
$$\dim(\hbar) = ML^2T^{-1} \qquad (\text{angular momentum} = n\hbar = rp) .$$

The only way to construct an energy from the three parameters is to take $\hbar$ to get the L^2 dependence, and multiply it by ω to complete the T^{-2} dependence. Therefore this much— that $E \propto \hbar\omega$—comes without knowing anything more about the Schrödinger equation than the parameters that occur in it. Of course, we already knew that $\hbar\omega$ is dimensionally an energy, but in fact it is the *only* combination of the available parameters that is so. (Had we chosen m and the spring constant k as the oscillator parameters, we would have found $E \propto \hbar\sqrt{k/m}$.)

(b) Example: hydrogen—The Schrödinger equation for hydrogen, where $V(r) = -ke^2/r$, is

$$-\frac{\hbar^2}{2m}\,\nabla^2\psi - \frac{ke^2}{r}\,\psi = E\psi \ .$$

Here the parameters are $\hbar$, m, $k \equiv 1/4\pi\epsilon_0$, and e; and now e brings in a fourth dimension, that of charge. However, since k and e^2 appear nowhere else in the equation, we can treat ke^2 as a single constant. Then since $V = -ke^2/r$ is an energy, the dimensions of ke^2 are those of length times energy,

$$\dim(ke^2) = ML^3T^{-2} \ .$$

To construct a quantity having the dimensions of energy from $\hbar$, m, and ke^2, we write

$$\dim(E) = [\dim(\hbar)]^a\ [\dim(m)]^b\ [\dim(ke^2)]^c \ ,$$

or

$$ML^2T^{-2} = (ML^2T^{-1})^a M^b (ML^3T^{-2})^c \ .$$

Here a, b, and c are numbers to be obtained by matching the powers of M, L, and T on the two sides of the equation. These constraints are

$$1 = a + b + c \ , \qquad 2 = 2a + 3c \ , \qquad \text{and} \quad -2 = -a - 2c \ .$$

Adding the last two equations gives $a + c = 0$; and then $c = +2$, $a = -2$, and $b = +1$. Thus the dependence of the eigenvalues on the parameters has to be

$$E \propto -(\hbar)^{-2}(m)^{+1}(ke^2)^{+2} = -\frac{mk^2e^4}{\hbar^2} \ ,$$

just as we got from the Bohr model in Sec. 1·4. We needed only the list of parameters in the Schrödinger equation, not the equation itself, to get this result. The minus sign is a reminder that, for $V(r) = -ke^2/r$, the bound states have negative energies. Thus if, say, m increases, the energies are more negative (sink lower).

(c) Uses of dimensional results—Here are some uses of dimensional results:

(1) *To estimate the energies and sizes of ground states*—If we multiply either of the above results, $\hbar\omega$ or $-mk^2e^4/\hbar^2$, by $1/2$, we get exactly the ground-state energy of the corresponding system (Sec. 2·8). In general, the combination of parameters that makes an energy gives an order-of-magnitude estimate of the ground-state energy. In the same way, for these and other simple power-law potentials, only one combination of parameters makes a length, which can serve as an estimate of the size of the ground state.

(2) *To see how energies and sizes "scale" when a parameter changes*—The energy of the hydrogen atom is proportional to the square of the product of the charges of the electron and proton. Thus the bound-state energies of a ionized atom with only one electron left and a nuclear charge Ze are Z^2 times those of hydrogen. For example, $E_1 = 4\times(-13.6) = -54.4\,\text{eV}$ for singly ionized helium. The energies are also proportional to the electron mass m, which properly (Sec. 1·4(d)) is the reduced mass, $\mu = mM/(m + M)$. The proton mass M is so much larger than m that for hydrogen μ is very nearly equal to m. But for *positronium*, the e^+e^- atom already discussed in Sec. 1·4(d), μ is $\frac{1}{2}m$, the bound-state energies are one-half those of hydrogen, and $E_1 = -6.8\,\text{eV}$.

(3) *To see how energies and sizes depend on a quantum number*—Some of the relations we have seen or shall see are

$$E_n = n^2 \frac{\pi^2 \hbar^2}{2mL^2}, \quad E_n = -\frac{1}{n^2} \frac{mk^2 e^4}{2\hbar^2}, \quad E_n = \left(n + \tfrac{1}{2}\right)\hbar\omega, \quad |\mathbf{L}| = \sqrt{\ell(\ell+1)}\,\hbar, \quad L_z = m_\ell \hbar ,$$

where here $\mathbf{L}$ is angular momentum. In each equation, the quantum number, n or ℓ or m_ℓ, to leading order goes with $\hbar$. (We put this relationship on a firmer footing in Sec. 2.) Therefore, having determined from a dimensional argument where $\hbar$ goes, we have a good idea where the associated quantum number goes. Thus from the earlier results of this section, we get

$$E_n \approx n\hbar\omega \quad \text{and} \quad E_n \approx -\frac{1}{n^2} \frac{mk^2 e^4}{\hbar^2}$$

for the oscillator and the hydrogen atom. We come close to the true eigenvalues,

$$E_n = \left(n + \tfrac{1}{2}\right)\hbar\omega \quad \text{and} \quad E_n = -\frac{1}{n^2} \frac{mk^2 e^4}{2\hbar^2} ,$$

using nothing more from the Schrödinger equations than the parameters in them and an approximate rule, "n goes with $\hbar$."

Caution is needed in applying the n-goes-with-$\hbar$ rule. The equation

$$E_n = -\frac{1}{n^2} \frac{\hbar^2}{2ma_0^2} ,$$

given in the first paragraph of this section, apparently violates it. This is because a factor $\hbar^4$ is buried in a_0^2. To get the right result, all factors of $\hbar$ have to be made explicit.

We may already know the eigenvalues for hydrogen and for the oscillator, but we now have a way to learn something quickly should we encounter potential energies such as $V(x) = b|x|$ or $V(r) = cr^4$.

(d) Remarks—Dimensional analysis and scaling methods have a distinguished ancestry that includes Lord Rayleigh and Fourier and goes all the way back to Galileo. However, much of the full power of the methods was developed by E. Buckingham in the 1910s. The examples above are very elementary. When there are more parameters, there will likely be more than one way to construct a quantity of interest: There is more than one way to construct an energy when, say, $V(r) = ar^2 - b/r$. Nevertheless, dimensional analysis will still lead to a reduction of the number of "degrees of freedom"—always a blessing. Then scaling permits, for example, the experimental testing of the design of an airplane wing, or an understanding of cratering due to explosives or asteroid impacts, using values of the parameters that are far from the "real" values but are experimentally accessible. And dimensional methods are invaluable when, as in fluid mechanics, the equations are known but are in many situations too difficult to solve; or when, as often in biomechanics, there is no equation from which to read the relevant parameters. The art then—and in many engineering applications—is to decide which of a multitude of parameters are important and which are not. For the atom, the art was Bohr's.

3·2. FITTING WAVELENGTHS IN A WELL

The Schrödinger equations for the infinite square well, the harmonic oscillator, the hydrogen atom, and a few other potential-energy wells, can be solved in closed form. However, there are wells of interest for which no closed-form solutions exist. We need other ways to find energy eigenvalues than solving the Schrödinger equation directly. In this section, we "invent" such a method, but it will usually only give approximate results.

(a) The idea We work from what we already know about the infinite square well and the oscillator. Figure 1 shows the lowest eigenstates for these wells. For the oscillator, we have so far found only the ground-state solution (Sec. 2·8), but all we need here are the qualitative features of the eigenfunctions—their symmetries, numbers of nodes, and curvatures in various regions of x (Sec. 2·9). The numbers of oscillations of eigenfunctions increase with energy: larger E means larger p means smaller $\lambda = h/p$ means more wavelengths fit in. For the oscillator, the oscillations increase (1) because the wavelengths get shorter, and (2) because the oscillations extend farther in x as the classical turning points—the values of x at which $E = V(x)$—move out. Note again that the oscillator eigenfunctions are oscillatory between the turning points and exponential outside them.

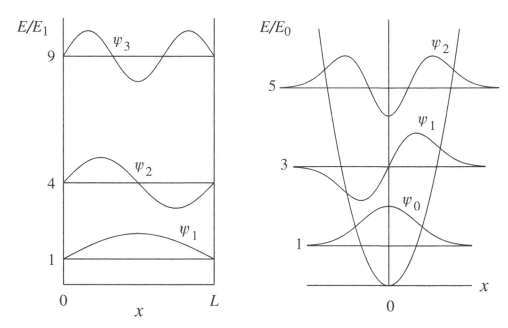

Figure 1. The lowest energy levels and eigenfunctions for the infinite square well and the harmonic oscillator.

We begin by calculating the number of de Broglie wavelengths that fit between the turning points of a well as a function of an *arbitrary* value of the energy. Then we find what numbers of wavelengths correspond to the energy eigenvalues we already know for the infinite square well and the oscillator. In this way, we "tune" the method, and then apply it to other wells. We cannot always expect to get exact results.

The approach here will be decidedly heuristic (Wikipedia: "heuristic ... [an] approach to problem solving, learning, or discovery that employs a practical method not guaranteed to be optimal or perfect, but sufficient for the immediate goals."). See also the end-of-section remarks. The usual derivation of the recipe we get involves much more mathematics.

(b) Example: the infinite square well—We consider again the infinite square well. The de Broglie wavelength λ is related to the momentum p by $\lambda = h/p$; and in the well $E = p^2/2m$. Thus

$$\lambda = \frac{h}{p} = \frac{2\pi\hbar}{\sqrt{2mE}} .$$

We calculate the number of wavelengths in the well for *any* E, not just for those "correct" values of E that correspond to an integral number of half-wavelengths in the well. The infinitesimal *fraction* of a wavelength in the infinitesimal distance dx is dx/λ. (If, say, $\lambda = 1$ mm, then $dx = 1\,\mu$m contains 1/1000th of a wavelength.) Then

$$\text{the number of wavelengths between 0 and } L \; = \int_0^L \frac{dx}{\lambda} = \frac{L}{\lambda} = \frac{\sqrt{2mE}}{2\pi\hbar} L ,$$

because here λ (for a given E) is constant. We are just adding up infinitesimal fractions.

To fit properly in the well and satisfy the boundary conditions, the number of wavelengths between the turning points, here at $x = 0$ and L, has to be $n/2$, where $n = 1, 2, 3, \ldots$ (see Fig. 1). Equating the right end of the above equation to $n/2$, we get

$$E_n = n^2 \frac{\pi^2\hbar^2}{2mL^2}$$

for the quantized energies. Which is of course the right answer.

So far there is little new—we made much the same argument in Sec. 2·4. However, we now have a two-step recipe for more general application:

(1) Find the number of wavelengths between the turning points for any value of E. (We only calculate between turning points because a wave function only *oscillates* between those points.)

(2) Set this result equal to the number of wavelengths we *want* between those points.

Think of slowly raising up the E line, like a high-jump bar, on a $V(x)$ diagram. As we go up, the wavelength gets shorter everywhere between the turning points, and at certain heights the E line "clicks" into a value that gives a proper number of wavelengths.

(c) Example: the oscillator—When $V(x)$ is not constant, then neither is the de Broglie wavelength. However, in regions of x where $E > V(x)$, we can still define a "local" wavelength. Since now $p^2/2m = E - V(x)$, we now have

$$\lambda(x) = \frac{h}{p(x)} = \frac{2\pi\hbar}{\sqrt{2m(E - V(x))}} .$$

With this varying $\lambda(x)$, the number of wavelengths turning points x_1 and x_2 is

$$\boxed{\int_{x_1}^{x_2} \frac{dx}{\lambda(x)} = \frac{1}{2\pi\hbar} \int_{x_1}^{x_2} \sqrt{2m(E - V(x))}\, dx .}$$

For the oscillator, $V(x) = \frac{1}{2}m\omega^2 x^2$, and the turning points $\pm a$ are where $E = \frac{1}{2}m\omega^2 a^2$. Thus $a = \sqrt{2E/m\omega^2}$, and

$$\text{the number of wavelengths between } -a \text{ and } +a \; = \frac{1}{2\pi\hbar} \int_{-a}^{+a} \sqrt{2m\left(E - \frac{1}{2}m\omega^2 x^2\right)}\, dx .$$

Here the limits of integration as well as the integrand are functions of E.

A change of variable, $y \equiv x/a$, changes the limits to ± 1 and the integral to

$$\frac{\sqrt{2mE}}{2\pi\hbar} \, a \int_{-1}^{+1} \sqrt{1-y^2} \, dy = \frac{E}{\pi\hbar\omega} \int_{-1}^{+1} \sqrt{1-y^2} \, dy = \frac{E}{2\hbar\omega} \, .$$

The integral itself is $\pi/2$; see Fig. 2. Thus the number of wavelengths between the turning points is here proportional to E.

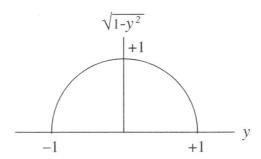

Figure 2. The area under the half circle is $\pi/2$.

The next question is: What do we now set $E/2\hbar\omega$ equal to? That is, how many wavelengths do we *want* between the turning points? The wave functions for the infinite square well are entirely oscillatory, and those oscillations bring the functions all the way back to the x axis at $x = 0$ and L. The wave functions for the oscillator, however, are finally brought back to the x axis by the decaying exponentials outside the turning points (Fig. 1). Thus, for the oscillator ground state, we don't need a whole half wavelength between the turning points; but we do need at least a bit of a wavelength in the $E > V(x)$ region to turn the wave function over to connect with the decaying exponentials. Suppose we minimize our maximum error by choosing midway between half a wavelength and none at all. Setting the number of wavelengths, $E/2\hbar\omega$, equal to $1/4$, we get $E_0 = \hbar\omega/2$. Which is the exact answer.

Instead of *guessing* the number of wavelengths that we need, we could have (and would have, had we got an incorrect result) *used* the known value of E_0 to get the number. But there is often something to be learned by first trying to puzzle out what a reasonable answer might be. Good guessing takes understanding.

To get each higher energy eigenvalue, we simply add another half wavelength, so that $E/2\hbar\omega = 1/4, 3/4, 5/4, \ldots$ (Fig. 1). This gives $E_n = (n+1/2)\hbar\omega$, where $n = 0, 1, 2, \ldots$ This too is exact for all n, as we shall prove in Chap. 6. (Note again that for the oscillator the numbering begins with 0.)

(d) Example: a bouncing ball—We apply the method to a ball bouncing elastically on a floor. (Experiments have been done with bouncing neutrons.) With the floor at $z = 0$, the potential energy is (see Fig. 3)

$$V(z) = \begin{cases} mgz & \text{for } z \geq 0 \\ \infty & \text{for } z < 0 \, . \end{cases}$$

The turning points are at $z = 0$ and $z = E/mg \equiv z_0$. The number of wavelengths between them as a function of E is

$$\int_0^{z_0} \frac{dz}{\lambda(z)} = \frac{1}{2\pi\hbar} \int_0^{z_0} \sqrt{2m(E - mgz)} \, dz \, .$$

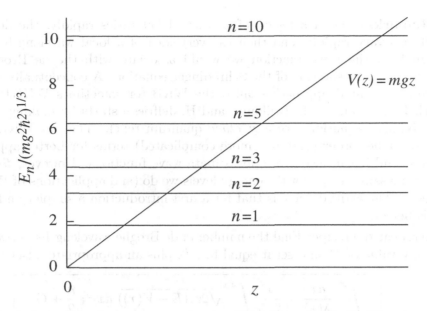

Figure 3. A few of the lowest energy levels of the bouncing ball. The approximate and true values are too close to be seen separately at this scale.

With a change of variable, $y \equiv z/z_0$, this becomes

$$\frac{\sqrt{2mE}}{2\pi\hbar} \frac{E}{mg} \int_0^1 \sqrt{1-y}\; dy = \frac{1}{2\pi g\hbar} \sqrt{\frac{2E^3}{m}} \left(-\tfrac{2}{3}\right)(1-y)^{3/2}\bigg|_0^1 = \frac{1}{3\pi g\hbar} \sqrt{\frac{2E^3}{m}}\;.$$

For the ground state, the best guess for the fraction of a wavelength to best fit between the turning points would be neither $1/4$ nor $1/2$ but instead $3/8$ (why?). For the nth state, our best guess is that $n/2 - 1/8 = 3/8, 7/8, 11/8, \ldots$, wavelengths, where $n = 1, 2, 3, \ldots$ Thus

$$\frac{1}{3\pi g\hbar} \sqrt{\frac{2E_n^3}{m}} \approx \frac{1}{2}\left(n - \frac{1}{4}\right)\;,$$

or

$$E_n \approx \frac{1}{2}\left[3\pi\left(n - \frac{1}{4}\right)\right]^{2/3} (mg^2\hbar^2)^{1/3}\;.$$

The dimensional part could have been found from a dimensional argument.

We compare the numerical factors with the known true values, obtained numerically, for a few values of n:

n =	1	2	3	5	10
$\frac{1}{2}\left[3\pi\left(n - \frac{1}{4}\right)\right]^{2/3}$ =	1.842	3.240	4.379	6.304	10.182
True =	1.856	3.245	4.382	6.305	10.183

Figure 3 shows these energy levels. The larger the value of n, the better the result. This is typical of the method, because the larger the value of n, the less it matters if our choice of what fraction of a wavelength to fit between turning points is exactly right. But even for $n = 1$ we get within 1% of the true value. The method does not always work so well.

(e) Remarks—(1) In a region of x where $V(x)$ varies rapidly, the de Broglie wavelength will also vary rapidly, and then the very idea of a local wavelength becomes problematic. In fact, the wave function we would associate with the de Broglie wavelength would not be an exact solution of the Schrödinger equation. A conceptually and mathematically more complicated approach—called the WKB (or sometimes WKBJ) method, after G. Wentzel, H.A. Kramers, L. Brillouin, and H. Jeffries—starts by making approximations to the Schrödinger equation (see any *other* quantum text). The *full* WKB method leads to a series of higher-order (and ever more complicated) terms for better approximations of energy levels, and it can also give approximate wave functions. However, *first-order* WKB gets exactly the same recipe for the energy levels we do (and applications of WKB often stop at first order). The attitude here is that for a first introduction a simpler and more intuitive approach is better.

(2) To repeat the recipe: Find the number of de Broglie wavelengths between the turning points for any value of E, and set it equal to $n/2$ plus an appropriate offset O:

$$\int_{x_1}^{x_2} \frac{dx}{\lambda(x)} = \frac{1}{2\pi\hbar} \int_{x_1}^{x_2} \sqrt{2m(E - V(x))} \, dx = \frac{n}{2} + O \ .$$

Then solve for the energies E_n. Note that multiplying the equation through by $\hbar$ gives, to leading order, the "n goes with $\hbar$" rule surmised in Sec. 1 from a sampling of equations.

(3) For the infinite square well, with two impenetrable boundaries and $n = 1, 2, 3, \ldots$, we get the required $1/2$, $2/2$, $3/2$, ... wavelengths in the well with an offset $O = 0$. For the oscillator, with two "soft" turning points and $n = 0, 1, 2, \ldots$, we get the desired $1/4$, $3/4$, $5/4$, ... wavelengths with $O = +1/4$; if n here had started with 1, O would have been *minus* $1/4$. For the bouncing ball, with one turning point hard, the other soft, and $n = 1, 2, 3, \ldots$, we get the desired $3/8$, $7/8$, $11/8$, ..., wavelengths with $O = -1/8$.

(4) All three of these examples have an infinite number of levels ($0 \le V(x) < \infty$). The method is not the tool of choice for shallow wells. For example, for a finite square well with, say, five bound states, the offsets would be smallest for the lowest levels and larger for levels near the top. What would we use for the offset?

3·3. GUESSING THE GROUND-STATE WAVE FUNCTION

(a) The idea—This method, called the variational method, is used mainly to get a good approximation for the ground-state energy E_1 of a system for which the Hamiltonian is $\hat{H}$. The idea is simple. If we knew the true ground-state wave function $\psi_1(x)$, then calculating $\hat{H}\psi_1$ would get us E_1, because $\hat{H}\psi_1 = E_1\psi_1$. Or, to say almost the same thing,

$$\int_{-\infty}^{+\infty} \psi_1^* \left(\hat{H}\psi_1\right) dx = E_1 \int_{-\infty}^{+\infty} \psi_1^* \psi_1 \, dx = E_1 \ .$$

But if we don't know $\psi_1(x)$, we *guess* a *trial* function $\psi_T(x)$, using what we know about the system and ground-state wave functions to make, as close as we can, $\psi_T \simeq \psi_1$. Then we calculate

$$E_T = \int_{-\infty}^{+\infty} \psi_T^* \left(\hat{H}\psi_T\right) dx \ .$$

62

The closer ψ_T is to ψ_1, the closer E_T will be to E_1. In fact, as we shall see below, E_T is always an upper limit on E_1. There is more to the method, but first an example.

(b) Example: the infinite square well—A guess at the ground-state wave function should have no nodes, vanish at the walls at $x = 0$ and L, and be symmetric about the midpoint of the well, because $V(x)$ is symmetric about that point. We try

$$\psi_T(x) = c\,x(L - x)\,, \quad 0 \le x \le L\,,$$

which satisfies all three conditions. The first thing is to find the normalization constant c:

$$\int_0^L |\psi_T(x)|^2 dx = |c|^2 \int_0^L x^2(L - x)^2 dx = |c|^2 \frac{L^5}{30} = 1\,,$$

so $|c|^2 = 30/L^5$. Figure 4 shows the normalized $\psi_T(x)$ as well as the exact $\psi_1(x) = \sqrt{2/L}\,\sin(\pi x/L)$. The Hamiltonian is just $\hat{H} = \hat{p}^2/2m = -(\hbar^2/2m)\,d^2/dx^2$, and

$$\hat{H}\psi_T(x) = -\frac{\hbar^2}{2m}\frac{d^2}{dx^2}\big[cx(L - x)\big] = +c\frac{\hbar^2}{m}\,.$$

The approximate value for E_1 is then

$$E_T = \int_0^L c^* x(L - x)\,c\,\frac{\hbar^2}{m}\,dx = \frac{30}{L^5}\frac{\hbar^2}{m}\frac{L^3}{6} = \frac{5\hbar^2}{mL^2}\,.$$

This is only 1.3% larger than the exact value, $E_1 = \pi^2\hbar^2/2mL^2 = 4.935\,\hbar^2/mL^2$. The answer is so good because the guess was good: $\psi_T(x)$ is (remarkably) close to the true sine wave.

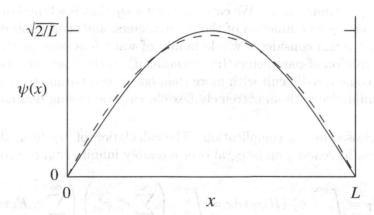

Figure 4. The dashed line is the trial $\psi_T(x)$, the solid line is the exact $\psi_1(x)$.

(c) Theory—Let $\hat{H}$ be the Hamiltonian for a particle in a potential-energy well. Let the (unknown) eigenvalues of $\hat{H}$ be $E_1 < E_2 < E_3 < \cdots$, and let the corresponding normalized (unknown) eigenfunctions be ψ_1, ψ_2, ψ_3, ... (For simplicity, we shall here assume that the energy levels are non-degenerate; but see below.) What we do know is that the only possible result of a measurement of the energy is one of those (unknown) eigenvalues; and that any possible state of the particle is some linear superposition of the those (unknown) eigenfunctions.

Now let ψ_T be a trial wave function that is normalized and satisfies the boundary conditions. The expansion of ψ_T in terms of the (unknown) set of true eigenfunctions ψ_n is the superposition

$$\psi_T = \sum_{n=1}^{\infty} c_n \psi_n \ .$$

The interpretation is that $|c_n|^2$ is the probability we would, on measuring the energy, get E_n, and the sum of all those probabilities, $\sum |c_n|^2$, is one. For example, if $c_1 = c_2 = 1/\sqrt{2}$, then a measurement of the energy would give E_1 or E_2 with equal probability; and no other value would be found.

The average or mean or *expectation value* of E_T in the state ψ_T is just the weighted sum over the various possible outcomes E_n, where the weights are those probabilities $|c_n|^2$:

$$E_T = \sum_{n=1}^{\infty} E_n |c_n|^2 \geq \sum_{n=1}^{\infty} E_1 |c_n|^2 = E_1 \ ,$$

or $E_T \geq E_1$. A perfect guess, $\psi_T = \psi_1$, gets us E_1. Any component of higher eigenfunctions in ψ_T adds something to E_T. The weighted sum E_T cannot be *less* that E_1.

There is another factor to our advantage. Suppose, for example, our guess at ψ_T has $c_2 = 0.1$, and $c_n = 0$ for $n > 2$. Then $|c_1|^2 = 0.99$ and the weighted sum that gives E_T is

$$E_T = 0.99 \, E_1 + 0.01 \, E_2 \ .$$

The 10% "contamination" of ψ_2 in ψ_T produces only 1% contamination of E_2 in E_T.

We can calculate E_T for several different trial wave functions, and the lowest value we get is our best upper limit on E_1. We can construct a ψ_T that is a function of free parameters, calculate E_T from ψ_T as a function of those parameters, and then minimize E_T with respect to them. Thus we can consider a whole family of wave functions at the same time. This is where the variation-of-parameters (or *variational*) method gets its name. By hand, the calculations become too difficult with more than one or two parameters, but a computer can calculate and minimize with an extremely flexible ψ_T, one having dozens of free parameters built into it.

We have glossed over a complication. The calculation of E_T from the expansion $\psi_T = \sum c_n \psi_n$ involves, in general, an integral over a doubly infinite sum of products $\psi_m^* \psi_n$:

$$E_T = \int_{-\infty}^{+\infty} \psi_T^* \left(\hat{H} \psi_T \right) dx = \int_{-\infty}^{+\infty} \left(\sum_{m=1}^{\infty} c_m^* \psi_m^* \right) \left(\sum_{n=1}^{\infty} c_n E_n \psi_n \right) dx$$

$$= \sum_{m=1}^{\infty} \sum_{n=1}^{\infty} c_m^* c_n E_n \int_{-\infty}^{+\infty} \psi_m^* \psi_n \, dx \ .$$

How can this reduce to $E_T = \sum_{n-1}^{\infty} E_n |c_n|^2$? Because:

(1) When $E_m \neq E_n$, then $\int \psi_m^* \psi_n \, dx = 0$ automatically. We prove this in Chap. 5, but a more limited result from linear algebra may be familiar: "Eigenvectors that belong to distinct eigenvalues of a symmetric matrix are orthogonal."

(2) If $E_m = E_n$ for some m and n (degeneracy), then ψ_m and ψ_n can be *chosen* to make $\int \psi_m^* \psi_n \, dx = 0$. Figure 2·9 showed the lowest energy levels for the one-, two-, and three-dimensional infinite square wells; for the latter two, most of the levels are degenerate. The lowest such levels in the 2D column, the $(n_x, n_y) = (1, 2)$ and $(2, 1)$ states, are orthogonal (why?). (Degeneracy *requires*, as here, two or more quantum numbers per state. Thus in systems with degenerate levels, m and n in the earlier equations would stand for more than one quantum number.)

To conclude, all the cross terms in the above—those with $m \neq n$—are automatically zero or can be made zero, and what remains is the singly infinite sum of diagonal terms, $\sum_{n=1}^{\infty} E_n |c_n|^2 = E_T$. And that being so, $E_T \geq E_1$.

(d) Example: hydrogen—We take $\psi_T(r) = c e^{-br}$ for a trial wave function, where b is the parameter to be varied. Because this $\psi_T(r)$ includes the true ground-state wave function (with $b = 1/a_0$, where a_0 is the Bohr radius), we expect, after we minimize $E_T(b)$ with respect to b, to get $E_T = E_1 = -13.6$ eV.

There are three integrals to do: for the normalization constant c, and for the kinetic- and potential-energy terms of the Hamiltonian. This integral,

$$\boxed{\int_0^{\infty} r^n e^{-ar} \, dr = \frac{n!}{a^{n+1}} \, ,}$$

will be needed not only here but often later. The normalization integral is

$$\int_{space} |\psi_T(r)|^2 r^2 \sin\theta \, dr \, d\theta \, d\phi = 4\pi |c|^2 \int_0^{\infty} e^{-2br} r^2 \, dr = 4\pi |c|^2 \frac{2}{(2b)^3} = \frac{\pi |c|^2}{b^3} = 1 \, ,$$

so $|c|^2 = b^3/\pi$. The potential-energy part of the integral of $\psi_T(\hat{H}\psi_T)$, with $V(r) = -ke^2/r$, is

$$-4\pi ke^2 |c|^2 \int_0^{\infty} e^{-2br} r \, dr = -4\pi \, ke^2 \frac{b^3}{\pi} \frac{1}{(2b)^2} = -ke^2 b \, .$$

The kinetic-energy integrand is $-(\hbar^2/2m)\psi_T^* \nabla^2 \psi_T$. The integral would be easy to calculate here, but a transformation makes things simpler and is often useful. So we digress (you could skip).

A brief mathematical digression—We begin in one dimension. Differentiating the product of $\psi^*(x)$ and $d\psi/dx$, we get

$$\frac{d}{dx}\left(\psi^* \frac{d\psi}{dx}\right) = \psi^* \frac{d^2\psi}{dx^2} + \frac{d\psi^*}{dx} \frac{d\psi}{dx} \, .$$

Integrating the left side from $-\infty$ to $+\infty$, we get

$$\int_{-\infty}^{+\infty} \frac{d}{dx}\left(\psi^* \frac{d\psi}{dx}\right) dx = \psi^* \frac{d\psi}{dx}\bigg|_{+\infty} - \psi^* \frac{d\psi}{dx}\bigg|_{-\infty} \, .$$

This is surely zero if $\psi(x)$ vanishes as $x \to \pm\infty$, as it must if ψ is normalizable. Therefore, the integral from $-\infty$ to $+\infty$ of the *right* side of the first equation of this paragraph is also zero, and so

$$\int_{-\infty}^{+\infty} \psi^* \frac{d^2\psi}{dx^2} \, dx = -\int_{-\infty}^{+\infty} \frac{d\psi^*}{dx} \frac{d\psi}{dx} \, dx = -\int_{-\infty}^{+\infty} \left|\frac{d\psi}{dx}\right|^2 dx \, .$$

65

(This is an equality of integrals, not of integrands.) The kinetic-energy integrand has got $\psi^* d^2\psi/dx^2$, but it is usually easier to calculate $d\psi/dx$ and then $|d\psi/dx|^2$ than to calculate $d^2\psi/dx^2$ and then multiply it by ψ^*.

The generalization to three dimensions begins with the vector identity

$$\nabla \cdot (\psi^* \nabla \psi) = \psi^* \nabla^2 \psi + \nabla \psi^* \cdot \nabla \psi \; .$$

Here ∇ is the vector differential operator *del*, and $\nabla\psi$ is the gradient of ψ. Now integrate the left side over all space, and use Gauss's theorem to convert the volume integral to a surface integral over a sphere S at $r = \infty$:

$$\int_{\text{space}} \nabla \cdot (\psi^* \nabla \psi) \, d^3\mathbf{r} = \oint_S (\psi^* \nabla \psi) \cdot d\mathbf{S} \; .$$

Here $d^3\mathbf{r}$ is an infinitesimal volume element and $d\mathbf{S}$ is a surface-area element (at infinity), directed outward. The surface integral is zero if $\psi(\mathbf{r})$ vanishes sufficiently rapidly as $r \to \infty$, as it must if ψ is normalizable. Therefore, the volume integral is also zero, and then so is the integral over all space of the *right* side of the first equation of this paragraph. So

$$\int_V \psi^* \nabla^2 \psi \, d^3\mathbf{r} = -\int_V \nabla \psi^* \cdot \nabla \psi \, d^3\mathbf{r} = -\int_V |\nabla \psi|^2 d^3\mathbf{r} \; .$$

The kinetic-energy integrand has got $\psi^* \nabla^2 \psi$, but it is usually easier to calculate $\nabla\psi$ and then $|\nabla\psi|^2$ than to calculate $\nabla^2\psi$ and then multiply it by ψ^*.

Back to the kinetic-energy integral and E_T—We have

$$\nabla \psi_T = c \frac{\partial(e^{-br})}{\partial r} \hat{\mathbf{r}} = -b \, \psi_T \, \hat{\mathbf{r}} \; ,$$

where $\hat{\mathbf{r}}$ is the unit radial vector. Then $|\nabla\psi|^2 = b^2 |\psi_T|^2$, and the kinetic-energy part of the integral of $\psi_T(\hat{H}\psi_T)$ is

$$-\frac{\hbar^2}{2m} \int_{space} \psi_T^* \nabla^2 \psi_T \, d^3\mathbf{r} = +\frac{\hbar^2}{2m} b^2 \int_{space} |\psi_T|^2 d^3\mathbf{r} = \frac{\hbar^2 b^2}{2m} \; ,$$

because ψ_T is normalized. The sum of the kinetic- and potential-energy integrals is

$$E_T(b) = \frac{\hbar^2 b^2}{2m} - ke^2 b \; .$$

To minimize $E_T(b)$, we differentiate it with respect to b and set the result equal to zero:

$$\frac{\partial E_T}{\partial b} = \frac{\hbar^2 b}{m} - ke^2 = 0 \; .$$

Thus E_T is smallest when the variational parameter b is

$$b = \frac{mke^2}{\hbar^2} = \frac{1}{a_0} \; ,$$

where a_0 is the Bohr radius. Putting this back into $E_T(b)$, we get

$$E_T(1/a_0) = \frac{\hbar^2}{2m}\left(\frac{mke^2}{\hbar^2}\right)^2 - ke^2\frac{mke^2}{\hbar^2} = -\frac{mk^2e^4}{2\hbar^2} = E_1 \ .$$

We get the exact E_1 because the functional form we chose for $\psi_T(r)$ includes the true $\psi_1(r)$. Figure 5 shows $E_T(b)$ as a function of b.

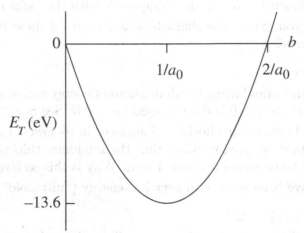

Figure 5. The function $E_T(b)$ is a parabola with its minimum at $b = 1/a_0$.

3·4. USEFUL INTEGRALS

There are two sets of integrals that occur over and over in the following problems and in later chapters. It will save a lot of time to have them at hand. The last problem of this chapter shows how to get $\int_0^\infty x^n e^{-ax}dx$. We derived $\int_{-\infty}^\infty e^{-ax^2}dx = \sqrt{\pi/a} \equiv I_0$ in Sec. 2·8(b); and Prob. 2·10 was to use Feynman's trick to get $I_n \equiv \int_{-\infty}^\infty x^n e^{-ax^2}dx$ from I_0 when n is even. The integrals below are over the half range 0 to ∞. The e^{-ax} integrals are usually over this range (why?). When n is *even*, the e^{-ax^2} integrals over $-\infty$ to $+\infty$ are just double those given below. (What are they when n is odd?) And what is called a in this table is likely to be 2a in your integrals, as the integrands will usually involve $|\psi(x)|^2$.

n	$\int_0^\infty x^n e^{-ax}dx$	$\int_0^\infty x^n e^{-ax^2}dx$	
0	$1/a$	$I_0/2$	
1	$1/a^2$	$1/2a$	
2	$2/a^3$	$I_0/4a$	$I_0 \equiv \sqrt{\pi/a}$
3	$6/a^4$	$1/2a^2$	
4	$24/a^5$	$3I_0/8a^2$	

PROBLEMS

1. *Characteristic lengths.*

(a) Find a combination of the parameters in the Schrödinger equation for the harmonic oscillator that has the dimension of length. Compare the result with the classical amplitude of oscillation when the energy is $\hbar\omega/2$.

(b) Find a combination of the parameters in the Schrödinger equation for the hydrogen atom that has the dimension of length. Compare it with the Bohr radius.

(c) How would you expect the characteristic lengths of these two systems to vary with the quantum number n?

2. *A bouncing particle.*

(a) Construct quantities having the dimensions of energy and of length when the potential energy is $V(z) = mgz$ for $z \geq 0$ and $V(z) = \infty$ for $z < 0$. See Sec. 2(d).

(b) An electron bounces on a horizontal surface. In its lowest energy states, the electron is held near the surface by gravity. Show that the minimum thickness of the layer in which the electron is likely to be found is about 1 mm. Why is this so large?

Experiments have been done with very low-energy ("ultracold") bouncing neutrons.

3. *Energy levels for $V(x) = Ax^4$.*

Use the method of fitting wavelengths in a well to show that the energy eigenvalues for $V(x) = Ax^4$ are approximately

$$E_n \approx \left(3\sqrt{\frac{\pi}{2}}\frac{\Gamma(\frac{3}{4})}{\Gamma(\frac{1}{4})}\right)^{4/3}(n-\tfrac{1}{2})^{4/3}\left(\frac{A\hbar^4}{m^2}\right)^{1/3} = 1.3765\,(n-\tfrac{1}{2})^{4/3}\left(\frac{A\hbar^4}{m^2}\right)^{1/3}.$$

Here Γ is the gamma function, and $n = 1, 2, \ldots$ All needed mathematics can be found in, for example, the *Mathematical Handbook of Formulas and Tables* by M.R. Spiegel, S. Lipschutz, and J. Liu (3rd ed., Schaum's Outline Series).

The true values of E_n for $n = 1$, 3, and 5, in units of $(A\hbar^4/m^2)^{1/3}$, are 0.668, 4.697, and 10.24. Show that the above equation for E_n is 18% off for $n = 1$ but is already within 1% for $n = 3$.

A dimensional analysis would give the dimensional factor here as well as the approximate dependence on n. Fitting wavelengths replaces n with $n - \frac{1}{2}$ and gives the approximate numerical factor 1.3765.

4. *Another bound state?*

Problem 2·14(a) was to show that $\psi_1(x) = \mathrm{sech}(ax)$ is an (unnormalized) eigenfunction of the Schrödinger equation for the potential energy

$$V(x) = -\frac{\hbar^2 a^2}{m}\,\mathrm{sech}^2(ax)\,,$$

and that $E_1 = \frac{1}{2}V(0)$. This is the only bound state of a particle of mass m in this well.

Find the largest number of de Broglie wavelengths that would fit in this well. How many wavelengths would there have to be for a second bound state?

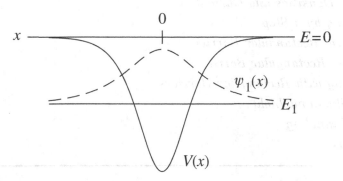

5. *The three-dimensional oscillator.*

Use $\psi_T(r) = ce^{-br^2}$ to get an upper limit on the ground-state energy for the three-dimensional oscillator, where $V(r) = \frac{1}{2}m\omega^2 r^2$. You should get the true value, $\frac{3}{2}\hbar\omega$.

6. *The $V(x) = Ax^4$ well again.*

Use $\psi_T(x) = ce^{-bx^2}$ to get an upper limit on the ground-state energy for $V(x) = Ax^4$. Compare the limit with the true value given in Prob. 3. It is about 2% too high.

7. *The bouncing ball again.*

Use $\psi_T(z) = cze^{-bz^2}$ to get an upper limit on the ground-state energy for $V(z) = mgz$ for $z > 0$, $V(z) = \infty$ otherwise. Compare with the approximate and true values given in Sec. 2. The result is less than 0.5% too high.

8. *The integral $\int_0^\infty x^n e^{-ax} dx = n!/a^{n+1}$.*

We used the integral $\int x^n e^{-ax} dx$ over the range 0 to ∞ in Sec. 3, and will need it over ranges both infinite and finite in later chapters. The integral $\int e^{-x} dx$ is of course $-e^{-x}$.

The indefinite integral of $x^n e^{-x}$ must be of the form $f(x)e^{-x}$, with $f(x)$ such that the derivative of $f(x)e^{-x}$ is $x^n e^{-x}$. Taking the derivative and factoring out the common factor e^{-x}, we are left with

$$f'(x) - f(x) = x^n .$$

(a) For $n = 1$, put $f(x) = c_0 + c_1 x$ into the equation for $f(x)$ and find that $c_1 = c_0 = -1$. Thus $\int xe^{-x} dx = -(1+x)e^{-x}$ and $\int_0^\infty xe^{-x} dx = 1$.

(b) For $n = 2$, put $f(x) = c_0 + c_1 x + c_2 x^2$ into the equation and find that $c_2 = -1$ and $c_1 = c_0 = -2$.

(c) For $n = 3$, put $f(x) = c_0 + c_1 x + c_2 x^2 + c_3 x^3$ into the equation and find the c_i's. Notice how in each case $c_n = -1$ and the coefficients cascade down to make $c_0 = -n!$

(d) Make a table of the integrals $\int_0^\infty x^n e^{-x} dx$ and $\int_0^1 x^n e^{-x} dx$ for $n = 0$, 1, 2, and 3.

(e) Given $\int x^n e^{-x} dx$, how do you get $\int x^n e^{-ax} dx$?

4. SCATTERING IN ONE DIMENSION

4.1. *Particle Densities and Currents*

4.2. *Scattering by a Step*

4.3. *A General Rectangular Barrier*

4.4. *A Simple Rectangular Barrier*

4.5. *Designing with Rectangular Barriers*

4.6. *Thin Films and Light*

4.7. *Weak Tunneling*

    Problems

Chapters 2 and 3 were almost entirely about the bound states of a particle, the energies of which are quantized. This chapter is about the scattering of a beam of *free* particles incident upon a potential-energy step or well or barrier, or a series of such. The incident particles travel in the $+x$ direction, and the changes in potential energy vary only with x. Thus the problem is one dimensional—to find what fraction of the incident particles are reflected back toward $-\infty$ and what fraction continue on toward $+\infty$. These fractions are also the probabilities that a given particle will be reflected or transmitted. With $V(-\infty) = 0$, the energy E of a particle can have any positive value.

Nearly all the problems solved here will be for series of regions of $V(x)$ that have sharp boundaries and are constant between those boundaries: "rectangular" potential energies. One reason for this restriction is its mathematical tractability. Another is that the results, when written in terms of wave numbers k, carry over directly to light incident normally on glass, water, soap bubbles, and the designer multilayered thin films on high-quality lenses, screens, mirrors, and other optical devices. Regions of constant potential energy with $E > V$ are to particles what regions of constant index of refraction are to light. This is the case even though here the particle beams will be nonrelativistic, which light certainly is not. The common element is wave numbers, $k = 2\pi/\lambda$. Waves act like waves.

In Sec. 2·5, on the finite square well, we found that the wave function and its derivative are continuous even across an abrupt (but finite) change of $V(x)$. In all but simplest cases, the efficient mathematical way for handling this double requirement at a boundary is with a 2-by-2 matrix. Then getting across a series of boundaries or regions can be reduced to simply multiplying a series of such matrices. And one way to handle regions of $V(x)$ that vary continuously is to slice them into very narrow regions, with an "infinitesimal" matrix for each slice, and use a computer to do the multiplying. Computers are good at that.

Applications of this chapter would include electronic devices, solar-energy collectors, and such, but that would take us too far afield; the short section on thin films and light will have to do for practical applications here. Chapter 16 takes up the considerably more complicated subject of scattering in three dimensions.

4·1. PARTICLE DENSITIES AND CURRENTS

(a) Infinite plane waves—The very first quantum-mechanical wave function we saw was in Sec. 2·1(c),

$$\Psi(x,t) = Ae^{i(px-Et)/\hbar} = Ae^{i(kx-\omega t)} .$$

This represented a particle with momentum $p = \hbar k$ and energy $E = \hbar\omega$. If $k > 0$, the particle is moving in the $+x$ direction. As was noted in Sec. 2·2(c), a wave function with an exact momentum p (or wavelength $\lambda = h/p$) must be infinite in extent. And then $\int_{-\infty}^{+\infty} |\Psi(x,t)|^2 \, dx = |A|^2 \int_{-\infty}^{+\infty} dx$ is infinite, and Ψ cannot be normalized. Nevertheless, idealized plane waves of monochromatic light are used throughout much of every textbook on optics. In this chapter, we reinterpret the above $\Psi(x,t)$ to be the wave function for a very long, steady *beam* of particles, all having the same momentum—a *plane wave* (a wave with a wide, planar front), not only long but also many wavelengths wide in the y and z directions.

To change the meaning of $\Psi(x,t)$, we change the meaning of A. We let $|\Psi(x,t)|^2$ be the *density* of particles in the beam,

$$\int_{\text{unit volume}} |\Psi(x,t)|^2 \, d(vol) = |A|^2 = \text{number of particles per unit volume of the beam} .$$

The physics problem is: What fractions of a beam of particles are reflected and transmitted when the beam, streaming in the $+x$ direction, is incident upon changes of $V(x)$? In addition to the measure of the density of particles, we shall need a measure of the flux (or current) of particles. We get this by using an analogy with conservation of charge.

(b) Conservation of charge—Charge is conserved—the net amount of charge in a volume changes only if there is a net flow of current across its boundary. Conservation of charge can be derived directly from Maxwell's equations. Figure 1 shows a volume of uniform cross section S. Within the volume, the charge density $\rho = \rho(x,t)$ and the current density $\mathbf{J} = J(x,t)\,\hat{\mathbf{x}}$ vary only with x and t, and any flow of charge is parallel to the x axis. The total charge between the planes at $x = a$ and b at time t is

$$Q_{ab}(t) = \int_a^b \rho(x,t)S \, dx .$$

Here $\rho(x,t)$ is in, say, coulombs per cubic meter.

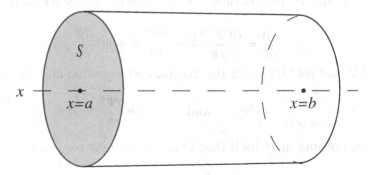

Figure 1. A volume in which the charge and current densities are $\rho(x,t)$ and $\mathbf{J} = J(x,t)\,\hat{\mathbf{x}}$.

Any change of Q_{ab} is due to currents across the boundaries, and here the relevant boundaries are the planes at a and b. Therefore,

$$\frac{dQ_{ab}}{dt} = \frac{d}{dt}\left(\int_a^b \rho(x,t)S\,dx\right) = -S\big[J(b,t) - J(a,t)\big]\,,$$

where $J(x,t)$ is in amperes per square meter. Now d/dt can be moved inside the integral as a *partial* derivative because the differentiation and integration are with respect to independent variables. And the difference between the current densities at a and b can be written as the integral of infinitesimal changes, $dJ = (\partial J/\partial x)\,dx$, from a to b at time t. Thus we have

$$\frac{dQ_{ab}}{dt} = \int_a^b \frac{\partial \rho(x,t)}{\partial t}\,S\,dx \qquad \text{and} \qquad J(b,t) - J(a,t) = \int_a^b dJ = \int_a^b \frac{\partial J(x,t)}{\partial x}\,dx\,.$$

Then the charge-conservation equation $dQ_{ab}/dt = -S\big[J(b,t) - J(a,t)\big]$ becomes

$$\int_a^b \left(\frac{\partial \rho}{\partial t} + \frac{\partial J}{\partial x}\right)dx = 0\,.$$

Charge conservation requires this to be true for any a and b, and mathematics then requires the integrand to be everywhere zero,

$$\frac{\partial \rho}{\partial t} + \frac{\partial J}{\partial x} = 0\,.$$

The integral equation for charge conservation over a volume translates into a differential equation at every point in space. In three dimensions, $\rho = \rho(x,y,z,t)$ and $\mathbf{J} = \mathbf{J}(x,y,z,t)$, and the differential equation becomes

$$\frac{\partial \rho}{\partial t} + \frac{\partial J_x}{\partial x} + \frac{\partial J_y}{\partial y} + \frac{\partial J_z}{\partial z} = \frac{\partial \rho}{\partial t} + \nabla \cdot \mathbf{J} = 0\,,$$

where $J_i = J_i(x,y,z,t)$, $\mathbf{J} = J_x\hat{\mathbf{x}} + J_y\hat{\mathbf{y}} + J_z\hat{\mathbf{z}}$, and $\nabla \cdot \mathbf{J}$ is the divergence of $\mathbf{J}$.

(c) **Conservation of particles**—Particles are not always conserved—they can be created or destroyed in reactions and decays. However, in this chapter particles will be conserved. We already have a particle density, $\rho(x,t) = |\Psi(x,t)|^2$, to correspond to the charge density $\rho(x,t)$. With charge conservation as a model, we *construct* a $J(x,t)$ that will fit the equation $\partial\rho/\partial t + \partial J/\partial x = 0$ and be our particle current density. The obvious place to begin is with $\partial\rho/\partial t$:

$$\frac{\partial \rho}{\partial t} = \frac{\partial(\Psi^*\Psi)}{\partial t} = \frac{\partial \Psi^*}{\partial t}\Psi + \Psi^*\frac{\partial \Psi}{\partial t}\,.$$

We replace $\partial\Psi/\partial t$ and $\partial\Psi^*/\partial t$ using the Schrödinger equation and its complex conjugate,

$$i\hbar\frac{\partial \Psi}{\partial t} = \hat{H}\Psi = -\frac{\hbar^2}{2m}\frac{\partial^2 \Psi}{\partial x^2} + V\Psi \qquad \text{and} \qquad -i\hbar\frac{\partial \Psi^*}{\partial t} = \hat{H}\Psi^* = -\frac{\hbar^2}{2m}\frac{\partial^2 \Psi^*}{\partial x^2} + V\Psi^*\,.$$

Note that there is nothing in $\hat{H}$ itself that requires complex conjugation. In particular, $V(x)$ is real. Thus

$$\frac{\partial \rho}{\partial t} = \frac{i}{\hbar}\big[(\hat{H}\Psi^*)\Psi - \Psi^*(\hat{H}\Psi)\big] = -\frac{i\hbar}{2m}\left[\frac{\partial^2 \Psi^*}{\partial x^2}\Psi - \Psi^*\frac{\partial^2 \Psi}{\partial x^2}\right] + \frac{i}{\hbar}\big[(V\Psi^*)\Psi - \Psi^*(V\Psi)\big]\,.$$

72

The terms with V are just the same three functions multiplied together in different order, and so they cancel. The rest of the equation may be rewritten like this:

$$\frac{\partial\rho}{\partial t} = -\frac{\partial}{\partial x}\left[\frac{i\hbar}{2m}\left(\frac{\partial\Psi^*}{\partial x}\Psi - \Psi^*\frac{\partial\Psi}{\partial x}\right)\right].$$

The quantity in the bracket is just what will make $\partial\rho/\partial t + \partial J/\partial x = 0$. Therefore the particle current density is

$$\boxed{J(x,t) = \frac{i\hbar}{2m}\left(\frac{\partial\Psi^*}{\partial x}\Psi - \Psi^*\frac{\partial\Psi}{\partial x}\right) = \frac{\hbar}{m}\,\mathrm{Im}\left(\Psi^*\frac{\partial\Psi}{\partial x}\right).}$$

The last equality follows because $c - c^* = (a + ib) - (a - ib) = 2ib = 2i\,\mathrm{Im}(c)$. Here $\mathrm{Im}(c)$ means b, not ib: $\mathrm{Im}(c)$ is real. And like $\rho(x,t)$, $J(x,t)$ is real.

(d) Simple examples—We consider a long steady beam in which all the particles have the same energy. Then the time-dependent factors $e^{-iEt/\hbar}$ and $e^{+iEt/\hbar}$ in $\Psi(x,t)$ and $\Psi^*(x,t)$ multiply to unity in both ρ and J. That is, when E is single-valued, neither ρ nor J has any time dependence. Thus in calculating ρ and J, we can work with $\psi(x)$ instead of $\Psi(x,t) = \psi(x)\,e^{-iEt/\hbar}$. And since then $\partial\rho/\partial t = 0$, so also $\partial J/\partial x = 0$, and J is independent of both of t and x. An idealization, but a very useful one.

The simplest example is $\psi(x) = Ae^{ikx}$, the spatial part of $Ae^{i(px-Et)/\hbar}$, where $k = p/\hbar$. Then J is

$$J = \frac{i\hbar}{2m}\left(\frac{\partial\psi^*}{\partial x}\psi - \psi^*\frac{\partial\psi}{\partial x}\right) = \frac{i\hbar}{2m}\left(-ikA^*A - ikA^*A\right) = \frac{\hbar k}{m}|A|^2.$$

This has the form of a speed (momentum $p = \hbar k$ over mass) times a density. The *classical* particle current density at a point where the population density is ρ and the speed is v is $J = v\rho$, so this result is perhaps not surprising. (Imagine a current of cars on a highway.)

Another example is $\psi(x) = Ce^{-qx}$, where q is real. This was the form of $\psi_+(x)$ outside the finite square well for a bound state (Sec. 2·5). Then J is

$$J = \frac{i\hbar}{2m}\left(-qC^*C + qC^*C\right)e^{-2qx} = 0.$$

Thus a nonzero ψ does not imply a nonzero J.

Here is a summary of simple cases (see Prob. 1). The coefficients A, B, C, and D will in general be complex.

$$
\begin{array}{ll}
\psi(x) = Ae^{\pm ikx} & J(x) = \pm(\hbar k/m)|A|^2 \\[4pt]
\psi(x) = Ae^{ikx} + Be^{-ikx} & J(x) = (\hbar k/m)(|A|^2 - |B|^2) \\[4pt]
\psi(x) = Ce^{\pm qx} & J(x) = 0 \\[4pt]
\psi(x) = Ce^{-qx} + De^{+qx} & J(x) = (2\hbar q/m)\,\mathrm{Im}(C^*D).
\end{array}
$$

The current for $\psi(x) = Ae^{ikx} + Be^{-ikx}$ is simply the difference of currents in opposite directions, with no cross term. In contrast, the current for $\psi(x) = Ce^{-qx} + De^{+qx}$ is zero if either term is absent, but is nonzero due to interference of the two terms unless C and D are relatively real.

4·2. SCATTERING BY A STEP

Figure 2 shows a step in potential energy of height V at $x = 0$. A beam of particles, all with the same energy $E > 0$, is incident from the left. We consider the $E > V$ and $E < V$ cases in turn. Note that $V < 0$ (a step down) is a sub-case of $E > V$. Classically, when $E > V$ all the particles would be slowed down or sped up at the step (as V is positive or negative), but all would continue on to the right. And when $E < V$, all of them would bounce back to the left. But now we are doing a *wave* mechanics: When $E > V$, some of the incident particles will be reflected, just as some of a beam of light incident upon an abrupt change of index of refraction is reflected.

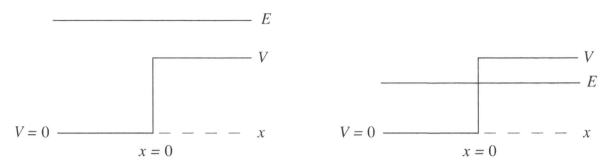

Figure 2. Steps in V, with $E > V$ and $E < V$. *Up* in the figures is in energy, not in space: In (a), the particles are not going *over* anything in space.

Here in outline is what we do:

(1) We write down the form of the wave functions in the two regions of x.

(2) We relate the amplitudes of the reflected and transmitted waves to the amplitude of the incident wave using the continuity of the wave function and its derivative at the step (Sec. 2·5).

(3) We find the fractions of particles reflected and transmitted from the ratios of the particle currents. These fractions are also the probabilities that a given particle is reflected or transmitted.

(a) Energy greater than step height—The Schrödinger equation is

$$\psi'' + \frac{2m}{\hbar^2}(E - V)\psi = 0 \ .$$

The wave functions and their derivatives in the $x < 0$ and $x > 0$ regions are

$$\psi_0(x) = A_0 e^{ik_0 x} + B_0 e^{-ik_0 x} \qquad d\psi_0/dx = ik(A_0 e^{ik_0 x} - B_0 e^{-ik_0 x})$$
$$\psi_1(x) = A_1 e^{ik_1 x} \qquad d\psi_1/dx = ik_1 A_1 e^{ik_1 x} \ ,$$

where k_0 and k_1 are real and positive:

$$k_0^2 \equiv \frac{2mE}{\hbar^2} > 0 \qquad \text{and} \qquad k_1^2 \equiv \frac{2m(E - V)}{\hbar^2} > 0 \ .$$

There is no term $B_1 e^{-ik_1 x}$ in $\psi_1(x)$ because there is no beam coming in from the far right. (Sunlight incident upon the sea is partly reflected and partly transmitted, but no light is coming up from the bottom of the sea.)

The boundary conditions $\psi_0(x) = \psi_1(x)$ and $d\psi_0/dx = d\psi_1/dx$ at $x = 0$ are

$$A_0 + B_0 = A_1 \qquad \text{and} \qquad A_0 - B_0 = \frac{k_1}{k_0} A_1 \;.$$

Adding the two equations, we get

$$2A_0 = \left(1 + \frac{k_1}{k_0}\right) A_1 \;, \qquad \text{or} \qquad \frac{A_1}{A_0} = \frac{2k_0}{k_0 + k_1} \;.$$

Then from $A_0 + B_0 = A_1$, we get

$$\frac{B_0}{A_0} = \frac{A_1}{A_0} - 1 = \frac{k_0 - k_1}{k_0 + k_1} \;.$$

The *fractions* R and T of particles reflected and transmitted are the ratios of *currents* (not of densities), and those currents were found in Sec. 1(d):

$$R = \frac{\hbar k_0/m \, |B_0|^2}{\hbar k_0/m \, |A_0|^2} = \left|\frac{B_0}{A_0}\right|^2 = \left(\frac{k_0 - k_1}{k_0 + k_1}\right)^2 \;,$$

and

$$T = \frac{\hbar k_1/m \, |A_1|^2}{\hbar k_0/m \, |A_0|^2} = \frac{k_1}{k_0}\left|\frac{A_1}{A_0}\right|^2 = \frac{4k_0 k_1}{(k_0 + k_1)^2} \;.$$

And $R + T = 1$ (check this)—a particle is either reflected or transmitted.

Now since $k_1^2/k_0^2 = (E - V)/E = 1 - V/E$, we get

$$\boxed{T = \frac{4k_0 k_1}{(k_0 + k_1)^2} = \frac{4k_1/k_0}{(1 + k_1/k_0)^2} = \frac{4\sqrt{1 - V/E}}{\left(1 + \sqrt{1 - V/E}\right)^2} \;.}$$

Figure 3 shows R and T as functions of E/V. The fraction of particles transmitted grows rapidly with E/V: $T = 71.2\%$ for $E/V = 1.1$, 92.8% for $E/V = 1.5$, and 97.1% for $E/V = 2$ (check this).

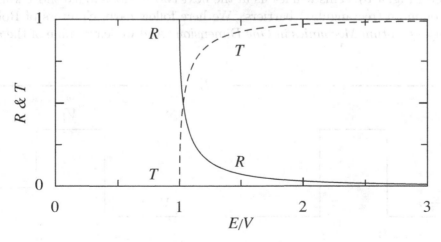

Figure 3. The probabilities R and T of reflection and transmission at a step as a function of E/V.

(b) Energy less than step height—When E is less than V, the form of the wave function $\psi_0(x)$ for $x < 0$ is unchanged, but now

$$\psi_1(x) = A_1 e^{-q_1 x} \qquad \text{and} \qquad d\psi_1/dx = -q_1 A_1 e^{-q_1 x} ,$$

where $q_1^2 \equiv 2m(V - E)/\hbar^2 > 0$, and q_1 is real and positive. Applying the boundary conditions at $x = 0$, we get

$$A_0 + B_0 = A_1 \qquad \text{and} \qquad A_0 - B_0 = \frac{iq_1}{k_0} A_1 .$$

Adding the equations, we now get

$$2 A_0 = \left(1 + \frac{iq_1}{k_0}\right) A_1 , \qquad \text{or} \qquad \frac{A_1}{A_0} = \frac{2k_0}{k_0 + iq_1} ,$$

and then

$$\frac{B_0}{A_0} = \frac{A_1}{A_0} - 1 = \frac{k_0 - iq_1}{k_0 + iq_1} .$$

The fraction of particles reflected is

$$R = \left|\frac{B_0}{A_0}\right|^2 = \left|\frac{k_0 - iq_1}{k_0 + iq_1}\right|^2 = \left(\frac{k_0 - iq_1}{k_0 + iq_1}\right)\left(\frac{k_0 + iq_1}{k_0 - iq_1}\right) = 1 .$$

The incident and reflected currents are equal in magnitude, and the *net* current in the $x < 0$ region is zero. In Sec. 1(d), we found that the current associated with $\psi(x) = C e^{-qx}$ is zero. Thus $T = 0$, and again $R + T = 1$. The wave function penetrates into the $x > 0$ region, but the current there, in the steady state considered, is zero. An analogous phenomenon is the total internal reflection of light.

4·3. A GENERAL RECTANGULAR BARRIER

Figure 4(a) shows the simplest barrier, but in this section we shall work from the slightly more general Fig. 4(b). This will let us in the next two sections make short work of problems with one or more (rectangular) barriers. We here follow early chapters of Robert Gilmore, *Elementary Quantum Mechanics in One Dimension*. And we leave some of the algebra to the reader.

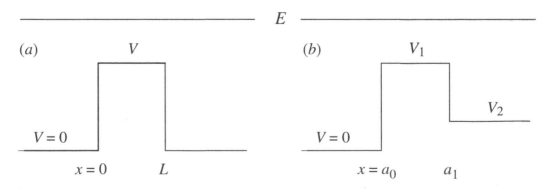

Figure 4. (a) A potential-energy barrier. (b) A slight generalization of (a).

(a) Getting across boundaries (E >V)—Figure 4(b) shows the x axis divided at $x = a_0$ and a_1 into regions over which the potential energies have the constant values 0, V_1, and V_2. In the first two of these regions, the wave functions and their derivatives are

$$x < a_0 : \quad \psi_0(x) = A_0 e^{ik_0 x} + B_0 e^{-ik_0 x} \qquad d\psi_0/dx = ik_0(A_0 e^{ik_0 x} - B_0 e^{-ik_0 x})$$

$$a_0 < x < a_1 : \quad \psi_1(x) = A_1 e^{ik_1 x} + B_1 e^{-ik_1 x} \qquad d\psi_1/dx = ik_1(A_1 e^{ik_1 x} - B_1 e^{-ik_1 x}) \ .$$

The wave numbers here are given by $k_0^2 = 2mE/\hbar^2$ and $k_1^2 = 2m(E-V_1)/\hbar^2$. For the present, we take $E > V_1$, in which case k_0 and k_1 are both real and positive. We no longer exclude the $e^{-ik_1 x}$ term in $\psi_1(x)$ as we did in Sec. 2(a), because now $\psi_1(x)$ only applies over a *finite* range; reflections will occur at $x = a_1$ as well as at a_0.

The two $x < a_0$ equations may be written as a single 2-by-2 matrix equation,

$$\begin{pmatrix} \psi_0 \\ d\psi_0/dx \end{pmatrix} = \begin{pmatrix} e^{ik_0 x} & e^{-ik_0 x} \\ ik_0 e^{ik_0 x} & -ik_0 e^{-ik_0 x} \end{pmatrix} \begin{pmatrix} A_0 \\ B_0 \end{pmatrix} .$$

The equations for ψ_1 and $d\psi_1/dx$ for $a_0 < x < a_1$ have exactly the same form: Just boost every subscript from 0 to 1. Then the continuity equations $\psi_0 = \psi_1$ and $d\psi_0/dx = d\psi_1/dx$ at the $x = a_0$ boundary may be written

$$\begin{pmatrix} e^{ik_0 a_0} & e^{-ik_0 a_0} \\ ik_0 e^{ik_0 a_0} & -ik_0 e^{-ik_0 a_0} \end{pmatrix} \begin{pmatrix} A_0 \\ B_0 \end{pmatrix} = \begin{pmatrix} e^{ik_1 a_0} & e^{-ik_1 a_0} \\ ik_1 e^{ik_1 a_0} & -ik_1 e^{-ik_1 a_0} \end{pmatrix} \begin{pmatrix} A_1 \\ B_1 \end{pmatrix} .$$

For the present, we assume that $E > V_2$, as in Fig. 4(b), so that $k_2^2 \equiv 2m(E - V_2)/\hbar^2 > 0$ and k_2 is real. Then the continuity equations over at $x = a_1$ have exactly the same form as those at a_1; again, just boost every subscript in the previous equation by one:

$$\begin{pmatrix} e^{ik_1 a_1} & e^{-ik_1 a_1} \\ ik_1 e^{ik_1 a_1} & -ik_1 e^{-ik_1 a_1} \end{pmatrix} \begin{pmatrix} A_1 \\ B_1 \end{pmatrix} = \begin{pmatrix} e^{ik_2 a_1} & e^{-ik_2 a_1} \\ ik_2 e^{ik_2 a_1} & -ik_2 e^{-ik_2 a_1} \end{pmatrix} \begin{pmatrix} A_2 \\ B_2 \end{pmatrix} .$$

And we keep the $e^{-ik_2 x}$ term in $\psi_2(x)$, to allow for the possibility that there are more steps in V farther to the right, and more reflections.

Matrices are a wonderful invention for organizing parallel relations in a concise way, but they can take some getting used to. This section is less complicated than it looks, because there are a lot of repetitions: All five of the 2-by-2 matrices so far have exactly the same form. We need the inverse of that matrix. The matrix itself can be factored like this,

$$\begin{pmatrix} e^{ika} & e^{-ika} \\ ike^{ika} & -ike^{-ika} \end{pmatrix} = \begin{pmatrix} 1 & 1 \\ ik & -ik \end{pmatrix} \begin{pmatrix} e^{ika} & 0 \\ 0 & e^{-ika} \end{pmatrix} ,$$

which you should check. And each of those factor matrices has an inverse,

$$\begin{pmatrix} 1 & 1 \\ ik & -ik \end{pmatrix}^{-1} = \frac{1}{2} \begin{pmatrix} 1 & -i/k \\ 1 & i/k \end{pmatrix} \qquad \text{and} \qquad \begin{pmatrix} e^{ika} & 0 \\ 0 & e^{-ika} \end{pmatrix}^{-1} = \begin{pmatrix} e^{-ika} & 0 \\ 0 & e^{ika} \end{pmatrix} ,$$

which you should check.

The inverse of the product $\hat{P}\hat{Q}$ of two matrices is $\hat{Q}^{-1}\hat{P}^{-1}$, in that reverse order,

$$(\hat{Q}^{-1}\hat{P}^{-1})\,\hat{P}\hat{Q} = \hat{Q}^{-1}(\hat{P}^{-1}\hat{P})\hat{Q} = \hat{Q}^{-1}\hat{Q} = \hat{I}\,.$$

Therefore, the inverse of our original matrix is

$$\begin{pmatrix} e^{ika} & e^{-ika} \\ ike^{ika} & -ike^{-ika} \end{pmatrix}^{-1} = \frac{1}{2}\begin{pmatrix} e^{-ika} & 0 \\ 0 & e^{ika} \end{pmatrix}\begin{pmatrix} 1 & -i/k \\ 1 & i/k \end{pmatrix} = \frac{1}{2}\begin{pmatrix} e^{-ika} & -ik^{-1}e^{-ika} \\ e^{ika} & ik^{-1}e^{ika} \end{pmatrix}.$$

Then

$$\frac{1}{2}\begin{pmatrix} e^{-ika} & -ik^{-1}e^{-ika} \\ e^{ika} & ik^{-1}e^{ika} \end{pmatrix}\begin{pmatrix} e^{ika} & e^{-ika} \\ ike^{ika} & -ike^{-ika} \end{pmatrix} = \begin{pmatrix} 1 & 0 \\ 0 & 1 \end{pmatrix},$$

which you should check!

Our aim is to eliminate the "middleman" coefficients, (A_1, B_1), in order to relate the (A_0, B_0) coefficients directly to the (A_2, B_2) coefficients. We go back to the continuity equation at a_0 that relates the (A_0, B_0) and (A_1, B_1) coefficients,

$$\begin{pmatrix} e^{ik_0 a_0} & e^{-ik_0 a_0} \\ ik_0 e^{ik_0 a_0} & -ik_0 e^{-ik_0 a_0} \end{pmatrix}\begin{pmatrix} A_0 \\ B_0 \end{pmatrix} = \begin{pmatrix} e^{ik_1 a_0} & e^{-ik_1 a_0} \\ ik_1 e^{ik_1 a_0} & -ik_1 e^{-ik_1 a_0} \end{pmatrix}\begin{pmatrix} A_1 \\ B_1 \end{pmatrix}.$$

We multiply this equation from the left by the inverse of the matrix on the left (the matrix with k_0's). All that will remain on the left is the (A_0, B_0) vector:

$$\begin{pmatrix} A_0 \\ B_0 \end{pmatrix} = \frac{1}{2}\begin{pmatrix} e^{-ik_0 a_0} & -ik_0^{-1}e^{-ik_0 a_0} \\ e^{ik_0 a_0} & ik_0^{-1}e^{ik_0 a_0} \end{pmatrix}\begin{pmatrix} e^{ik_1 a_0} & e^{-ik_1 a_0} \\ ik_1 e^{ik_1 a_0} & -ik_1 e^{-ik_1 a_0} \end{pmatrix}\begin{pmatrix} A_1 \\ B_1 \end{pmatrix} = \mathcal{T}_{01}\begin{pmatrix} A_1 \\ B_1 \end{pmatrix}.$$

And by boosting all the subscripts by one, we get the corresponding equation for the boundary at a_1,

$$\begin{pmatrix} A_1 \\ B_1 \end{pmatrix} = \frac{1}{2}\begin{pmatrix} e^{-ik_1 a_1} & -ik_1^{-1}e^{-ik_1 a_1} \\ e^{ik_1 a_1} & ik_1^{-1}e^{ik_1 a_1} \end{pmatrix}\begin{pmatrix} e^{ik_2 a_1} & e^{-ik_2 a_1} \\ ik_2 e^{ik_2 a_1} & -ik_2 e^{-ik_2 a_1} \end{pmatrix}\begin{pmatrix} A_2 \\ B_2 \end{pmatrix} = \mathcal{T}_{12}\begin{pmatrix} A_2 \\ B_2 \end{pmatrix}.$$

The $\mathcal{T}_{01}$ and $\mathcal{T}_{12}$ matrices (were we to write them out) are called transfer matrices. They "transfer" the (A, B) vector from one region to the next across the boundary between them. We could then write

$$\begin{pmatrix} A_0 \\ B_0 \end{pmatrix} = \mathcal{T}_{01}\begin{pmatrix} A_1 \\ B_1 \end{pmatrix} = \mathcal{T}_{01}\mathcal{T}_{12}\begin{pmatrix} A_2 \\ B_2 \end{pmatrix} = \mathcal{T}_{02}\begin{pmatrix} A_2 \\ B_2 \end{pmatrix},$$

where $\mathcal{T}_{02} = \mathcal{T}_{01}\mathcal{T}_{12}$. The *full* transfer matrix, $\mathcal{T}_{02}$, would be the product of the four matrices in the previous two set-off equations, in the order in which they occur. But the transfer matrices are not quite what we are after here.

(b) Getting across a region with E > V—The product of the *middle two* of the four matrices that go into $\mathcal{T}_{02}$—the only two of the four that involve k_1 is

$$\begin{pmatrix} e^{ik_1 a_0} & e^{-ik_1 a_0} \\ ik_1 e^{ik_1 a_0} & -ik_1 e^{-ik_1 a_0} \end{pmatrix}\frac{1}{2}\begin{pmatrix} e^{-ik_1 a_1} & -ik_1^{-1}e^{-ik_1 a_1} \\ e^{ik_1 a_1} & ik_1^{-1}e^{ik_1 a_1} \end{pmatrix} = \begin{pmatrix} \cos\phi_1 & -k_1^{-1}\sin\phi_1 \\ k_1 \sin\phi_1 & \cos\phi_1 \end{pmatrix},$$

where $\phi_1 \equiv k_1(a_1 - a_0) = k_1 L_1$, and $L_1 = a_1 - a_0$ is the width of region 1. *This* matrix,

$$\mathcal{M}_1 \equiv \begin{pmatrix} \cos\phi_1 & -k_1^{-1}\sin\phi_1 \\ k_1\sin\phi_1 & \cos\phi_1 \end{pmatrix},$$

gets us across a *region*. The *transfer* matrices $\mathcal{T}_{01}$ and $\mathcal{T}_{12}$ would have got us across boundaries. This is the first main result of this section. It applies when $E > V_1$.

The $\mathcal{M}_1$ matrix itself is not complicated. The determinant of $\mathcal{M}_1$ equals one. The wave number k_1 is by definition $2\pi/\lambda_1$, where λ_1 is the de Broglie wavelength in region 1. Therefore L_1/λ_1 is the number of de Broglie wavelengths in the distance L_1, and $\phi_1 = k_1 L_1 = 2\pi L_1/\lambda_1$ is the change in phase (in radians) of the wave function from a_0 to a_1. If $\phi_1 = 2\pi$, then $\mathcal{M}_1 = \hat{I}$, the 2-by-2 identity matrix. What is $\mathcal{M}_1$ if $\phi_1 = \pi$?

With the two k_1 matrices reduced to one, the matrix product that relates the (A_0, B_0) and (A_2, B_2) coefficients is down from four pieces to three,

$$\begin{pmatrix} A_0 \\ B_0 \end{pmatrix} =$$

$$\frac{1}{2}\begin{pmatrix} e^{-ik_0 a_0} & -ik_0^{-1}e^{-ik_0 a_0} \\ e^{ik_0 a_0} & ik_0^{-1}e^{ik_0 a_0} \end{pmatrix}\begin{pmatrix} \cos\phi_1 & -k_1^{-1}\sin\phi_1 \\ k_1\sin\phi_1 & \cos\phi_1 \end{pmatrix}\begin{pmatrix} e^{ik_2 a_1} & e^{-ik_2 a_1} \\ ik_2 e^{ik_2 a_1} & -ik_2 e^{-ik_2 a_1} \end{pmatrix}\begin{pmatrix} A_2 \\ B_2 \end{pmatrix}.$$

And now without loss of generality we can set the left edge, a_0, of the first barrier (no matter what else may lie beyond a_1) to be at $x = 0$. Then

$$\begin{pmatrix} A_0 \\ B_0 \end{pmatrix} = \frac{1}{2}\begin{pmatrix} 1 & -i/k_0 \\ 1 & i/k_0 \end{pmatrix}\begin{pmatrix} \cos\phi_1 & -k_1^{-1}\sin\phi_1 \\ k_1\sin\phi_1 & \cos\phi_1 \end{pmatrix}\begin{pmatrix} e^{ik_2 a_1} & e^{-ik_2 a_1} \\ ik_2 e^{ik_2 a_1} & -ik_2 e^{-ik_2 a_1} \end{pmatrix}\begin{pmatrix} A_2 \\ B_2 \end{pmatrix},$$

where $\phi_1 = k_1 L_1$. This is the second main result, to be used in the next two sections. But first ...

(c) **Getting across a region with $E < V$**—If in Fig. 4(b) E is less than V_1, then ψ_1 and $d\psi_1/dx$ between $x = a_0$ and a_1 are

$$\psi_1(x) = A_1 e^{-q_1 x} + B_1 e^{+q_1 x} \quad \text{and} \quad d\psi_1/dx = -q_1(A_1 e^{-q_1 x} - B_1 e^{+q_1 x}),$$

where $q_1^2 \equiv 2m(V_1 - E)/\hbar^2 > 0$. The $E > V$ equations for this same region were

$$\psi_1(x) = A_1 e^{ik_1 x} + B_1 e^{-ik_1 x} \quad \text{and} \quad d\psi_1/dx = ik_1(A_1 e^{ik_1 x} - B_1 e^{-ik_1 x}).$$

Replacing ik_1 with $-q_1$ (or k_1 with iq_1) in these $E > V_1$ equations turns them into the $E < V_1$ equations. Thus by replacing k_1 with iq_1 in the $E > V_1$ region-crossing matrix $\mathcal{M}_1$, we get the region-crossing matrix for $E < V_1$. For this, we need

$$\cos(k_1 L_1) \rightarrow \cos(iq_1 L_1) = \cosh(q_1 L_1) \quad \text{and} \quad \sin(k_1 L_1) \rightarrow \sin(iq_1 L_1) = i\sinh(q_1 L_1).$$

For example, $\cos\delta = \frac{1}{2}(e^{i\delta} + e^{-i\delta}) \longrightarrow \cos(i\delta) = \frac{1}{2}(e^{-\delta} + e^{\delta}) = \cosh\delta$. With these substitutions, the matrix that takes us from a_0 to a_1 when $E < V_1$ is

$$\begin{pmatrix} \cos\phi_1 & -k_1^{-1}\sin\phi_1 \\ k_1\sin\phi_1 & \cos\phi_1 \end{pmatrix} \longrightarrow \mathcal{M}_1 \equiv \begin{pmatrix} \cosh\Phi_1 & -q_1^{-1}\sinh\Phi_1 \\ -q_1\sinh\Phi_1 & \cosh\Phi_1 \end{pmatrix},$$

where $\Phi_1 \equiv q_1 L_1$. This is the third main result. The determinant is again one.

4·4. A SIMPLE RECTANGULAR BARRIER

(a) **A single barrier with E > V**—The last equation in Sec. 3(b) is

$$\begin{pmatrix} A_0 \\ B_0 \end{pmatrix} = \frac{1}{2} \begin{pmatrix} 1 & -i/k_0 \\ 1 & i/k_0 \end{pmatrix} \begin{pmatrix} \cos\phi_1 & -k_1^{-1}\sin\phi_1 \\ k_1\sin\phi_1 & \cos\phi_1 \end{pmatrix} \begin{pmatrix} e^{ik_2 a_1} & e^{-ik_2 a_1} \\ ik_2 e^{ik_2 a_1} & -ik_2 e^{-ik_2 a_1} \end{pmatrix} \begin{pmatrix} A_2 \\ B_2 \end{pmatrix}.$$

Figure 4(a) showed a single barrier with $a_0 = 0$, $a_1 = L$, $V_0 = V_2 = 0$, and $V_1 = V$. Then $k_2 = k_0$; and B_2 is 0 because there are no particles coming in from the right. With these simplifications, the equation reduces to

$$\begin{pmatrix} A_0 \\ B_0 \end{pmatrix} = \frac{1}{2} \begin{pmatrix} 1 & -i/k_0 \\ 1 & i/k_0 \end{pmatrix} \begin{pmatrix} \cos\phi_1 & -k_1^{-1}\sin\phi_1 \\ k_1\sin\phi_1 & \cos\phi_1 \end{pmatrix} \begin{pmatrix} 1 \\ ik_0 \end{pmatrix} A_2 e^{ik_0 L},$$

where $\phi_1 = k_1 L$ is the phase change from $x = 0$ to L. There are two equations here: the "top" one relates A_0 to A_2, the bottom one B_0 to A_2. We only need the top one. The second matrix times the column vector is

$$\begin{pmatrix} \cos\phi_1 & -k_1^{-1}\sin\phi_1 \\ k_1\sin\phi_1 & \cos\phi_1 \end{pmatrix} \begin{pmatrix} 1 \\ ik_0 \end{pmatrix} = \begin{pmatrix} \cos\phi_1 - i(k_0/k_1)\sin\phi_1 \\ k_1\sin\phi_1 + ik_0\cos\phi_1 \end{pmatrix}.$$

The top row, $(1, -i/k_0)$, of the first matrix times this column vector and the remaining factors $\frac{1}{2}$ and $A_2 e^{ik_0 L}$ is

$$A_0 = \left[\cos\phi_1 - \frac{i}{2}\left(\frac{k_1}{k_0} + \frac{k_0}{k_1}\right)\sin\phi_1\right] A_2 e^{ik_0 L},$$

which you should check. Then the fraction T of particles transmitted through the barrier is given by

$$\frac{1}{T} = \left|\frac{A_0}{A_2}\right|^2 = \cos^2\phi_1 + \frac{(k_0^2 + k_1^2)^2}{4k_0^2 k_1^2}\sin^2\phi_1 = 1 + \frac{(k_0^2 - k_1^2)^2}{4k_0^2 k_1^2}\sin^2\phi_1.$$

Since $k_0^2 \equiv 2mE/\hbar^2$ and $k_1^2 \equiv 2m(E - V)/\hbar^2$, we can write this more neatly as

$$\frac{1}{T} = 1 + \frac{V^2}{4E(E - V)}\sin^2\phi_1 = 1 + \frac{1}{4\epsilon(\epsilon - 1)}\sin^2\phi_1,$$

where $\epsilon \equiv E/V$. All of which you should ... check! In summary,

$$T = \frac{1}{1 + Z} \quad \text{and} \quad R = 1 - T = \frac{Z}{1 + Z}, \quad \text{where} \quad Z \equiv \frac{1}{4\epsilon(\epsilon - 1)}\sin^2\phi_1.$$

Both Z and ϵ are dimensionless; and, again, $\phi_1 = k_1 L = 2\pi L/\lambda_1$ is the phase change across the barrier. The equations also apply to scattering from a well ($V < 0$), in which case $\epsilon = E/V$ is negative. This $1/(1 + Z)$ way of writing T, with differing equations for Z, will be used again.

When $\sin\phi_1 = 0$, Z is zero, T is one, and *no* particles are reflected. This occurs when $\phi_1 = k_1 L = n\pi$, where $n = 1, 2, 3, \ldots$ In terms of the de Broglie wavelength $\lambda_1 = 2\pi/k_1$ in the barrier region, the condition is $L = n\lambda_1/2$; that is, an integral number of half wavelengths fit into L. From $k_1^2 = 2m(E - V)/\hbar^2$, we get

$$E_n - V = \frac{\hbar^2 k_1^2}{2m} = n^2 \frac{\pi^2 \hbar^2}{2mL^2} \, ,$$

for the energies that give 100% transmission. The same condition, $L = n\lambda_1/2$, gave the energy levels, $E_n = n^2(\pi^2\hbar^2/2mL^2)$, of the infinite square well.

We rewrite T in terms of the ratio $\epsilon = E/V$ and a parameter β, defined as

$$\boxed{\beta^2 \equiv \frac{2m}{\hbar^2}VL^2 \, .}$$

A higher and/or wider barrier makes β larger. Like ϵ, β is dimensionless, and it contains all the parameters of the problem in the only combination that matters. The phase angle is given by

$$\phi_1^2 = (k_1 L)^2 = \frac{2m}{\hbar^2}(E - V)L^2 = \frac{2m}{\hbar^2}VL^2\,(\epsilon - 1) = \beta^2(\epsilon - 1) \, .$$

Then

$$\boxed{T = \frac{1}{1 + Z} \, , \qquad \text{where} \quad Z = \frac{1}{4\epsilon(\epsilon - 1)} \sin^2 \phi_1 = \frac{1}{4\epsilon(\epsilon - 1)} \sin^2(\beta\sqrt{\epsilon - 1}) \, .}$$

Figure 5 shows T as a function of ϵ for two values of β. For each value, the maxima occur at energies in the ratios $1:4:9:\ldots$ above $\epsilon = 1$.

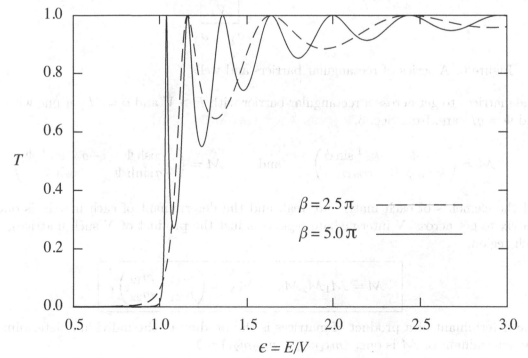

$$\beta = 2.5\,\pi$$
$$\beta = 5.0\,\pi$$

Figure 5. The fraction T of particles transmitted at a rectangular barrier as a function of $\epsilon = E/V$ for $\beta = 2.5\pi$ and $\beta = 5\pi$ rad, where $\beta^2 \equiv 2mVL^2/\hbar^2$.

(b) A single barrier with E < V—In Sec. 3(c), we went from $E > V$ to $E < V$ by replacing k_1 with iq_1. In particular, this turned $\sin(k_1 L)$ into $i\sinh(q_1 L)$, or $\sin\phi_1$ into $i\sinh\Phi_1$, where $q_1^2 \equiv 2m(V-E)/\hbar^2$ and $\Phi_1 = q_1 L$. Making these same changes in the T in the last set-off equation, and with $\epsilon = E/V$ and $\beta^2 \equiv 2mVL^2/\hbar^2$ as before, we get

$$T = \frac{1}{1+Z'}, \qquad \text{where} \quad Z' = \frac{1}{4\epsilon(1-\epsilon)}\sinh^2(q_1 L) = \frac{1}{4\epsilon(1-\epsilon)}\sinh^2(\beta\sqrt{1-\epsilon}).$$

Classically, when $E < V$, all the particles bounce back at $x = 0$. But continuity of the *wave* function at $x = 0$ requires that the wave leak into the classically forbidden region. And then continuity again over at $x = L$ requires that some remnant of the wave leak out the far side. This is *quantum tunneling*. For the values of β used in Fig. 5, T is very small until E is nearly as large as V.

4·5. DESIGNING WITH RECTANGULAR BARRIERS

(a) Barriers in series—Figure 6 shows a series of rectangular barriers and wells.

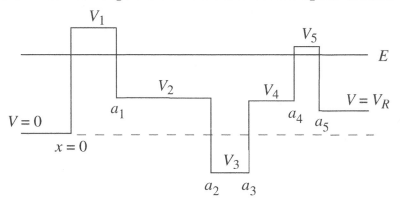

Figure 6. A series of rectangular barriers and wells.

The matrices to get across a rectangular barrier with $E > V$ and $\phi = kL$, or one with $E < V$ and $\Phi = qL$, are, from Sec. 3,

$$\mathcal{M} = \begin{pmatrix} \cos\phi & -k^{-1}\sin\phi \\ k\sin\phi & \cos\phi \end{pmatrix} \qquad \text{and} \qquad \mathcal{M} = \begin{pmatrix} \cosh\Phi & -q^{-1}\sinh\Phi \\ -q\sinh\Phi & \cosh\Phi \end{pmatrix}.$$

All the elements of each matrix are real, and the determinant of each matrix is one. The matrix to get across N intermediate regions is just the product of N such matrices, one for each region,

$$\mathcal{M} = \mathcal{M}_1\mathcal{M}_2\mathcal{M}_3\cdots\mathcal{M}_N = \begin{pmatrix} m_{11} & m_{12} \\ m_{21} & m_{22} \end{pmatrix}.$$

The determinant of a product of matrices is the product of the individual determinants, so the determinant of $\mathcal{M}$ is one: $(m_{11}m_{22} - m_{12}m_{21}) = 1$.

To the left of $x = 0$ in Fig. 6, $V = 0$ and $k = k_0$. The N barriers run from $x = 0$ to $x = L_1 + L_2 + \cdots + L_N \equiv L$, where $L_1 = a_1 - a_0$, etc. To the right of $x = L$, $V = V_R$

and $k = k_R$. And there is no wave coming in from the far right, so $B_R = 0$. With obvious substitutions in the second set-off equation of Sec. 4, we get

$$\begin{pmatrix} A_0 \\ B_0 \end{pmatrix} = \frac{1}{2} \begin{pmatrix} 1 & -i/k_0 \\ 1 & i/k_0 \end{pmatrix} \begin{pmatrix} m_{11} & m_{12} \\ m_{21} & m_{22} \end{pmatrix} \begin{pmatrix} 1 \\ ik_R \end{pmatrix} A_R e^{ik_R L}.$$

The $\mathcal{M}$ matrix times the column vector is

$$\begin{pmatrix} m_{11} & m_{12} \\ m_{21} & m_{22} \end{pmatrix} \begin{pmatrix} 1 \\ ik_R \end{pmatrix} = \begin{pmatrix} m_{11} + ik_R\, m_{12} \\ m_{21} + ik_R\, m_{22} \end{pmatrix}.$$

The top row, $(1, -i/k_0)$, of the first matrix times this column vector and the factors $\frac{1}{2}$ and $A_R e^{ik_R L}$ is

$$A_0 = \frac{1}{2} \left[(m_{11} + m_{22}\, k_R/k_0) + i\,(k_R\, m_{12} - m_{21}/k_0) \right] A_R e^{ik_R L}.$$

The incident and transmitted currents are $(\hbar k_0/m)|A_0|^2$ and $(\hbar k_R/m)|A_R|^2$, so

$$T = \frac{k_R}{k_0} \frac{|A_R|^2}{|A_0|^2} = \frac{4k_R/k_0}{(m_{11} + m_{22}\, k_R/k_0)^2 + (k_R\, m_{12} - m_{21}/k_0)^2} \quad \text{for } k_R \neq k_0 .$$

This is an important result! If $V_R = 0$, then $k_R = k_0$ and T simplifies a bit to

$$T = \frac{|A_R|^2}{|A_0|^2} = \frac{4}{(m_{11} + m_{22})^2 + (k_0\, m_{12} - m_{21}/k_0)^2} \quad \text{for } k_R = k_0 .$$

In the following, these will be called the $k_R \neq k_0$ and $k_R = k_0$ equations for T. In the latter case, the sum of the *cross* terms in squaring out the terms in the denominator is $2(m_{11}m_{22} - m_{12}m_{21}) = 2$. (Why $= 2$?) Thus the whole denominator is equal to $\left(2 + m_{11}^2 + m_{22}^2 + k_0^2 m_{12}^2 + m_{21}^2/k_0^2\right)$. However, this does not seem to simplify further algebra.

We can now do some designing.

(b) Design: 100% transmission at a step—In Sec. 2(a), we found that the transmission probability for getting past a step V_R high when $E > V_R$ is

$$T = \frac{4k_0 k_R}{(k_0 + k_R)^2},$$

where k_0 and k_R are the wave numbers in the two regions. This is always less than 1 (see Fig. 3). Figure 7 shows a rectangular barrier V_1 high and L_1 wide in front of a step V_R high. Can we choose V_1 and L_1 so that *all* the particles with a specified energy $E > V_R$ get through?

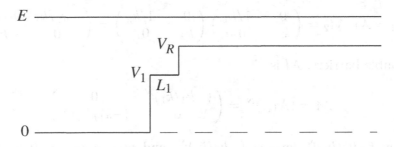

Figure 7. Here $E = 4V_R/3$. If $V_1 = E/2$ and $L_1 = \lambda_1/4$, then $T = 1$.

83

The answer is yes. There are two conditions. The first is that $\phi_1 = k_1 L_1 = \frac{1}{2}\pi$, where $k_1^2 = 2m(E - V_1)/\hbar^2$. Since $k = 2\pi/\lambda$, this makes $L_1 = \lambda_1/4$, where λ_1 is the de Broglie wavelength in region 1. And with $\phi_1 = \frac{1}{2}\pi$ the $\mathcal{M}$ matrix is very simple:

$$\mathcal{M} = \begin{pmatrix} \cos\phi_1 & -k_1^{-1}\sin\phi_1 \\ k_1\sin\phi_1 & \cos\phi_1 \end{pmatrix} \longrightarrow \begin{pmatrix} 0 & -1/k_1 \\ k_1 & 0 \end{pmatrix}.$$

Setting $m_{11} = m_{22} = 0$, $m_{12} = -1/k_1$, and $m_{21} = k_1$ in the $k_R \neq k_0$ equation for T (the first boxed equation on p. 83), we get

$$T = \frac{4k_R/k_0}{(k_R/k_1 + k_1/k_0)^2}.$$

Now let $a = k_R/k_1$ and $b = k_1/k_0$. If $a = b$, then $(a+b)^2 = a^2 + 2ab + b^2 = 4ab$. So if $k_R/k_1 = k_1/k_0$, then the denominator in T is $4k_R/k_0$, and $T = 1$. The *two* conditions for 100% transmission at a step are

$$\boxed{(1)\ \phi_1 = k_1 L_1 = \tfrac{1}{2}\pi \ (\text{or}\ L_1 = \tfrac{1}{4}\lambda_1), \quad \text{and} \quad (2)\ k_1 = \sqrt{k_0 k_R}\ .}$$

(c) Design: a particle mirror—Figure 8 shows a series of N identical double barriers, with heights V_1 and V_2 and widths L_1 and L_2. Can we choose the parameters so that nearly all the particles with a specified energy E are *reflected*?

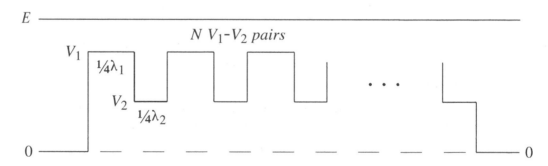

Figure 8. A series of N double barriers to make a particle mirror.

Yes, again. The conditions are, (1) make $\phi_1 = k_1 L_1$ and $\phi_2 = k_2 L_2$ both equal to $\frac{1}{2}\pi$, and (2) make N large. The first condition is equivalent to making $L_i = \lambda_i/4$ in each region. For *one* double-barrier pair, $\mathcal{M}$ is

$$\mathcal{M}_{12} = \mathcal{M}_1 \mathcal{M}_2 = \begin{pmatrix} 0 & -1/k_1 \\ k_1 & 0 \end{pmatrix} \begin{pmatrix} 0 & -1/k_2 \\ k_2 & 0 \end{pmatrix} = \begin{pmatrix} -k_2/k_1 & 0 \\ 0 & -k_1/k_2 \end{pmatrix}.$$

For the N double barriers, $\mathcal{M}$ is

$$\mathcal{M} = (\mathcal{M}_{12})^N = \begin{pmatrix} (-k_2/k_1)^N & 0 \\ 0 & (-k_1/k_2)^N \end{pmatrix}.$$

Setting $m_{11} = (-k_2/k_1)^N$, $m_{22} = (-k_1/k_2)^N$, and $m_{12} = m_{21} = 0$ in the second boxed equation on p. 83, we get

84

$$\boxed{T = \frac{4}{\left[(-k_2/k_1)^N + (-k_1/k_2)^N\right]^2} \cdot}$$

Suppose $V_1 > V_2$, as in Fig. 8. Then $k_2 > k_1$ (the momentum $p_2 = \hbar k_2$ is larger where V is smaller), and $(k_2/k_1)^{2N}$ will grow rapidly with N and make T small.

In Fig. 8, $V_1 = 3E/4$ and $V_2 = E/2$, but for a simpler example, let $V_2 = 0$. Then $k_1 = k_0/2$, $k_2 = k_0$, and $k_2/k_1 = 2$. Then $R = 1 - T$ is

$N =$	1	2	3	5	10
$R\ (\%) =$	36.0	77.85	93.94	99.61	99.9996

$N = 10$ makes a pretty good mirror.

(d) Design: $0.2 \times 0.2 = 1$—Problem 4 is to find $\mathcal{M}$ and T for a δ-function barrier $V(x) = g\,\delta(x)$, where $g > 0$. With a subscript added to indicate they are for one δ barrier, the answers are

$$\mathcal{M}_1 = \begin{pmatrix} 1 & 0 \\ -2mg/\hbar^2 & 1 \end{pmatrix} \quad \text{and} \quad T_1 = \frac{1}{1 + Z_1}, \quad \text{where} \quad Z_1 = \frac{mg^2}{2E\hbar^2} = \left(\frac{mg}{\hbar^2 k_0}\right)^2.$$

Suppose the parameters are such that $Z_1 = 4$. Then $T_1 = 0.2$, and 20% of the particles get through the barrier. Now add an identical barrier, $V(x) = g\delta(x - L)$, a distance L downstream. What fraction of particles will get past both barriers? One might guess that $T_2 = 0.2 \times 0.2$, or 4%, but that is not the way of waves. In fact, we can choose the separation L between the δ's to make $T = 100\%$.

For brevity, let $A \equiv 2mg/\hbar^2$ in $\mathcal{M}_1$ above. Then the matrix $\mathcal{M}_2$ for two δ's and the intermediate region where $V = 0$ is

$$\mathcal{M}_2 = \begin{pmatrix} 1 & 0 \\ -A & 1 \end{pmatrix} \begin{pmatrix} \cos\phi & -k_0^{-1}\sin\phi \\ k_0\sin\phi & \cos\phi \end{pmatrix} \begin{pmatrix} 1 & 0 \\ -A & 1 \end{pmatrix},$$

where $\phi = k_0 L$. The multiplication gives

$$\mathcal{M}_2 = \begin{pmatrix} \cos\phi + Ak_0^{-1}\sin\phi & -k_0^{-1}\sin\phi \\ (k_0 - A^2 k_0^{-1})\sin\phi - 2A\cos\phi & \cos\phi + Ak_0^{-1}\sin\phi \end{pmatrix}.$$

(See Prob. 9.) Putting these m_{ij} elements into the $k_R = k_0$ equation for T at the end of Sec. 5(a), we eventually get

$$\boxed{T_2 = \frac{1}{1 + Z_2}, \quad \text{where} \quad Z_2 = \gamma^2\left(\cos\phi + \frac{\gamma}{2}\sin\phi\right)^2 \cdot}$$

and $\gamma \equiv A/k_0 = 2mg/\hbar^2 k_0$. And if $\tan\phi = -2/\gamma$, then $T_2 = 1$.

From the short-hand definitions used above, which are

$$Z_1 = \left(\frac{mg}{\hbar^2 k_0}\right)^2 \quad \text{and} \quad \gamma = \frac{A}{k_0} = 2\frac{mg}{\hbar^2 k_0},$$

we get $\gamma = 2\sqrt{Z_1}$. If, as in the example earlier, $Z_1 = 4$, then $\gamma = 4$. So if $\tan\phi = -2/\gamma = -0.5$, then T_2 is 1. The angles for this are $\phi = 153.43°$ or $333.44°$ (second and fourth quadrants), and all the repeats (mod $360°$) as ϕ spins through $360°$, $720°$, ... Figure 9 shows T_2 as a function of ϕ.

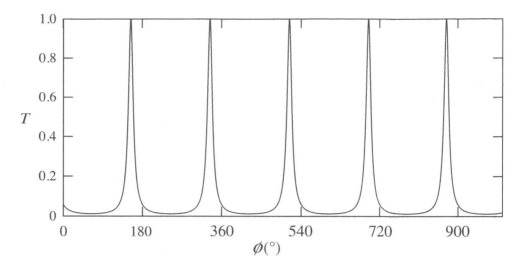

Figure 9. The transmission T_2 for two δ-function barriers a distance L apart as a function of the phase angle $\phi = k_0 L$ when the transmission for each barrier alone is $T_1 = 0.2$. Since k_0 is fixed by the other parameters and T_1, ϕ is proportional to L.

4·6. THIN FILMS AND LIGHT

The mathematics for light incident normally on abrupt changes of index of refraction in transparent media is very similar to that for particles incident normally on abrupt changes of V in regions with $E > V$. Waves are waves, and in either case the wave or field functions are of the form $Ae^{i(kx-\omega t)}$. (The relation between k and ω is of course different for light and particles; see below.) The electric and magnetic fields of light oscillate perpendicularly to the direction the light is going. Thus for normally incident light, the fields oscillate parallel to the interface planes. The boundary conditions are continuity of those fields across the interface, not continuity of a scalar wave function and its derivative. But the resulting equations for the applications to be described below are almost identical in form.

The loss of intensity of light by reflection at a typical air-glass interface is about 4%. An 8% loss with a single lens (there is a loss at each side) might be acceptable; but losses at every interface of a complicated optical device might be prohibitive. A large industry is devoted to designing and fabricating "multilayer thin films" to improve the optical characteristics of lenses, screens, mirrors, filters, and other optical devices. The films are durable transparent materials evaporated on components under carefully controlled conditions. Sophisticated applications can involve dozens of layers—each layer being usually a quarter- or half-wavelength thin. Losses at interfaces can be greatly reduced, and filters can be made that select on wavelength in bands either narrow or wide for either transmission or reflection.

The speed of light in vacuum is $c = \lambda_0 \nu_0$, where λ_0 and ν_0 are the wavelength and frequency in vacuum. In a transparent medium such as glass, the speed is $v = \lambda \nu_0$ (the frequency does not change). The index of refraction is defined by $n = c/v = \lambda_0/\lambda$ ($n = 1$ in vacuum); or since $k = 2\pi/\lambda$, n is equal to k/k_0. Then $k_1/k_2 = n_1/n_2$ for two transparent media with indices n_1 and n_2. The great simplification here is this: In terms of ratios of wave numbers, the equations for T for particles and for light in the examples below are identical. Therefore, in going from particles to light, we can simply replace ratios like k_1/k_2 with n_1/n_2. Of course, the wave numbers are related to the relevant parameters differently: $k_1/k_2 = n_1/n_2$ for light, but $k_1/k_2 = \sqrt{(E - V_1)/(E - V_2)}$ for particles.

In the following, n_0 and k_0 are those of the first medium, which would usually but not always be vacuum or air. And the light is normally incident on the films. The results were known in optics long before the classical wall between waves and particles was breached by quantum mechanics.

(1) At a sharp interface, such as between air and glass, with indices n_0 and n_R, the transmission is

$$T = \frac{4 n_0 n_R}{(n_0 + n_R)^2} \qquad \left[\text{see in Sec. 2(a),} \quad T = \frac{4 k_0 k_1}{(k_0 + k_1)^2}\right. .$$

If, for example, $n_0 = 1$ and $n_R = 1.5$, then $T = 0.96$.

(2) To get $T = 1$ for an incident wavelength of λ_0 at that interface, coat the second medium with a thin film with index n_1 and width L_1, where

$$\begin{aligned} n_1 &= \sqrt{n_0 n_R} \\ L_1 &= \lambda_1/4 \end{aligned} \qquad \left[\text{see in Sec. 5(b),} \quad \begin{aligned} k_1 &= \sqrt{k_0 k_R} \\ L_1 &= \lambda_1/4 \end{aligned}\right. .$$

For the example in (1), make $n_1 = \sqrt{1.5} = 1.225$. (A practical matter in applying the equations here and in (3) is that durable transparent materials with the desired indices of refraction may not exist.)

(3) Or get $T = 1$ by using two thin films, where

$$\begin{aligned} n_2/n_1 &= \sqrt{n_R/n_0} \\ L_i &= \lambda_i/4, \ i = 1,2 \end{aligned} \qquad \left[\text{see in Prob. 7,} \quad \begin{aligned} k_2/k_1 &= \sqrt{k_R/k_0} \\ L_i &= \lambda_i/4, \ i = 1,2 \end{aligned}\right. .$$

(4) For a thin film, like a soap-bubble film, with index n_1 and width L in a medium with index n_0,

$$\frac{1}{T} = 1 + \frac{(n_0^2 - n_1^2)^2}{4 n_0^2 n_1^2} \sin^2 \phi_1 \qquad \left[\text{see in Sec. 4(a),} \quad \frac{1}{T} = 1 + \frac{(k_0^2 - k_1^2)^2}{4 k_0^2 k_1^2} \sin^2 \phi_1\right. .$$

where $\phi_1 = 2\pi L/\lambda_1$ is the phase change across the film.

(5) For a better mirror than one can make with metal, use N double layers with indices n_1 and n_2, where the widths are $L_1 = \frac{1}{4}\lambda_1$ and $L_2 = \frac{1}{4}\lambda_2$. Then

$$T = \frac{4}{\left[(-n_2/n_1)^N + (-n_1/n_2)^N\right]^2} \qquad \left[\text{see in Sec. 5(c),} \quad T = \frac{4}{\left[(-k_2/k_1)^N + (-k_1/k_2)^N\right]^2}\right. .$$

The chapter "Theory of Multilayer Films" in F.L. Pedrotti, L.S. Pedrotti, and L.M. Pedrotti, *Introduction to Optics* (3rd Ed.), has several nice figures of transmittances and reflectances as a function of wavelength.

4·7. WEAK TUNNELING

(a) An approximation—Section 4(b) was about a single rectangular barrier V high, L wide, and with $E < V$. There, with $\epsilon \equiv E/V$ and $q^2 \equiv 2m(V - E)/\hbar^2$, we had

$$T = \frac{1}{1 + Z'}, \qquad \text{where} \quad Z' = \frac{1}{4\epsilon(1 - \epsilon)} \sinh^2(qL) .$$

If qL is at all large, then

$$\sinh(qL) = \frac{1}{2}\left(e^{qL} - e^{-qL}\right) \approx \frac{1}{2}\,e^{qL} \gg 1 \;.$$

For example, even with $qL = 2$, e^{qL} is 55 times e^{-qL}. So when qL is large,

$$Z' = \frac{1}{4\epsilon(1-\epsilon)}\sinh^2(qL) \approx \frac{1}{16\epsilon(1-\epsilon)}\,e^{2qL} \gg 1 \;,$$

and

$$T = \frac{1}{1 + Z'} \approx \frac{1}{Z'} \approx 16\,\epsilon(1-\epsilon)\,e^{-2qL} \ll 1 \;.$$

The factor $16\epsilon(1-\epsilon)$ is bound between 0 to 4 (why?), and it is not a very sensitive function of ϵ. On the other hand, the factor e^{-2qL} *is* a very sensitive function of its exponent. For example, with ϵ fixed, doubling the width L reduces T by a factor e^{-2qL}. So if e^{-2qL} is 10^{-2}, a mere doubling of the width reduces T by a factor of 100. Since nearly all the sensitivity of T is in that exponential, the approximation $T \approx e^{-2qL}$ is sometimes used for rough calculations when qL is large.

If qL is really large, so that T is *really* small, why even bother? Because Nature has many *really* slow processes, such as nuclear decays with lifetimes of thousands or millions or billions of years, and these are of great interest—for example, in dating rocks. And some of the decays can be modeled as quantum tunneling through high and/or wide barriers.

Now suppose E is less than V across some region over which V is not constant but varies with x. Figure 10(a) shows an example, with $V(x) = V_0 - Ax$ for $x > 0$. If E is less than V_0, then $E < V(x)$ between $x = 0$ and L, where $L = L(E)$. We make an approximation analogous to the fitting-wavelengths-in-a-well method of Sec. 3·2. There, for bound states, we summed the differential bits of wavelength between the classical turning points (the region over which the wave function is oscillatory),

$$\int_{x_1}^{x_2} \frac{dx}{\lambda(x)} = \frac{1}{2\pi\hbar}\int_{x_1}^{x_2}\sqrt{2m(E - V(x))}\,dx \;.$$

Here, we sum the differential bits of $q(x)$ between the turning points 0 and L in the same way,

$$qL \longrightarrow \int_0^L q(x)\,dx = \frac{1}{\hbar}\int_0^L \sqrt{2m(V(x) - E)}\,dx \;.$$

For $V(x) = V_0 - Ax$ for $x > 0$, we get L from $E = V_0 - AL$, or $L = (V_0 - E)/A$. Then

$$\int_0^L q(x)\,dx = \frac{1}{\hbar}\int_0^L \sqrt{2m(V_0 - Ax - E)}\,dx = \frac{1}{\hbar}\int_0^L \sqrt{2mA(L - x)}\,dx \;.$$

With the substitution $y = L - x$, we get

$$\int_0^L q(x)\,dx = \frac{1}{\hbar}\sqrt{2mA}\int_0^L y^{1/2}\,dy = \frac{2}{3\hbar}\sqrt{2mAL^3} \;.$$

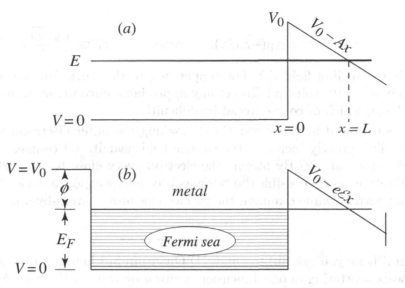

Figure 10. (a) A barrier of variable height. (b) The Fermi sea of conduction electrons in a block of metal. A strong electric field $\mathcal{E}$ is set up between the metal and an electrode.

(b) Field emission of electrons—Problem 17 of Chap. 2 was about the conduction electrons in a block of metal such as copper. To summarize: There is about one conduction electron per metal atom, and these electrons move about almost freely, like the molecules of a gas in a box, through the fixed neutralizing background of the positive ions. At normal temperatures, these free electrons stack up in the lowest energy levels, according to the Pauli exclusion principle, as shown in Fig. 10(b). The depth of this "Fermi sea" is the *Fermi energy* E_F, which for copper is about 7 eV. The difference $\phi = V_0 - E_F$ is the minimum energy an incident photon would have to have in order to eject one of the most loosely bound conduction electrons; ϕ is the work function in Einstein's photo-electric equation $K_{max} = h\nu - \phi$ (Sec. 1·1). For copper, $\phi = 4.7$ eV.

In Fig. 10(b), there is a strong electric field $\mathcal{E}$ between a flat side of the metal and a flat electrode. The potential energy $V(x)$ between the block and the electrode is

$$V(x) = V_0 - e\mathcal{E}x \ .$$

A conduction electron with energy E in the metal sees a triangular potential-energy barrier. A tunneling current to the outside will involve the electrons with energies near the top of the Fermi sea; for electrons lower in the sea, the barrier is both higher and wider and the transmission probability would be far smaller. We simply take over our earlier result, with $A = e\mathcal{E}$ and $V_0 - E = V_0 - E_F = \phi$, the work function. The barrier extends from $x = 0$ to $x = L$, where L is

$$L = (V_0 - E_F)/e\mathcal{E} = \phi/e\mathcal{E} \ .$$

Our earlier result becomes

$$\int_0^L q(x)\, dx = \frac{2}{3\hbar}\sqrt{2mAL^3} \ \longrightarrow \ \frac{2}{3\hbar}\frac{\sqrt{2m\phi^3}}{e\mathcal{E}}$$

89

Then

$$T \approx e^{-2qL} = \exp(-\mathcal{E}_0/\mathcal{E}), \qquad \text{where} \qquad \mathcal{E}_0 \equiv \frac{4}{3}\frac{\sqrt{2m}}{e\hbar}\phi^{3/2}.$$

How large is the scaling field $\mathcal{E}_0$? For copper, where the work function ϕ is 4.7 eV, $\mathcal{E}_0$ is approximately 6×10^{10} volts/m! To get any appreciable current, we would have to make $\mathcal{E}$ nearly this large, which of course would be difficult!

This is got around in two ways: (1) By making the outer electrode have an extremely sharp point. This greatly increases the electric field near it. Of course, then the geometry is not one-dimensional. (2) By placing the electrode very close to the surface of the metal. These modifications make possible the "atomic-force microscope," a device that by measuring the tunneling current while scanning the surface can probe, atom-by-atom, the structure of the surface.

———————

If on an iPhone you ask Siri to "find '1D Quantum Mechanics' in the Apps," you get an app about wave scattering in one dimension, written by Damon English. As you run a finger back and forth on the lower plot or up and down on the upper plot, the energy E follows, and much about the incident, reflected, and transmitted waves is shown. There are a number of pre-set barriers to choose from. Perhaps start with a double barrier, and notice how things change rapidly as you go through peaks in transmission; the lower plot shows where those peaks are at. Tap "?" for information. Tap "$|\Psi(x,t)|^2$" to see the scattering of a gaussian pulse. Explore!

PROBLEMS

1. *Probability current densities.*

(a) Show that the current density for $\psi(x) = Ae^{ikx} + Be^{-ikx}$ is $J = (\hbar k/m)(|A|^2 - |B|^2)$.

(b) Show that the current density for $\psi(x) = Ce^{-qx} + De^{+qx}$ is $J = (2\hbar q/m)\,\text{Im}(C^*D)$.

(c) Find J for $\psi(x) = Cx + D$.

2. *Cases.*

(a) What (nonzero) minimum width of a 3-eV-deep rectangular potential well gives 100% transmission of 5-eV electrons?

(b) What minimum width of a 3-eV-high rectangular potential barrier gives 100% transmission of 5-eV electrons?

(c) A potential barrier is 3 eV high and 0.3 nm wide. What fraction of 1-eV electrons will be transmitted? What fraction of 2-eV electrons will be transmitted?

3. *A rectangular barrier with $E = V$.*

A barrier is V high and L wide; elsewhere $V = 0$. Particles of energy exactly $E = V$ are incident.

(a) Show that the $\mathcal{M}$ matrix is

$$\mathcal{M} = \begin{pmatrix} 1 & -L \\ 0 & 1 \end{pmatrix}$$

by letting $E \to V$ in the $\mathcal{M}$ matrices for $E > V$ and for $E < V$.

(b) Show that the probability of transmission is $T = 1/(1 + Z)$, where $Z = mVL^2/2\hbar^2$.

(c) For what value of L/λ_0, where λ_0 is the de Broglie wavelength of the incident particles, is $T = 1/2$?

4. *A δ-function barrier.*

Particles of energy E are incident upon a δ-function barrier $V(x) = g\,\delta(x)$, where $g > 0$.

(a) By letting $V \to \infty$ and $VL \to g$ in the $\mathcal{M}$ matrix for $E < V$, show that the $\mathcal{M}$ matrix here is

$$\mathcal{M} = \begin{pmatrix} 1 & 0 \\ -2mg/\hbar^2 & 1 \end{pmatrix}.$$

(b) Show that the probability of transmission is $T = 1/(1 + Z)$, where $Z = mg^2/2E\hbar^2$.

5. *Doubling the width of a barrier.*

Particles of energy $E = V/2$ are incident upon a rectangular barrier of height V, and the transmission probability is T_1. The width of the barrier is then doubled and the transmission probability is T_2. Find T_2 in terms of T_1. Check that if $T_1 = 1/2$ then $T_2 = 1/9$. What in this case does T become if the width is again doubled?

6. *The anti-reflecting step.*

The conditions to get $T = 1$ with the arrangement in Fig. 7 are $L_1 = \lambda_1/4$ and $k_1 = \sqrt{k_0 k_R}$. Show from the latter that
$$V_1 = E\left[1 - \sqrt{1 - V_R/E}\right].$$

In Fig. 7, $V_R = 3E/4$. Is V_1 properly drawn?

7. *Another anti-reflector.*

Put *two* quarter-wavelength regions, V_1 and V_2 high, in front of a step V_R. Show that for a given incident energy E the condition for $T = 1$ is

$$\frac{k_2}{k_1} = \sqrt{\frac{k_R}{k_0}}.$$

8. *Mirrors.*

(a) In the numerical example at the end of Sec. 5(c), we took $V_1 = 3E/4$ and $V_2 = 0$ and got $R = 36\%$ for $N = 1$. But $N = 1$ and $V_2 = 0$ leaves just a single V_1 barrier. Check that you get $R = 36\%$ with the equation at the end of Sec. 4(a) with $\phi_1 = \frac{1}{2}\pi$ and $\epsilon = 4/3$.

(b) Make a table like the one at the end of Sec. 5(c) with V_1 still $3E/4$, but now with $V_2 = E/2$, as in Fig. 8.

(c) Would it matter if V_1 and V_2 were interchanged, with $E/2$ coming before $3E/4$?

9. *Algebra for the double-δ barrier.*

Work through the algebra for the double-δ barrier that was only sketched in Sec. 5(d). Get

$$T_2 = \frac{1}{1 + Z_2}, \qquad \text{where} \qquad Z_2 = \gamma^2\left(\cos\phi + \frac{\gamma}{2}\sin\phi\right)^2,$$

in which $\phi = k_0 L$ and $\gamma = 2mg/\hbar^2 k_0$.

91

10. *Loss of light by reflection at interfaces.*

The equation for R for light incident normally on an interface between transparent media, such as between air and water or glass, is

$$R = \left(\frac{n - n'}{n + n'} \right)^2 ,$$

where n and n' are the indices of refraction of the two media.

What is R at an air-water ($n = 1$, $n' = 1.33$), air-glass ($n' = 1.50$), and air-diamond ($n' = 2.42$) interface? What values of E/V for particles correspond to these cases?

11. *Escape of a trapped particle.*

A particle bounces back and forth between a rigid wall at $x = 0$ and a δ-function barrier $V(x) = g\delta(x - L)$. If g is large and the energy E is small, the probability of transmission T_1 in a single approach will be small, and then $T_1 \simeq 2E\hbar^2/mg^2$ (see Prob. 4). Treat the back-and-forth motion of the particle classically, and find the number of collisions N it will take to make $N \times T_1 = 1/2$. Thus find the "half-life" $\tau_{1/2}$ of this decay. How does it scale with E?

5. MATHEMATICAL FORMALISM

5.1. *Vector Spaces. Dirac Notation*
5.2. *States as Vectors*
5.3. *Operators*
5.4. *Successive Operations. Commutators*
5.5. *Operators as Matrices*
5.6. *Expectation Values*
5.7. *More Theorems*
5.8. *Revised Rules*
     Problems

We develop much of the mathematical formalism of quantum mechanics. As Eugene Wigner said, "The miracle of the appropriateness of the language of mathematics for the formulation of the laws of physics is a wonderful gift which we neither understand nor deserve." We begin by introducing in an informal way a unifying notational scheme invented by Paul Adrien Maurice Dirac. Our first use of Dirac notation is to extend the domain of vectors to include the states that describe a quantum system. This lets us carry over the visual language of vector geometry (length, direction, orthogonality, orthogonal unit axes, components, and so on) to those states. A vector to represent the general state of a harmonic oscillator will require an infinite number of axes and components (Chapter 6); a vector to represent the general spin state of a spin-1/2 particle will require just two (Chapter 10).

Whereas the states of quantum systems are represented by functions or vectors, the physical properties we can measure ("observables") are represented by operators. Thus far we have used the position and momentum operators $\hat{x}$ and $\hat{p}$, and the energy operator $\hat{H}$ constructed from them. The operators that represent observables are both linear and Hermitian. These properties (to be defined) lead to two easy-to-prove and extremely useful results: (1) The eigenvalues of linear, Hermitian operators are real; (2) the eigenstates that belong to eigenvalues that differ are orthogonal. A crucial property of the operators for observables is that the order of the operations can matter: The operators do not all *commute* with one another. Most fundamentally, the operators $\hat{x}$ and $\hat{p}$ do not commute.

Full understanding of the formalism will only come from using it. In Chapters 6 and 8, we use abstract algebras of operator commutation relations to solve the oscillator and angular-momentum problems. In Chapter 7, we find from the algebras that the exact values of two observables can only be known simultaneously if the corresponding operators commute, and that there are limits on what we can know if the operators do not commute.

5·1. VECTOR SPACES. DIRAC NOTATION

(a) Ordinary vectors—In ordinary three-dimensional (xyz) space, we can set up a rectangular coordinate system and write any vector $\mathbf{a}$ in the space in terms of three unit (i.e., normalized) vectors, $\hat{\mathbf{x}}$, $\hat{\mathbf{y}}$, and $\hat{\mathbf{z}}$, that lie along the orthogonal axes,

$$\mathbf{a} = a_x\hat{\mathbf{x}} + a_y\hat{\mathbf{y}} + a_z\hat{\mathbf{z}} \ .$$

The *orthonormal* vectors $\hat{\mathbf{x}}$, $\hat{\mathbf{y}}$, and $\hat{\mathbf{z}}$ form a *basis* for the space, and a_x, a_y, and a_z are the *components* of $\mathbf{a}$ in that basis. The length or magnitude a of $\mathbf{a}$ is the positive square root of the scalar product of $\mathbf{a}$ with itself. Because $\hat{\mathbf{x}} \cdot \hat{\mathbf{y}} = \hat{\mathbf{x}} \cdot \hat{\mathbf{z}} = \hat{\mathbf{y}} \cdot \hat{\mathbf{z}} = 0$ (the unit vectors are orthogonal), we get

$$\mathbf{a} \cdot \mathbf{a} = (a_x\hat{\mathbf{x}} + a_y\hat{\mathbf{y}} + a_z\hat{\mathbf{z}}) \cdot (a_x\hat{\mathbf{x}} + a_y\hat{\mathbf{y}} + a_z\hat{\mathbf{z}}) = a_x^2 + a_y^2 + a_z^2 = a^2 \ .$$

If $a = 1$, $\mathbf{a}$ is normalized. The scalar product of two vectors $\mathbf{a}$ and $\mathbf{b}$ is

$$\mathbf{a} \cdot \mathbf{b} = a_xb_x + a_yb_y + a_zb_z = ab\cos\theta \ ,$$

where θ is the angle between $\mathbf{a}$ and $\mathbf{b}$. If $\mathbf{a}$ and $\mathbf{b}$ are perpendicular (or if either or both of a and b are zero), then $\mathbf{a} \cdot \mathbf{b} = 0$.

It is common and useful to write vectors using their components as the elements of column vectors. In order to more easily generalize, we label the axes 1, 2, 3 instead of x, y, z. The orthonormal basis vectors are then

$$\hat{\mathbf{1}} = \begin{pmatrix} 1 \\ 0 \\ 0 \end{pmatrix}, \quad \hat{\mathbf{2}} = \begin{pmatrix} 0 \\ 1 \\ 0 \end{pmatrix}, \quad \hat{\mathbf{3}} = \begin{pmatrix} 0 \\ 0 \\ 1 \end{pmatrix}, \quad \text{and then} \quad \mathbf{a} = \begin{pmatrix} a_1 \\ a_2 \\ a_3 \end{pmatrix} .$$

Here we have used the two fundamental operations of vector algebra: (1) Multiplication of a vector $\mathbf{a}$ by a scalar c, and (2) addition of two vectors $\mathbf{a}$ and $\mathbf{b}$. The operations are defined by

$$c\mathbf{a} \equiv \begin{pmatrix} c\,a_1 \\ c\,a_2 \\ c\,a_3 \end{pmatrix} \quad \text{and} \quad \mathbf{a} + \mathbf{b} \equiv \begin{pmatrix} a_1 + b_1 \\ a_2 + b_2 \\ a_3 + b_3 \end{pmatrix} .$$

Any linear sum of vectors in the space, such as $c_1\mathbf{a} + c_2\mathbf{b} + \cdots$, is a vector in the same space, because its three components are defined by these two operations.

With vectors as skinny matrices, we can use ordinary matrix multiplication to calculate scalar products if we tilt the first vector on its side. Thus

$$\mathbf{a} \cdot \mathbf{b} = \begin{pmatrix} a_1 & a_2 & a_3 \end{pmatrix} \begin{pmatrix} b_1 \\ b_2 \\ b_3 \end{pmatrix} = a_1b_1 + a_2b_2 + a_3b_3 \ .$$

The scalar product of, say, $\hat{\mathbf{3}}$ with an arbitrary vector $\mathbf{a}$ picks out its third component:

$$\hat{\mathbf{3}} \cdot \mathbf{a} = \begin{pmatrix} 0 & 0 & 1 \end{pmatrix} \begin{pmatrix} a_1 \\ a_2 \\ a_3 \end{pmatrix} = a_3 \ .$$

94

(b) Dirac notation—Dirac invented a notation that, at the simplest level, can be used to distinguish row from column vectors. In this notation, $\mathbf{a} \cdot \mathbf{b}$ is written $\langle a|b \rangle$. This is the scalar product of the *bra* (row) vector $\langle a|$ and the *ket* (column) vector $|b\rangle$; the product is a *bra(c)ket*. A scalar product—a bracket—is just a number (perhaps with dimensions), a simpler object than the vectors that make it up. Dirac notation makes marking vectors with bold print or an arrow overhead superfluous. The unit column vectors are $|1\rangle$, $|2\rangle$, and $|3\rangle$, the unit row vectors are $\langle 1|$, $\langle 2|$, and $\langle 3|$, and they are orthonormal: $\langle i|j \rangle = \delta_{ij}$, where $i, j = 1, 2, 3$ and $\delta_{ij} = 1$ if $i = j$ and 0 if $i \neq j$. The third component of $|a\rangle$ is $a_3 = \langle 3|a \rangle$.

If we multiply the vectors that represent bras and kets in the opposite order, we get a *square matrix* instead of a scalar. For example,

$$|1\rangle\langle 3| = \begin{pmatrix} 1 \\ 0 \\ 0 \end{pmatrix} \begin{pmatrix} 0 & 0 & 1 \end{pmatrix} = \begin{pmatrix} 0 & 0 & 1 \\ 0 & 0 & 0 \\ 0 & 0 & 0 \end{pmatrix}.$$

A square matrix multiplying a vector—*operating* on the vector—makes another vector. For example,

$$(|1\rangle\langle 3|)\,|a\rangle = \begin{pmatrix} 0 & 0 & 1 \\ 0 & 0 & 0 \\ 0 & 0 & 0 \end{pmatrix} \begin{pmatrix} a_1 \\ a_2 \\ a_3 \end{pmatrix} = \begin{pmatrix} a_3 \\ 0 \\ 0 \end{pmatrix} = a_3|1\rangle = |1\rangle(\langle 3|a\rangle) \ .$$

Of more common use are the operators $|i\rangle\langle i|$ made of unit vectors where $i = j$. For example,

$$(|1\rangle\langle 1|)\,|a\rangle = \begin{pmatrix} 1 \\ 0 \\ 0 \end{pmatrix} \begin{pmatrix} 1 & 0 & 0 \end{pmatrix} \begin{pmatrix} a_1 \\ a_2 \\ a_3 \end{pmatrix} = \begin{pmatrix} 1 & 0 & 0 \\ 0 & 0 & 0 \\ 0 & 0 & 0 \end{pmatrix} \begin{pmatrix} a_1 \\ a_2 \\ a_3 \end{pmatrix} = \begin{pmatrix} a_1 \\ 0 \\ 0 \end{pmatrix} = a_1|1\rangle \ .$$

Operating on any vector, $|1\rangle\langle 1|$ picks out—*projects* out—the part of a vector (not just the component a_1, but $a_1|1\rangle$) along the 1 axis: $|1\rangle\langle 1|$ is a *projection operator*. Similarly, $|2\rangle\langle 2|$ and $|3\rangle\langle 3|$ would project out the parts of a vector along the 2 and 3 axes; and adding the three projections together would give the whole vector back again. That is,

$$|1\rangle\langle 1| + |2\rangle\langle 2| + |3\rangle\langle 3| = \begin{pmatrix} 1 & 0 & 0 \\ 0 & 1 & 0 \\ 0 & 0 & 1 \end{pmatrix} = \hat{I} \ ,$$

where $\hat{I}$ is the 3-by-3 identity matrix.

(c) Three generalizations—(1) The first generalization is to allow any number of dimensions, not just three. This just changes the number of elements in the vectors from three to N, the number of terms in summations for scalar products from three to N, the number of elements in square matrices from 3^2 to N^2, etc. In the N-dimensional space, we can set up orthonormal basis ket vectors $|i\rangle$, where $i = 1$ through N. The first such ket, $|1\rangle$, will have a 1 as the top element of its column and 0's elsewhere. Basis ket $|2\rangle$ will have a 1 as its second element and 0's elsewhere. And so on. The ket $|a\rangle$ is the column of components $a_1, a_2, \ldots a_N$. The fifth component of $|a\rangle$ is $a_5 = \langle 5|a\rangle$. And as with $N = 3$,

$$|1\rangle\langle 1| + |2\rangle\langle 2| + \cdots + |N\rangle\langle N| = \sum_{i=1}^{N} |i\rangle\langle i| = \hat{I} \ ,$$

where $\hat{I}$ is the N-by-N identity matrix.

(2) The second generalization is to vectors with complex components, and to be useful this requires a change in the relation between a ket $|a\rangle$ and the corresponding bra $\langle a|$. Consider the (provisional) scalar product $\langle a|a\rangle$ of a ket with components 1 and i,

$$\langle a|a\rangle = \begin{pmatrix} 1 & i \end{pmatrix} \begin{pmatrix} 1 \\ i \end{pmatrix} = 1 - 1 = 0 \ .$$

This tells us nothing about the magnitude of $|a\rangle$. To get the magnitude $|c|$ of a complex number $c = c_r + ic_i$, where the real numbers c_r and c_i are its real and imaginary parts, we know to multiply c by its complex conjugate,

$$|c|^2 = c^*c = (c_r - ic_i)(c_r + ic_i) = c_r^2 + c_i^2 \geq 0 \ .$$

This suggests that in going from the ket $|a\rangle$, with complex components a_1, a_2, ..., to the corresponding bra $\langle a|$ (or vice versa), we complex conjugate all the components. That is,

$$\boxed{\ |a\rangle = \begin{pmatrix} a_1 \\ a_2 \\ \vdots \end{pmatrix} \qquad \longleftrightarrow \qquad \langle a| = \begin{pmatrix} a_1^* & a_2^* & \cdots \end{pmatrix} \ .\ }$$

(Note that we do not put an asterisk on $\langle a|$.) The scalar product $\langle a|a\rangle$ is now

$$\langle a|a\rangle = \begin{pmatrix} a_1^* & a_2^* & \dots \end{pmatrix} \begin{pmatrix} a_1 \\ a_2 \\ \vdots \end{pmatrix} = \sum_{i=1}^{N} |a_i|^2 \ ,$$

which is real, and is positive for all nonzero vectors. Then the length or magnitude a of $|a\rangle$ is $\sqrt{\langle a|a\rangle}$. In the above example where the components were 1 and i, $a = \sqrt{\langle a|a\rangle} = \sqrt{2}$. This change in the definition of the scalar product makes it *useful*.

The scalar product of two different vectors will now, in general, be a complex number, and $\langle a|b\rangle = \langle b|a\rangle^*$:

$$\boxed{\ \langle a|b\rangle = \begin{pmatrix} a_1^* & a_2^* & \dots \end{pmatrix} \begin{pmatrix} b_1 \\ b_2 \\ \vdots \end{pmatrix} = a_1^*b_1 + a_2^*b_2 + \cdots = (b_1^*a_1 + b_2^*a_2 + \cdots)^* = \langle b|a\rangle^* \ .\ }$$

Here we have used basic properties of complex numbers: $(c_1 + c_2)^* = c_1^* + c_2^*$, $(c^*)^* = c$, and $(c_1^*c_2)^* = c_1c_2^* = c_2^*c_1$. We can write $\langle a|b\rangle$ in a slightly different way. The ith component of $|a\rangle$ is $a_i = \langle i|a\rangle$, and $a_i^* = \langle i|a\rangle^* = \langle a|i\rangle$. Then

$$\langle a|b\rangle = \sum_{i=1}^{N} a_i^* b_i = \sum_{i=1}^{N} \langle a|i\rangle\langle i|b\rangle = \langle a| \left(\sum_{i=1}^{N} |i\rangle\langle i| \right) |b\rangle = \langle a|\hat{I}|b\rangle \ .$$

We can insert the identity $\hat{I} = \sum |i\rangle\langle i|$ into the middle of a scalar product.

(3) The third generalization is to function spaces. The functions of most interest will of course be our complex wave functions. In Dirac notation, ψ becomes $|\psi\rangle$ and ψ^* becomes $\langle\psi|$ (but see below). The scalar product of two functions ϕ and ψ is

$$\langle\phi|\psi\rangle = \int_{-\infty}^{+\infty} \phi^*(x)\,\psi(x)\,dx\ .$$

Here a sum of products of two functions over a *continuous variable* x replaces a sum of products of components of two vectors over a *discrete index* i: $\int$ replaces $\sum$. In either case, the result is a scalar.

Writing $\psi \to |\psi\rangle$ for a function is like writing $\mathbf{a} \to |a\rangle$ for a vector. But what then corresponds to the index i in $a_i = \langle i|a\rangle$ for a vector? The answer, implicit in the integral above, is x: $\psi(x)$ becomes $\langle x|\psi\rangle$ and $\phi^*(x)$ becomes $\langle\phi|x\rangle$. The discrete index i goes over to the continuous index x, and

$$\langle a|b\rangle = \sum_{i=1}^{N}\langle a|i\rangle\langle i|b\rangle \qquad \text{goes over to} \qquad \langle\phi|\psi\rangle = \int_{-\infty}^{+\infty}\langle\phi|x\rangle\langle x|\psi\rangle dx\ .$$

The probability density $|\psi(x)|^2 = \psi^*(x)\psi(x)$ in Dirac notation is $\langle\psi|x\rangle\langle x|\psi\rangle$. Certainly, $\langle\psi|\psi\rangle$ can't be the density, because it is the *integral* of $|\psi(x)|^2$ over the range of x, and is just a scalar. (As a matter of notation, we shall almost always favor $\psi(x)$ over $\langle x|\psi\rangle$.)

If $\langle\phi|\psi\rangle = 0$, the functions are orthogonal. If $\langle\psi|\psi\rangle = 1$, ψ is normalized. If $\langle\psi|\psi\rangle \neq 1$ but is finite, we can normalize $|\psi\rangle$ by dividing it by $\sqrt{\langle\psi|\psi\rangle}$. And just as for vectors, $\langle\phi|\psi\rangle^* = \langle\psi|\phi\rangle$. This follows from properties of complex functions, such as $(\psi^*)^* = \psi$, $(\phi^*\psi)^* = \phi\psi^* = \psi^*\phi$, and $\int f^*(x)\,dx = \left(\int f(x)\,dx\right)^*$.

(d) Linear vector spaces—The vectors and functions we have been discussing belong to linear vector spaces. The fundamental operations in a linear vector space are those noted at the start of the chapter: (1) multiplication of a vector or function by a scalar c, and (2) addition of two vectors or functions. Let $|f\rangle$, $|g\rangle$, and $|h\rangle$ be either vectors or functions. The two operations satisfy the following rules:

(1) $|f\rangle + |g\rangle = |g\rangle + |f\rangle$.

(2) $|f\rangle + (|g\rangle + |h\rangle) = (|f\rangle + |g\rangle) + |h\rangle = |f\rangle + |g\rangle + |h\rangle$.

(3) There is a unique zero vector or function, $|0\rangle$, such that $|f\rangle + |0\rangle = |f\rangle$ for all $|f\rangle$.

(4) For each $|f\rangle$, there is a unique $|-f\rangle$, such that $|f\rangle + |-f\rangle = |0\rangle$.

(5) Multiplication by 1 leaves $|f\rangle$ unchanged: $1|f\rangle = |f\rangle$.

(6) $c_1 c_2|f\rangle = c_1(c_2|f\rangle)$.

(7) $c(|f\rangle + |g\rangle) = c|f\rangle + c|g\rangle$.

(8) $(c_1 + c_2)|f\rangle = c_1|f\rangle + c_2|f\rangle$.

The rules license us to do what comes naturally. If any rule does not seem quite natural, write it out with ordinary three-component column vectors and with functions.

Here are four simple relations for calculating with our thrice-generalized scalar products that are used over and over, whether they be of vectors or of functions:

$$\langle f|cg\rangle = c\langle f|g\rangle \qquad\qquad \langle f|g+h\rangle = \langle f|g\rangle + \langle f|h\rangle$$

$$\langle cf|g\rangle = c^*\langle f|g\rangle \qquad\qquad \langle f+g|h\rangle = \langle f|h\rangle + \langle g|h\rangle \ .$$

These relations follow at once from the properties of the summations and integrations that define the scalar products. You might write them out in both vector and function forms.

Dirac notation is an abstract notation that encompasses both vectors and functions and brings some simplicity and unity to thinking about them. When the vectors or functions represent states of a quantum system, we shall often call them *states*. In the next sections, we go with little formality back and forth between Dirac notation and the ordinary notations of vectors and functions. This quickly becomes second nature.

5·2. STATES AS VECTORS

(a) **A Fourier sine series**—In Sec. 2·4, we found the energy eigenvalues and normalized eigenfunctions of a particle in an infinite square well. They are

$$E_n = n^2 \frac{\pi^2\hbar^2}{2mL^2} \qquad \text{and} \qquad \psi_n(x) = \sqrt{\frac{2}{L}}\sin\left(\frac{n\pi x}{L}\right),$$

where $0 \le x \le L$ and $n = 1, 2, 3, \ldots$ These eigenfunctions are not only normalized, they are also orthogonal. That is,

$$\int_0^L \psi_m^*(x)\,\psi_n(x)\,dx = \frac{2}{L}\int_0^L \sin\left(\frac{m\pi x}{L}\right)\sin\left(\frac{n\pi x}{L}\right)\,dx = \delta_{mn}\ ,$$

where $\delta_{mn} = 1$ if $m = n$ and 0 if $m \neq n$. Look up the integral. In the next section, we prove the orthogonality of energy eigenstates with $E_m \neq E_n$ *without* doing the integral.

Let $\psi(x) = \Psi(x, t=0)$ be the state of a particle in the well at $t = 0$. In general, $\psi(x)$ is a *linear superposition* of any or all of the energy eigenfunctions,

$$\psi(x) = \sum_{n=1}^{\infty} c_n\psi_n(x)\ ,$$

where $\sum_n |c_n|^2 = 1$ (see below). (Problem 2·2 was about a superposition of ψ_1 and ψ_2.) Given a $\psi(x)$, we can get the values of the expansion coefficients c_n by doing integrals. For example,

$$\int_0^L \sqrt{\frac{2}{L}}\sin\left(\frac{5\pi x}{L}\right)\psi(x)\,dx = \int_0^L \psi_5^*(x)\left(\sum_{n=1}^{\infty} c_n\psi_n(x)\right)dx = c_5\ .$$

The last step follows because $\psi_5(x)$ is orthogonal to all the other $\psi_n(x)$. The c_n are just the coefficients of a sine series—the Fourier sine series—of $\psi(x)$ over $0 \le x \le L$.

The physical interpretation of, say, c_5 is this: $|c_5|^2$ is the probability that a measurement of the energy will yield E_5. The sum of the probabilities for all the mutually exclusive possible outcomes of a measurement of the energy has to be one. This interpretive constraint *normalizes* the expansion coefficients: $\sum_n |c_n|^2 = 1$. And *after* the measurement, if we get E_5, then $|c_5| = 1$ and the state is ψ_5, not ψ (unless $|c_5|$ was 1 to begin with). Measurement "collapses the wave function."

(b) States as vectors—The ket notation for a normalized energy eigenstate ψ_n of the infinite square well, the harmonic oscillator, or some other physical system, is $|n\rangle$. In more than one dimension, we will need more than one quantum number, but that can wait. The notation $|n\rangle$ promotes the important information—the quantum number of the eigenstate—from subscript to full-size character. Thus $|5\rangle$ represents the eigenstate with energy E_5, $\psi_5(x)$ is $\langle x|5\rangle$, and c_5 in a superposition state $\psi(x)$ is $\langle 5|\psi\rangle$. The superposition state $\Psi(t{=}0) = \psi$ at $t = 0$ is $|\psi\rangle = \sum_n c_n(0)|n\rangle$. The orthogonality (not yet proved) of states with energies $E_m \neq E_n$ is $\langle m|n\rangle = \delta_{mn}$.

Since the energy eigenkets have all the characteristics of orthonormal unit vectors, we sometimes represent them as orthonormal unit column vectors,

$$|1\rangle = \begin{pmatrix} 1 \\ 0 \\ 0 \\ \vdots \end{pmatrix}, \quad |2\rangle = \begin{pmatrix} 0 \\ 1 \\ 0 \\ \vdots \end{pmatrix}, \quad \text{etc.,} \quad \text{and then} \quad |\psi\rangle = \begin{pmatrix} c_1(0) \\ c_2(0) \\ c_3(0) \\ \vdots \end{pmatrix}.$$

If as here the system being considered is the infinite square well, where did the sine functions go? They are now implicit, just as in writing an ordinary vector $\mathbf{a} = a_x\hat{\mathbf{x}} + a_y\hat{\mathbf{y}} + a_z\hat{\mathbf{z}}$ as a column vector with components a_x, a_y, and a_z, we put $\hat{\mathbf{x}}$, $\hat{\mathbf{y}}$, and $\hat{\mathbf{z}}$ out of sight. *Place in the column vector would tell which sine function is indicated.* The notation is abstract but unifying, the same for the infinite square well, the harmonic oscillator, and other systems.

Each component of a full state vector evolves with its own time dependence, $e^{-iE_n t/\hbar}$. Thus starting from $|\psi\rangle = |\Psi(t{=}0)\rangle$, we get

$$|\Psi(t)\rangle = \sum_{n=1}^{\infty} c_n(0)|n\rangle \, e^{-iE_n t/\hbar} = \begin{pmatrix} c_1(0)\, e^{-iE_1 t/\hbar} \\ c_2(0)\, e^{-iE_2 t/\hbar} \\ c_3(0)\, e^{-iE_3 t/\hbar} \\ \vdots \end{pmatrix} = \begin{pmatrix} c_1(t) \\ c_2(t) \\ c_3(t) \\ \vdots \end{pmatrix},$$

where $c_n(t) = c_n(0)e^{-iE_n t/\hbar}$. Going from ket to bra, the column of $c_n(t)$'s becomes a row of $c_n^*(t)$'s. Then

$$\langle\Psi(t)|\Psi(t)\rangle = \begin{pmatrix} c_1^*(0)\, e^{+iE_1 t/\hbar} & c_2^*(0)\, e^{+iE_2 t/\hbar} & \cdots \end{pmatrix} \begin{pmatrix} c_1(0)\, e^{-iE_1 t/\hbar} \\ c_2(0)\, e^{-iE_2 t/\hbar} \\ \vdots \end{pmatrix} = \sum_{n=1}^{\infty} |c_n(0)|^2 \, .$$

The wave function once normalized stays normalized.

In summary, vectors (columns of coefficients c_n) and functions (wave functions Ψ) are two ways of representing quantum states, and we go freely back and forth between them. And again: The vector language lets us use the familiar geometrical language of length, direction, orthogonality, orthonormal axis vectors, components, and so on. Dirac notation is especially simplifying for scalar products: $\langle 5|\psi\rangle$, say, or $\langle\phi|\psi\rangle$ are more concise than the sum or integral for which each stands.

The quantum states exist in a *Hilbert space*. A Hilbert space is a linear vector space of normalizable functions of one or more independent variables such as x or (r, θ, ϕ). (Vectors of infinite length do not belong to a Hilbert space.) Since the functions will in general be complex, the complex form of the scalar product is needed. A vector that represents a particular $\Psi(t)$ will sweep out a path in Hilbert space as time advances.

5·3. OPERATORS

An *observable* is a physical property that one can measure, such as the energy of a particle or a component of its position, momentum, or angular momentum. In quantum mechanics, physical observables are represented by operators. We have already seen the operators

$$\hat{x} = x \ , \quad \hat{p}_x = -i\hbar \frac{\partial}{\partial x} \ , \quad \hat{V}(x) = V(x) \ , \quad \hat{H} = \frac{\hat{p}_x^2}{2m} + \hat{V} = -\frac{\hbar^2}{2m} \frac{\partial^2}{\partial x^2} + V(x) \ .$$

The basic operators $\hat{x}$ and $\hat{p}_x$ are used to construct the more complicated operators such as $\hat{H}$ and (in Chap. 8) angular momentum. The operators are then used to set up equations like $\hat{H}\psi_n = E_n\psi_n$. The operators that represent physical observables in quantum mechanics are *linear* and *Hermitian*.

(a) Linear operators—An operator $\hat{A}$ operating on a state $|\psi\rangle$ produces another state, $|\hat{A}\psi\rangle$; that is, $\hat{A}|\psi\rangle = |\hat{A}\psi\rangle$. A *linear* operator $\hat{A}$ satisfies two conditions,

$$\hat{A}|c\psi\rangle = c\hat{A}|\psi\rangle \qquad \text{and} \qquad \hat{A}(|\psi\rangle + |\phi\rangle) = \hat{A}|\psi\rangle + \hat{A}|\phi\rangle \ ,$$

where c is an arbitrary constant. The two conditions can be lumped together,

$$\boxed{\hat{A}(c_1|\psi\rangle + c_2|\phi\rangle) = c_1\hat{A}|\psi\rangle + c_2\hat{A}|\phi\rangle \ .}$$

Multiplying by a constant, multiplying by a function, taking a derivative, are linear operations that change one function into another. For example,

$$\frac{d}{dx}\big[c_1\psi(x) + c_2\phi(x)\big]dx = c_1\frac{d\psi(x)}{dx} + c_2\frac{d\phi(x)}{dx} \ .$$

Multiplying a column vector by a square matrix is a linear operation that produces another column vector. Linear algebra is all about such operations.

Squaring is an example of an operation that is not linear; in general,

$$(c\psi)^2 \neq c(\psi)^2 \quad \text{and} \quad (\psi_1 + \psi_2)^2 \neq \psi_1^2 + \psi_2^2 \ .$$

Taking the sine or the logarithm of a function are not linear operations: $\sin(c\psi) \neq c\sin\psi$. Of course we use squared functions, sines, cosines, exponentials, and other functions in quantum mechanics all the time. Here is the difference: Multiplying a wave function ψ by x^2 or $\sin\theta$ is a *linear* operation on ψ. Taking ψ^2 or $\sin\psi$ is *not* a linear operation on ψ. The operators that represent *physical observables* are linear operations on *quantum states*.

(b) Hermitian operators—Calculations will often involve the scalar product $\langle\phi|\hat{A}\psi\rangle$ of states $|\phi\rangle$ and $|\hat{A}\psi\rangle$. Since $|\hat{A}\psi\rangle = \hat{A}|\psi\rangle$, then also $\langle\phi|\hat{A}\psi\rangle = \langle\phi|\hat{A}|\psi\rangle$ which is more symmetrical. In this sandwich of $\hat{A}$ between $\langle\phi|$ and $|\psi\rangle$, $\hat{A}$ operates on the state that follows it, here $|\psi\rangle$.

There is an "adjoint" operator $\hat{A}^\dagger$ (read "A-dagger") that makes the scalar product of $|\hat{A}^\dagger\phi\rangle$ with $|\psi\rangle$ equal to the scalar product of $|\phi\rangle$ with $|\hat{A}\psi\rangle$. That is, $\hat{A}^\dagger$ is defined by

$$\boxed{\langle\hat{A}^\dagger\phi|\psi\rangle \equiv \langle\phi|\hat{A}\psi\rangle. \qquad \text{And if } \hat{A}^\dagger = \hat{A}, \text{ then } \hat{A} \text{ is Hermitian.}}$$

If $\hat{A}$ is Hermitian, then $\langle\hat{A}^\dagger\phi|\psi\rangle = \langle\hat{A}\phi|\psi\rangle = \langle\phi|\hat{A}\psi\rangle$, and we can freely move it back and forth in the bra(c)ket.

To find the adjoint $\hat{A}^{\dagger}$ of an operator $\hat{A}$, we manipulate scalar products. For example, let $f(x)$ be any *real* function of the position x, such as x itself or a potential energy $V(x)$. Then

$$\langle\phi|f\psi\rangle = \int_{-\infty}^{+\infty} \phi^*(x)\, f(x)\psi(x)\, dx = \int_{-\infty}^{+\infty} [f(x)\phi(x)]^*\,\psi(x)\, dx = \langle f\phi|\psi\rangle \ .$$

We could move $f(x)$ over to the front part of the integrand without change because $f^* = f$. Thus for any real function $f(x)$, $f^{\dagger} = f$ and f is Hermitian.

Any real constant is Hermitian, but a complex constant c is not. The rules at the end of Sec. 1 tell us that $\langle\phi|c\psi\rangle = c\langle\phi|\psi\rangle$, whereas $\langle c\phi|\psi\rangle = c^*\langle\phi|\psi\rangle \neq \langle\phi|c\psi\rangle$ if $c^* \neq c$. However, $\langle c^*\phi|\psi\rangle = c\langle\phi|\psi\rangle = \langle\phi|c\psi\rangle$, because $(c^*)^* = c$. Therefore $c^{\dagger} = c^*$, and in particular $i^{\dagger} = -i$.

Another operator that is not Hermitian is $\partial/\partial x$. We shall assume that the functions ϕ and ψ vanish as $x \to \pm\infty$, as they must if they are normalizable. Then integration by parts gives

$$\int_{-\infty}^{+\infty} \phi^* \frac{\partial\psi}{\partial x}\, dx = \phi^*\psi\Big|_{-\infty}^{+\infty} - \int_{-\infty}^{+\infty} \left(\frac{\partial\phi^*}{\partial x}\right)\psi\, dx = \int_{-\infty}^{+\infty} \left(-\frac{\partial\phi}{\partial x}\right)^*\psi\, dx \ .$$

Thus $(\partial/\partial x)^{\dagger} = -\partial/\partial x$. However, the operator $\hat{p}_x = -i\hbar\,\partial/\partial x$ is Hermitian even though it is the product of operators i and $\partial/\partial x$ that are not: minus $\times$ minus $=$ plus.

(c) Real eigenvalues, orthogonal eigenstates—The proofs are simple, the theorems are among the most important in mathematical physics. Let a_n and $|\psi_n\rangle$ be the eigenvalues and normalized eigenstates of the Hermitian operator $\hat{A}$: $\hat{A}|\psi_n\rangle = a_n|\psi_n\rangle$. (We use $|\psi_n\rangle$ instead of $|n\rangle$ here because $|\hat{A}n\rangle$ is poor notation.)

(1) *The eigenvalues of $\hat{A}$ are real.* We have

$$\langle\psi_n|\hat{A}\psi_n\rangle = \langle\psi_n|a_n\psi_n\rangle = a_n\langle\psi_n|\psi_n\rangle = a_n \ ,$$

and

$$\langle\hat{A}\psi_n|\psi_n\rangle = \langle a_n\psi_n|\psi_n\rangle = a_n^*\langle\psi_n|\psi_n\rangle = a_n^* \ .$$

Since by hypothesis $\hat{A}$ is Hermitian, the left ends of these lines are equal, and therefore so are the right ends: $a_n^* = a_n$, and a_n is real.

The eigenvalues of Hermitian operators are real.
---

(2) *Eigenstates of $\hat{A}$ with eigenvalues a_m and a_n are orthogonal if $a_m \neq a_n$.* We have

$$\langle\psi_m|\hat{A}\psi_n\rangle = \langle\psi_m|a_n\psi_n\rangle = a_n\langle\psi_m|\psi_n\rangle \ ,$$

and

$$\langle\hat{A}\psi_m|\psi_n\rangle = \langle a_m\psi_m|\psi_n\rangle = a_m^*\langle\psi_m|\psi_n\rangle = a_m\langle\psi_m|\psi_n\rangle \ .$$

Since $\hat{A}$ is Hermitian, the left ends of these lines are equal, and therefore so are the right ends. Subtracting the right ends from one another, we get

$$(a_m - a_n)\langle\psi_m|\psi_n\rangle = 0 \ .$$

If $a_m \neq a_n$, then $\langle\psi_m|\psi_n\rangle = 0$.

Eigenstates of Hermitian operators with different eigenvalues are orthogonal.

It is hard to imagine a more useful theorem that the second one. If the eigenvalues E_m and E_n of a Hamiltonian $\hat{H}$ are not equal, the corresponding eigenfunctions $|m\rangle$ and $|n\rangle$ are orthogonal: $\langle m|n\rangle = 0$. For example, the infinite set of eigenfunctions of the infinite square well are orthogonal,

$$\langle m|n\rangle = \frac{2}{L} \int_0^L \sin\left(\frac{m\pi x}{L}\right) \sin\left(\frac{n\pi x}{L}\right) dx = 0 \qquad \text{if } m \neq n \ .$$

An infinite set of integrals are equal to zero—for *free*! We do not even need to know what the eigenfunctions *are*! And there are like orthogonalities of the eigenfunctions of *any* Hermitian operator $\hat{A}$.

In a course on linear algebra, one learns that the eigenvalues of symmetric matrices are real, and that the eigenvectors of such matrices that belong to distinct eigenvalues are orthogonal. Symmetric matrices are a subclass of Hermitian operators, as we shall see below.

5·4. SUCCESSIVE OPERATIONS. COMMUTATORS

Sums and products of linear operators are themselves linear operators, and are used to construct Hamiltonians and angular-momentum operators from simpler operators. Sums of operators are easy, products require some care. The operators here will all be linear. Not all are Hermitian.

(a) **Successive operations**—The operator $\hat{A} + \hat{B}$ operating on $|\psi\rangle$ means $\hat{A}|\psi\rangle + \hat{B}|\psi\rangle$. The first rule of linear vector spaces, given in Sec. 1(d), was that $|f\rangle + |g\rangle = |g\rangle + |f\rangle$: Which order the states are summed does not matter,

$$(\hat{A} + \hat{B})|\psi\rangle = \hat{A}|\psi\rangle + \hat{B}|\psi\rangle = \hat{B}|\psi\rangle + \hat{A}|\psi\rangle = (\hat{B} + \hat{A})|\psi\rangle \ .$$

The adjoint of a sum is the sum of the adjoints,

$$\langle (\hat{A} + \hat{B})^\dagger \phi|\psi\rangle = \langle \phi|(\hat{A} + \hat{B})\psi\rangle = \langle \phi|\hat{A}\psi\rangle + \langle \phi|\hat{B}\psi\rangle = \langle \hat{A}^\dagger \phi|\psi\rangle + \langle \hat{B}^\dagger \phi|\psi\rangle \ ,$$

so $(\hat{A} + \hat{B})^\dagger = \hat{A}^\dagger + \hat{B}^\dagger$. If $\hat{A}$ and $\hat{B}$ are Hermitian, then so is $\hat{A} + \hat{B}$.

The operator $\hat{A}\hat{B}$ operating on $|\psi\rangle$ means operate on $|\psi\rangle$ with $\hat{B}$, *then* operate on $\hat{B}|\psi\rangle$ with $\hat{A}$:

$$\hat{A}\hat{B}|\psi\rangle = \hat{A}(\hat{B}|\psi\rangle) = \hat{A}|\hat{B}\psi\rangle = |\hat{A}\hat{B}\psi\rangle \ .$$

Considered as a single operator, the adjoint $(\hat{A}\hat{B})^\dagger$ of $\hat{A}\hat{B}$ is defined by

$$\langle (\hat{A}\hat{B})^\dagger \phi|\psi\rangle = \langle \phi|\hat{A}\hat{B}\psi\rangle \ .$$

However, if we peel $\hat{A}$ and $\hat{B}$ over one by one, we get

$$\langle \phi|\hat{A}\hat{B}\psi\rangle = \langle \hat{A}^\dagger \phi|\hat{B}\psi\rangle = \langle \hat{B}^\dagger \hat{A}^\dagger \phi|\psi\rangle \ ,$$

so $(\hat{A}\hat{B})^\dagger = \hat{B}^\dagger \hat{A}^\dagger$. Then of course $(\hat{A}\hat{B}\hat{C})^\dagger = \hat{C}^\dagger \hat{B}^\dagger \hat{A}^\dagger$, and so on. The "rules" here for manipulating sums and products of operators are used repeatedly.

Now suppose that $\hat{A}$ and $\hat{B}$ are Hermitian operators: $\hat{A}^\dagger = \hat{A}$ and $\hat{B}^\dagger = \hat{B}$. Then $(\hat{A}\hat{B})^\dagger = \hat{B}^\dagger \hat{A}^\dagger = \hat{B}\hat{A}$. Is the operator $\hat{A}\hat{B}$ Hermitian? Not unless $\hat{B}\hat{A} = \hat{A}\hat{B}$; not unless $\hat{A}$ and $\hat{B}$ *commute*.

(b) Commutators—The order in which product operators operate on a function often matters. Let $\psi(x)$ be an arbitrary wave function. Then

$$\hat{x}\hat{p}_x\psi = -i\hbar x\frac{\partial\psi}{\partial x}, \quad \text{but} \quad \hat{p}_x\hat{x}\,\psi = -i\hbar\frac{\partial(x\psi)}{\partial x} = -i\hbar x\frac{\partial\psi}{\partial x} - i\hbar\psi .$$

The most basic of operators, $\hat{x}$ and $\hat{p}_x$, *do not commute*. Since $(\hat{x}\hat{p}_x)^\dagger = \hat{p}_x^\dagger\hat{x}^\dagger = \hat{p}_x\hat{x} \neq \hat{x}\hat{p}_x$, neither of $\hat{x}\hat{p}_x$ or $\hat{p}_x\hat{x}$ is Hermitian. The operator that is the difference of operators $\hat{A}\hat{B}$ and $\hat{B}\hat{A}$ is the *commutator* of $[\hat{A}, \hat{B}]$,

$$\boxed{[\hat{A}, \hat{B}] \equiv \hat{A}\hat{B} - \hat{B}\hat{A} .}$$

For $\hat{x}$ and $\hat{p}_x$, it is

$$[\hat{x}, \hat{p}_x]\,\psi \equiv (\hat{x}\hat{p}_x - \hat{p}_x\hat{x})\psi = i\hbar\psi .$$

Since $[\hat{x}, \hat{p}_x]$ operating on any ψ always multiplies it by $i\hbar$, we can write

$$\boxed{[\hat{x}, \hat{p}_x] = i\hbar .}$$

(Is $[\hat{x}, \hat{p}_x]$ Hermitian?) In general, however, the effect of a commutator $[\hat{A}, \hat{B}]$ of operators that do not commute depends on the state it operates on.

The kinetic-energy term $\hat{K} = \hat{p}_x^2/2m$ is Hermitian because $(\hat{p}_x^2)^\dagger = \hat{p}_x^\dagger\hat{p}_x^\dagger = \hat{p}_x\hat{p}_x = \hat{p}_x^2$. And we saw earlier that the potential energy $V(x)$ is Hermitian. Therefore the Hamiltonian operator $\hat{H} = \hat{K} + \hat{V}$ is Hermitian and its eigenvalues are real. The position operators $\hat{x}$ and $\hat{y}$ commute because x and y are ordinary functions, which have the same effect applied in either order: $xy\psi = yx\psi$. The momentum operators $\hat{p}_x$ and $\hat{p}_y$ commute because x and y are independent variables, in which case $\partial^2\psi/\partial x\partial y = \partial^2\psi/\partial y\partial x$.

Perhaps more familiar is the fact that matrices do not in general commute. For example,

$$\begin{pmatrix} 0 & 1 \\ 1 & 0 \end{pmatrix}\begin{pmatrix} 1 & 0 \\ 0 & -1 \end{pmatrix}\begin{pmatrix} a \\ b \end{pmatrix} = \begin{pmatrix} -b \\ a \end{pmatrix}, \quad \text{but} \quad \begin{pmatrix} 1 & 0 \\ 0 & -1 \end{pmatrix}\begin{pmatrix} 0 & 1 \\ 1 & 0 \end{pmatrix}\begin{pmatrix} a \\ b \end{pmatrix} = \begin{pmatrix} b \\ -a \end{pmatrix}.$$

The first matrix interchanges the two components of a vector, the second one changes the sign of the bottom component. The order matters. The commutator of the matrices is

$$\begin{pmatrix} 0 & 1 \\ 1 & 0 \end{pmatrix}\begin{pmatrix} 1 & 0 \\ 0 & -1 \end{pmatrix} - \begin{pmatrix} 1 & 0 \\ 0 & -1 \end{pmatrix}\begin{pmatrix} 0 & 1 \\ 1 & 0 \end{pmatrix} = \begin{pmatrix} 0 & -2 \\ 2 & 0 \end{pmatrix}.$$

These matrices play starring roles in Chap. 10, on spin angular momentum.

The basic rules for reducing compound commutators to simpler form are (see Prob. 2)

$$\boxed{\begin{aligned} [\hat{A}, \hat{B} + \hat{C}] &= [\hat{A}, \hat{B}] + [\hat{A}, \hat{C}] & [\hat{A}\hat{B}, \hat{C}] &= \hat{A}[\hat{B}, \hat{C}] + [\hat{A}, \hat{C}]\hat{B} \\ [\hat{A} + \hat{B}, \hat{C}] &= [\hat{A}, \hat{C}] + [\hat{B}, \hat{C}] & [\hat{A}, \hat{B}\hat{C}] &= \hat{B}[\hat{A}, \hat{C}] + [\hat{A}, \hat{B}]\hat{C} \end{aligned}}$$

The $[\hat{A}\hat{B}, \hat{C}]$ rule says: Pull the *first* of the product of operators $\hat{A}\hat{B}$ out *front*, to multiply $[\hat{B}, \hat{C}]$; then pull the *second* of the product operators out *back*, to multiply $[\hat{A}, \hat{C}]$. The "out-front" and "out-back" orderings are necessary because, for example, $\hat{A}$ might not commute with $[\hat{B}, \hat{C}]$. To reduce $[\hat{A}\hat{B}\hat{C}, \hat{D}]$ or $[\hat{A}\hat{B}, \hat{C}\hat{D}]$, you would use the basic rules repeatedly.

Half of Chap. 6 (on the harmonic oscillator) and most of Chap. 8 (on angular momentum) start with the basic commutator relations (in one and three dimensions). In Chap. 7, we prove that if $\hat{A}$ and $\hat{B}$ do not commute, then (a trivial exception aside) they have no simultaneous eigenstates and the observables A and B cannot simultaneously be known exactly. The famous example is $\Delta x \, \Delta p_x \geq \frac{1}{2}\hbar$ for the root-mean-square uncertainties of x and p_x for *any* quantum state; it follows at once from $[\hat{x}, \hat{p}_x] = i\hbar$. Just as important is the case where two (or more) operators *do* commute. Then the operators *do* have simultaneous eigenstates, with exact eigenvalues. But that too can wait until Chap. 7.

5·5. OPERATORS AS MATRICES

It will often be useful to represent the operators for physical observables as square matrices, especially when the eigenvalues of an operator are discrete and equally spaced. This is the case in Chap. 6, where we find matrices for the operators $\hat{H}$, $\hat{x}$, and $\hat{p}$ for the simple harmonic oscillator, and in Chap. 8, where we find matrices for the angular-momentum operators. The matrices are then used in later chapters.

Suppose we have an orthonormal set of vectors that can serve as a basis for representing the states of a system. Let the set be $|n\rangle$, $n = 1, 2, 3, \ldots$, where $\langle 1| = (1, 0, 0, 0, \ldots)$, $\langle 2| = (0, 1, 0, 0, \ldots)$, etc. The recipe for the matrix for an operator $\hat{A}$ in this basis is:

(1) Calculate the vectors $\hat{A}|n\rangle$ for each n.

(2) Stack those vectors up in order side-by-side like the pickets of a picket fence,

$$\hat{A} = \begin{pmatrix} \vdots & \vdots & \vdots & \\ \hat{A}|1\rangle & \hat{A}|2\rangle & \hat{A}|3\rangle & \cdots \\ \vdots & \vdots & \vdots & \end{pmatrix}.$$

That's it—*the picket-fence recipe*.

To see this, consider a 2-by-2 matrix operating on a 2-component vector,

$$\begin{pmatrix} a & b \\ c & d \end{pmatrix} \begin{pmatrix} x \\ y \end{pmatrix} = \begin{pmatrix} ax + by \\ cx + dy \end{pmatrix} = \begin{pmatrix} a \\ c \end{pmatrix} x + \begin{pmatrix} b \\ d \end{pmatrix} y .$$

The result is a weighted sum of *the columns of the matrix*, where the weights are *the components of the vector*. If $x = 1$ and $y = 0$, you get the first column of the matrix; if $x = 0$ and $y = 1$, you get the second column. This is a useful way to think about matrix multiplication for any number of dimensions. (Write out $\hat{B}v$, where $\hat{B}$ is 3-by-3 and the components of v are x, y, z.)

Now consider $\hat{A}|\psi\rangle$, where $|\psi\rangle = \sum c_n |n\rangle$. Our operators are *linear* operators, so that

$$\hat{A}|\psi\rangle = \hat{A}\left(\sum_n c_n |n\rangle\right) = \sum_n c_n \hat{A}|n\rangle .$$

That is, $\hat{A}$ operating on $|\psi\rangle$ is the weighted sum of the $\hat{A}|n\rangle$, where the weights are the components c_n of $|\psi\rangle$. This is exactly what the recipe for $\hat{A}$—make the columns of $\hat{A}$ be the $\hat{A}|n\rangle$—will get when it operates on an arbitrary $|\psi\rangle$ with components c_n. If $c_1 = 1$ and all other $c_n = 0$, then you get column 1 of $\hat{A}$ (which had better be $\hat{A}|1\rangle$); and so on.

However, if all we need are a few of the elements $A_{mn} \equiv \langle m|\hat{A}|n\rangle$ of a matrix—perhaps just the diagonal elements—they are probably better calculated one by one.

Here is the matrix representation of the Hamiltonian for the infinite square well, using the energy eigenstates as the basis vectors. Since $\hat{H}|n\rangle = E_n|n\rangle$, we get at once

$$\hat{H} = \begin{pmatrix} E_1 & 0 & 0 & \cdots \\ 0 & E_2 & 0 & \cdots \\ 0 & 0 & E_3 & \cdots \\ \vdots & \vdots & \vdots & \ddots \end{pmatrix} = E_1 \begin{pmatrix} 1 & 0 & 0 & \cdots \\ 0 & 4 & 0 & \cdots \\ 0 & 0 & 9 & \cdots \\ \vdots & \vdots & \vdots & \ddots \end{pmatrix}.$$

The matrix for $\hat{H}$ is diagonal because the basis vectors are the eigenstates of $\hat{H}$.

The matrix elements of $\hat{A}$ and of its adjoint $\hat{A}^\dagger$ are related by

$$\langle m|\hat{A}|n\rangle = \langle \psi_m|\hat{A}\psi_n\rangle = \langle \hat{A}^\dagger\psi_m|\psi_n\rangle = \langle \psi_n|\hat{A}^\dagger\psi_m\rangle^* = \langle n|\hat{A}^\dagger|m\rangle^*,$$

or

$$\boxed{A_{mn} = (A^\dagger_{nm})^*, \qquad \text{or} \qquad A^\dagger_{nm} = A^*_{mn}.}$$

To get the $\hat{A}^\dagger$ matrix from the $\hat{A}$ matrix, reflect $\hat{A}$ across its main diagonal and complex conjugate all the elements. If $\hat{A}^\dagger = \hat{A}$ ($\hat{A}$ is Hermitian), then $A^\dagger_{nm} = A_{nm} = A^*_{mn}$: The elements on the main diagonal are real and elements symmetrically placed across that diagonal are complex conjugates of one another. Later chapters will have uses for operators (and their matrices) that are *not* Hermitian.

5·6. EXPECTATION VALUES

In Sec. 2, we wrote the general superposition state $\Psi(x,t)$ for a particle in the infinite square well as a vector. In Sec. 5, we wrote the Hamiltonian $\hat{H}$ as a matrix. The action of $\hat{H}$ operating on Ψ is

$$\hat{H}\Psi(x,t) = \begin{pmatrix} E_1 & 0 & 0 & \cdots \\ 0 & E_2 & 0 & \cdots \\ 0 & 0 & E_3 & \cdots \\ \vdots & \vdots & \vdots & \ddots \end{pmatrix} \begin{pmatrix} c_1\,e^{-iE_1t/\hbar} \\ c_2\,e^{-iE_2t/\hbar} \\ c_3\,e^{-iE_3t/\hbar} \\ \vdots \end{pmatrix} = \begin{pmatrix} E_1c_1\,e^{-iE_1t/\hbar} \\ E_2c_2\,e^{-iE_2t/\hbar} \\ E_3c_3\,e^{-iE_3t/\hbar} \\ \vdots \end{pmatrix},$$

where $c_1 = c_1(t=0)$, etc. Then

$$\langle \Psi(x,t)|\hat{H}\Psi(x,t)\rangle = \begin{pmatrix} c_1^*\,e^{+iE_1t/\hbar} & c_2^*\,e^{+iE_2t/\hbar} & \cdots \end{pmatrix} \begin{pmatrix} E_1c_1\,e^{-iE_1t/\hbar} \\ E_2c_2\,e^{-iE_2t/\hbar} \\ \cdots \end{pmatrix} = \sum_{n=1}^{\infty} E_n\,|c_n|^2.$$

This is the sum of the eigenvalues E_n, each weighted by the probability $|c_n|^2$ of getting that value on making a measurement of the energy.

One measurement of the energy would produce *one* of the values E_n—and then the particle would be in the corresponding energy eigenstate $\Psi_n(x,t)$. The weighted quantity above is the *expectation value*—the expected average or mean value of the energy if *many* measurements are made, each on an *identical* system, starting *always* in the original state $\Psi(x,t)$. If between measurements the system is not returned to the original $\Psi(x,t)$, repeated measurements will simply repeatedly get the value of E obtained on the first measurement. (Why?)

The expectation value $\langle A \rangle$ of the physical quantity represented by the operator $\hat{A}$ is

$$\langle A \rangle \equiv \langle \Psi | \hat{A} \Psi \rangle = \langle \Psi | \hat{A} | \Psi \rangle \,,$$

a sandwich, with Ψ the bread and $\hat{A}$ the filling. First operate on $|\Psi\rangle$ with $\hat{A}$, then calculate the scalar product of $\langle \Psi |$ with $\hat{A} |\Psi\rangle$. This ordering is the only one that makes sense. When the operator is a matrix and Ψ is a column vector, then $\hat{A} |\Psi\rangle$ is a column vector and $\langle \Psi |$ is a row vector, and $\langle A \rangle = \langle \Psi | \hat{A} | \Psi \rangle$ is a scalar. No other ordering of $\langle \Psi |$, $\hat{A}$, and $|\Psi\rangle$ fits the pieces together to make a scalar.

When calculating, say, $\langle V \rangle = \langle \Psi | V | \Psi \rangle$ with an integral instead of a matrix, we can write the integrand as $\Psi^* V \Psi$ or as $V |\Psi|^2$, because each is just a product of the same three ordinary functions. However, in getting $\langle E \rangle$ or $\langle p \rangle$, which involve differential operators, writing the integrand as $\hat{H}(\Psi^* \Psi)$ or $\hat{p}(\Psi^* \Psi)$ leads to nonsense—the integrand must be $\Psi^*(\hat{A}\Psi)$.

5·7. MORE THEOREMS

(a) Time dependence—In Sec. 2, we showed that $\Psi(x,t)$ once normalized stays normalized. We show it again (and more generally), this time by proving directly that

$$\frac{d}{dt}\langle \Psi | \Psi \rangle = \frac{d}{dt}\left(\int_{-\infty}^{+\infty} \Psi^*(x,t)\Psi(x,t)\,dx \right) = 0 \,.$$

The differentiation and integration here are with respect to independent variables t and x. Therefore we can (by the rules of calculus) move the differentiation inside the integral as partial differentiation:

$$\frac{d}{dt}\left(\int_{-\infty}^{+\infty} \Psi^*(x,t)\Psi(x,t)\,dx \right) = \int_{-\infty}^{+\infty} \left[\left(\frac{\partial \Psi^*}{\partial t} \right)\Psi + \Psi^*\left(\frac{\partial \Psi}{\partial t} \right) \right] dx \,.$$

Then, just as we did in Sec. 4·1(c), we replace the partial derivatives using the Schrödinger equation $i\hbar\,\partial\Psi/\partial t = \hat{H}\Psi$ and its complex conjugate $-i\hbar\,\partial\Psi^*/\partial t = \hat{H}\Psi^*$:

$$\frac{i}{\hbar}\int_{-\infty}^{+\infty}\left[(\hat{H}\Psi^*)\Psi - \Psi^*(\hat{H}\Psi) \right]dx = \frac{i}{\hbar}\left(\langle \hat{H}\Psi | \Psi \rangle - \langle \Psi | \hat{H}\Psi \rangle \right) = \frac{i}{\hbar}\left(\langle \Psi | \hat{H}\Psi \rangle - \langle \Psi | \hat{H}\Psi \rangle \right) = 0 \,.$$

The next-to-last step follows because $\hat{H}$ is Hermitian. And so $d(\langle \Psi | \Psi \rangle)/dt = 0$.

The time rate of change of the expectation value $\langle A \rangle \equiv \langle \Psi | \hat{A} | \Psi \rangle$ of an observable A is

$$\frac{d}{dt}\langle A \rangle = \frac{d}{dt}\left(\int_{-\infty}^{+\infty} \Psi^*(\hat{A}\Psi)\,dx \right).$$

We can allow for possible explicit t-dependence in $\hat{A}$, such as there would be for the potential energy $V(t) = q\mathcal{E}(t)$ of a charge q in an oscillating electric field $\mathcal{E}(t)$. We get

$$\frac{d}{dt}\langle A \rangle = \int_{-\infty}^{+\infty}\left[\left(\frac{\partial \Psi^*}{\partial t} \right)(\hat{A}\Psi) + \Psi^*\left(\frac{\partial \hat{A}}{\partial t} \right)\Psi + \Psi^*\left(\hat{A}\frac{\partial \Psi}{\partial t} \right) \right] dx$$

$$= \frac{i}{\hbar}\langle \hat{H}\Psi | \hat{A}\Psi \rangle + \langle \Psi | \frac{\partial \hat{A}}{\partial t} | \Psi \rangle - \frac{i}{\hbar}\langle \Psi | \hat{A}\hat{H}\Psi \rangle$$

$$\frac{d}{dt}\langle A \rangle = \frac{i}{\hbar}\langle \Psi | [\hat{H}, \hat{A}] | \Psi \rangle + \left\langle \frac{\partial \hat{A}}{\partial t} \right\rangle \,.$$

Often $\hat{A}$ will have no explicit time dependence, so that $\partial\hat{A}/\partial t = 0$; $\hat{x}$ and $\hat{p}_x$ are examples. Then there are two simple cases in which $d\langle A\rangle/dt = 0$, and $\langle A\rangle$ is constant in time.

(1) If $[\hat{H},\hat{A}] = 0$, then it follows at once from the boxed equation that $d\langle A\rangle/dt = 0$. For example, since $[\hat{H},\hat{H}] = 0$, the mean value of the energy stays constant—the energy is *conserved*. The same will hold for other observables A that commute with $\hat{H}$.

(2) If Ψ is an eigenstate $\Psi_n(t) = \psi_n e^{-iE_n t}$ of $\hat{H}$, then

$$\langle\psi_n|[\hat{H},\hat{A}]|\psi_n\rangle = \langle\hat{H}\psi_n|\hat{A}\psi_n\rangle - \langle\psi_n|\hat{A}\hat{H}\psi_n\rangle = (E_n - E_n)\langle\psi_n|\hat{A}|\psi_n\rangle = 0 ,$$

even if $[\hat{H},\hat{A}]$ is not zero (the $e^{\mp iE_n t}$ factors cancel), so again $d\langle A\rangle/dt = 0$. For example, neither of $[\hat{H},\hat{x}]$ or $[\hat{H},\hat{p}_x]$ is zero, but in an energy eigenstate $\langle x\rangle$ and $\langle p_x\rangle$ are both constants.

(b) Ehrenfest relations—To find how $\langle x\rangle$ varies in time when $\Psi(t)$ is a superposition of energy eigenstates, we need $[\hat{H},\hat{x}]$. Using relations $[\hat{A}\hat{B},\hat{C}] = \hat{A}[\hat{B},\hat{C}] + [\hat{A},\hat{C}]\hat{B}$ and $[\hat{x},\hat{p}_x] = i\hbar$ from Sec. 4(b), we get

$$[\hat{H},\hat{x}] = \frac{1}{2m}[\hat{p}_x^2,\hat{x}] + [\hat{V}(x),\hat{x}] = \frac{\hat{p}_x}{2m}[\hat{p}_x,\hat{x}] + [\hat{p}_x,\hat{x}]\frac{\hat{p}_x}{2m} = -\frac{i\hbar}{m}\hat{p}_x .$$

Then from the last boxed equation, we get

$$\boxed{\frac{d\langle x\rangle}{dt} = \frac{i}{\hbar}\langle\Psi|[\hat{H},\hat{x}]|\Psi\rangle = \frac{\langle p_x\rangle}{m} .}$$

This has the same *form* as the classical relation $dx/dt = v_x = p_x/m$. Similarly, one can show that

$$\boxed{\frac{d\langle p_x\rangle}{dt} = \frac{i}{\hbar}\langle\Psi|[\hat{H},\hat{p}_x]|\Psi\rangle = -\left\langle\frac{\partial V}{\partial x}\right\rangle ,}$$

which has the same form as the classical relation $dp_x/dt = F_x = -\partial V/\partial x$. These are examples of *Ehrenfest's theorem*: Quantum-mechanical expectation values obey the classical equations of motion.

(c) The virial theorem—We showed at the end of part (a) that when ψ_n is an eigenstate of $\hat{H}$ then

$$\langle\psi_n|[\hat{H},\hat{A}]|\psi_n\rangle = 0 .$$

Let $\hat{A} = \hat{x}\hat{p}_x$ and $\hat{H} = \hat{p}_x^2/2m + \hat{V}(x)$. Using again $[\hat{A}\hat{B},\hat{C}] = \hat{A}[\hat{B},\hat{C}] + [\hat{A},\hat{C}]\hat{B}$ and $[\hat{x},\hat{p}_x] = i\hbar$, and taking care to let the $\hat{p}_x$ operator "bite" on a function (as we did in deriving $[\hat{x},\hat{p}_x] = i\hbar$ in Sec. 4(b)), we eventually get

$$[\hat{H},\hat{x}\hat{p}_x] = \hat{x}[\hat{H},\hat{p}_x] + [\hat{H},\hat{x}]\hat{p}_x = i\hbar\hat{x}\frac{\partial\hat{V}}{\partial x} - i\hbar\frac{\hat{p}_x^2}{m} .$$

Now since $\langle\psi_n|[\hat{H},\hat{x}\hat{p}_x]|\psi_n\rangle = 0$ in an eigenstate ψ_n, the $\langle\psi_n|\cdots|\psi_n\rangle$ bracket of the right side of the equation is equal to zero too. From this, we get

$$\boxed{\left\langle\frac{p_x^2}{2m}\right\rangle_n = \langle K\rangle_n = \frac{1}{2}\left\langle x\frac{\partial V}{\partial x}\right\rangle_n .}$$

This is the *virial theorem* (in one dimension). The subscript n is a reminder that the expectation value is for an energy eigenstate ψ_n, not for a superposition state.

107

For a power-law potential, $V(x) = ax^b$, we get $x\,\partial V/\partial x = b\,V(x)$, and therefore

$$\langle K \rangle_n = \frac{b}{2} \langle V \rangle_n \; .$$

For the harmonic oscillator, $V(x) = \frac{1}{2}m\omega^2 x^2$, and $\langle K \rangle_n = \langle V \rangle_n$. In an eigenstate, the total energy is, on average, shared equally between the kinetic and potential forms.

5·8. REVISED RULES

These following rules replace the provisional rules given in Secs. 2·2 and 2·3.

Rule 1—The state of a system is completely specified by a state function Ψ, which is a function of space (and spin) coordinates and time. Anything knowable about the system can be obtained from this function, but what is knowable is less complete than what is assumed in classical mechanics.

Rule 2—The Schrödinger equation $i\hbar\,\partial\Psi/\partial t = \hat{H}\Psi$, where $\hat{H}$ is the Hamiltonian operator, tells how Ψ evolves in time—as long as the sytem is undisturbed. Measurement can cause an abrupt change in Ψ.

Rule 3—For every physical observable there is a linear, Hermitian operator. The only possible result of an accurate measurement of a observable A is one of the eigenvalues of the corresponding operator $\hat{A}$.

Rule 4—If the result of a measurement of A is the eigenvalue a_n, then immediately afterwards the system is in the eigenstate (or a superposition of the eigenstates) that belong to a_n. Conversely, if the system is in an eigenstate of $\hat{A}$ when a measurement is made, the result of the measurement will be the eigenvalue that belongs to that eigenstate.

Rule 5—Any (renormalized) linear superposition of possible states of a system is a possible state of the system.

In classical physics, a particle has at any instant a definite position, momentum, and energy. Classical physics puts no theoretical limits on how well those properties can be measured. And in principle measurements can be made that leave the system completely undisturbed. This is clearly not true in quantum mechanics. For example, in any of the energy eigenstates of a particle in an infinite square well (or in any other well), there is only a probability distribution for the particle's position. And after a measurement of its position (with some small error; perfect measurements of a continuous variable are not possible), an expansion of the new wave function in terms of energy eigenstates would require a superposition of a number of them. The particle no longer has a definite energy—a situation unheard of in classical physics. We return to such questions in Chap. 7.

PROBLEMS

1. *Simple matrix operators.*

The space is ordinary 3-dimensional space, with the usual orthonormal vectors $|1\rangle$, $|2\rangle$, and $|3\rangle$. What is the effect of multiplying an arbitrary vector $|a\rangle$ from the left by $|3\rangle\langle 3|$? What matrix operating on $|a\rangle$ would give the "projection" of $|a\rangle$ on the 23 (i.e., the yz) plane? Write this matrix in terms of $|i\rangle\langle j|$ matrices. Write in terms of the unit vectors a matrix that interchanges the first and third components of a vector while doubling the length of the second component.

2. *Commutator identities.*

(a) Prove the following operator identities:
$$[\hat{A}, \hat{B} + \hat{C}] = [\hat{A}, \hat{B}] + [\hat{A}, \hat{C}]$$
$$[\hat{A}\hat{B}, \hat{C}] = \hat{A}[\hat{B}, \hat{C}] + [\hat{A}, \hat{C}]\hat{B}$$
$$[\hat{A}, \hat{B}\hat{C}] = \hat{B}[\hat{A}, \hat{C}] + [\hat{A}, \hat{B}]\hat{C} .$$

(b) Reduce the commutators $[\hat{p}, \hat{x}^2]$, $[\hat{p}^2, \hat{x}]$, and $[\hat{p}^2, \hat{x}^2]$ to the simplest forms possible.

3. *On being Hermitian.*

(a) Prove that the expectation value $\langle A \rangle = \langle \psi|\hat{A}|\psi\rangle$ of a Hermitian operator $\hat{A}$ is real.

(b) Prove that the expectation value of the square of a Hermitian operator, $\langle A^2 \rangle = \langle \psi|\hat{A}^2|\psi\rangle$, is nonnegative.

(c) Prove that $(\hat{B}^\dagger)^\dagger = \hat{B}$, even if $\hat{B}$ is not Hermitian.

(d) Is $\hat{B}^\dagger\hat{B}$ Hermitian even if $\hat{B}$ is not? If $\hat{B}$ and $\hat{C}$ are Hermitian, which of $\hat{B}\hat{C}$, $(\hat{B}\hat{C} + \hat{C}\hat{B})$, $(\hat{B}\hat{C} - \hat{C}\hat{B})$, $i(\hat{B}\hat{C} + \hat{C}\hat{B})$, and $i(\hat{B}\hat{C} - \hat{C}\hat{B})$ are necessarily Hermitian?

4. *Proofs, etc.*

(a) Prove the second of the Ehrenfest relations given in Sec. 7(b).

(b) Fill in the details of the proof of the virial theorem given in Sec. 7(c).

(c) One can formulate quantum mechanics in "momentum (p) space" as well as in "configuration (x) space." Then, instead of $\Psi(x, t)$, we have $\Phi(p, t)$. The commutation relation $[\hat{x}, \hat{p}] = i\hbar$ remains unchanged. In momentum space, $\hat{p} = p$. So what then will the operator $\hat{x}$ be? Guess and then check to see if anything needs fixing.

5. *Exercises.*

The operators $\hat{A}$, $\hat{B}$, and $\hat{C}$ are linear and Hermitian. Operator $\hat{D}$ is linear but not necessarily Hermitian.

(a) Operator $\hat{A}$ has eigenstates ψ_1 and ψ_2 with eigenvalues a_1 and a_2, where $a_1 \neq a_2$. Prove that if $[\hat{A}, \hat{B}] = 0$, then $\langle \psi_1|\hat{B}|\psi_2\rangle = 0$.

(b) Operator $\hat{C}$ has an eigenstate ψ_1 with eigenvalue $c_1 = 1$. Operator $\hat{D}$ has the following commutation relation with $\hat{C}$: $[\hat{C}, \hat{D}] = \hat{D}\hat{C}^2 + \hat{D}$. Show that $\psi_2 = \hat{D}\psi_1$ is an eigenstate of $\hat{C}$, and find the eigenvalue c_2. Then show that $\psi_3 = \hat{D}\psi_2$ is also an eigenstate of $\hat{C}$, and find the eigenvalue c_3.

6. *Matrix eigenvalue problems.*

The Pauli matrices $\hat{\sigma}_x$, $\hat{\sigma}_y$, and $\hat{\sigma}_z$, for which we shall eventually have much use, are defined as

$$\hat{\sigma}_x = \begin{pmatrix} 0 & 1 \\ 1 & 0 \end{pmatrix}, \quad \hat{\sigma}_y = \begin{pmatrix} 0 & -i \\ i & 0 \end{pmatrix}, \quad \hat{\sigma}_z = \begin{pmatrix} 1 & 0 \\ 0 & -1 \end{pmatrix}.$$

Find the eigenvalues λ and the corresponding normalized eigenvectors of each of these matrices. That is, solve

$$\begin{pmatrix} 0 & 1 \\ 1 & 0 \end{pmatrix} \begin{pmatrix} c \\ d \end{pmatrix} = \lambda \begin{pmatrix} c \\ d \end{pmatrix},$$

etc. For each eigenvector, choose c to be real and nonnegative. Check that the two eigenvectors for each operator are orthogonal.

7. *The general 2-by-2 eigenvalue problem.*

Any 2-by-2 Hermitian matrix may be written as

$$a\hat{I} + b_x\hat{\sigma}_x + b_y\hat{\sigma}_y + b_z\hat{\sigma}_z = a\hat{I} + \mathbf{b} \cdot \hat{\boldsymbol{\sigma}} = \begin{pmatrix} a + b_z & b_x - ib_y \\ b_x + ib_y & a - b_z \end{pmatrix},$$

where the parameters a, b_x, b_y, and b_z are real, and $\mathbf{b} = (b_x, b_y, b_z) = b\,\hat{\mathbf{n}}$ is an ordinary vector of magnitude b in the direction $\hat{\mathbf{n}}$.

(a) Show that the eigenvalues of this matrix are $\lambda_\pm = a \pm (b_x^2 + b_y^2 + b_z^2)^{1/2} = a \pm b$. Therefore, to get the eigenvalues of *any* 2-by-2 Hermitian matrix, all you need to do is identify a and b.

(b) Show that the normalized eigenstates belonging to the eigenvalues $a \pm b$ are, aside from an arbitrary overall phase factor,

$$|{+}n\rangle = \begin{pmatrix} \cos\frac{1}{2}\theta \\ e^{+i\phi}\sin\frac{1}{2}\theta \end{pmatrix} \quad \text{and} \quad |{-}n\rangle = \begin{pmatrix} \sin\frac{1}{2}\theta \\ -e^{+i\phi}\cos\frac{1}{2}\theta \end{pmatrix},$$

where θ and ϕ are the polar and azimuthal angles of $\hat{\mathbf{n}}$. To get the normalized eigenvectors of *any* 2-by-2 Hermitian matrix, simply identify the angles θ and ϕ using $\mathbf{b} = (b_x, b_y, b_z)$.

(c) Show that the states $|{+}n\rangle$ and $|{-}n\rangle$ are orthogonal, and that

$$|{+}n\rangle\langle{+}n| + |{-}n\rangle\langle{-}n| = \hat{I}.$$

6. THE HARMONIC OSCILLATOR

6.1. *The Classical Oscillator*

6.2. *The Quantum Oscillator: Series Solution*

6.3. *The Operator Solution*

6.4. *States as Vectors, Operators as Matrices*

    Problems

The restoring force acting on an object displaced from a position of stable equilibrium is often, for small enough displacements, proportional to the displacement x; hence the importance of Hooke's law $F(x) = -kx$ and the harmonic oscillator. Of course, since $F(x)$ becomes infinite as $x \to \pm\infty$, it can only be valid over a limited range of x. Nevertheless, the oscillator approximation is one of the most important models in both classical and quantum physics. "Essentially, all models are wrong, but some are useful" (George E.P. Box).

We solve the Schrödinger equation for the oscillator potential energy, $V(x) = \frac{1}{2}m\omega^2 x^2$, in two very different ways. The first is the series-solution method taught in courses on differential equations—the method that led to the special functions and polynomials carrying the names of such as Bessel, Legendre, Laguerre, and (relevant for this chapter) Hermite. The second method is more exotic: The eigenvalues and eigenfunctions are found, almost by magic, from the commutator relation $[\hat{x}, \hat{p}] = i\hbar$ and the form of the Hamiltonian, which is quadratic in $\hat{x}$ and $\hat{p}$. Each method of solution will be used again—the operator method for angular momentum, the series-solution method for the radial equation of hydrogen. Solving differential equations with the series method is not everyone's favorite mathematics. However, if you slight Section 2, you still need to read Section 2(a), on "rescaling" the Schrödinger equation.

With the eigenvalues and eigenfunctions in hand, we write general states of an oscillator as vectors and the operators as matrices. Applications of the physics (most of them beyond the scope of this book) include molecular vibrations, specific heats of solids and gases, and quantization of the electromagnetic field.

6·1. THE CLASSICAL OSCILLATOR

A brief review: Figure 1 shows the potential energy for the simple harmonic oscillator, the parabolic $V(x) = \frac{1}{2}m\omega^2 x^2$ we introduced in Sec. 2·8. The horizontal line represents a particular total energy E. Classically, a particle with energy E is restricted to the range of x between the turning points $\pm x_0$, where $E = V(\pm x_0) = \frac{1}{2}m\omega^2 x_0^2$, or $x_0 = \sqrt{2E/m\omega^2}$. Within the allowed range, the difference $E - V(x)$ is the kinetic energy $K = p^2/2m$. The greater the energy, the wider the range of motion.

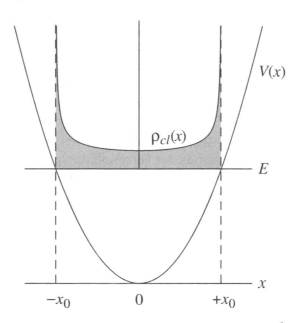

Figure 1. The harmonic oscillator potential energy $V(x) = \frac{1}{2}m\omega^2 x^2$, the turning points $\pm x_0$ for an arbitrary energy E, and the classical probability density $\rho_{\text{cl}}(x)$ (measured from the E line) for a particle with this energy.

The force on m is $F(x) = -dV/dx = -m\omega^2 x$; $m\omega^2$ is the Hooke's-law spring constant (often written k). Newton's second law here is $m\ddot{x} = -m\omega^2 x$, and the general solution is $x(t) = A\cos\omega t + B\sin\omega t$. If at $t = 0$ the particle is at rest at $x = +x_0$, then its position and velocity at later times are $x(t) = x_0\cos\omega t$ and $v(t) = dx/dt = -\omega x_0\sin\omega t$.

We shall want to compare the classical position probability density $\rho_{cl}(x)$ with—when we have found them—the quantum-mechanical densities $|\psi_n(x)|^2$. Imagine taking snapshots of the oscillating particle at 1000 random times: the particle would be found most often where it is moving slowly, near $\pm x_0$. The probability $dP(x)$ that it would be found in dx at x is proportional to the time dt it spends in dx, which is inversely proportional to the speed with which it crosses dx. From $v = dx/dt$, we get $dP(x) \propto dt = dx/v$. The normalized probability distribution is (see Prob. 1)

$$dP(x) = \rho_{cl}(x)\,dx = \frac{dx}{\pi\sqrt{x_0^2 - x^2}}$$

for $-x_0 < x < +x_0$; see Fig. 1. The integral $\int_{-x_0}^{+x_0} \rho_{cl}(x)\,dx$ is one, even though $\rho_{cl}(x)$ rises to infinity at $x = \pm x_0$.

112

Now $V(x) = \frac{1}{2}m\omega^2 x^2$ implies, unphysically, that $F(x)$ becomes infinite as $x \to \pm\infty$. Clearly, the oscillator $V(x)$ can only be an approximation over a limited range of x. Consider a smoothly varying $V(x)$ with a minimum at $x = a$, such as shown in Fig. 2. The Taylor series expansion of $V(x)$ about $x = a$ is

$$V(x) = V(a) + \frac{(x-a)}{1!}\frac{dV}{dx}\Big|_a + \frac{(x-a)^2}{2!}\frac{d^2V}{dx^2}\Big|_a + \frac{(x-a)^3}{3!}\frac{d^3V}{dx^3}\Big|_a + \cdots$$

At a, $dV/dx = 0$ (why?). Close enough to a, $(x-a)^2$ is much larger than $(x-a)^n$ for $n \geq 3$. Therefore, provided that $d^2V/dx^2 \neq 0$ at a, close enough to a we get

$$V(x) \approx V(a) + \frac{1}{2}k(x-a)^2 \, ,$$

where $k \equiv d^2V/dx^2|_a$. The approximating parabola has the same values of V, dV/dx, and d^2V/dx^2 as the actual $V(x)$ at $x = a$. The curvature of $V(x)$ at a is the effective spring constant k, and the angular frequency for oscillations in the parabolic approximation is $\omega = \sqrt{k/m}$.

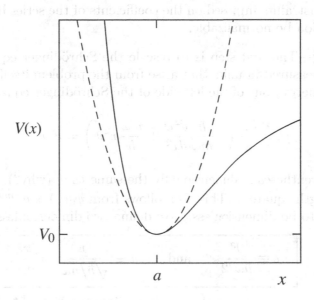

Figure 2. A smoothly varying potential energy and its oscillator approximation.

The range of applicability of the oscillator approximation can be overstated—not every smoothly varying $V(x)$ can usefully be approximated near a minimum with a parabola. For example, for $V(x) = bx^4$, d^2V/dx^2 at $x = 0$ is zero, and we have no spring constant. And sometimes a minimum in $V(x)$ is so shallow and/or asymmetric that the oscillator approximation would give completely wrong results.

6·2. THE QUANTUM OSCILLATOR: SERIES SOLUTION

The time-independent Schrödinger equation for the harmonic oscillator is

$$\hat{H}\psi = \left(\frac{\hat{p}^2}{2m} + \frac{1}{2}m\omega^2\hat{x}^2\right)\psi = -\frac{\hbar^2}{2m}\frac{d^2\psi}{dx^2} + \frac{1}{2}m\omega^2 x^2\psi = E\psi \, .$$

In Sec. 2·8, we found that the ground-state energy and normalized wave function are

$$E_0 = \frac{1}{2}\hbar\omega \quad \text{and} \quad \psi_0(x) = \left(\frac{m\omega}{\pi\hbar}\right)^{1/4} e^{-m\omega x^2/2\hbar} .$$

Now we find all the eigenvalues and eigenfunctions. Since $V(x)$ rises to infinity, we expect to find an infinite number of bound states. Since $V(x) = V(-x)$, we expect the eigenfunctions to alternate between even and odd as the energy increases.

Here are the steps toward the solution:

(a) We rescale the equation, so that all harmonic oscillator problems reduce to the same simple form.

(b) We investigate the behavior of the solutions of the rescaled equation as $x \to 0$ and $x \to \pm\infty$.

(c) We factor the $x \to \pm\infty$ ("asymptotic") behavior out of the rescaled equation to get a reduced equation.

(d) We try a series solution for the reduced equation that has the $x \to 0$ behavior built into it.

(e) We find the constraints imposed on the coefficients of the series by the requirement that the wave function be normalizable.

(a) Rescaling—The first step is to rescale the Schrödinger equation so that energies and distances are measured in units that arise from the problem itself. We pull a factor $\frac{1}{2}\hbar\omega$ (the ground-state energy) out of the left side of the Schrödinger equation, to get

$$\frac{\hbar\omega}{2}\left(-\frac{\hbar}{m\omega}\frac{d^2\psi}{dx^2} + \frac{m\omega}{\hbar}x^2\psi\right) = E\psi .$$

The quantity in parentheses is dimensionally the same as ψ (why?), and therefore $\hbar/m\omega$ is dimensionally a length squared. (This also follows from $\psi_0(x) \propto e^{-m\omega x^2/2\hbar}$; the argument of an exponential has to be dimensionless.) We define two dimensionless quantities,

$$\epsilon \equiv \frac{E}{\hbar\omega/2} \quad \text{and} \quad y \equiv \frac{x}{\sqrt{\hbar/m\omega}} = \frac{x}{\sigma_0} ,$$

where $\sigma_0 \equiv \sqrt{\hbar/m\omega}$ ($x = \pm\sigma_0$ are the turning points for $E_0 = \frac{1}{2}\hbar\omega$). In terms of ϵ and y, the Schrödinger equation is

$$-\frac{d^2\psi}{dy^2} + y^2\psi = \epsilon\psi ,$$

where now $\psi = \psi(y)$. The kinetic-, potential-, and total-energy terms are now $-\psi''$, $y^2\psi$, and $\epsilon\psi$. In these dimensionless units, the ground-state energy is $\epsilon_0 = 1$, and the ground-state wave function is

$$\psi_0(y) = \pi^{-1/4} e^{-y^2/2} ,$$

where $\int_{-\infty}^{+\infty} |\psi_0(y)|^2 \, dy = 1$; we proved that $\int e^{-y^2} dy = \sqrt{\pi}$ in Sec. 2·8(b). The turning points, where $\epsilon_0 = y^2$, are at $y = \pm 1$.

114

The rescaled Schrödinger equation leaves behind all the particulars of the problem—the values of m and ω and tiny numbers such as $\hbar \approx 10^{-34}$ J s. But once the tidier generic problem is solved, we can always go back to those particulars using $E = \epsilon \times \frac{1}{2}\hbar\omega$ and $x = y \times \sqrt{\hbar/m\omega}$. And the conversion factors here give us contextual measures of whether a quantity is large or small. Is 2 eV a large energy? Compare it to $\frac{1}{2}\hbar\omega$. Is 3 nm a small distance? Compare it to $\sqrt{\hbar/m\omega}$.

(b) Approximate behavior at small y and at large y—We follow the usual series-solution method to solve the rescaled equation,

$$\psi'' + (\epsilon - y^2)\psi = 0 ,$$

but the path is not straight. First, we look for the leading-order behavior of $\psi(y)$ when $y \to 0$ and when $y \to \pm\infty$. Since ϵ is positive (why?), very close to the origin $\epsilon \gg y^2$, and $\psi'' + \epsilon\psi \approx 0$. The solution in this limit is

$$\psi(y \to 0) \approx A\cos(\sqrt{\epsilon}\,y) + B\sin(\sqrt{\epsilon}\,y) \approx A + B'y ,$$

(where $B' = \sqrt{\epsilon}\,B$) because here y is small. The point, to be used below, is that $\psi(y)$ is well-behaved at $y = 0$. There are no terms like $1/y$ or $1/y^2$ that would cause $\psi(y)$ to blow up at the origin.

For $y \to \pm\infty$, the y^2 term swamps the ϵ term, and now $\psi'' \approx y^2\psi$. One *asymptotic* solution (that is, to highest power in y) is $\psi(y) = \exp(\pm y^2/2)$, for then

$$\psi'(y) = \pm y\, e^{\pm y^2/2} \quad \text{and} \quad \psi'' = (y^2 \pm 1)\, e^{\pm y^2/2} \approx y^2\psi(y) .$$

The last step follows because here $y^2 \gg 1$. And we throw away $\psi(y) = e^{+y^2/2}$ because it blows up as $y \to \pm\infty$.

We could multiply the asymptotic solution $e^{-y^2/2}$ by any power of y, say y^n, and it would still be an *asymptotic* solution. This is because, in taking derivatives of $y^n e^{-y^2/2}$, each derivative of y^n *lowers* the power of y, whereas each derivative of $e^{-y^2/2}$ *raises* the power of y. Thus the highest-order term in the second derivative of $y^n e^{-y^2/2}$ is $y^{n+2}e^{-y^2/2}$, which is just what we need to satisfy $\psi'' \approx y^2\psi$ *asymptotically*.

By the same argument, $\psi_n(y) = h_n(y)\, e^{-y^2/2}$, where $h_n(y)$ is a finite polynomial in y whose highest-order term is y^n, is an asymptotic solution. And no matter how large (but *finite*) n is, $e^{-y^2/2}$ will always eventually overwhelm $h_n(y)$ and force $\psi(y)$ to vanish as $y \to \pm\infty$. That is, of course, a requirement if the wave function is to be normalizable.

(c) A new equation—The next step is to use what we have learned about $\psi(y)$ at large and small values of y to construct solutions of the Schrödinger equation for the whole range $-\infty < y < +\infty$. First, we factor the asymptotic behavior $e^{-y^2/2}$ out of $\psi(y)$:

$$\boxed{\psi(y) = h(y)\, e^{-y^2/2} .}$$

Below, we intend to write $h(y)$ as a power series in y. We cannot write $\psi(y)$ itself as a power series because positive powers of y would blow up as $y \to \pm\infty$, and negative powers of y would blow up at the origin—and $\psi(x)$ would not be normalizable.

So we put $\psi(y) = h(y)\, e^{-y^2/2}$ in the Schrödinger equation. We need ψ'':

$$\psi'(y) = (h' - yh)\, e^{-y^2/2} , \qquad \text{and then} \quad \psi'' = (h'' - 2yh' - h + y^2 h)\, e^{-y^2/2} .$$

Putting ψ and ψ'' into $\psi'' + (\epsilon - y^2)\psi = 0$ and cancelling the common factor $e^{-y^2/2}$, we get

$$\boxed{h'' - 2yh' + (\epsilon - 1)h = 0 .}$$

If we solve this equation, we have solved the Schrödinger equation. And if we look for a moment (always look!), we can *see* two solutions,

$$h_0(y) = 1 \text{ with } \epsilon_0 = 1, \text{ making } E_0 = \tfrac{1}{2}\hbar\omega \text{ and } \psi_0(y) \propto e^{-y^2/2} ;$$

$$h_1(y) = y \text{ with } \epsilon_1 = 3, \text{ making } E_1 = \tfrac{3}{2}\hbar\omega \text{ and } \psi_1(y) \propto y\, e^{-y^2/2} .$$

The first wave function has even parity and no nodes, the second has odd parity and one node. These are, of course, the ground- and first-excited-state solutions.

(d) Series solution—The two solutions found above suggest that higher-energy solutions will involve higher powers of y. We try a power series for $h(y)$,

$$\boxed{h(y) = \sum_{i=0} b_i\, y^i .}$$

The series is to apply for the full range of y, $-\infty$ to $+\infty$. The coefficients b_i are to be determined; and for the moment we allow all nonnegative powers of y. No *negative* powers of y are included, because any such terms would cause $h(y)$, and thus $\psi(y)$, to blow up at $y = 0$, where we have shown that $\psi(y)$ is well-behaved.

We need h' and h'':

$$h'(y) = \sum_{i=0} i\, b_i\, y^{i-1}, \qquad h''(y) = \sum_{i=0} i(i-1)\, b_i\, y^{i-2} .$$

Putting h, h', and h'' into $h'' - 2yh' + (\epsilon - 1)h = 0$, we get

$$\sum_{i=0} i(i-1)\, b_i\, y^{i-2} + \sum_{i=0}(\epsilon - 1 - 2i)\, b_i\, y^i = 0 .$$

In the first sum, the series really begins with the $i = 2$ term because the factors i and $(i-1)$ kill the $i = 0$ and $i = 1$ terms. Let us, in this sum, shift i (a "dummy index") by 2, letting $i \to i + 2$. This shifts the sum as follows:

$$\sum_{i=2} i(i-1)\, b_i\, y^{i-2} \longrightarrow \sum_{i=0}(i+2)(i+1)\, b_{i+2}\, y^i ,$$

and the new sum starts up with the $i = 0$ term. What have we changed? Nothing! Write a few terms out. The two sides give exactly the same thing:

$$2b_2 y^0 + 6b_3 y^1 + 12b_4 y^2 + \cdots$$

Shifting the variable in an integral, instead of the index in a sum, would probably cause little unease.

The reason for shifting the index is that now the powers of y in the two series in the equation for $h(y)$ march in step and may be combined into a single series,

$$\sum_{i=0}\left[(i+2)(i+1)\, b_{i+2} + (\epsilon - 1 - 2i)\, b_i\right] y^i = 0 .$$

Now the different powers of y $(1, y, y^2, \ldots)$ are *linearly independent functions*, and therefore the equation is only satisfied if the coefficient of each and every power of y is equal to zero. This gives a *recurrence relation* for the b_i,

$$b_{i+2} = \frac{(2i + 1 - \epsilon)}{(i+1)(i+2)}\, b_i \,.$$

Thus (if we know ϵ) b_0 begets b_2, which in turn begets b_4, etc. And b_1 begets b_3, which in turn begets b_5, etc. Of all the b_i, only b_0 and b_1 are independent.

(e) Normalizability—As $y \to \pm\infty$, the highest (that is, largest-i) powers of y in the series for $h(y)$ will dominate. Suppose the series does not terminate; that is, suppose there is no upper limit on i. The recurrence relation shows that for large i the ratio of the coefficients b_{i+2} and b_i approaches $2/i$. Consider now the expansion of e^{y^2},

$$e^{y^2} = 1 + \frac{y^2}{1!} + \frac{y^4}{2!} + \cdots + \frac{y^j}{(j/2)!} + \frac{y^{j+2}}{(j/2+1)!} + \cdots,$$

where j is even. The ratio of coefficients of y^{j+2} and y^j is $1/(j/2 + 1) \approx 2/j$. Therefore, if the series for $h(y)$ does not terminate, $h(y)$ will behave like e^{y^2} at large y. But then the series would overpower the factor $e^{-y^2/2}$ we took out of $\psi(y)$ in writing $\psi(y) = h(y)\, e^{-y^2/2}$. This is physically unacceptable because the bound states of a particle have to be normalizable (the single particle is somewhere). Thus the series for $h(y)$ can only have a finite number of terms. At large enough y, $e^{-y^2/2}$ will kill any *finite* series.

The numerator $(2i + 1 - \epsilon)$ in the recurrence relation for the b_i suggests the answer. Let $\epsilon = 2n + 1$, where n is a nonnegative integer. Then when $i = 0, 1, 2, \ldots$, reaches n, the numerator will equal zero and so will the coefficients b_{n+2}, b_{n+4}, etc. However, the choice $\epsilon = 2n + 1$ can only kill (when i reaches n) *even* powers of y if n is even, or only *odd* powers of y if n is odd. Therefore, either the odd series (starting with b_1) or the even series (starting with b_0) has to be killed at its start, and there are two ways to get physically acceptable solutions:

(1) Set $\epsilon = 2n + 1$ with n any *even* nonnegative integer, $n = 0, 2, 4, \ldots$, and set $b_1 = 0$. Then $h(y)$ and $\psi(y)$ will be *even* functions of y (have even parity).

(2) Set $\epsilon = 2n + 1$ with n any *odd* positive integer, $n = 1, 3, 5, \ldots$, and set $b_0 = 0$. Then $h(y)$ and $\psi(y)$ will be *odd* functions of y (have odd parity).

(f) Eigenvalues and eigenfunctions—Thus, once again, the eigenfunctions in a symmetric $V(x)$ alternate with even and odd parity as the energy increases. The n-even and n-odd sets of solutions together give all the eigenvalues,

$$E_n = \epsilon_n \frac{\hbar\omega}{2} = (2n + 1)\frac{\hbar\omega}{2} = \left(n + \frac{1}{2}\right)\hbar\omega \,,$$

where $n = 0, 1, 2, \ldots$ And the recurrence relation for the b_i gives the polynomials $h_n(y)$. For $b_1 = 0$ and $n = 0$, so that $\epsilon_0 = 1$, we get

$$b_2 = \frac{(2 \times 0 + 1 - 1)}{1 \times 2}\, b_0 = 0 \,,$$

117

and $h_0(y) = b_0$, as seen by inspection earlier. For $b_0 = 0$ and $n = 1$, so that $\epsilon_1 = 3$, we get

$$b_3 = \frac{(2 \times 1 + 1 - 3)}{2 \times 3} b_1 = 0 \ ,$$

and $h_1(y) = b_1 y$, as seen by inspection earlier. For $b_1 = 0$ and $n = 2$, so that $\epsilon_2 = 5$, we get

$$b_2 = \frac{(2 \times 0 + 1 - 5)}{1 \times 2} b_0 = -2b_0, \qquad b_4 = \frac{(2 \times 2 + 1 - 5)}{3 \times 4} b_2 = 0 \ ,$$

and $h_2(y) = b_0(1 - 2y^2)$. And so on for larger n.

Putting $\epsilon = 2n + 1$ into the differential equation $h'' - 2yh' + (\epsilon - 1)h = 0$ for $h(y)$, we get

$$h'' - 2yh' + 2nh = 0 \ ,$$

where $n = 0, 1, 2, \ \ldots$ This is the Hermite equation (the Hermite of Hermitian operators). The *Hermite polynomials* $H_n(y)$ are defined (by convention) so that the highest power, y^n, is multiplied by 2^n and is positive. The first few are

$$H_0(y) = 1 \qquad\qquad H_1(y) = 2y$$
$$H_2(y) = 4y^2 - 2 \qquad\qquad H_3(y) = 8y^3 - 12y$$
$$H_4(y) = 16y^4 - 48y^2 + 12 \qquad H_5(y) = 32y^5 - 160y^3 + 120y \ .$$

As with other special functions and polynomials, there are many relations among the Hermite polynomials. They may be found in any good mathematical reference.

It is simpler to calculate the normalization constants using the methods of the next section. The normalized eigenfunctions are

$$\boxed{\ \psi_n(y) = \frac{1}{\sqrt{2^n n! \sqrt{\pi}}} H_n(y) \, e^{-y^2/2} = \frac{1}{\sqrt{2^n n!}} H_n(y) \, \psi_0(y) \ . \ }$$

In terms of x, the normalized eigenfunctions are

$$\psi_n(x) = \frac{1}{\sqrt{2^n n!}} \frac{1}{(\pi \sigma_0^2)^{1/4}} H_n(x/\sigma_0) \, e^{-x^2/2\sigma_0^2} = \frac{1}{\sqrt{2^n n!}} H_n(x/\sigma_0) \, \psi_0(x) \ ,$$

where (again) $\sigma_0 \equiv \sqrt{\hbar/m\omega}$ is the turning-point distance for $E_0 = \frac{1}{2}\hbar\omega$. Except for those normalization constants, the solution is complete.

Figure 3(a) shows the wave functions $\psi_n(y)$ for $n = 0$ through 5, and Fig. 3(b) shows the corresponding probability density distributions $|\psi_n(y)|^2$. Figure 3(c) shows the probability density for $n = 10$. For all $n > 0$, the Hermite polynomials grow without limit as $y \to \pm\infty$, but they are eventually killed off by the exponential factor $e^{-y^2/2}$, which is common to all the wave functions.

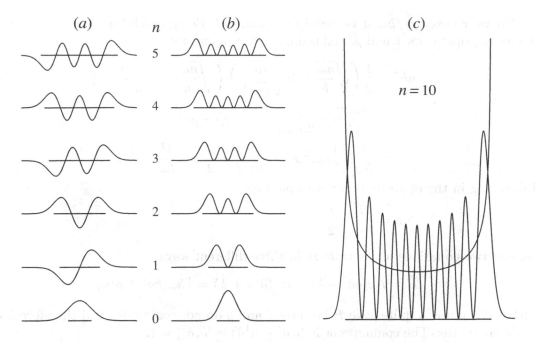

Figure 3. (a) The oscillator wave functions $\psi_n(y)$ versus y for $n = 0$ through 5. (b) The corresponding probability densities $|\psi_n(y)|^2$. The bars on the y axes indicate the classically allowed regions. (c) The classical and quantum probability densities for $n = 10$, not to the same scale as in (b).

6·3. THE OPERATOR SOLUTION

(a) The tools—We solve the oscillator all over again using an entirely different method, one to be used again for angular momentum. Everything now will come from the commutation relation $[\hat{x}, \hat{p}] = i\hbar$ and the fact that the oscillator Hamiltonian is quadratic in the operators $\hat{x}$ and $\hat{p}$. The *algebraic* quadratic $a^2 + b^2$ may be factored $(a + ib)(a - ib)$, and we try to factor $\hat{H}$ in the same way. First, we pull out an $\hbar\omega$ out of $\hat{H}$ and put the $\hat{x}^2$ term first:

$$\hat{H} = \frac{\hat{p}^2}{2m} + \frac{1}{2}m\omega^2\hat{x}^2 = \hbar\omega\left(\frac{m\omega\hat{x}^2}{2\hbar} + \frac{\hat{p}^2}{2m\omega\hbar}\right).$$

Then the $(a \pm ib)$-like factors of the expression in the parentheses are

$$\hat{a} \equiv \frac{1}{\sqrt{2}}\left(\sqrt{\frac{m\omega}{\hbar}}\,\hat{x} + \frac{i\hat{p}}{\sqrt{m\omega\hbar}}\right) \quad \text{and} \quad \hat{a}^\dagger \equiv \frac{1}{\sqrt{2}}\left(\sqrt{\frac{m\omega}{\hbar}}\,\hat{x} - \frac{i\hat{p}}{\sqrt{m\omega\hbar}}\right).$$

These new operators are linear operators, because they are linear combinations of the linear operators $\hat{x}$ and $\hat{p}$. They are not self adjoint (because $i^\dagger = -i$), but they are adjoints of one another. In terms of the dimensionless length variable $y \equiv x/\sqrt{\hbar/m\omega}$ introduced in Sec. 2(a), they are (with $\hat{p} = -i\hbar\,d/dx$)

$$\hat{a} = \frac{1}{\sqrt{2}}\left(y + \frac{d}{dy}\right) \quad \text{and} \quad \hat{a}^\dagger = \frac{1}{\sqrt{2}}\left(y - \frac{d}{dy}\right).$$

In Sec. 5·3(b), we showed that $(d/dy)^\dagger = -d/dy$. We shall need these reduced forms later.

119

Do we recover $\hat{H}/\hbar\omega$ if we multiply $\hat{a}$ and $\hat{a}^\dagger$? Being careful of the order of the non-commuting operators $\hat{x}$ and $\hat{p}$, and using $[\hat{x}, \hat{p}] = i\hbar$, we get

$$\hat{a}\hat{a}^\dagger = \frac{1}{2}\left(\sqrt{\frac{m\omega}{\hbar}}\,\hat{x} + \frac{i\hat{p}}{\sqrt{m\omega\hbar}}\right)\left(\sqrt{\frac{m\omega}{\hbar}}\,\hat{x} - \frac{i\hat{p}}{\sqrt{m\omega\hbar}}\right)$$

$$= \frac{m\omega\,\hat{x}^2}{2\hbar} + \frac{\hat{p}^2}{2m\omega\hbar} - \frac{i}{2\hbar}(\hat{x}\hat{p} - \hat{p}\hat{x})$$

$$= \frac{1}{\hbar\omega}\left(\frac{1}{2}m\omega^2\,\hat{x}^2 + \frac{\hat{p}^2}{2m}\right) + \frac{1}{2} = \frac{\hat{H}}{\hbar\omega} + \frac{1}{2}\;.$$

Multiplying in the opposite order, we would get

$$\hat{a}^\dagger\hat{a} = \frac{\hat{H}}{\hbar\omega} - \frac{1}{2}\;.$$

The last two equations let us write $\hat{H}$ in three different ways,

$$\hat{H} = \hbar\omega(\hat{a}\hat{a}^\dagger - \tfrac{1}{2}) = \hbar\omega(\hat{a}^\dagger\hat{a} + \tfrac{1}{2}) = \tfrac{1}{2}\hbar\omega(\hat{a}\hat{a}^\dagger + \hat{a}^\dagger\hat{a})\;.$$

Neither $\hat{a}\hat{a}^\dagger$ nor $\hat{a}^\dagger\hat{a}$ quite gave $\hat{H}$ because $\hat{x}$ and $\hat{p}$ do not commute. And therefore $\hat{a}$ and $\hat{a}^\dagger$ do not commute: The commutator is $(\hat{a}\hat{a}^\dagger - \hat{a}^\dagger\hat{a}) = [\hat{a}, \hat{a}^\dagger] = 1$.

We also need the commutators of $\hat{H}$ with $\hat{a}$ and $\hat{a}^\dagger$. Using $\hat{H} = \hbar\omega(\hat{a}^\dagger\hat{a} + \tfrac{1}{2})$, we get

$$[\hat{H}, \hat{a}] = \hbar\omega(\hat{a}^\dagger\hat{a} + \tfrac{1}{2})\hat{a} - \hat{a}(\hat{a}^\dagger\hat{a} + \tfrac{1}{2})\hbar\omega = \hbar\omega(\hat{a}^\dagger\hat{a} - \hat{a}\hat{a}^\dagger)\hat{a}$$

$$= -\hbar\omega\,\hat{a}\;,$$

since $[\hat{a}, \hat{a}^\dagger] = 1$. In the same way, we would get $[\hat{H}, \hat{a}^\dagger] = +\hbar\omega\,\hat{a}^\dagger$. We now have our tools, and put them in a box,

$$\boxed{\begin{array}{c} \hat{H} = \hbar\omega(\hat{a}\hat{a}^\dagger - \tfrac{1}{2}) = \hbar\omega(\hat{a}^\dagger\hat{a} + \tfrac{1}{2}) = \tfrac{1}{2}\hbar\omega(\hat{a}\hat{a}^\dagger + \hat{a}^\dagger\hat{a}) \\[2mm] [\hat{H}, \hat{a}^\dagger] = +\hbar\omega\hat{a}^\dagger, \qquad [\hat{H}, \hat{a}] = -\hbar\omega\hat{a}, \qquad [\hat{a}, \hat{a}^\dagger] = 1\;. \end{array}}$$

The hard work is done.

(b) The eigenvalues—Forget that we already know the eigenvalues. We assume the existence of a normalized eigenstate $|E\rangle$ of $\hat{H}$, labeled by its unknown energy E,

$$\hat{H}|E\rangle = E|E\rangle, \quad \text{where} \quad \langle E|E\rangle = 1\;.$$

Consider the state $\hat{a}^\dagger|E\rangle$. Is it an eigenstate of $\hat{H}$? Using the tool $[\hat{H}, \hat{a}^\dagger] = +\hbar\omega\hat{a}^\dagger$, we get

$$\hat{H}(\hat{a}^\dagger|E\rangle) = \hat{a}^\dagger\hat{H}|E\rangle + \hbar\omega\,\hat{a}^\dagger|E\rangle = (E + \hbar\omega)\,(\hat{a}^\dagger|E\rangle)\;.$$

We used the commutation relation to get $\hat{H}$ next to $|E\rangle$, where we know what it does. This is one of the main uses of commutation relations: to exchange an order of operations. The equation says that $\hat{a}^\dagger|E\rangle$ is indeed an eigenstate of $\hat{H}$, but that its eigenvalue is $E + \hbar\omega$, not E. That is,

$$\hat{a}^\dagger|E\rangle = C_E^+\,|E + \hbar\omega\rangle\;.$$

The constant C_E^+ is to allow for the possibility that $\hat{a}^\dagger|E\rangle$ is not the *normalized* state $|E+\hbar\omega\rangle$. We insist that our kets be normalized, that $\langle E+\hbar\omega|E+\hbar\omega\rangle = 1$. We shall find C_E^+ below.

In the same way, using $[\hat{H}, \hat{a}] = -\hbar\omega\hat{a}$, we would get

$$\hat{H}(\hat{a}|E\rangle) = (E - \hbar\omega)(\hat{a}\,|E\rangle) ,$$

so that $\hat{a}|E\rangle = C_E^-|E{-}\hbar\omega\rangle$. The operators $\hat{a}^\dagger$ and $\hat{a}$ are called raising and lowering operators because they raise or lower the energy.

In the above, $|E\rangle$ was *any* eigenstate of $\hat{H}$. Starting with any $|E\rangle$, we can get a new state with an energy higher or lower by $\hbar\omega$. In this way, we can generate a whole ladder of states, up and down, using the operators repeatedly, as in Fig. 4. (So $\hat{a}^\dagger$ and $\hat{a}$ are also called ladder operators.) But the ladder has to have a bottom rung, one with $E_0 > 0$. No state can have an energy below the minimum of $V(x)$, which for the oscillator is $V(0) = 0$. Let $|E_0\rangle$ be that lowest rung. The only way that $\hat{a}|E_0\rangle$ is *not* a new state with an even lower energy is for $\hat{a}|E_0\rangle$ to be zero; that is, $\hat{a}|E_0\rangle = 0$.

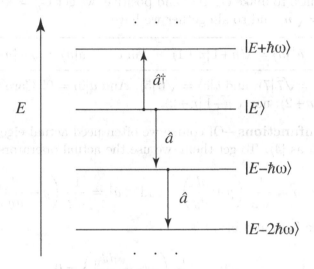

Figure 4. The ladder of states generated by the ladder operators $\hat{a}^\dagger$ and $\hat{a}$.

Now, again, $\hat{a}$ and $\hat{a}^\dagger$ are linear operators, and a *linear* operator operating on zero gives zero. Therefore,

$$\hat{a}^\dagger(\hat{a}|E_0\rangle) = 0 .$$

But $\hat{a}^\dagger\hat{a} = \hat{H}/\hbar\omega - 1/2$ (see the tool box), so that

$$\hat{a}^\dagger\hat{a}\,|E_0\rangle = \left(\frac{\hat{H}}{\hbar\omega} - \frac{1}{2}\right)|E_0\rangle = 0 ,$$

or (!)

$$\hat{H}|E_0\rangle = \tfrac{1}{2}\hbar\omega\,|E_0\rangle .$$

The energy of the lowest state is $E_0 = \tfrac{1}{2}\hbar\omega$! (Right here, I always think of the rabbit jumping out of the magician's hat.) And since the rungs of the ladder are separated by $\hbar\omega$, *all* the eigenvalues are given by $E_n = (n + \tfrac{1}{2})\hbar\omega$, where $n = 0, 1, 2, \ldots$ And now we label the eigenkets in the most economical way, with the quantum number: $|E_n\rangle \to |n\rangle$:

$$\boxed{\hat{H}|n\rangle = (n + \tfrac{1}{2})\hbar\omega\,|n\rangle .}$$

121

(c) The constants $C_n^\pm$—We relabel the constants $C_E^\pm$ as $C_n^\pm$. They are the constants such that

$$\hat{a}^\dagger|n\rangle = C_n^+ |n{+}1\rangle \qquad \text{and} \qquad \hat{a}|n\rangle = C_n^- |n{-}1\rangle \,,$$

where the kets $|n\rangle$, $|n{+}1\rangle$, and $|n{-}1\rangle$ are all normalized. We have

$$\langle n|\hat{a}\hat{a}^\dagger|n\rangle = \langle \psi_n|\hat{a}\hat{a}^\dagger|\psi_n\rangle = \langle \hat{a}^\dagger\psi_n|\hat{a}^\dagger\psi_n\rangle = |C_n^+|^2 \langle n{+}1|n{+}1\rangle = |C_n^+|^2 \,.$$

(For clarity, we again briefly revert from $|n\rangle$ to $|\psi_n\rangle$, as $|\hat{a}^\dagger n\rangle$ is poor notation.) Now we found earlier that $\hat{a}\hat{a}^\dagger = \hat{H}/\hbar\omega + 1/2$, and also that $\hat{H}|n\rangle = (n + \frac{1}{2})\hbar\omega|n\rangle$. Therefore,

$$|C_n^+|^2 = \langle n|\hat{a}\hat{a}^\dagger|n\rangle = \langle n|(\hat{H}/\hbar\omega + 1/2)|n\rangle = (n + 1)\langle n|n\rangle = n + 1 \,.$$

With a choice of phase to make C_n^+ real and positive, we get $C_n^+ = \sqrt{n+1}$. In the same way, we would get $C_n^- = \sqrt{n}$, and so altogether we have

$$\boxed{\hat{a}^\dagger|n\rangle = \sqrt{n+1}\,|n{+}1\rangle \qquad \text{and} \qquad \hat{a}|n\rangle = \sqrt{n}\,|n{-}1\rangle \,.}$$

For example, $\hat{a}^\dagger|6\rangle = \sqrt{7}\,|7\rangle$, and $\hat{a}|6\rangle = \sqrt{6}\,|5\rangle$. And $\hat{a}|0\rangle = 0$! Care is needed: For example, $\hat{a}^\dagger|n{+}1\rangle$ is $\sqrt{n+2}\,|n{+}2\rangle$, not $\sqrt{n+1}\,|n{+}2\rangle$.

(d) The eigenfunctions—Of course we often need actual eigenfunctions, not just abstract symbols such as $|4\rangle$. To get them, we use the actual operators we started this section with,

$$\hat{a} = \frac{1}{\sqrt{2}} \left(y + \frac{d}{dy} \right) \qquad \text{and} \qquad \hat{a}^\dagger = \frac{1}{\sqrt{2}} \left(y - \frac{d}{dy} \right) \,.$$

Then $\hat{a}|0\rangle = 0$ made flesh and blood is

$$\hat{a}|0\rangle = \frac{1}{\sqrt{2}} \left(y\psi_0 + \frac{d\psi_0}{dy} \right) = 0 \,.$$

But the function $\psi_0(y)$ whose derivative is $-y$ times itself is $e^{-y^2/2}$, which normalized is $\psi_0(y) = \pi^{-1/4}e^{-y^2/2}$. (We proved that $\int e^{-y^2}\,dy = \sqrt{\pi}$ in Sec. 2·8.)

To get higher states, we use $\hat{a}^\dagger|n\rangle = \sqrt{n+1}\,|n{+}1\rangle$ repeatedly. Thus $\hat{a}^\dagger|0\rangle = \sqrt{1}\,|1\rangle$ means

$$\psi_1(y) = \frac{1}{\sqrt{1}}\frac{1}{\sqrt{2}} \left(y - \frac{d}{dy} \right) \psi_0(y) = \frac{1}{\sqrt{2\sqrt{\pi}}} 2y\,e^{-y^2/2} \,,$$

and so on. The eigenfunctions so generated are *already normalized*, because we started with $|0\rangle$ normalized and the kets in $\hat{a}^\dagger|n\rangle = \sqrt{n+1}\,|n{+}1\rangle$ are normalized. (That was the point of introducing the $C_n^\pm$.) The general solution is

$$\boxed{\psi_n(y) = \frac{1}{\sqrt{2^n n!}\sqrt{\pi}} H_n(y)\,e^{-y^2/2} = \frac{1}{\sqrt{2^n n!}} H_n(y)\,\psi_0(y) \,.}$$

The operator $(y - d/dy)$ used repeatedly on $e^{-y^2/2}$ "generates" the Hermite polynomials: $(y - d/dy)^n e^{-y^2/2} = H_n(y)\,e^{-y^2/2}$. In the normalization constants here, where does the $\sqrt{2^n}$ come from? Where does the $\sqrt{n!}$ come from?

6·4. STATES AS VECTORS, OPERATORS AS MATRICES

(a) States as vectors—The energy eigenstates $|n\rangle$, the solutions of $\hat{H}|n\rangle = E_n|n\rangle$, form an infinite orthonormal set that we can represent with infinite column vectors,

$$|0\rangle = \begin{pmatrix} 1 \\ 0 \\ 0 \\ \vdots \end{pmatrix}, \quad |1\rangle = \begin{pmatrix} 0 \\ 1 \\ 0 \\ \vdots \end{pmatrix}, \quad |2\rangle = \begin{pmatrix} 0 \\ 0 \\ 1 \\ \vdots \end{pmatrix}, \quad \cdots$$

The corresponding bra vectors are $\langle 0| = (1, 0, 0, \ldots)$, $\langle 1| = (0, 1, 0, \ldots)$, ... These vectors obviously satisfy $\langle m|n\rangle = \delta_{mn}$. We can use them as a basis for expanding any state that is square-integrable over $-\infty < x < +\infty$:

$$|\psi\rangle = |\Psi(t=0)\rangle = \sum_{n=0}^{\infty} c_n|n\rangle = \begin{pmatrix} c_0 \\ c_1 \\ c_2 \\ \vdots \end{pmatrix}, \quad |\Psi(t)\rangle = \sum_{n=0}^{\infty} c_n e^{-iE_n t/\hbar}|n\rangle = \begin{pmatrix} c_0\, e^{-iE_0 t/\hbar} \\ c_1\, e^{-iE_1 t/\hbar} \\ c_2\, e^{-iE_2 t/\hbar} \\ \vdots \end{pmatrix}.$$

The components $c_n = c_n(0)$ of $|\psi\rangle$ with respect to the basis states $|n\rangle$ are the scalar products $c_n = \langle n|\psi\rangle$. For example,

$$c_5 = \langle 5|\psi\rangle = \int_{-\infty}^{+\infty} \psi_5^*(x)\,\psi(x)\,dx \ .$$

Were we to measure the energy of the oscillator, $|c_5|^2$ is the probability we would get E_5.

We might have to do some integrals to get the c_n, but once we have got them we can do further calculations using matrix methods. For example, $|\psi\rangle$ is normalized if

$$\int_{-\infty}^{+\infty} |\psi(x)|^2\,dx = \langle\psi|\psi\rangle = \begin{pmatrix} c_0^* & c_1^* & c_2^* & \cdots \end{pmatrix} \begin{pmatrix} c_0 \\ c_1 \\ c_2 \\ \vdots \end{pmatrix} = \sum_{n=0}^{\infty} |c_n|^2 = 1 \ .$$

The scalar product of two functions $|\psi\rangle = \sum c_n|n\rangle$ and $|\phi\rangle = \sum d_n|n\rangle$ is $\langle\psi|\phi\rangle = \sum c_n^* d_n$.

(b) Operators as matrices—In Sec. 5·5, we showed that the matrix representation for an operator $\hat{A}$ in an orthonormal basis $|n\rangle$ is found as follows:

(1) Find the vectors $\hat{A}|n\rangle$ for all n.

(2) Stack them up in order side-by-side like the pickets of a picket fence.

Thus for the oscillator, where $\hat{H}|n\rangle = (n + \tfrac{1}{2})\hbar\omega|n\rangle$, the matrix representation of $\hat{H}$ is diagonal:

$$\hat{H} = \frac{\hbar\omega}{2} \begin{pmatrix} 1 & 0 & 0 & \cdots \\ 0 & 3 & 0 & \cdots \\ 0 & 0 & 5 & \cdots \\ \vdots & \vdots & \vdots & \ddots \end{pmatrix}.$$

Of course $\hat{H}$ is diagonal, because we are using the eigenstates of $\hat{H}$ as the basis vectors.

Using $\hat{a}|n\rangle = \sqrt{n}\,|n-1\rangle$, we get $\hat{a}\,|0\rangle = 0$, $\hat{a}\,|1\rangle = \sqrt{1}\,|0\rangle$, $\hat{a}\,|2\rangle = \sqrt{2}\,|1\rangle$, etc., and the picket-fence recipe for $\hat{a}$ is

$$\hat{a} = \begin{pmatrix} 0 & \sqrt{1} & 0 & 0 & \cdots \\ 0 & 0 & \sqrt{2} & 0 & \cdots \\ 0 & 0 & 0 & \sqrt{3} & \cdots \\ \vdots & \vdots & \vdots & \vdots & \ddots \end{pmatrix}.$$

The first column is the *zero* vector, not to be confused with the $n = 0$ state $|0\rangle$, which has a one at the top. And using $\hat{a}^{\dagger}|n\rangle = \sqrt{n+1}\,|n+1\rangle$, we get $\hat{a}^{\dagger}|0\rangle = \sqrt{1}\,|1\rangle$, $\hat{a}^{\dagger}|1\rangle = \sqrt{2}\,|2\rangle$, $\hat{a}^{\dagger}|2\rangle = \sqrt{3}\,|3\rangle$, etc., and the matrix for $\hat{a}^{\dagger}$ is

$$\hat{a}^{\dagger} = \begin{pmatrix} 0 & 0 & 0 & 0 & \cdots \\ \sqrt{1} & 0 & 0 & 0 & \cdots \\ 0 & \sqrt{2} & 0 & 0 & \cdots \\ 0 & 0 & \sqrt{3} & 0 & \cdots \\ \vdots & \vdots & \vdots & \vdots & \ddots \end{pmatrix}.$$

The matrices for $\hat{a}$ and $\hat{a}^{\dagger}$ are not self adjoint but are adjoints of one another.

The operators $\hat{a}$ and $\hat{a}^{\dagger}$ were defined in Sec. 3(a) in terms of $\hat{x}$ and $\hat{p}$ as

$$\hat{a} \equiv \frac{1}{\sqrt{2}}\left(\sqrt{\frac{m\omega}{\hbar}}\,\hat{x} + \frac{i\hat{p}}{\sqrt{m\omega\hbar}}\right), \qquad \hat{a}^{\dagger} \equiv \frac{1}{\sqrt{2}}\left(\sqrt{\frac{m\omega}{\hbar}}\,\hat{x} - \frac{i\hat{p}}{\sqrt{m\omega\hbar}}\right).$$

Adding and subtracting these equations, we solve for $\hat{x}$ and $\hat{p}$:

$$\hat{x} = \sqrt{\frac{\hbar}{2m\omega}}\,(\hat{a}+\hat{a}^{\dagger}) = \sqrt{\frac{\hbar}{2m\omega}} \begin{pmatrix} 0 & \sqrt{1} & 0 & 0 & \cdots \\ \sqrt{1} & 0 & \sqrt{2} & 0 & \cdots \\ 0 & \sqrt{2} & 0 & \sqrt{3} & \cdots \\ 0 & 0 & \sqrt{3} & 0 & \cdots \\ \vdots & \vdots & \vdots & \vdots & \ddots \end{pmatrix}$$

and

$$\hat{p} = i\sqrt{\frac{m\omega\hbar}{2}}\,(\hat{a}^{\dagger}-\hat{a}) = i\sqrt{\frac{m\omega\hbar}{2}} \begin{pmatrix} 0 & -\sqrt{1} & 0 & 0 & \cdots \\ \sqrt{1} & 0 & -\sqrt{2} & 0 & \cdots \\ 0 & \sqrt{2} & 0 & -\sqrt{3} & \cdots \\ 0 & 0 & \sqrt{3} & 0 & \cdots \\ \vdots & \vdots & \vdots & \vdots & \ddots \end{pmatrix}.$$

These matrices are self adjoint (note the i in $\hat{p}$), because $\hat{x}$ and $\hat{p}$ are Hermitian operators. They satisfy all the operator relations, such as $[\hat{x},\hat{p}] = i\hbar\hat{I}$, $\hat{H} = \hat{p}^2/2m + \frac{1}{2}m\omega^2\hat{x}^2$, etc.

The diagonal elements $\langle n|\hat{x}|n\rangle$ and $\langle n|\hat{p}|n\rangle$ of the $\hat{x}$ and $\hat{p}$ matrices are zero. From the functional point of view, these expectation values are zero because the integrands $x|\psi_n|^2$ and $-i\hbar\psi_n^*\,d\psi_n/dx$ are odd functions of x, to be integrated over even intervals of x.

(c) Matrix elements—Sometimes we just need a few matrix elements—maybe, say, just the diagonal elements of the matrix for $\hat{x}^2$. Squaring $\hat{x} = \sqrt{\hbar/2m\omega}\,(\hat{a} + \hat{a}^\dagger)$, and being careful to preserve the order of the noncommuting operators $\hat{a}$ and $\hat{a}^\dagger$ in expanding $(\hat{a}+\hat{a}^\dagger)^2$, we get

$$\langle n|\hat{x}^2|n\rangle = \frac{\hbar}{2m\omega}\langle n|(\hat{a}+\hat{a}^\dagger)^2|n\rangle = \frac{\hbar}{2m\omega}\langle n|(\hat{a}^2 + \hat{a}\hat{a}^\dagger + \hat{a}^\dagger\hat{a} + \hat{a}^{\dagger 2})|n\rangle .$$

The $\langle n|\hat{a}^2|n\rangle$ and $\langle n|\hat{a}^{\dagger 2}|n\rangle$ pieces are zero (why?). One of the operator relations in the tool box in Sec. 3(a) is

$$\hat{H} = \tfrac{1}{2}\hbar\omega(\hat{a}\hat{a}^\dagger + \hat{a}^\dagger\hat{a}) .$$

Then, since $\hat{H}|n\rangle = (n + \tfrac{1}{2})\hbar\omega\,|n\rangle$, we get

$$\langle n|\hat{x}^2|n\rangle = \frac{\hbar}{2m\omega}\langle n|(\hat{a}\hat{a}^\dagger + \hat{a}^\dagger\hat{a})|n\rangle = \frac{\hbar}{2m\omega}\langle n|(2n + 1)|n\rangle = \left(n + \frac{1}{2}\right)\frac{\hbar}{m\omega} .$$

From this, we get $\langle n|\hat{V}|n\rangle = \tfrac{1}{2}m\omega^2\langle n|\hat{x}^2|n\rangle = \tfrac{1}{2}E_n$; and then $\langle n|\hat{K}|n\rangle = E_n - \langle n|\hat{V}|n\rangle = \tfrac{1}{2}E_n$. In the oscillator energy eigenstates, the expectation values of the kinetic and potential energies are equal. We found this same result in Sec. 5·7(c) using the virial theorem.

Much of the work of perturbation theory, the subject of Chaps. 13 and 15, is calculating matrix elements. However, we already have the the full array of matrix elements for $\hat{x}$ and for $\hat{p}$—for the *harmonic oscillator*. The matrix representations above are *only* for the oscillator. Matrix elements for a different physical system will be different. We shall find, however, that the matrix representations of the operators for angular momenta are similar (Chap. 8). There, depending on the magnitude of the angular momentum, the matrices will be 2-by-2, or 3-by-3, or 4-by-4, ..., not infinite.

PROBLEMS

1. *The classical probability distribution.*

Show that the normalized position probability distribution $dP(x)$ between the turning points $\pm x_0$ of the classical oscillator is

$$dP(x) = \rho_{cl}(x)\,dx = \frac{dx}{\pi\sqrt{x_0^2 - x^2}} .$$

2. *The isotropic 3-dimensional oscillator.*

The potential energy for the 3-dimensional isotropic oscillator is

$$V(x, y, z) = \tfrac{1}{2}m\omega^2(x^2 + y^2 + z^2) .$$

Write down the energy eigenvalues E in terms of quantum numbers n_x, n_y, and n_z (see Sec. 2·7(a)). Show that the number of linearly independent eigenfunctions for a given $n \equiv n_x + n_y + n_z$ is $N = \tfrac{1}{2}(n+1)(n+2)$. This is the *degeneracy* of the level: $N = 1, 3, 6, 10, \ldots$ for $n = 0, 1, 2, 3, \ldots$ (Think of a clever way to count; N is the sum of the integers 1 through $n + 1$.)

3. *The oscillator in an electric field.*

Show that the addition of a term $-\gamma x$ to the oscillator potential energy $\frac{1}{2}m\omega^2 x^2$ results in a parabola having the same form as $\frac{1}{2}m\omega^2 x^2$ alone but with the minimum displaced sideways in x and down in energy. Then write down the new eigenvalues.

For example, an electric field $\mathcal{E}$ in the x direction acting on a particle with charge q would add a potential energy $-q\mathcal{E}x$ to the oscillator $\frac{1}{2}m\omega^2 x^2$.

4. *Leftover details.*

(a) Show that the highest-order term in the second derivative of $y^n e^{-y^2/2}$ is $y^{n+2}e^{-y^2/2}$.

(b) Prove that $\hat{H}(\hat{a}|E\rangle) = (E - \hbar\omega)(\hat{a}|E\rangle)$.

(c) Prove that $\hat{a}|n\rangle = \sqrt{n}\,|n-1\rangle$.

(d) We can shift an index of summation by any amount we want, provided we also shift the range of summation appropriately. Make the shift $i \to i + 5$ in

$$\sum_{i=0} i(i-1)\, b_i\, y^{i-2} \; ,$$

and show that the first three terms of the two sums are the same.

5. *Normalized eigenfunctions and Hermite polynomials.*

(a) Use $\hat{a}^\dagger|n\rangle = \sqrt{n+1}\,|n+1\rangle$ to get the normalized eigenfunctions $\psi_n(y)$ through $n = 4$.

(b) Use the relations for y and d/dy in terms of $\hat{a}$ and $\hat{a}^\dagger$ to find these "recurrence relations" for the Hermite polynomials:

$$H_{n+1}(y) = 2yH_n(y) - 2nH_{n-1}(y) \qquad \text{and} \qquad \frac{dH_n(y)}{dy} = 2nH_{n-1}(y) \; .$$

6. *An uncertainty relation.*

In Sec. 4, we found the expectation values of $\hat{x}$, $\hat{p}$, and $\hat{x}^2$ for any oscillator energy eigenstate $|n\rangle$. Find the values of $\langle n|\hat{p}^2|n\rangle$, and then write down $(\Delta x)^2 \equiv \langle(x - \langle x\rangle)^2\rangle$ and $(\Delta p)^2 \equiv \langle(p - \langle p\rangle)^2\rangle$. Here Δx and Δp are the root-mean-square uncertainties in x and p in these states (see Chap. 7). Show that $\Delta x\,\Delta p \geq \hbar/2$ for any oscillator energy eigenstate. Does the equal sign ever apply?

7. *A matrix element.*

Show that $\langle n|\hat{x}^4|n\rangle = (6n^2 + 6n + 3)(\hbar/2m\omega)^2$.

8. $[\hat{x}, \hat{p}] = i\hbar\hat{I}$ *with matrices.*

By multiplying the matrices, show that the upper left 4-by-4 corner of $[\hat{x}, \hat{p}]$ is $i\hbar\hat{I}$. (You need to start with a little more than the 4-by-4 corners of $\hat{x}$ and $\hat{p}$.)

9. *The Morse potential.*

The figure below shows the Morse potential energy (P.M. Morse, 1929),

$$V(r) = V_0\left[e^{-2(r-r_0)/b} - 2e^{-(r-r_0)/b}\right] \; .$$

(The figure is drawn with $b = r_0/5$.) This function is used to model the vibrational states of diatomic molecules, such as CO or HCl. The Schrödinger equation for $V(r)$ can be solved exactly (but not easily). The bound-state eigenvalues are

$$E_n = -V_0 + (n+1/2)\hbar\omega - \frac{[(n+1/2)\hbar\omega]^2}{4V_0} \ ,$$

where $\omega \equiv \sqrt{2V_0/\mu b^2}$, $\mu = m_1 m_2/(m_1 + m_2)$ is the reduced mass, and $n = 0, 1, 2, \ldots$ As a function of $(n + \frac{1}{2})\hbar\omega$, E_n rises from $-V_0$ to a maximum of zero and then falls toward minus infinity. The only physically meaningful values of n are those before the maximum is reached.

(a) Verify the main features of the figure. In particular, show that $V(+\infty)$ is zero; that the minimum value of $V(r)$ is $-V_0$ at $r = r_0$; and that if $b \ll r_0$ then $V(0) \gg V_0$. This last is important because the model $V(r)$ is not infinite at $r = 0$, whereas physically the atoms cannot sit on top of one another.

(b) Show that d^2V/dr^2 at r_0 is $\mu\omega^2$. Thus the oscillator approximation to the Morse potential gives the first two terms of the exact expression for E_n.

(c) Half way up from the bottom of the well, show that the classical turning points are at $r = r_0 - 0.535b$ and $r_0 + 1.228b$. What value of b would make the distance between the turning points at this energy be $r_0/10$? And what then is V at $r = 0$?

(d) How does the number of bound states depend on the parameters V_0, b, and μ?

(e) The levels of the Morse potential get closer together as n increases; the levels of the oscillator approximation do not. Suppose the parameters of $V(r)$ are such that the number of bound states is large. Show that this number is twice the number of negative-energy states for the oscillator approximation to $V(r)$.

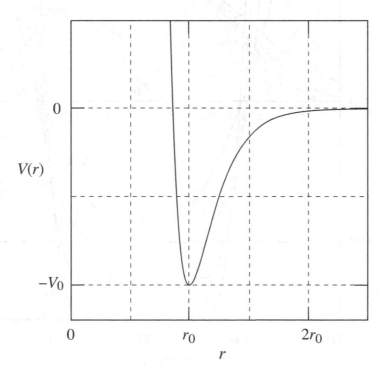

10. *A family of states and potential energies.*

(a) Consider an (unnormalized) wave function of the form $\psi(y) = e^{-f(y)}$. Show that

$$-\psi'' = \left[f'' - (f')^2\right]\psi .$$

(b) Let $f(y) = \frac{1}{2}y^2 + \lambda y^4$. For $\lambda = 0$, $\psi_\lambda(y)$ is the (unnormalized) oscillator ground state $\psi_0(y) = e^{-y^2/2}$. And $\psi_\lambda(y)$ will be normalizable for $\lambda > 0$ (why?). Calculate ψ'' and arrange the result in the form of a Schrödinger equation for $\psi_\lambda(y)$,

$$-\psi_\lambda'' + \left[(1 - 12\lambda)y^2 + 8\lambda y^4 + 16\lambda^2 y^6\right]\psi_\lambda = 1\,\psi_\lambda .$$

The function in the square brackets is a continuous infinity of potential energies $V_\lambda(y)$, all of which have $V_\lambda(0) = 0$ and a ground-state energy of $\epsilon_0 = 1$. The figure shows the functions $V_\lambda(y)$ and $\psi_\lambda(y)$ for $\lambda = 0.0$, 0.1, 0.2, and 0.3. The (unnormalized) wave functions are measured from the $\epsilon_0 = 1$ line.

This is from J.P. Killingbeck, *Microcomputer Quantum Mechanics*, p. 98. See also Probs. 2·14 and 9·7, which also reverse matters by starting with a wave function and getting $V(x)$ and the lowest eigenvalue.

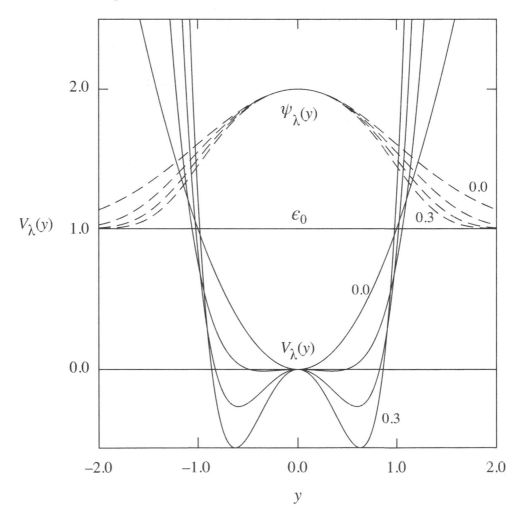

11. *Two coupled oscillators.*

This problem is from parts of D. Saxon, *Elementary Quantum Mechanics*, Chap. VIII, Secs. 2, 4, and 7.

(a) The kinetic-energy terms of a Hamiltonian for two oscillators, each of mass m, is

$$-\frac{\hbar^2}{2m}\left(\frac{\partial^2\psi}{\partial x_1^2} + \frac{\partial^2\psi}{\partial x_2^2}\right),$$

where $\psi = \psi(x_1, x_2)$. Show that the coordinate transformations $X \equiv \frac{1}{2}(x_1 + x_2)$ and $x \equiv x_1 - x_2$ transforms those terms to

$$-\frac{\hbar^2}{2M}\frac{\partial^2\psi}{\partial X^2} - \frac{\hbar^2}{2\mu}\frac{\partial^2\psi}{\partial x^2},$$

where $M = 2m$, the total mass, and $\mu = \frac{1}{2}m$, the reduced mass. And now $\psi = \psi(X, x)$. The rule for changing the variables here (the chain rule) is

$$\frac{\partial\psi}{\partial x_1} = \frac{\partial X}{\partial x_1}\frac{\partial\psi}{\partial X} + \frac{\partial x}{\partial x_1}\frac{\partial\psi}{\partial x}, \quad \text{etc.}$$

(b) Now suppose that the total potential-energy is

$$V(x_1, x_2) = \frac{1}{2}m\omega^2 x_1^2 + \frac{1}{2}m\omega^2 x_2^2 + \frac{1}{2}k(x_1 - x_2)^2.$$

Without the term that couples the oscillators, there would just be two independent oscillators. Make the same $(x_1, x_2) \to (X, x)$ and $(m_1, m_2) \to (M, m)$ transforms as in (a), and show that here too there is separation.

(c) Thus without further calculation, write down the energy levels:

$$E_{N,n} = (N + \tfrac{1}{2})\hbar\omega + (n + \tfrac{1}{2})\hbar\omega',$$

where $\omega'^2 = \omega^2 + k/\mu$.

(d) Draw the energy-level diagram for the lowest six levels. Assume that k/μ is small compared to ω^2 (the coupling is weak) and is positive. Label the levels with the values of the quantum numbers (N, n).

12. *A different coupling.*

(a) Suppose that the coupling term in the previous problem is instead $c\,x_1 x_2$. Show that now

$$E_{N,n} = (N + \tfrac{1}{2})\hbar\omega_+ + (n + \tfrac{1}{2})\hbar\omega_-,$$

where $\omega_\pm = \omega^2 \pm c/m$.

(b) Draw the energy-level diagram for the lowest six levels, and label the levels with the values of (N, n). Assume that c/m is small compared to ω^2 and is positive.

7. UNCERTAINTY RELATIONS. SIMULTANEOUS EIGENSTATES

7.1. *Heisenberg Uncertainty Relations*

7.2. *The Schwarz Inequality*

7.3. *Proof of Uncertainty Relations*

7.4. *Fourier Transforms. Momentum Space*

7.5. *Time and Energy*

7.6. *When Operators Commute*

Problems

Most of this chapter is about whether the values of two physical properties of a particle (observables) can be precisely known at the same instant. That depends on whether the corresponding operators commute or not. When two operators do not commute, there is an "uncertainty relation," which sets a limit on how well the values can be simultaneously known. The famous example is $\Delta x \, \Delta p_x \geq \frac{1}{2}\hbar$, where x and p_x are the position and momentum of a particle along a rectangular axis, and Δx and Δp_x are the root-mean-square (standard-deviation) measures of the uncertainties in the values. Those uncertainties depend on the state of the system. For example, as n increases, the oscillator energy eigenfunctions $\psi_n(x)$ spread wider in x, and both Δx and Δp_x are larger.

The absence of exact knowledge of the state of a system is not somehow due to short-comings of our experimental arrangements. Werner Heisenberg put it perfectly: "...in the strong formulation of the causal law 'If we know exactly the present, we can predict the future' it is not the conclusion but rather the premise which is false. We *cannot* know, as a matter of principle, the present in all its details."

In Section 3, we derive the uncertainty relation for the observables A and B from the commutator $[\hat{A}, \hat{B}]$ of their operators. Then $\Delta x \, \Delta p_x \geq \frac{1}{2}\hbar$ follows at once from $[\hat{x}, \hat{p}_x] = i\hbar$. In the next chapter, we come to the more complicated $[\hat{L}_x, \hat{L}_y] = i\hat{L}_z$ (and cyclic permutations of x, y, and z), where $\hat{L}_x$, $\hat{L}_y$, and $\hat{L}_z$ represent the components of the orbital angular momentum $\mathbf{L}$ of a particle or system. A trivial case $\mathbf{L} = 0$ aside, we find that we can only know the exact value of one component of $\mathbf{L}$ at a time—there will always be some unknowability about which way $\mathbf{L}$ is pointing.

In the last section, we consider the case that two or more operators *do* all commute with one another. Then there do exist states that are simultaneously eigenstates of all the members of the set. In Chapter 8, on angular momentum, we find the simultaneous eigenstates of $\hat{L}_z$ and $\hat{L}^2 = L_x^2 + L_y^2 + L_z^2$. Then any state of orbital angular momentum is a superposition of those *basis* states: $\hat{L}_z$ and $\hat{L}^2$ are *compatible* operators and the set is *complete*. In Chapter 9, on the hydrogen atom, we find simultaneous eigenstates of $\hat{L}_z$, $\hat{L}^2$, and the Hamiltonian $\hat{H}$. Then any spatial state of hydrogen is a superposition of those *basis* states: The operators are *compatible* and the set is *complete*.

7·1. HEISENBERG UNCERTAINTY RELATIONS

(a) **Uncertainties**—The measure used in quantum mechanics for the spread of values around a mean is the root-mean-square (rms) deviation, also called the standard deviation—a measure used widely in many fields. First one calculates the mean, which for the physical quantity A in the quantum state Ψ is

$$\langle A \rangle \equiv \langle \Psi | \hat{A} | \Psi \rangle \, ,$$

where $\hat{A}$ is the operator for A. The integration is over the spatial coordinates, and the result $\langle A \rangle$ may still be a function of time. Then one calculates the mean of the squared deviation from the mean,

$$(\Delta A)^2 \equiv \langle \Psi | (\hat{A} - \langle A \rangle)^2 | \Psi \rangle \, .$$

The rms deviation is ΔA—the root of the mean of the square of the deviation. One can calculate mean values *after* measurements, with actual data: The mean grade on the exam was $\langle g \rangle = 73$ and the standard deviation was $\Delta g = 16$. Or one can calculate with an a priori probability distribution: The mean value of many measurements of the position x of a particle, each measurement being made on a system whose wave function is the *same* $\Psi(x, t)$ (*not* the wave functions *after* the measurements), is expected to be about

$$\langle x \rangle = \langle \Psi | \hat{x} | \Psi \rangle = \int_{-\infty}^{+\infty} \Psi^*(x, t) \, x \, \Psi(x, t) \, dx \, .$$

In an energy eigenstate, where factors $e^{iEt/\hbar}$ and $e^{-iEt/\hbar}$ multiply to one, $\langle x \rangle$ is a constant. The rms spread of positions for these measurements is expected to be about Δx, where

$$(\Delta x)^2 = \langle \Psi | (\hat{x} - \langle x \rangle)^2 | \Psi \rangle \, .$$

In quantum mechanics, $\langle A \rangle$ is the *expectation value*, and ΔA is the *uncertainty*.

We can write $(\Delta A)^2$ in another way:

$$\begin{aligned}
(\Delta A)^2 &= \langle \Psi | (\hat{A} - \langle A \rangle)^2 | \Psi \rangle \\
&= \langle \Psi | (\hat{A}^2 - 2\hat{A}\langle A \rangle + \langle A \rangle^2) | \Psi \rangle \\
&= \langle A^2 \rangle - 2\langle A \rangle^2 + \langle A \rangle^2 \\
&= \langle A^2 \rangle - \langle A \rangle^2 \, .
\end{aligned}$$

This way, we can calculate $\langle A^2 \rangle$ in parallel with $\langle A \rangle$, without having to first get $\langle A \rangle$. We could sum grades and squares of grades at the same time, and then $(\Delta g)^2$ follows at once. We could pull $\langle A \rangle$ out of the outer brackets, because as far as further sums or integrations are concerned $\langle A \rangle$ is already fixed: $\langle \Psi | \langle A \rangle | \Psi \rangle = \langle A \rangle$.

In Sec. 3, we prove the following theorem about uncertainties. In a state Ψ, the product of uncertainties of the physical observables represented by the operators $\hat{A}$ and $\hat{B}$ is bounded by

$$\Delta A \, \Delta B \geq \frac{1}{2} |\langle \Psi | [\hat{A}, \hat{B}] | \Psi \rangle| = \frac{1}{2} |\langle \Psi | \hat{C} | \Psi \rangle| \, ,$$

where $[\hat{A}, \hat{B}] = i\hat{C}$. Often $\hat{C}$ is, as in $[\hat{x}, \hat{p}] = i\hbar$, simpler than $[\hat{A}, \hat{B}]$. And factoring i out of the right-hand side makes $\hat{C}$ Hermitian (show this), as is $\hbar$.

(b) First examples—Let $\hat{A} = \hat{x}$ and $\hat{B} = \hat{p}_x$. Since $[\hat{x}, \hat{p}] = i\hbar$, $\hat{C}$ is $\hbar$, and

$$\boxed{\Delta x\, \Delta p_x \geq \frac{1}{2}\, |\langle \Psi | \hbar | \Psi \rangle| = \frac{1}{2}\hbar \;.}$$

This says that we can never simultaneously know both x and p_x of a particle exactly. The better we know the one, the worse we can know the other. This is a fundamental limitation unimagined in classical mechanics—and one that caused Einstein great unease. To him, a particle certainly must have an exact position and momentum. In a 1935 paper that for decades was largely ignored, he acknowledged the remarkable successes of quantum mechanics, but insisted that it could not be *complete*: There must be a still more fundamental theory. We return to Einstein's deeply thoughtful objections and to the experiments they inspired 40 years later in Chap. 12. (Quantum mechanics wins!)

At the macroscopic level, we do not see the "fuzziness" implied by the uncertainty relation because $\hbar = 1.05 \times 10^{-34}$ J s is so very small. The mass of a speck of soot of radius 10^{-6} m and density twice that of water is 8.4×10^{-15} kg. If its uncertainty in position is $\Delta x = 10^{-7}$ m, its uncertainty in speed is only $\Delta v_x \geq 6.3 \times 10^{-14}$ m/s. However, for an electron ($m = 9.11 \times 10^{-31}$ kg) with Δx equal to the Bohr radius ($a_0 = 5.29 \times 10^{-11}$ m), the uncertainty in speed is $\Delta v_x \geq 1.09 \times 10^6$ m/s. Check these numbers.

The right side of the uncertainty relation for Δx and Δp_x is always $\frac{1}{2}\hbar$. For other non-commuting operators $\hat{A}$ and $\hat{B}$, the right side is usually more complicated, and is dependent upon the state of the system. But if $\hat{A}$ and $\hat{B}$ do commute, the right side is zero, and then there exist states that are simultaneously eigenstates of both $\hat{A}$ and $\hat{B}$. All this in what follows.

7·2. THE SCHWARZ INEQUALITY

We see to a preliminary matter first. Figure 1(a) shows two ordinary vectors **a** and **b**. The scalar product is $\mathbf{a} \cdot \mathbf{b} = ab\cos\theta$, where a is the length of **a**, etc. Then

$$a\,b \geq ab\,|\cos\theta| = |\mathbf{a} \cdot \mathbf{b}|$$

because $|\cos\theta| \leq 1$. Very simple—but to make an analogy we give a geometrical proof. We write **a** as the sum of vectors parallel and perpendicular to **b**, $\mathbf{a} = \mathbf{a}_{\parallel} + \mathbf{a}_{\perp}$. By the Pythagorean theorem, $a_{\perp}^2 = a^2 - a_{\parallel}^2 \geq 0$, where $a^2 = \mathbf{a} \cdot \mathbf{a}$, etc. Also since $\mathbf{a} \cdot \mathbf{b} = (a\cos\theta)\,b = a_{\parallel}b$, then $a_{\parallel} = (\mathbf{a} \cdot \mathbf{b})/b$.

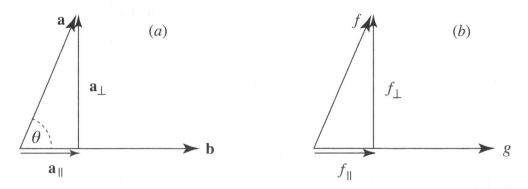

Figure 1. (a) Two ordinary vectors. (b) Two functions.

Therefore,

$$a_\perp^2 = a^2 - \frac{(\mathbf{a} \cdot \mathbf{b})^2}{b^2} \geq 0, \qquad \text{or} \qquad a^2 b^2 = (\mathbf{a} \cdot \mathbf{a})(\mathbf{b} \cdot \mathbf{b}) \geq (\mathbf{a} \cdot \mathbf{b})^2 .$$

The square root of this is $ab \geq |\mathbf{a} \cdot \mathbf{b}|$, as before. In Dirac notation, the inequality is

$$\langle a|a \rangle \langle b|b \rangle \geq |\langle a|b \rangle|^2 .$$

The equal sign holds only if $a_\perp = 0$, in which case $\mathbf{a}$ and $\mathbf{b}$ are colinear and therefore linearly dependent; that is, then $\mathbf{a} = c\,\mathbf{b}$, where c is a constant.

We prove an analogous relation for functions $f(x)$ and $g(x)$ (x might stand for more than one spatial dimension). The scalar product is, as usual,

$$\langle f|g \rangle = \int_{\text{range of } x} f^*(x)\, g(x)\, dx = \langle g|f \rangle^* .$$

The part of $f(x)$ that is orthogonal to $g(x)$, analogous to $\mathbf{a}_\perp$ for vectors, is

$$|f_\perp \rangle = |f \rangle - \frac{\langle g|f \rangle}{\langle g|g \rangle} |g \rangle .$$

(Show that $\langle g|f_\perp \rangle = 0$.) Then

$$\langle f_\perp|f_\perp \rangle = \langle f|f \rangle - \frac{\langle f|g \rangle \langle g|f \rangle}{\langle g|g \rangle} - \frac{\langle g|f \rangle \langle f|g \rangle}{\langle g|g \rangle} + \frac{\langle f|g \rangle \langle g|f \rangle \langle g|g \rangle}{\langle g|g \rangle^2} \geq 0 .$$

The ≥ 0 at the right is because $\langle f_\perp|f_\perp \rangle = \int |f_\perp|^2 dx$ is nonnegative. The last two terms cancel, and what remains is

$$\langle f|f \rangle \langle g|g \rangle \geq |\langle f|g \rangle|^2 .$$

This is the *Schwarz inequality*. The equal sign holds only if $f_\perp(x) = 0$, in which case $f(x)$ lies "along" $g(x)$ and they are linearly dependent; that is, then $f(x) = c\,g(x)$.

7·3. PROOF OF UNCERTAINTY RELATIONS

(a) The proof—We derive the uncertainty principle in one swoop, and explain each step immediately afterwards. The line of proof follows that in R. Shankar, *Principles of Quantum Mechanics*, Chap. 9. $\hat{A}$ and $\hat{B}$ are Hermitian operators representing physical observables, and $[\hat{A}, \hat{B}] = i\hat{C}$. The left side here is $(\Delta A)^2 (\Delta B)^2$ all the way down.

$$(\Delta A)^2 (\Delta B)^2 = \langle \Psi|(\hat{A} - \langle A \rangle)^2 \Psi \rangle \langle \Psi|(\hat{B} - \langle B \rangle)^2 \Psi \rangle \tag{1}$$

$$= \langle \Psi|\hat{\mathcal{A}}^2 \Psi \rangle \langle \Psi|\hat{\mathcal{B}}^2 \Psi \rangle \tag{2}$$

$$= \langle \hat{\mathcal{A}}\Psi|\hat{\mathcal{A}}\Psi \rangle \langle \hat{\mathcal{B}}\Psi|\hat{\mathcal{B}}\Psi \rangle \tag{3}$$

$$(\Delta A)^2 (\Delta B)^2 \geq |\langle \hat{\mathcal{A}}\Psi|\hat{\mathcal{B}}\Psi \rangle|^2 \tag{4}$$

$$\geq |\langle \Psi|\hat{\mathcal{A}}\hat{\mathcal{B}}\Psi \rangle|^2 \tag{5}$$

$$\geq \frac{1}{4} |\langle \Psi|([\hat{\mathcal{A}}, \hat{\mathcal{B}}]_+ + [\hat{A}, \hat{B}])|\Psi \rangle|^2 \tag{6}$$

$$\geq \frac{1}{4} |\langle \Psi|[\hat{\mathcal{A}}, \hat{\mathcal{B}}]_+|\Psi \rangle + i\langle \Psi|\hat{C}\Psi \rangle|^2 \tag{7}$$

$$\geq \frac{1}{4} |\langle \Psi|[\hat{\mathcal{A}}, \hat{\mathcal{B}}]_+|\Psi \rangle|^2 + \frac{1}{4} |\langle \Psi|\hat{C}\Psi \rangle|^2 \tag{8}$$

$$(\Delta A)^2 (\Delta B)^2 \geq \frac{1}{4} |\langle \Psi|\hat{C}\Psi \rangle|^2 . \tag{9}$$

133

The square root of this is the uncertainty relation for two general Hermitian operators,

$$\Delta A \, \Delta B \geq \frac{1}{2}|\langle \Psi | \hat{C} \Psi \rangle| \ .$$

(1) This is the definition of $(\Delta A)^2 (\Delta B)^2$.

(2) For brevity, we write $\hat{\mathcal{A}} \equiv \hat{A} - \langle A \rangle$ and $\hat{\mathcal{B}} \equiv \hat{B} - \langle B \rangle$. Just shorthand.

(3) $\hat{\mathcal{A}} \equiv \hat{A} - \langle A \rangle$ is Hermitian because $\hat{A}^\dagger = \hat{A}$, and $\langle A \rangle$ is real; and similarly for $\hat{\mathcal{B}}$.

(4) Using the Schwarz inequality $\langle f|f \rangle \langle g|g \rangle \geq |\langle f|g \rangle|^2$, with $f = \hat{\mathcal{A}}\Psi$ and $g = \hat{\mathcal{B}}\Psi$. The first indent is for the first inequality.

(5) $\hat{\mathcal{A}}$ is Hermitian.

(6) Using the obvious identity $\hat{\mathcal{A}}\hat{\mathcal{B}} = \frac{1}{2}(\hat{\mathcal{A}}\hat{\mathcal{B}} + \hat{\mathcal{B}}\hat{\mathcal{A}}) + \frac{1}{2}(\hat{\mathcal{A}}\hat{\mathcal{B}} - \hat{\mathcal{B}}\hat{\mathcal{A}})$, and abbreviating the "+" term as $[\hat{\mathcal{A}}, \hat{\mathcal{B}}]_+$.

(7) Using $[\hat{\mathcal{A}}, \hat{\mathcal{B}}] = [\hat{A}, \hat{B}]$ (why?), and $[\hat{A}, \hat{B}] = i\hat{C}$; and also $(\hat{\mathcal{O}}_1 + \hat{\mathcal{O}}_2)\Psi = \hat{\mathcal{O}}_1\Psi + \hat{\mathcal{O}}_2\Psi$.

(8) $\langle \Psi | [\hat{\mathcal{A}}, \hat{\mathcal{B}}]_+ | \Psi \rangle$ and $\langle \Psi | \hat{C} \Psi \rangle$ are real because $[\hat{\mathcal{A}}, \hat{\mathcal{B}}]_+$ and $\hat{C}$ are Hermitian; and $|a + ib|^2 = a^2 + b^2$, where a and b are real. (Why is $\langle \Psi | [\hat{\mathcal{A}}, \hat{\mathcal{B}}]_+ | \Psi \rangle$ Hermitian?)

(9) Discarding a nonnegative piece from the right side, which if anything makes it smaller. The second inequality, the second indent.

(b) Conditions for equality—For the $\geq$ in uncertainty relations to reduce to $=$, two separate conditions are required First, the Schwarz inequality used at step 4 must be "saturated," or

$$\langle \hat{\mathcal{A}}\Psi | \hat{\mathcal{A}}\Psi \rangle \langle \hat{\mathcal{B}}\Psi | \hat{\mathcal{B}}\Psi \rangle = |\langle \hat{\mathcal{A}}\Psi | \hat{\mathcal{B}}\Psi \rangle|^2 \ .$$

This means (see Sec. 2) that the functions $\hat{\mathcal{A}}\Psi$ and $\hat{\mathcal{B}}\Psi$ must be linearly dependent: $\hat{\mathcal{A}}\Psi = c\,(\hat{\mathcal{B}}\Psi)$. Secondly, the nonnegative piece we discarded at step 9 must be zero, or

$$\langle \Psi | [\hat{\mathcal{A}}, \hat{\mathcal{B}}]_+ | \Psi \rangle = \langle \Psi | (\hat{A} - \langle A \rangle)(\hat{B} - \langle B \rangle) + (\hat{B} - \langle B \rangle)(\hat{A} - \langle A \rangle) | \Psi \rangle = 0 \ .$$

The oscillator provides examples. Problem 6·6 was to show by direct calculation that the uncertainly relation $\Delta x \, \Delta p_x \geq \frac{1}{2}\hbar$ is satisfied for all the oscillator energy eigenstates $|n\rangle$, and that in fact the equal sign holds for the ground state $|0\rangle$. Problem 3 of this chapter is to show that *all* those oscillator eigenstates satisfy the second of the above two conditions, but that only the ground state also satisfies the first one. In Chap. 8, analogous results will be found for angular-momentum eigenstates.

(c) Another example—The $\Delta x \, \Delta p_x \geq \frac{1}{2}\hbar$ relation has a sharp lower limit: No matter what the state, the product of those uncertainties cannot be smaller than $\frac{1}{2}\hbar$ (and nearly always the product is larger). Usually things are more complicated. For example, for any of the energy eigenstates of the harmonic oscillator, $\Delta x \neq 0$ but $\Delta E = 0$, and so $\Delta x \, \Delta E = 0$. Is there a problem here? Well, since $[\hat{x}, \hat{V}(x)] = 0$, we get

$$[\hat{x}, \hat{H}] = \frac{1}{2m}[\hat{x}, \hat{p}_x^2] = \frac{i\hbar}{m}\,\hat{p}_x \ ,$$

so that

$$\Delta x \, \Delta E \geq \frac{\hbar}{2m}|\langle \Psi | \hat{p}_x | \Psi \rangle| = \frac{h}{2m}\langle p_x \rangle \ .$$

For consistency, $\langle p_x \rangle$ must be zero in an energy eigenstate. To see that this is indeed so, consider the Ehrenfest relation $d\langle x \rangle/dt = \langle p_x \rangle/m$, found in Sec. 5·7(b). Here $\langle x \rangle = \int x \, |\Psi_n(x,t))|^2 \, dx$ is constant in an energy eigenstate because $|\Psi_n(x,t))|^2 = |\psi_n(x))|^2$ is time independent. Thus $d\langle x \rangle/dt = 0$, and then so is $\langle p_x \rangle$, and everything is all right. In a superposition of states with different energies, ΔE would be nonzero, and $\langle x \rangle$ and $\langle p_x \rangle$ would be time dependent. (See yet again Prob. 2·2.)

Thus the fact that $\hat{A}$ and $\hat{B}$ do not commute does not force $\Delta A \, \Delta B > 0$. Nevertheless, for x and E, there exists no state in which both are known with certainty at the same time. In Chap. 8, we find a (trivial) example where all of ΔA, ΔB, and $\langle \Psi | \hat{C} | \Psi \rangle$ are zero, even though the operators $\hat{A}$ and $\hat{B}$ do not commute.

7·4. FOURIER TRANSFORMS. MOMENTUM SPACE

(a) Position and momentum—The first quantum-mechanical wave function we saw was

$$\Psi(x,t) = Ae^{i(px - Et)/\hbar} = Ae^{i(kx - \omega t)} = Ae^{ikx}e^{-i\omega t} \ .$$

This represents a particle (or, as in Chap. 4, a plane wave of them) with exact momentum $p = \hbar k$ and energy $E = \hbar \omega$. We noted in Sec. 4·1(a) that although this wave function is not normalizable, it is a useful idealization—and then we used it. However, by superimposing a continuum of functions of this form, we can construct a normalizable $\Psi(x,t)$. Suppose that at $t = 0$ we want a specific $\psi(x)$. The recipe is

$$\boxed{\Psi(x,0) = \psi(x) = \frac{1}{\sqrt{2\pi}} \int_{-\infty}^{+\infty} \phi(k) \, e^{ikx} \, dk \ .}$$

For each value of the wave number k, $\phi(k)$ tells how much of e^{ikx} goes into the make-up of $\psi(x)$. But given $\psi(x)$, what should the function $\phi(k)$ here be? It is

$$\boxed{\phi(k) = \frac{1}{\sqrt{2\pi}} \int_{-\infty}^{+\infty} \psi(x) \, e^{-ikx} \, dx \ .}$$

Note that the first integral is over k and the second one is over x; and that the first integrand has e^{ikx}, while the second one has e^{-ikx}. Physically, $\phi(k)$ gives (because $p = \hbar k$) the probability amplitude for momentum, just as $\psi(x)$ is the probability amplitude for position. Thus $\phi(k)$ is called the *momentum-space* wave function, and $|\phi(k)|^2$ gives the probability distribution of momentum. The narrower one makes $\psi(x)$, the broader one makes $\phi(k)$; and no function exists that makes $\Delta x \, \Delta k < 1/2$.

The integral transform of $\psi(x)$ in the second box is called the *Fourier transform*, and the transform of $\phi(k)$ in the first box is called the *inverse Fourier transform*. It does not really matter much which we call Fourier and which inverse Fourier. Sometimes the two are simply called the $+i$ (with e^{ikx}) and $-i$ transforms. The two transforms are sometimes written with the 2π differently apportioned between the two integrals.

We simply state two central (and remarkable) properties of Fourier transforms:

(1) If $\phi(k)$ is the Fourier transform (FT) of $\psi(x)$, then $\psi(x)$ is the inverse transform (IFT) of $\phi(k)$. Schematically,

$$\psi(x) \xrightarrow{\text{FT}} \phi(k) \xrightarrow{\text{IFT}} \psi(x) \ .$$

(2) If either one of $\psi(x)$ or $\phi(k)$ is normalized, then so is the other. That is,

$$\langle\psi|\psi\rangle = \int_{-\infty}^{+\infty} |\psi(x)|^2 \, dx = 1 \qquad \text{implies} \qquad \langle\phi|\phi\rangle = \int_{-\infty}^{+\infty} |\phi(k)|^2 \, dk = 1 \ ,$$

and vice versa. A good reference for all matters concerning the properties of Fourier transforms, as well as for many of its applications, is R.N. Bracewell, *The Fourier Integral and Its Applications*, 3rd ed. A nice feature is a 19-page pictorial dictionary of transform pairs.

Since $e^{-ikx} = \cos kx - i \sin kx$, we can write the transform of $\psi(x)$ as

$$\phi(k) = \frac{1}{\sqrt{2\pi}} \int_{-\infty}^{+\infty} \psi(x) \, e^{-ikx} \, dx = \frac{1}{\sqrt{2\pi}} \int_{-\infty}^{+\infty} [\psi(x) \cos kx - i\psi(x) \sin kx] \, dx \ .$$

And we can break down any $\psi(x)$ over $-\infty < x < +\infty$ into a sum of even and odd parts about $x = 0$: $\psi(x) = \psi_E(x) + \psi_O(x)$. Then since $\cos kx$ and $\sin kx$ are even and odd functions about $x = 0$, and the integral is over a symmetric range of x, the integral is

$$\phi(k) = \frac{1}{\sqrt{2\pi}} \int_{-\infty}^{+\infty} [\psi_E(x) \cos kx - i\psi_O(x) \sin kx] \, dx \ .$$

Therefore,

$$\phi(k) = \phi_E(k) = \frac{2}{\sqrt{2\pi}} \int_0^{+\infty} \psi_E(x) \cos kx \, dx \qquad \text{when } \psi(x) \text{ is even} \ ,$$

$$\phi(k) = \phi_O(k) = -\frac{2i}{\sqrt{2\pi}} \int_0^{+\infty} \psi_O(x) \sin kx \, dx \qquad \text{when } \psi(x) \text{ is odd} \ .$$

When $\psi(x)$ is neither even nor odd, its Fourier transform is the sum, $\phi_E(k) + \phi_O(k)$.

(b) Example: the double slit—We look back at the experiment of Sec. 1·2, where a beam of neutrons was incident upon a double slit. The top row of Fig. 2 below shows schematically aperture screens for two very narrow slits, for one slit of finite width, and for two slits of finite width. The middle row shows the amplitudes at a distance behind the aperture screens. These are amplitudes for Fraunhofer diffraction, derived in elementary optics using Huygen's principle. The bottom row shows the intensity patterns—the squared amplitudes—on the screen.

The patterns can be found most simply and powerfully using the Fourier transform. Let y be the coordinate perpendicular to both the beam axis, x, and to the slits. The two slits are centered at $y = +a$ and $-a$, and the widths of the finite slits are $2b$. Then the aperture functions for the three cases in Fig. 2 are

$$\psi_1(y) = \delta(y - a) + \delta(y + a)$$
$$\psi_2(y) = \begin{cases} 1 & \text{for } -b < y < b \\ 0 & \text{otherwise} \end{cases}$$
$$\psi_3(y) = \begin{cases} 1 & \text{for } |y \pm a| < b \\ 0 & \text{otherwise} \ . \end{cases}$$

The apertures "chop" the broad incident wave transversely, and thereby induce distributions of transverse momenta. If the observing screen were directly behind the aperture screen, the patterns would of course simply be images of the apertures; the screens would "record" the probability distributions $|\psi(y)|^2$. But if the observing screen lies far beyond the aperture screen, the transverse momentum distributions have time to "develop," and the intensity patterns in Fig. 2 are pictures of those distributions.

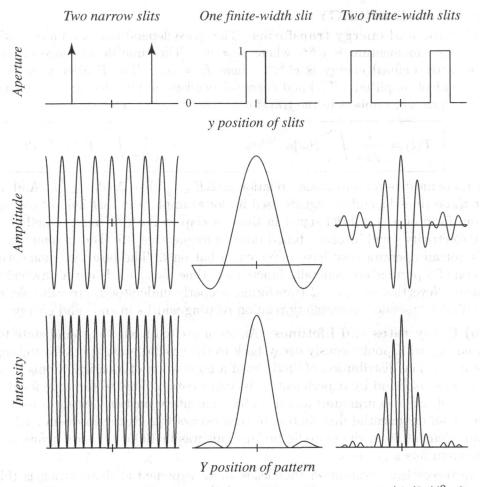

Figure 2. Apertures, amplitudes $\phi_i(k_y)$, and intensity patterns $|\phi_i(k_y)|^2$, for two very narrow slits, one finite-width slit, and two finite-width slits. In the last, the ratio of slit separation to width is 4:1. Note that $\phi_3(k_y)$ is the product of $\phi_1(k_y)$ and $\phi_2(k_y)$.

Since we are interested here in patterns, not in absolute quantities (which would depend on the intensity of the incoming wave), we ignore multiplicative constants. Since all three apertures are symmetrical about $y = 0$, we use the cosine transform. We get

$$\phi_1(k_y) \propto \int_0^\infty \delta(y - a) \cos ky \, dy = \cos ka$$

$$\phi_2(k_y) \propto \int_0^b \cos ky \, dy = b \frac{\sin kb}{kb}$$

$$\phi_3(k_y) \propto \int_{a-b}^{a+b} \cos ky \, dy = b \cos ka \frac{\sin kb}{kb} \, ,$$

and $\phi_3(k_y) = \phi_1(k_y) \times \phi_2(k_y)$. It could hardly be simpler. The second row of Fig. 2 shows these amplitudes, the third row the intensities. If the two narrow slits were closer together, the fringes would be farther apart. If the single slit was narrower, the diffraction pattern would be broader. Narrower in y, broader in p_y—your uncertainty principle at work.

7·5. TIME AND ENERGY

(a) Time-and-energy transforms—The space dependence of an amplitude with perfectly defined momentum is e^{ikx}, where $p = \hbar k$. The time dependence of an amplitude with perfectly defined energy is $e^{-i\omega t}$, where $E = \hbar \omega$. The Fourier transforms relating time-dependent amplitudes $T(t)$ and energy-dependent amplitudes $b(\omega)$ are given by simply changing signs and symbols in the transforms relating $\psi(x)$ and $\phi(k)$:

$$T(t) = \frac{1}{\sqrt{2\pi}} \int_{-\infty}^{\infty} b(\omega) e^{-i\omega t} d\omega, \qquad b(\omega) = \frac{1}{\sqrt{2\pi}} \int_{-\infty}^{\infty} T(t) e^{+i\omega t} dt .$$

Surely there must be an uncertainty relation $\Delta t \Delta E \geq \frac{1}{2}\hbar$ like $\Delta x \Delta p \geq \frac{1}{2}\hbar$. And indeed there is—for those time-dependent signals used in, for example, the vast field of communications. The transform of a very brief signal in time—a clap or flash or burst, whether of sound or light or electricity—will involve a broad range of frequencies, etc. But in quantum mechanics, time is not an operator—we have never put a hat on t. The position, momentum, energy, and so on of a particle are generally functions of time, but time is not a physical property of a particle. Nevertheless, the ω-t transforms, properly understood, are valuable, and we can usually find a "pseudo" uncertainty relation relating widths in time and energy.

(b) Decay rates and lifetimes—An atom excited from its ground state to a state of higher energy will spontaneously decay back to the ground state (perhaps through intermediate states). The distribution of lifetimes of a population of identical atoms in a particular decay process is found by experiment to be exponential. This is also true for decays of unstable nuclei (such as uranium) and unstable elementary particles (such as pions). The first to observe an exponential distribution in such decays was Rutherford—in nuclear decays of thorium (element 90 in the periodic table). This was in 1900, 11 years before he discovered that an atom *has* a nucleus.

The theoretical assumption that leads to an exponential distribution is this: The infinitesimal number of decays $dN(t)$ that occur in the next infinitesimal time interval dt is proportional to the number of atoms (or nuclei, or particles) $N(t)$ that are still "alive" at t. So let $dN(t) = -AN(t)\, dt$, where A is a constant that will depend on the particular decay process. Starting with N_0 excited atoms at $t = 0$, we get

$$\int_{N_0}^{N(t)} \frac{dN}{N} = -A \int_0^t dt, \quad \text{or} \quad N(t) = N_0\, e^{-At} ,$$

the exponential decay law. Exponential growth or decay also often describes approximately the time dependence of populations, values of investments, losses in epidemics, and many other phenomena—at least over limited time intervals.

The relation $dN(t) = -AN(t)\, dt$ also says that $|dN(t)|$ atoms *have* lifetimes between t and $t + dt$. Weighting the lifetimes with the numbers of atoms that live that long, and dividing by the total number of atoms, we get the average or mean lifetime τ,

$$\tau = \langle t \rangle = -\frac{1}{N_0} \int_0^\infty t\, dN = +A \int_0^\infty t\, e^{-At}\, dt = \frac{1}{A} .$$

Thus the mean life τ is $1/A$, and $N(t) = N_0 e^{-t/\tau}$. Lifetimes can vary from less than 10^{-20} second to billions of years. It is left for a problem to show that $(\Delta t)^2 = \langle t^2 \rangle - \langle t \rangle^2 = \tau^2$, so that the rms spread of a lifetime is $\Delta t = \tau$. To summarize,

138

$$dN(t) = -\frac{1}{\tau}N(t)\,dt, \quad N(t) = N_0\,e^{-t/\tau}, \quad \text{and} \quad \Delta t = \tau .$$

(c) The distribution in energy—At $t = 0$ an atom is excited to a state with energy E_0. Since now $N_0 = 1$, $N(t) = e^{-t/\tau}$ is the probability that the atom is still in the excited state at t. The normalized time-dependent amplitude $T(t)$ for being in the excited state is

$$|T(t)|^2 = \frac{1}{\tau}\,e^{-t/\tau}, \quad \text{where} \quad \int_0^\infty |T(t)|^2\,dt = 1 ,$$

and $t \geq 0$. Aside from an arbitrary phase, $T(t)$ is

$$T(t) = \frac{1}{\sqrt{\tau}}\,e^{-t/2\tau}\,e^{-i\omega_0 t} .$$

The factor $e^{-i\omega_0 t}$ is the usual time-dependent phase of a quantum state, where $\omega_0 = E_0/\hbar$.

The Fourier transform of $T(t)$ is

$$b(\omega) = \frac{1}{\sqrt{2\pi}} \int_{-\infty}^\infty T(t)\,e^{+i\omega t}\,dt = \frac{1}{\sqrt{2\pi\tau}} \int_0^\infty e^{-t/2\tau}\,e^{i(\omega - \omega_0)t}\,dt$$

$$= \frac{1}{\sqrt{2\pi\tau}}\,\frac{1}{[1/2\tau - i(\omega - \omega_0)]} .$$

With $\omega_0 = E_0/\hbar$, $\omega = E/\hbar$, and $\Gamma \equiv \hbar/\tau$, we write this in terms of energy:

$$b(E) = \sqrt{\frac{2\tau}{\pi}} \times \frac{1}{1 - 2\tau i(E - E_0)/\hbar} = \sqrt{\frac{2\hbar}{\pi\Gamma}} \times \frac{1}{1 - i(E - E_0)/\tfrac{1}{2}\Gamma} .$$

The units of Γ are those of energy. The probability distribution in E is

$$|b(E)|^2 = \frac{2\hbar}{\pi\Gamma} \times \frac{1}{1 + [(E - E_0)/\tfrac{1}{2}\Gamma]^2} = \frac{2\hbar}{\pi\Gamma} \times BW(E) .$$

Here $BW(E)$ is the *Breit-Wigner resonance function*,

$$BW(E) = \frac{1}{1 + [(E - E_0)/\tfrac{1}{2}\Gamma]^2} , \quad \text{or} \quad BW(\epsilon) = \frac{1}{1 + \epsilon^2} ,$$

where $\epsilon \equiv (E - E_0)/\tfrac{1}{2}\Gamma$; ϵ measures $E - E_0$ (plus and minus) in units of $\tfrac{1}{2}\Gamma$.

Figure 3 shows $BW(\epsilon)$. The maximum value is 1 at $\epsilon = 0$ (or $E = E_0$), and the value is down by half when $\epsilon = \pm 1$ (or $E - E_0 = \pm\tfrac{1}{2}\Gamma$). Thus Γ is the full width at half maximum (FWHM) of the Breit-Wigner function. The mean-squared deviation $(\Delta\epsilon)^2$ of ϵ from zero is infinite: The integrand for the calculation is $\epsilon^2 BW(\epsilon) = \epsilon^2/(1 + \epsilon^2)$, and this does not vanish as $\epsilon \to \pm\infty$. So then $\Delta E \propto \Delta\epsilon$ is infinite too.

Therefore, with no proper rms ΔE to be had, we use the FWHM in its place, and call it δE. And since δE equals $\Gamma = \hbar/\tau$, and the rms Δt equals τ, we get $\delta E \equiv \Gamma = \hbar/\tau$, or

$$\tau\,\delta E = \Delta t\,\delta E = \hbar .$$

This certainly looks like an uncertainty relation, but δE here is not an rms uncertainty, and time is not a particle property. Nevertheless, $\Delta t = \tau$ and the FWHM δE are natural and well-defined measures, and the relation $\tau\,\delta E = \hbar$ is widely used.

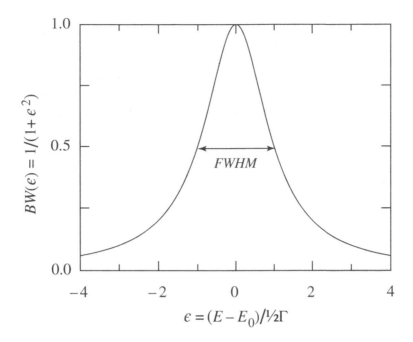

Figure 3. The Breit-Wigner resonance function $BW(\epsilon) = 1/(1 + \epsilon^2)$, where $\epsilon \equiv (E - E_0)/\frac{1}{2}\Gamma$. The full width at half maximum (FWHM) is Γ.

(d) Examples—Here are two important applications of $\tau\,\delta E = \hbar$:

(1) A state with a finite lifetime does not have an exact energy, and so the frequency of the transition to a lower state is not infinitely sharp. Spectral lines have a width, as in Fig. 3. Since $\hbar = 6.58 \times 10^{-16}$ eV s, atomic decays with lifetimes of order 10^{-8} s (not untypical) will have widths of order 10^{-7} eV. This is of course nearly always very small compared to the energy difference between the states. We calculate a hydrogen decay lifetime in Chap. 15.

(2) If a lifetime is so short we cannot measure it, δE will be large, so perhaps we can measure that. Elementary-particle decays with lifetimes of order 10^{-22} s (not untypical) will have widths of order 10 MeV. For such short-lived particles, it is in fact FWHM's that are measured and tabulated, not lifetimes—and then the states are often called resonances. Some resonances are produced in particle-scattering reactions like $a + b \to$ resonance $\to a + b$. The reaction rate (or "cross section") is measured as a function of center-of-mass (c.m.) energy E. A peak in the rate at E_0, with a shape as a function of c.m. energy like that in Fig. 3, is likely a resonance. Chapter 16 is about reactions and cross sections.

A partial parallel: Figure 5 in Chap. 4 showed the fraction T of particles transmitted by a rectangular barrier V high and L wide as a function of E/V for two values of a parameter β. If we did not know one or the other of V or L, we could find it by varying E and measuring the values at which *all* the particles were transmitted ($T = 1$). The form of the equation for T was

$$T = \frac{1}{1+Z}, \qquad \text{in form just like} \quad BW(\epsilon) = \frac{1}{1+\epsilon^2} \,.$$

The $T = 1$ peaks occurred when Z was 0; the $BW = 1$ peak occurs when ϵ is 0. However, Z was a more complicated function of energy than ϵ^2, and it contained a multitude of peaks.

(e) The Breit-Wigner amplitude—The Breit-Wigner resonance function shown in Fig. 3 is the magnitude squared of the right side of

$$\sqrt{\frac{\pi\Gamma}{2\hbar}} \times b(E) = \frac{1}{1 - i(E - E_0)/\frac{1}{2}\Gamma} = \frac{1}{1 - i\epsilon} \ .$$

We multiply it by i (for why, see below), which does not change the magnitude squared. The amplitude is then

$$\frac{i}{1 - i\epsilon} = \frac{i(1 + i\epsilon)}{1 + \epsilon^2} = \frac{i - \epsilon}{1 + \epsilon^2} \ .$$

As $(E - E_0)$ ranges from $-\infty$ to $+\infty$, so does ϵ. (Actually, there is usually a lower limit for E; see Chap. 16.) Figure 4 shows that the amplitude traces a circle on the complex plane (see the problems). The reason for inserting the i is to stand the circle upright, with its main diameter along the imaginary axis; otherwise, it would lie along the real axis. Upright, the circle can describe (as already mentioned) the energy dependence of an amplitude for a *resonance* in a *partial wave* in the elastic scattering of two particles, $a + b \rightarrow a + b$. The imaginary part of the amplitude for elastic scattering is strictly positive. But that is for Chap. 16.

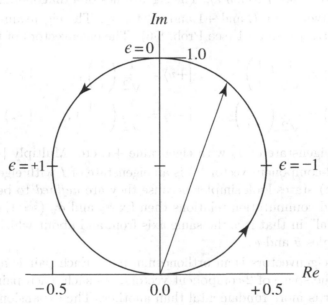

Figure 4. The Breit-Wigner amplitude $i/(1 - i\epsilon) = (i - \epsilon)/(1 + \epsilon^2)$. As ϵ runs from $-\infty$ to $+\infty$, the amplitude, starting with zero length at the origin, traverses the circle in the counterclockwise direction and comes back to the origin. The magnitude of the amplitude (the length of the straight arrow) is $1/\sqrt{1 + \epsilon^2}$. The magnitude squared, $1/(1 + \epsilon^2)$, is the Breit-Wigner function $BW(\epsilon)$ shown in Fig. 3.

7·6. WHEN OPERATORS COMMUTE

(a) A simple system—We begin with four 2-by-2 Hermitian matrices,

$$\hat{I} = \begin{pmatrix} 1 & 0 \\ 0 & 1 \end{pmatrix}, \qquad \hat{\sigma}_x = \begin{pmatrix} 0 & 1 \\ 1 & 0 \end{pmatrix}, \qquad \hat{\sigma}_y = \begin{pmatrix} 0 & -i \\ i & 0 \end{pmatrix}, \qquad \hat{\sigma}_z = \begin{pmatrix} 1 & 0 \\ 0 & -1 \end{pmatrix}.$$

The first one is the 2-by-2 identity matrix. The other three are called the Pauli spin matrices, because (see below) $\hat{s}_i = \frac{1}{2}\hbar\hat{\sigma}_i$, $i = x, y, z$, are the operators for the three components of the angular momentum of a spin-1/2 particle. Now

$$\hat{\sigma}_x^2 = \hat{\sigma}_y^2 = \hat{\sigma}_z^2 = \hat{I}$$

(show this), so that $\hat{s}_x^2 + \hat{s}_y^2 + \hat{s}_z^2 = \hat{s}^2 = \frac{3}{4}\hbar^2\hat{I}$ is the operator for the total spin-angular-momentum squared. Electrons, protons, neutrons, and certain other particles, have spin 1/2, and the spin is as intrinsic to those particles as are their masses and charges. Angular momentum is the subject of the next chapter, and spin-1/2 gets the whole of Chap. 10 to itself. But we can talk here about measuring a component of the spin angular momentum of a particle. The set of four operators provides the simplest example for what follows.

Any 2-by-2 Hermitian matrix can be written as $a\hat{I} + b_x\hat{\sigma}_x + b_y\hat{\sigma}_y + b_z\hat{\sigma}_z$, where a, b_x, b_y, and b_z are all real (see Prob. 5·7). The eigenvalues of a diagonal matrix are the values on its diagonal: $+1$ (twice) for $\hat{I}$, and $+1$ and -1 for $\hat{\sigma}_z$. The eigenvalues of $\hat{\sigma}_x$ and $\hat{\sigma}_y$ are the same as those of $\hat{\sigma}_z$, $+1$ and -1 (see Prob. 5·6). The eigenvectors of the three operators are

$$|+x\rangle = \frac{1}{\sqrt{2}}\begin{pmatrix} 1 \\ 1 \end{pmatrix}, \qquad |+y\rangle = \frac{1}{\sqrt{2}}\begin{pmatrix} 1 \\ i \end{pmatrix}, \qquad |+z\rangle = \begin{pmatrix} 1 \\ 0 \end{pmatrix},$$

$$|-x\rangle = \frac{1}{\sqrt{2}}\begin{pmatrix} 1 \\ -1 \end{pmatrix}, \qquad |-y\rangle = \frac{1}{\sqrt{2}}\begin{pmatrix} 1 \\ -i \end{pmatrix}, \qquad |-z\rangle = \begin{pmatrix} 0 \\ 1 \end{pmatrix}.$$

Here $|+x\rangle$ is the eigenstate of $\hat{\sigma}_x$ with eigenvalue $+1$, etc. Multiply $|-y\rangle$ by $\hat{\sigma}_y$ and you get $-|-y\rangle$, etc. Any 2-component vector $|v\rangle$ is an eigenstate of $\hat{I}$ with eigenvalue $+1$: $\hat{I}|v\rangle = |v\rangle$. The $|+z\rangle$ and $|-z\rangle$ states look simpler because they are *defined* to be as given. That choice determines $\hat{\sigma}_z$, and commutation relations then fix $\hat{\sigma}_x$ and $\hat{\sigma}_y$ (see the next chapter). The z axis is only "special" in that it is the same axis from and about which we measure the polar and azimuthal angles θ and ϕ.

Each pair of eigenvectors is an orthonormal pair. Each pair is a minimal but *complete* set for spanning the space of 2-component vectors. As such, each pair is a possible *basis* for the space. No pair is more fundamental than another. The expansion of an arbitrary vector $|v\rangle$ in terms of $|+x\rangle$ and $|-x\rangle$ would be

$$|v\rangle = \hat{I}|v\rangle = \big(|+x\rangle\langle+x| + |-x\rangle\langle-x|\big)|v\rangle = |+x\rangle\langle+x|v\rangle + |-x\rangle\langle-x|v\rangle$$

(see Sec. 5·1). The components of $|v\rangle$ in this basis are $\langle+x|v\rangle$ and $\langle-x|v\rangle$.

The eigenvectors of, say, $\hat{\sigma}_x$ are not eigenvectors of $\hat{\sigma}_y$ or $\hat{\sigma}_z$. The magnitude of the scalar product of either member of one pair with any of the other four eigenvectors is $1/\sqrt{2}$. For example, $|\langle+x|-y\rangle| = |(1-i)|/2 = 1/\sqrt{2}$, and so on. Likewise,

$$|\langle+x|+y\rangle|^2 = |\langle+x|+z\rangle|^2 = |\langle+x|-z\rangle|^2 = 1/2 ,$$

and so on.

(b) Sequential measurements—An accurate measurement of a physical property can only get an eigenvalue of the corresponding operator (Sec. 5·8). And then the system is in the eigenstate (or if the state is degenerate, in a superposition of the eigenstates) that belong to the eigenvalue. Since $\hat{s}^2 = \frac{3}{4}\hbar^2\hat{I}$, a measurement of s^2 can only get $\frac{3}{4}\hbar^2$. Since $\hat{s}_z = \frac{1}{2}\hbar\hat{\sigma}_z$, a measurement of s_z can only get $+\frac{1}{2}\hbar$ or $-\frac{1}{2}\hbar$. If we get $-\frac{1}{2}\hbar$, then afterwards the state is $|-z\rangle$. If we quickly measure s_z again, before possible decays or outside influences can change the state, we get $-\frac{1}{2}\hbar$ again. Rapid remeasurements of either s^2 or s_z would not change the state.

But suppose we now measure s_x. We either get $+\frac{1}{2}\hbar$, and the state is then $|+x\rangle$, or we get $-\frac{1}{2}\hbar$, and the state is then $|-x\rangle$; the probability of each outcome is 1/2. And nothing then remains in the new state to "remember" that just before this measurement the state was $|-z\rangle$. A sequence of measurements of a component, each time along a different axis (say, z, x, z, y, x, ...) would give a completely random sequence of $+\frac{1}{2}\hbar$'s and $-\frac{1}{2}\hbar$'s. In Chap. 12, we shall see that this observation underlies quantum cryptography—and much else.

(c) Simultaneous eigenstates—Here is what we are getting at:

(1) If two operators commute, then there exists a set of simultaneous eigenstates. For $\hat{I}$ and $\hat{\sigma}_y$, the set is $|+y\rangle$ and $|-y\rangle$.

(2) If two operators do not commute, then there is not a set of simultaneous eigenstates. The operators $\hat{\sigma}_z$ and $\hat{\sigma}_x$ do not commute, and the eigenstates of one are not the eigenstates of the other. The commutation relations for the matrices are

$$[\hat{I}, \hat{\sigma}_i] = 0, \quad [\hat{\sigma}_x, \hat{\sigma}_y] = 2i\hat{\sigma}_z, \quad [\hat{\sigma}_y, \hat{\sigma}_z] = 2i\hat{\sigma}_x, \quad \text{and} \quad [\hat{\sigma}_z, \hat{\sigma}_x] = 2i\hat{\sigma}_y$$

(show this). Uncertainty relations follow and forbid simultaneous knowledge (and simultaneous *existence*) of exact values for more than one component of a spin.

In the next chapter, operators $\hat{L}_x$, $\hat{L}_y$, and $\hat{L}_z$ will represent the three components of the *orbital* angular momentum of a particle or system. The commutation relations are $[\hat{L}_x, \hat{L}_y] = i\hat{L}_z$ and the two relations obtained by cyclically permuting the indices x, y, and z. And $\hat{L}_x$, $\hat{L}_y$, and $\hat{L}_z$ all commute with $\hat{L}^2 \equiv \hat{L}_x^2 + \hat{L}_y^2 + \hat{L}_z^2$. Thus the simultaneous eigenfunctions of, say, $\hat{L}_z$ and $\hat{L}^2$ can serve as a basis for writing any state of orbital angular momentum.

Sometimes there are three or more mutually commuting operators. Chapter 9 is about hydrogen and other systems for which the potential energy depends only on r, and not on θ or ϕ: $V = V(r)$. The Hamiltonian $\hat{H}$ and the operators $\hat{L}^2$, and $\hat{L}_z$ then all commute with one another, and we find the simultaneous eigenstates of all three. The set could as well be $\hat{H}$, $\hat{L}^2$, and $\hat{L}_x$, or $\hat{H}$, $\hat{L}^2$, and $\hat{L}_y$, but again the z axis is the axis from and about which we measure the polar and azimuthal angles, and $\hat{L}_z$ is the usual choice.

PROBLEMS

The first two problems are back-of-the envelope estimates, really just dimensional arguments, as in Sec. 3·1. For them, ignore numerical factors.

1. *A beam of well-defined momentum chopped longitudinally.*

A long beam of particles, all having momentum $p_0 = mv_0$, is incident upon a closed shutter. The shutter is opened for a short time, so that a segment of length L of the beam passes through.

Make a rough order-of-magnitude estimate of the spread Δp in momentum values for the particles that get through the shutter. Then estimate, in terms of L, m, and $\hbar$, how long it takes the segment to become twice its original length L.

2. *A beam of well-defined momentum chopped transversely.*

A long beam of electrons, all having momentum p_0 in the x direction, is incident upon a slit of width D.

(a) For the electrons that pass through the slit, make a rough order-of-magnitude estimate of the spread Δp_y of momentum values transverse to the initial beam direction and to the long axis of the slit.

(b) The electrons strike a screen a distance $L \gg D$ beyond the slit. The width W of the pattern on the screen is of course large if D is large, and it is also large, due to the spread in p_y, if D is very small. By simply adding the contributions to W from the uncertainties in transverse position and momentum, and assuming that Δp_y is small compared to p_0 (the beam goes forward), estimate W in terms of D, L, and the de Broglie wavelength λ of the electrons.

3. *Getting the equal sign in $\Delta x \, \Delta p_x \geq \hbar/2$ for the oscillator.*

In deriving the general uncertainty relation, we saw that there are two conditions for the equal sign to hold. When $\hat{A} = \hat{x}$ and $\hat{B} = \hat{p}_x$, these conditions are

$$(\hat{x} - \langle x \rangle)|\psi\rangle = c\,(\hat{p}_x - \langle p_x \rangle)|\psi\rangle \tag{1}$$

and

$$\langle\psi|(\hat{x} - \langle x \rangle)(\hat{p}_x - \langle p_x \rangle) + (\hat{p}_x - \langle p_x \rangle)(\hat{x} - \langle x \rangle)|\psi\rangle = 0 \ . \tag{2}$$

In Chap. 6·4(b), we found that $\langle x \rangle$ and $\langle p_x \rangle$ are zero in any energy eigenstate $|n\rangle$ of the oscillator. Thus the inside of the bracket in (2) reduces to $\hat{x}\hat{p}_x + \hat{p}_x\hat{x}$ for these states.

(a) Show that (2) is satisfied for any eigenstate $|n\rangle$ of the oscillator.

(b) Show that (1) is satisfied for the ground state $|0\rangle$ but not for any other of the states. Now look back at Problem 6·6.

4. *Transforms of states of the infinite square well.*

Find the Fourier transforms $\phi(k)$ for the $n = 1$ and $n = 4$ states of the infinite square well, $0 < x < L$. Sketch the momentum probability distributions.

5. *Same-form Fourier transform pairs.*

(a) Find the transform $\phi_0(k)$ of the normalized oscillator ground state $\psi_0(x)$. Then find the inverse transform of $\phi_0(k)$. Show that if $\psi_0(x)$ gets narrower, $\phi_0(k)$ gets wider.

(b) Find $\phi_1(k)$ from $\psi_1(x)$.

In each case, you should find that $\phi(k)$ has the same *functional form* as the corresponding $\psi(x)$. In fact, all the $\psi_n(x)$–$\phi_n(k)$ pairs of *oscillator* eigenfunctions are same-form pairs. That is because x and p enter form-equally (both quadratically) in the Hamiltonian.

(c) Show that the Fourier transform of $\text{sech } x$ is a $\text{sech } k$ function.

6. *Exponential decays.*

(a) Prove that the rms spread of lifetimes, Δt, for an exponential decay is equal to the mean lifetime τ.

(b) The *half* life $\tau_{1/2}$ is the time for half of the decays to occur, $N_0 e^{-t/\tau_{1/2}} = \frac{1}{2} N_0$. Show that the mean and half-lives τ and $\tau_{1/2}$ (both of which are used) are related by

$$e^{\tau_{1/2}/\tau} = 2 \qquad \text{or} \qquad \tau_{1/2} = \ln 2 \times \tau = 0.693\,\tau .$$

(c) The time-dependent amplitude is $T(t) = \frac{1}{\sqrt{\tau}} e^{-t/2\tau} e^{-i\omega_0 t}$. Sketch on the complex plane the path of $\sqrt{\tau}\, T(t) = e^{-t/2\tau} e^{-i\omega_0 t}$ from $t = 0$ through two cycles. Take $e^{-t/2\tau}$ to equal 0.8 at the end of the first cycle.

7. *Exponential growth and decay.*

(a) A saving account earns 2% a year. How many years will it take to double in value? How long if the rate is 5%? Or 10%? Make up a simple rule for making quick estimates of these doubling times.

(b) One penny is put on the first square of a chess board (64 squares), two pennies on the second square, four pennies on the third square, ... How many will be put on the eighth square? On the 64th? If a penny is 1 mm thick, how high will the stack on the 64th square be?

(d) Scum on a pond doubles in area every 24 hours. In 30 days it will cover the whole pond. On which day does it first cover at least 1% of the pond? Make a quick guess (< 10 s) before figuring.

8. *The Breit-Wigner resonance function and amplitude.*

(a) Prove that the area under the Breit-Wigner function $1/(1 + \epsilon^2)$ is π, and thus show that $\int_{-\infty}^{+\infty} |b(\omega)|^2 d\omega = 1$.

(b) Prove that as ϵ runs from $-\infty$ to $+\infty$, the Breit-Wigner amplitude $(i + \epsilon)/(1 + \epsilon^2)$ traces the unit circle on the complex plane shown in Fig. 4.

(c) Prove that the magnitude of the amplitude is $1/\sqrt{1 + \epsilon^2}$. When squared, this is the resonance function $1/(1 + \epsilon^2)$ shown in Fig. 3.

9. *Uncertainty relations for spin 1/2.*

(a) Use the matrix operators given at the start of Sec. 6 to prove the commutation relations $[\hat{I}, \hat{\sigma}_z] = 0$ and $[\hat{\sigma}_x, \hat{\sigma}_y] = 2i\hat{\sigma}_z$. Cyclicly permuting subscripts would give $[\hat{I}, \hat{\sigma}_x]$, $[\hat{I}, \hat{\sigma}_y]$, $[\hat{\sigma}_y, \hat{\sigma}_z]$, and $[\hat{\sigma}_z, \hat{\sigma}_x]$. Thus $[\hat{s}^2, \hat{s}_i] = 0$ and $[\hat{s}_x, \hat{s}_y] = i\hbar \hat{s}_z$, where $\hat{s}_x = \frac{1}{2}\hbar\hat{\sigma}_x$, etc.

(b) Use the matrix operators to calculate the expectation values of $\hat{s}_x$, $\hat{s}_y$, $\hat{s}_z$, $\hat{s}_x^2$, $\hat{s}_y^2$, $\hat{s}_z^2$, and $\hat{s}^2$ for a particle in the $|+z\rangle$ state. That is, calculate $\langle +z|\hat{O}|+z\rangle$ for each of those seven operators.

(c) Thus write down the rms uncertainties Δs_x and Δs_y in the $|+z\rangle$ state. The general uncertainty principle $\Delta A\, \Delta B \geq \frac{1}{2}\,|\langle \Psi|[\hat{A}, \hat{B}]|\Psi\rangle|$ applied here reads

$$\Delta s_x\, \Delta s_y \geq \frac{\hbar}{2}\,|\langle +z|\,\hat{s}_z\,|+z\rangle| \ .$$

Which do you get, $>$ or $=$?

8. ANGULAR MOMENTUM

8.1. *Central Forces. Separation of Variables*
8.2. *Angular Momentum Commutation Relations*
8.3. *The Operator Solution*
8.4. *Certainty and Uncertainty*
8.5. *States as Vectors, Operators as Matrices (Again)*
8.6. *Differential Operators for Orbital Angular Momentum*
8.7. *Spherical Harmonics*
8.8. *Angular Momentum and the Oscillator*
    Problems

We begin the solution of three-dimensional problems for forces that are central, those for which $\mathbf{F}(\mathbf{r}) = F(r)\,\hat{\mathbf{r}}$. Then the potential energy is only a function of r, $V(\mathbf{r}) = V(r)$, and the Schrödinger equation in spherical coordinates separates into two equations, one involving only r and one involving only the angles θ and ϕ. Since the angular equation is independent of $V(r)$, its solutions apply for any $V(r)$. We solve the angular equation in this chapter. We solve the r-dependent equation for hydrogen, where $V(r) = -ke^2/r$, and for the isotropic harmonic oscillator, where $V(r) = \frac{1}{2}m\omega^2 r^2$, in the next.

The angular equation is an eigenvalue problem for angular momentum. To solve it, we develop operators for the components of the orbital angular momentum, $\mathbf{L} = \mathbf{r} \times \mathbf{p}$, and then use the same operator method we used in the latter half of Chapter 6 for the oscillator problem. "Whenever one encounters a set of operators satisfying similar commutation rules, one can play the same game" (Harry J. Lipkin, in *Lie Groups for Pedestrians*). The close parallels of the operator solutions of the oscillator and angular-momentum problems are summarized at the end of the chapter.

The operator method finds twice as many solutions as would be found in the series solution of the differential equation in θ and ϕ. Nature, not to be outdone by Mathematics, uses these extra solutions—most fundamentally as the intrinsic angular momenta (spins) of electrons, protons, neutrons, neutrinos, quarks, and other particles. Chapter 10 is all about spin.

We then show that a subset of the solutions obtained using the operator method are the solutions of the orbital angular equation we got from the Schrödinger equation. These solutions are the spherical harmonics, $Y_\ell^m(\theta, \phi)$. They play a large role in much of the rest of this text, and are widely used in other fields involving angular distributions in three dimensions.

147

8·1. CENTRAL FORCES. SEPARATION OF VARIABLES

(a) Angular momentum—Figure 1 shows a (classical) particle in motion about a center of force O. The position of the particle is $\mathbf{r}(t)$, its momentum is $\mathbf{p}(t)$, and the components of $\mathbf{p}$ parallel and perpendicular to $\mathbf{r}$ are $\mathbf{p}_\parallel$ and $\mathbf{p}_\perp$. The orbital angular momentum about O is $\mathbf{L} = \mathbf{r} \times \mathbf{p}$; its magnitude is $L = rp_\perp$. The torque about O is $\mathbf{N} = d\mathbf{L}/dt = \mathbf{r} \times \mathbf{F}$, where $\mathbf{F}$ is the force. If $\mathbf{F}$ is along $\mathbf{r}$, as it is for central forces, then $\mathbf{r} \times \mathbf{F} = 0$ and $\mathbf{L}$ is conserved. The kinetic energy is

$$K = \frac{p^2}{2m} = \frac{p_\parallel^2}{2m} + \frac{p_\perp^2}{2m} = \frac{p_\parallel^2}{2m} + \frac{L^2}{2mr^2} ,$$

where we have used $p_\perp = L/r$.

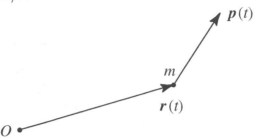

Figure 1. A particle of mass m in motion about the origin O of a central force.

To quantum mechanics. When the force is central, so that $V = V(r)$, we use spherical coordinates. In these coordinates, the Schrödinger equation is, as given in Sec. 2·7(b),

$$\hat{H}\psi = -\frac{\hbar^2}{2m}\nabla^2\psi + V(r)\psi = E\psi ,$$

or

$$-\frac{\hbar^2}{2m}\left[\frac{1}{r}\frac{\partial^2(r\psi)}{\partial r^2} + \frac{1}{r^2\sin\theta}\frac{\partial}{\partial\theta}\left(\sin\theta\frac{\partial\psi}{\partial\theta}\right) + \frac{1}{r^2\sin^2\theta}\frac{\partial^2\psi}{\partial\phi^2}\right] + V(r)\,\psi = E\psi ,$$

where $\psi = \psi(r,\theta,\phi)$.

Now compare the classical kinetic energy K with the kinetic-energy term $-(\hbar^2/2m)\nabla^2\psi$ of the Schrödinger equation. The angular-*momentum* part of the classical K, $L^2/2mr^2$, corresponds to the angular-*derivative* parts of $-(\hbar^2/2m)\nabla^2\psi$. We might therefore suppose the angular-momentum-squared operator, operating on ψ, to be

$$\boxed{\hat{L}^2\psi = -\hbar^2\left[\frac{1}{\sin\theta}\frac{\partial}{\partial\theta}\left(\sin\theta\frac{\partial\psi}{\partial\theta}\right) + \frac{1}{\sin^2\theta}\frac{\partial^2\psi}{\partial\phi^2}\right].}$$

And this identification is in fact correct (it could hardly be otherwise), as we return to in Sec. 6. For the moment we simply use $\hat{L}^2\psi$ as shorthand, and write the Schrödinger equation more compactly as

$$\hat{H}\psi = -\frac{\hbar^2}{2mr}\frac{\partial^2(r\psi)}{\partial r^2} + \frac{\hat{L}^2\psi}{2mr^2} + V(r)\,\psi = E\psi .$$

Since the *only* appearance of the variables θ and ϕ in the Hamiltonian operator is in the $\hat{L}^2$ term, it ought to be apparent that $\hat{L}^2$ commutes with $\hat{H}$: $[\hat{H}, \hat{L}^2] = 0$. And therefore at least the magnitude of the angular momentum is conserved.

(b) Separation—To separate variables, first r from θ and ϕ, we put the product $\psi(r, \theta, \phi) = R(r) Y(\theta, \phi)$ into the Schrödinger equation and divide by RY:

$$-\frac{\hbar^2}{2mrR}\frac{d^2(rR)}{dr^2} + \frac{\hat{L}^2 Y}{2mr^2 Y} + V(r) = E \, .$$

Multiplying through by $2mr^2$ and moving all r-dependent terms to the right side, we get

$$\frac{\hat{L}^2 Y}{Y} = 2mr^2 \left[\frac{\hbar^2}{2mrR}\frac{d^2(rR)}{dr^2} + E - V(r) \right] \, .$$

We have separation. There are only angles on the left, there is only r on the right.

Equating each side of the last equation to a separation constant $\lambda\hbar^2$, we get

$$\hat{L}^2 Y(\theta, \phi) = \lambda\hbar^2 \, Y(\theta, \phi)$$

for the angular equation (including $\hbar^2$ here makes λ dimensionless). With some rearrangement, the radial equation becomes

$$\boxed{\frac{d^2(rR)}{dr^2} + \frac{2m}{\hbar^2}\left[E - V(r) - \frac{\lambda\hbar^2}{2mr^2} \right] rR = 0 \, ,}$$

where $R = R(r)$. This is an ordinary differential equation, but it contains *two* separation constants, E and λ. We have to solve the angular equation—the work of this chapter—and find its eigenvalues λ before we can attack the radial equation. Therefore, we see little more of r until the next chapter, in which we solve the radial equation for the hydrogen atom and for the isotropic oscillator.

The angular equation is an eigenvalue equation for the operator $\hat{L}^2$ (in units of $\hbar^2$):

$$\frac{1}{\hbar^2}\hat{L}^2 Y = -\left[\frac{1}{\sin\theta}\frac{\partial}{\partial\theta}\left(\sin\theta\frac{\partial Y}{\partial\theta} \right) + \frac{1}{\sin^2\theta}\frac{\partial^2 Y}{\partial\phi^2} \right] = \lambda Y \, ,$$

This is independent of all the parameters in the problem: m, $\hbar$, and those that come with $V(r)$. Its solutions, the functions $Y(\theta, \phi)$, will apply to *any* central-force problem. One solution is apparent: $Y = 1$ (no angular dependence), with eigenvalue $\lambda = 0$. This is a state whose orbital angular momentum is zero.

It remains to separate the angles. Putting $Y(\theta, \phi) = P(\theta) \, \Phi(\phi)$ into the above equation, multiplying through by $\sin^2\theta$, and dividing by $P\Phi$, we get

$$-\frac{\sin\theta}{P}\frac{d}{d\theta}\left(\sin\theta\frac{dP}{d\theta} \right) - \frac{1}{\Phi}\frac{d^2\Phi}{d\phi^2} = \lambda\sin^2\theta \, ,$$

or

$$-\frac{1}{\Phi}\frac{d^2\Phi}{d\phi^2} = \frac{\sin\theta}{P}\frac{d}{d\theta}\left(\sin\theta\frac{dP}{d\theta} \right) + \lambda\sin^2\theta = \mu \, ,$$

where μ is a third separation constant. Thus

$$\frac{1}{\sin\theta}\frac{d}{d\theta}\left(\sin\theta\frac{dP}{d\theta} \right) + \left(\lambda - \frac{\mu}{\sin^2\theta} \right) P = 0 \qquad \text{and} \qquad \frac{d^2\Phi}{d\phi^2} + \mu\Phi = 0 \, .$$

The schematic

$$t \uparrow^E r \uparrow^\lambda \theta \uparrow^\mu \phi \, ,$$

read from left to right, summarizes the sequence of separations and separation constants.

We could now solve the angular equations directly, as differential equations, working backwards from the equation for $\Phi(\phi)$, where there is only one separation constant. However, there is another method, very like the operator method for the harmonic oscillator, that uses the commutation relations of the angular-momentum operators. This second method brings forth twice as many solutions as are found solving the differential equations, and these extra solutions *exist* in Nature. So we venture again into a blizzard of commutation relations, and find eigenstates and eigenvalues. But not until the very end do we find our way to the functional solutions $Y(\theta, \phi)$ of the differential equations in θ and ϕ.

8·2. ANGULAR MOMENTUM COMMUTATION RELATIONS

(a) **The angular-momentum operators**—The components of the orbital angular momentum $\mathbf{L} = \mathbf{r} \times \mathbf{p}$ are

$$L_x = yp_z - zp_y , \quad L_y = zp_x - xp_z , \quad L_z = xp_y - yp_x .$$

We can move from one to the next of these equations by cyclically permuting the variables x, y, and z: $x \to y \to z \to x$. This is because the y axis bears the same geometrical relation to the x axis as the z axis does to the y axis as the x axis does to the z axis; see Fig. 2. The components of $\mathbf{L}$ are also given by expanding the determinant

$$\mathbf{L} = \begin{vmatrix} \hat{\mathbf{x}} & \hat{\mathbf{y}} & \hat{\mathbf{z}} \\ x & y & z \\ p_x & p_y & p_z \end{vmatrix} .$$

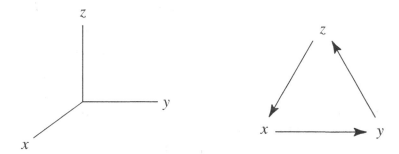

Figure 2. The y axis bears the same relation to the x axis as the z axis does to the y axis as the x axis does to the z axis.

To begin, we need the commutation relations of L_x, L_y, and L_z with one another. Since these operators are built from the fundamental operators $\hat{x}_i \equiv \hat{x}, \hat{y}, \hat{z}$ and $\hat{p}_i \equiv \hat{p}_x, \hat{p}_y, \hat{p}_z$, we need their commutation relations:

$[\hat{x}_i, \hat{x}_j] = 0$	because	$xy\psi = yx\psi$, etc.
$[\hat{p}_i, \hat{p}_j] = 0$	"	$\partial^2\psi/\partial x\partial y = \partial^2\psi/\partial y\partial x$, etc.
$[\hat{x}_i, \hat{p}_j] = i\hbar\,\delta_{ij}$	"	$[\hat{x}, \hat{p}_x] = i\hbar$, but $[\hat{x}, \hat{p}_y] = -i\hbar x\,\partial\psi/\partial y + i\hbar\partial(x\psi)/\partial y = 0$, etc.

Each of the six operators commutes with all but one of the other five.

(b) The basic commutators—Here and below, we make repeated use of the basic commutator identities given in Sec. 5·4(b),

$$[\hat{A} + \hat{B}, \hat{C}] = [\hat{A}, \hat{C}] + [\hat{B}, \hat{C}] \quad \text{and} \quad [\hat{A}\hat{B}, \hat{C}] = \hat{A}[\hat{B}, \hat{C}] + [\hat{A}, \hat{C}]\hat{B} .$$

The commutator of $\hat{L}_x$ and $\hat{L}_y$ is

$$[\hat{L}_x, \hat{L}_y] = \hat{L}_x\hat{L}_y - \hat{L}_y\hat{L}_x = (\hat{y}\hat{p}_z - \hat{z}\hat{p}_y)(\hat{z}\hat{p}_x - \hat{x}\hat{p}_z) - (\hat{z}\hat{p}_x - \hat{x}\hat{p}_z)(\hat{y}\hat{p}_z - \hat{z}\hat{p}_y) .$$

In writing out the (eight) terms, we can reorder as convenient operators that commute with one another, but need to keep the order of operators that do not commute (here $\hat{z}$ and $\hat{p}_z$). Thus we get

$$[\hat{L}_x, \hat{L}_y] = \hat{y}\hat{p}_x\hat{p}_z\hat{z} - \hat{x}\hat{y}\hat{p}_z^2 - \hat{z}^2\hat{p}_x\hat{p}_y + \hat{x}\hat{p}_y\hat{z}\hat{p}_z - \hat{y}\hat{p}_x\hat{z}\hat{p}_z + \hat{z}^2\hat{p}_x\hat{p}_y + \hat{x}\hat{y}\hat{p}_z^2 - \hat{x}\hat{p}_y\hat{p}_z\hat{z}$$

$$= \hat{x}\hat{p}_y(\hat{z}\hat{p}_z - \hat{p}_z\hat{z}) - \hat{y}\hat{p}_x(\hat{z}\hat{p}_z - \hat{p}_z\hat{z})$$

$$= i\hbar(\hat{x}\hat{p}_y - \hat{y}\hat{p}_x) = i\hbar\hat{L}_z .$$

Cyclically permuting variables, we get altogether

$$[\hat{L}_x, \hat{L}_y] = i\hbar\hat{L}_z , \quad [\hat{L}_y, \hat{L}_z] = i\hbar\hat{L}_x , \quad [\hat{L}_z, \hat{L}_x] = i\hbar\hat{L}_y .$$

And of course $[\hat{L}_x, \hat{L}_x] = 0$, etc. These are the fundamental angular-momentum commutation relations: the components L_x, L_y, and L_z of **L** do not commute with one another. Everything below follows from these relations.

Next, we define a total-angular-momentum-squared operator, $\hat{L}^2 \equiv \hat{L}_x^2 + \hat{L}_y^2 + \hat{L}_z^2$, and show that it commutes with the components of $\hat{\mathbf{L}}$. We get

$$[\hat{L}^2, \hat{L}_x] = [\hat{L}_x^2 + \hat{L}_y^2 + \hat{L}_z^2, \hat{L}_x]$$

$$= 0 + \hat{L}_y[\hat{L}_y, \hat{L}_x] + [\hat{L}_y, \hat{L}_x]\hat{L}_y + \hat{L}_z[\hat{L}_z, \hat{L}_x] + [\hat{L}_z, \hat{L}_x]\hat{L}_z$$

$$= i\hbar(-\hat{L}_y\hat{L}_z - \hat{L}_z\hat{L}_y + \hat{L}_z\hat{L}_y + \hat{L}_y\hat{L}_z)$$

$$= 0 .$$

By symmetry, $\hat{L}^2$ also commutes with $\hat{L}_y$ and $\hat{L}_z$.

(c) Ladder operators—Lastly, we introduce two operators, $\hat{L}_\pm = \hat{L}_x \pm i\hat{L}_y$. They may be thought of as an attempt to factor the right side of $\hat{L}^2 - \hat{L}_z^2 = \hat{L}_x^2 + \hat{L}_y^2$, like $a^2 + b^2 = (a + ib)(a - ib)$. They will play roles like $\hat{a}^\dagger$ and $\hat{a}$ did in the oscillator problem. And like $\hat{a}^\dagger$ and $\hat{a}$, they are not Hermitian but are adjoints of one another: $(\hat{L}_\pm)^\dagger = \hat{L}_\mp$. We have

$$[\hat{L}^2, \hat{L}_\pm] = [\hat{L}^2, \hat{L}_x] \pm i[\hat{L}^2, \hat{L}_y] = 0 ,$$

and

$$[\hat{L}_z, \hat{L}_+] = [\hat{L}_z, \hat{L}_x] + i[\hat{L}_z, \hat{L}_y] = i\hbar\hat{L}_y + i(-i\hbar\hat{L}_x) = +\hbar\hat{L}_+ .$$

Similarly, $[\hat{L}_z, \hat{L}_-] = -\hbar\hat{L}_-$. Also useful will be

$$\hat{L}_+\hat{L}_- = (\hat{L}_x + i\hat{L}_y)(\hat{L}_x - i\hat{L}_y) = \hat{L}_x^2 + \hat{L}_y^2 - i(\hat{L}_x\hat{L}_y - \hat{L}_y\hat{L}_x) = \hat{L}^2 - \hat{L}_z^2 + \hbar\hat{L}_z .$$

Similarly, $\hat{L}_-\hat{L}_+ = \hat{L}^2 - \hat{L}_z^2 - \hbar\hat{L}_z$. Lastly, we will need $\hat{L}_x$ and $\hat{L}_y$ in terms of $\hat{L}_+$ and $\hat{L}_-$:

$$\hat{L}_x = \frac{1}{2}(\hat{L}_+ + \hat{L}_-) \quad \text{and} \quad \hat{L}_y = \frac{i}{2}(\hat{L}_- - \hat{L}_+) .$$

This completes the basic setup.

151

8·3. THE OPERATOR SOLUTION

(a) Notation—In the following, we are going to use symbols $\hat{J}_x$, $\hat{J}_y$, $\hat{J}_z$, $\hat{J}^2$, and $\hat{J}_\pm$, instead of $\hat{L}_x$, $\hat{L}_y$, $\hat{L}_z$, $\hat{L}^2$, and $\hat{L}_\pm$, for the angular-momentum operators. As was already remarked, the operator method finds *more* solutions than we would get from solving the differential equations of Sec. 1—even though we got the commutation relations of Sec. 2 from the *orbital*-angular-momentum operators. The extra solutions exist in Nature, most fundamentally as the intrinsic angular momenta (spins) of electrons, protons, neutrons, neutrinos, quarks, and other particles. Therefore, we shall use J for general or unspecified angular momenta, reserve L for orbital angular momenta, and use S for spin angular momenta.

What follows is the same in spirit, although not in all particulars, as the operator solution of the harmonic oscillator. There everything followed from $[\hat{x}, \hat{p}] = i\hbar$ and the form of the Hamiltonian, quadratic in $\hat{x}$ and in $\hat{p}$. Here everything follows from the commutation relations for the components of angular momentum,

$$\boxed{[\hat{J}_x, \hat{J}_y] = i\hbar\hat{J}_z \ , \quad [\hat{J}_y, \hat{J}_z] = i\hbar\hat{J}_x \ , \quad [\hat{J}_z, \hat{J}_x] = i\hbar\hat{J}_y \ ,}$$

and the quadratic $\hat{J}_x^2 + \hat{J}_y^2$. We restate, using J instead of L, the relations developed in Sec. 2:

$$\boxed{\begin{array}{ll}
\hat{J}^2 \equiv \hat{J}_x^2 + \hat{J}_y^2 + \hat{J}_z^2 & \hat{J}_\pm \equiv \hat{J}_x \pm i\hat{J}_y \\[2mm]
\hat{J}_x = \dfrac{1}{2}(\hat{J}_+ + \hat{J}_-) & \hat{J}_y = \dfrac{i}{2}(\hat{J}_- - \hat{J}_+) \\[2mm]
[\hat{J}^2, \hat{J}_i] = 0, \ \ i =, x, y, z & [\hat{J}^2, \hat{J}_\pm] = 0 \\[2mm]
[\hat{J}_z, \hat{J}_\pm] = \pm\hbar\hat{J}_\pm & \hat{J}_\pm\hat{J}_\mp = \hat{J}^2 - \hat{J}_z^2 \pm \hbar\hat{J}_z \ .
\end{array}}$$

(b) Ladders of states—Because the operators for J^2 and J_z commute, the uncertainty principle places no limit on how well we can simultaneously know their values. We start by assuming there do exist simultaneous eigenstates of $\hat{J}^2$ and $\hat{J}_z$:

$$\hat{J}^2\,|\alpha, \beta\rangle = \alpha\,|\alpha, \beta\rangle \qquad \text{and} \qquad \hat{J}_z\,|\alpha, \beta\rangle = \beta\,|\alpha, \beta\rangle \ .$$

We label the (normalized) kets by the unknown eigenvalues α and β; α is the square of the length of $\mathbf{J}$ and β is the z component of $\mathbf{J}$. We cannot look for simultaneous eigenstates of, say, all of $\hat{J}^2$, $\hat{J}_y$, and $\hat{J}_z$ because $\hat{J}_y$ and $\hat{J}_z$ do not commute.

Consider $\hat{J}_+|\alpha, \beta\rangle$: Is it an eigenstate of $\hat{J}^2$ and of $\hat{J}_z$? (*Déjà vu*: Consider $\hat{a}^\dagger|E\rangle$: Is it an eigenstate of $\hat{H}$?) Using $[\hat{J}^2, \hat{J}_+] = 0$, which means $\hat{J}^2\hat{J}_+ = \hat{J}_+\hat{J}^2$, we get

$$\hat{J}^2(\hat{J}_+\,|\alpha, \beta\rangle) = \hat{J}_+\hat{J}^2\,|\alpha, \beta\rangle = \alpha\,(\hat{J}_+\,|\alpha, \beta\rangle) \ .$$

Using $[\hat{J}_z, \hat{J}_+] = +\hbar\hat{J}_+$, or $\hat{J}_z\hat{J}_+ = \hat{J}_+\hat{J}_z + \hbar\hat{J}_+$, we get

$$\hat{J}_z(\hat{J}_+\,|\alpha, \beta\rangle) = (\hat{J}_+\hat{J}_z + \hbar\hat{J}_+)\,|\alpha, \beta\rangle = (\beta + \hbar)(\hat{J}_+\,|\alpha, \beta\rangle) \ .$$

Thus $\hat{J}_+\,|\alpha, \beta\rangle$ is an eigenstate of $\hat{J}^2$ with an unchanged eigenvalue α; and it is also an eigenstate of $\hat{J}_z$ but with the eigenvalue raised by $\hbar$. That is,

$$\hat{J}_+\,|\alpha, \beta\rangle = C_{\alpha\beta}^+\,|\alpha, \beta + \hbar\rangle \ .$$

152

The constant $C_{\alpha\beta}^+$ is to allow for the possibility that $\hat{J}_+ |\alpha, \beta\rangle$ is not the *normalized* state $|\alpha, \beta + \hbar\rangle$; all kets here are normalized. Similarly, using $[\hat{J}^2, \hat{J}_-] = 0$ and $[\hat{J}_z, \hat{J}_-] = -\hbar \hat{J}_-$, we would get

$$\hat{J}_- |\alpha, \beta\rangle = C_{\alpha\beta}^- |\alpha, \beta - \hbar\rangle \ .$$

Thus $\hat{J}_+$ and $\hat{J}_-$ create new states $|\alpha, \beta + \hbar\rangle$ and $|\alpha, \beta - \hbar\rangle$ from $|\alpha, \beta\rangle$, just as $\hat{a}^\dagger$ and $\hat{a}$ created new states $|E + \hbar\omega\rangle$ and $|E - \hbar\omega\rangle$ from $|E\rangle$.

By repeated use of $\hat{J}_+$ and $\hat{J}_-$, we can create from any state $|\alpha, \beta\rangle$ a ladder of states, all with the same eigenvalue α of $\hat{J}^2$, but with different values of $\hat{J}_z$; see Fig. 3. However α, whatever its value, sets bounds on the range of values of β. The left side of

$$\langle \alpha, \beta | \hat{J}^2 | \alpha, \beta\rangle = \langle \alpha, \beta | \hat{J}_x^2 | \alpha, \beta\rangle + \langle \alpha, \beta | \hat{J}_y^2 | \alpha, \beta\rangle + \langle \alpha, \beta | \hat{J}_z^2 | \alpha, \beta\rangle$$

equals α. The last term on the right side equals β^2. The other two terms, being expectation values of the squares of Hermitian operators, are nonnegative. Thus $\alpha \geq \beta^2 \geq 0$. Not even in quantum mechanics can the projection β of $\mathbf{J}$ on an axis be larger than the length $\sqrt{\alpha}$ of $\mathbf{J}$. We conclude that the ladder of states having the "family" name α has a top rung with $\beta = \beta_{max}$ and a bottom rung with $\beta = \beta_{min}$, where $\alpha \geq \beta_{max}^2$ and $\alpha \geq \beta_{min}^2$.

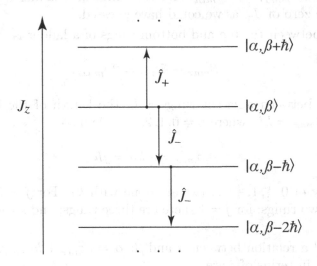

Figure 3. The ladder of states created by $\hat{J}_+$ and $\hat{J}_-$ from $|\alpha, \beta\rangle$.

(c) Eigenvalues—It follows that J_+ acting on the top rung, and J_- acting on the bottom rung, cannot create new states. That is,

$$\hat{J}_+ |\alpha, \beta_{max}\rangle = 0 \quad \text{and} \quad \hat{J}_- |\alpha, \beta_{min}\rangle = 0 \ .$$

Then, since $\hat{J}_+$ and $\hat{J}_-$ are linear operators, and linear operators operating on zero give zero, we get

$$\hat{J}_- \hat{J}_+ |\alpha, \beta_{max}\rangle = 0 \quad \text{and} \quad \hat{J}_+ \hat{J}_- |\alpha, \beta_{min}\rangle = 0 \ .$$

Using one of our operator identities, $\hat{J}_- \hat{J}_+ = \hat{J}^2 - \hat{J}_z^2 - \hbar \hat{J}_z$, we get

$$\hat{J}_- \hat{J}_+ |\alpha, \beta_{max}\rangle = (\hat{J}^2 - \hat{J}_z^2 - \hbar \hat{J}_z)|\alpha, \beta_{max}\rangle = (\alpha - \beta_{max}^2 - \hbar \beta_{max})|\alpha, \beta_{max}\rangle = 0 \ .$$

153

And using $\hat{J}_+\hat{J}_- = \hat{J}^2 - \hat{J}_z^2 + \hbar\hat{J}_z$, we get

$$\hat{J}_+\hat{J}_-|\alpha, \beta_{min}\rangle = (\hat{J}^2 - \hat{J}_z^2 + \hbar\hat{J}_z)|\alpha, \beta_{min}\rangle = (\alpha - \beta_{min}^2 + \hbar\beta_{min})|\alpha, \beta_{min}\rangle = 0 \ .$$

The factors $(\alpha - \beta_{max}^2 - \hbar\beta_{max})$ and $(\alpha - \beta_{min}^2 + \hbar\beta_{min})$ at the right ends of the last two equations must equal zero because the kets $|\alpha, \beta_{max}\rangle$ and $|\alpha, \beta_{min}\rangle$ they multiply are not zero. Equating the two resultant expressions for α, we get

$$\alpha = \beta_{max}^2 + \hbar\beta_{max} = \beta_{min}^2 - \hbar\beta_{min} \ .$$

This factors as

$$\beta_{max}^2 - \beta_{min}^2 + \hbar(\beta_{max} + \beta_{min}) = (\beta_{max} + \beta_{min})(\beta_{max} - \beta_{min} + \hbar) = 0 \ .$$

The solutions are

$$\beta_{min} = -\beta_{max} \qquad \text{or} \qquad \beta_{min} = \beta_{max} + \hbar \ .$$

We discard the second solution because it makes $\beta_{min} > \beta_{max}$, but β_{max} is *up* the ladder from β_{min}. Therefore $\beta_{max} = -\beta_{min}$, and the rungs of a ladder are symmetrically placed with respect to the zero of J_z, as we could have guessed.

The distance between the top and bottom rungs of a ladder is

$$\beta_{max} - \beta_{min} = 2\beta_{max} \ .$$

Since the distance between adjacent rungs is $\hbar$, the length of the ladder is some integral multiple k of $\hbar$: $2\beta_{max} = k\hbar$, where $k = 0, 1, 2, \ldots$ And so

$$\beta_{max} = \tfrac{1}{2}k\hbar = j\hbar \ ,$$

where j can be any of $0, \frac{1}{2}, 1, \frac{3}{2}, \ldots$ (we are done with k). For $j = 0$ there is one rung; for $j = 1/2$ there are two rungs; for $j = 1$ there are three rungs; and so on. There are an infinite number of ladders.

Above, we had a relation between α and β: $\alpha = \beta_{max}^2 + \hbar\beta_{max}$. Since $\beta_{max} = j\hbar$, the eigenvalues α of $\hat{J}^2$ in terms of j are

$$\alpha = \beta_{max}^2 + \hbar\beta_{max} = j(j+1)\hbar^2 \ .$$

For a given value of j, the eigenvalues β of $\hat{J}_z$ range in steps of size $\hbar$ from $\beta_{max} = j\hbar$ down to $\beta_{min} = -\beta_{max} = -j\hbar$. Letting $\beta = m\hbar$, and labeling kets by the dimensionless quantum numbers j and m, we finally have the simultaneous eigenstates of $\hat{J}^2$ and J_z,

$$\hat{J}^2|j,m\rangle = j(j+1)\hbar^2|j,m\rangle, \qquad \text{where } j = 0, \tfrac{1}{2}, 1, \tfrac{3}{2}, \ldots$$

$$\hat{J}_z|j,m\rangle = m\hbar|j,m\rangle, \qquad \text{where } m = j, j-1, \ldots, -j \ .$$

For a given value of j, there are $2j + 1$ values of m. Figure 4 shows the first few *angular-momentum multiplets*.

154

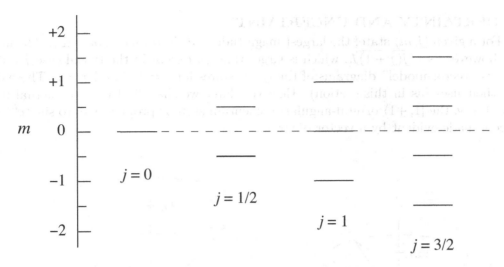

Figure 4. The angular-momentum multiplets with $j = 0, 1/2, 1,$ and $3/2$. There are $2j + 1$ rungs per ladder.

(d) Constants $C_{jm}^{\pm}$—We relabel the constants $C_{\alpha\beta}^{\pm}$ with the quantum numbers j and m. Then

$$\hat{J}_{\pm} \,|j, m\rangle = C_{jm}^{\pm} \,|j, m \pm 1\rangle \,,$$

where $|j, m\rangle$ and $|j, m \pm 1\rangle$ are normalized kets. The magnitudes squared, $|C_{jm}^{+}|^2$, are

$$\langle j, m | \hat{J}_{-} \hat{J}_{+} | j, m \rangle = |C_{jm}^{+}|^2 \langle j, m + 1 | j, m + 1 \rangle = |C_{jm}^{+}|^2 \,.$$

Using the identity $\hat{J}_{-} \hat{J}_{+} = \hat{J}^2 - \hat{J}_z^2 - \hbar \hat{J}_z$, we then get

$$|C_{jm}^{+}|^2 = \langle j, m | (\hat{J}^2 - \hat{J}_z^2 - \hbar J_z) | j, m \rangle = [j(j + 1) - m^2 - m]\hbar^2 = (j - m)(j + m + 1)\hbar^2 \,.$$

Choosing a phase to make C_{jm}^{+} real and non-negative, we have

$$C_{jm}^{+} = \sqrt{(j - m)(j + m + 1)} \, \hbar \,.$$

And replacing m with $-m$ turns C_{jm}^{+} into C_{jm}^{-}: $C_{jm}^{-} = \sqrt{(j + m)(j - m + 1)} \, \hbar$. Note that $C_{j,+j}^{+} = C_{j,-j}^{-} = 0$, so that $\hat{J}_{+} |j, +j\rangle = 0$ and $\hat{J}_{-} |j, -j\rangle = 0$ and the ladders have ends.

The four main results, with which one can do a great deal with angular momentum, are

$$
\boxed{
\begin{aligned}
\hat{J}^2 \,|j, m\rangle &= j(j + 1)\hbar^2 \,|j, m\rangle \qquad &&\text{where } j = 0, \tfrac{1}{2}, 1, \tfrac{3}{2}, \dots \\
\hat{J}_z \,|j, m\rangle &= m\hbar \,|j, m\rangle \qquad &&\text{where } m = j, j - 1, \dots, -j \\
\hat{J}_{+} \,|j, m\rangle &= \sqrt{(j - m)(j + m + 1)} \, \hbar \,|j, m + 1\rangle \\
\hat{J}_{-} \,|j, m\rangle &= \sqrt{(j + m)(j - m + 1)} \, \hbar \,|j, m - 1\rangle \,.
\end{aligned}
}
$$

The kets here are normalized. And each $|j, m\rangle$ state is orthogonal to every other one,

$$\boxed{\langle j', m' | j, m \rangle = \delta_{j'j} \delta_{m'm} \,.}$$

8·4. CERTAINTY AND UNCERTAINTY

For a given $|j, m\rangle$ state, the largest magnitude of J_z is $j\hbar$, when $m = \pm j$. The magnitude of $\mathbf{J}$, however, is $\sqrt{j(j+1)}\,\hbar$, which is larger than $j\hbar$ except in the trivial case $j = 0$. Figure 5 shows "vector-model" diagrams of the $|j, m\rangle$ states for $j = 1/2$ and $j = 1$. These diagrams have their uses (as in this section). However, later we shall find that the actual functional form of, say, the $|1, +1\rangle$ orbital-angular-momentum state is proportional to $\sin\theta\, e^{i\phi}$, which is scarcely made evident by a vector diagram.

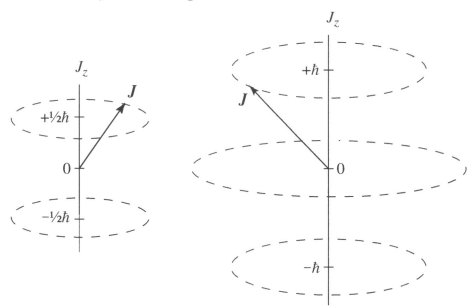

Figure 5. The $|j, m\rangle$ states for $j = 1/2$ and 1, and the $\mathbf{J}$ vectors for the $m = j$ states.

In any $|j, m\rangle$ state, $|\mathbf{J}|$ and J_z are known exactly, but where on a circle in Fig. 5 the tip of $\mathbf{J}$ would lie is unknown (and unknowable). Since $J_x^2 + J_y^2 = J^2 - J_z^2 = [j(j+1) - m^2]\,\hbar^2$, the radii of those circles are $\sqrt{[j(j+1) - m^2]}\,\hbar$. The angle between $\mathbf{J}$ and the z axis is

$$\cos\theta = \frac{m}{\sqrt{j(j+1)}} \, .$$

For the $|\frac{1}{2}, \pm\frac{1}{2}\rangle$ states, $\mathbf{J}$ is at $55°$ to the $\pm z$ directions; for the $|1, \pm1\rangle$ states it is at $45°$ to the $\pm z$ directions. In the classical limit of very large j, the angular momentum can point as close as you like to any specified direction.

Why can't $\mathbf{J}$ lie along the z (or any other specified) direction? Because if it did, then J_x and J_y would be zero, and all the uncertainties ΔJ_x, ΔJ_y, and ΔJ_z would be zero. The general uncertainty relation, derived in Sec. 7·3, states that if $[\hat{A}, \hat{B}] = i\hat{C}$, then in the state Ψ,

$$\Delta A\, \Delta B \geq \tfrac{1}{2}|\langle\Psi|\hat{C}|\Psi\rangle|$$

Here $[\hat{A}, \hat{B}] = i\hat{C}$ is $[\hat{J}_x, \hat{J}_y] = i\hbar\hat{J}_z$ and the state is $|j, m\rangle$, so

$$\Delta J_x\, \Delta J_y \geq \tfrac{1}{2}\hbar|\langle j, m|\hat{J}_z|j, m\rangle| \, ,$$

where $(\Delta J_x)^2 = \langle\hat{J}_x^2\rangle - \langle\hat{J}_x\rangle^2$, etc. It remains to show that the uncertainty relation is in fact satisfied for every $|j, m\rangle$ state.

In terms of the ladder operators $\hat{J}_\pm = \hat{J}_x \pm i\hat{J}_y$, the $\hat{J}_x$ and $\hat{J}_y$ operators are

$$\hat{J}_x = \frac{1}{2}(\hat{J}_+ + \hat{J}_-) \qquad \text{and} \qquad \hat{J}_y = \frac{i}{2}(\hat{J}_- - \hat{J}_+) .$$

Thus $\hat{J}_x |j, m\rangle$ is a linear superposition of the $|j, m+1\rangle$ and $|j, m-1\rangle$ states, each of which is orthogonal to $|j, m\rangle$. Then $\langle J_x \rangle = \langle j, m|\hat{J}_x|j, m\rangle = 0$, and $(\Delta J_x)^2 = \langle J_x^2 \rangle - \langle J_x \rangle^2$ reduces to $\langle J_x^2 \rangle$. Similarly, $(\Delta J_y)^2$ reduces to $\langle J_y^2 \rangle$, and

$$\begin{aligned}
(\Delta J_x)^2 + (\Delta J_y)^2 &= \langle j, m|(\hat{J}_x^2 + \hat{J}_y^2)|j, m\rangle \\
&= \langle j, m|(\hat{J}^2 - \hat{J}_z^2)|j, m\rangle \\
&= \left[j(j+1) - m^2 \right]\hbar^2 .
\end{aligned}$$

Furthermore, the symmetric relation of the x and y axes to the z axis in Fig. 5 makes it evident that $\Delta J_x = \Delta J_y$, so that $(\Delta J_x)^2 + (\Delta J_y)^2 = 2\,\Delta J_x \Delta J_y$.

Therefore, for any $|j, m\rangle$ state, the left side of the uncertainty relation $\Delta J_x \Delta J_y \geq \frac{1}{2}\hbar|\langle j, m|\hat{J}_z|j, m\rangle|$ is

$$\Delta J_x \Delta J_y = \tfrac{1}{2}\left[j(j+1) - m^2 \right]\hbar^2 .$$

The right side is

$$\tfrac{1}{2}\hbar|\langle j, m|\hat{J}_z|j, m\rangle| = \tfrac{1}{2}|m|\hbar^2 .$$

The largest value of $|m|$ is j, and this value both minimizes the left side ($\Delta J_x \Delta J_y$) and maximizes the right side ($\frac{1}{2}\hbar|\langle \hat{J}_z \rangle|$), and so it provides the strongest test of the uncertainty relation. But when $|m| = j$, the two sides of the uncertainty relation are equal to $j\hbar^2/2$. If $\mathbf{J}$ lay any closer to the z axis, $\Delta J_x \Delta J_y$ would be smaller and $\frac{1}{2}\hbar|\langle \hat{J}_z \rangle|$ would be larger, and the uncertainty principle would be violated. The top and bottom rungs of each ladder of $|j, m\rangle$ states "saturate" the angular-momentum uncertainty relation. Just like the bottom ($n = 0$) rung of the oscillator ladder made $\Delta x \Delta p = \hbar/2$.

8·5. STATES AS VECTORS, OPERATORS AS MATRICES (AGAIN)

It is often useful to represent angular-momentum states as column vectors and the operators as matrices, just as was done for oscillator states and operators in Sec. 6·4. We consider $j = 1/2$ and $j = 1$ in turn, and then it will be clear what to do for any value of j.

(a) j=1/2—When $j = 1/2$, there are just two simultaneous eigenstates of $\hat{J}^2$ and $\hat{J}_z$: $|j, m\rangle = |\frac{1}{2}, +\frac{1}{2}\rangle$ and $|\frac{1}{2}, -\frac{1}{2}\rangle$. We represent them with two-component vectors, called *spinors* (already seen at the end of Chap. 7),

$$|\tfrac{1}{2}, +\tfrac{1}{2}\rangle = \begin{pmatrix} 1 \\ 0 \end{pmatrix} \qquad \text{and} \qquad |\tfrac{1}{2}, -\tfrac{1}{2}\rangle = \begin{pmatrix} 0 \\ 1 \end{pmatrix} .$$

These orthonormal vectors act as basis states for writing any state $|\chi\rangle$ of a $j = 1/2$ system:

$$|\chi\rangle = c_+ |\tfrac{1}{2}, +\tfrac{1}{2}\rangle + c_- |\tfrac{1}{2}, -\tfrac{1}{2}\rangle = c_+ \begin{pmatrix} 1 \\ 0 \end{pmatrix} + c_- \begin{pmatrix} 0 \\ 1 \end{pmatrix} = \begin{pmatrix} c_+ \\ c_- \end{pmatrix} .$$

The state is normalized, $\langle \chi|\chi \rangle = 1$, when $|c_+|^2 + |c_-|^2 = 1$. Then the probability of getting $+\hbar/2$ on measuring J_z would be $|c_+|^2$.

The picket-fence recipe of Sec. 5·5 specified that the columns of the matrix representation of an operator $\hat{A}$ are what the operator does to the basis states,

$$\hat{A} = \begin{pmatrix} \overset{\uparrow}{\hat{A}|1\rangle} & \overset{\uparrow}{\hat{A}|2\rangle} & \cdots \\ \downarrow & \downarrow \end{pmatrix}.$$

The matrix operators for $j = 1/2$ will be 2-by-2 because there are only two basis states. Using $\hat{J}^2 |\frac{1}{2}, \pm\frac{1}{2}\rangle = j(j+1)\hbar^2 |\frac{1}{2}, \pm\frac{1}{2}\rangle$, we get

$$\hat{J}^2 = \frac{3\hbar^2}{4} \begin{pmatrix} 1 & 0 \\ 0 & 1 \end{pmatrix} = \frac{3\hbar^2}{4} \hat{I},$$

where $\hat{I}$ is the 2-by-2 identity matrix. Using $\hat{J}_z |\frac{1}{2}, \pm\frac{1}{2}\rangle = m\hbar |\frac{1}{2}, \pm\frac{1}{2}\rangle$, we get

$$\hat{J}_z = \frac{\hbar}{2} \begin{pmatrix} 1 & 0 \\ 0 & -1 \end{pmatrix}.$$

The matrices for $\hat{J}^2$ and $\hat{J}_z$ are diagonal in the $|j, m\rangle$ basis because the $|j, m\rangle$ states are eigenstates of those operators. Using $\hat{J}_+ |\frac{1}{2}, \pm\frac{1}{2}\rangle = \sqrt{(j-m)(j+m+1)}\, \hbar |\frac{1}{2}, \pm\frac{1}{2}\rangle$ and the corresponding equation for $\hat{J}_-$, we get

$$\hat{J}_+ = \hbar \begin{pmatrix} 0 & 1 \\ 0 & 0 \end{pmatrix} \qquad \text{and} \qquad \hat{J}_- = \hbar \begin{pmatrix} 0 & 0 \\ 1 & 0 \end{pmatrix},$$

which are adjoints of one another. And finally we get

$$\hat{J}_x = \frac{1}{2}(\hat{J}_+ + \hat{J}_-) = \frac{\hbar}{2} \begin{pmatrix} 0 & 1 \\ 1 & 0 \end{pmatrix} \qquad \text{and} \qquad \hat{J}_y = \frac{i}{2}(\hat{J}_- - \hat{J}_+) = \frac{\hbar}{2} \begin{pmatrix} 0 & -i \\ i & 0 \end{pmatrix},$$

which are Hermitian.

These matrix representations satisfy all the relations $[\hat{J}_x, \hat{J}_y] = i\hbar\hat{J}_z$, $\hat{J}^2 = \hat{J}_x^2 + \hat{J}_y^2 + \hat{J}_z^2$, $[\hat{J}_z, \hat{J}_+] = +\hbar\hat{J}_+$, and so on. The matrices for J_x, J_y, and J_z, without the factors $\hbar/2$, are called the *Pauli spin matrices*. We make much use of them in Chap. 10, which is entirely about spin-1/2 particles.

(b) **j=1**—Now there are three $|j, m\rangle$ states, $|1, +1\rangle$, $|1, 0\rangle$, and $|1, -1\rangle$, which we represent with three-component vectors:

$$|1, +1\rangle = \begin{pmatrix} 1 \\ 0 \\ 0 \end{pmatrix}, \qquad |1, 0\rangle = \begin{pmatrix} 0 \\ 1 \\ 0 \end{pmatrix}, \qquad \text{and} \qquad |1, -1\rangle = \begin{pmatrix} 0 \\ 0 \\ 1 \end{pmatrix}.$$

A general $j = 1$ state is the linear superposition of these states with coefficients c_+, c_0, and c_-, where $|c_+|^2 + |c_0|^2 + |c_-|^2 = 1$.

The representations of the $\hat{J}^2$ and $\hat{J}_z$ operators have their eigenvalues along the diagonal,

$$\hat{J}^2 = 2\hbar^2 \begin{pmatrix} 1 & 0 & 0 \\ 0 & 1 & 0 \\ 0 & 0 & 1 \end{pmatrix} \quad \text{and} \quad \hat{J}_z = \hbar \begin{pmatrix} +1 & 0 & 0 \\ 0 & 0 & 0 \\ 0 & 0 & -1 \end{pmatrix}.$$

The raising and lowering operators are easily shown to be

$$\hat{J}_+ = \hbar \begin{pmatrix} 0 & \sqrt{2} & 0 \\ 0 & 0 & \sqrt{2} \\ 0 & 0 & 0 \end{pmatrix} \quad \text{and} \quad \hat{J}_- = \hbar \begin{pmatrix} 0 & 0 & 0 \\ \sqrt{2} & 0 & 0 \\ 0 & \sqrt{2} & 0 \end{pmatrix},$$

which are adjoints of one another. And then from $\hat{J}_x = \frac{1}{2}(\hat{J}_+ + \hat{J}_-)$, etc. we get

$$\hat{J}_x = \frac{\hbar}{\sqrt{2}} \begin{pmatrix} 0 & 1 & 0 \\ 1 & 0 & 1 \\ 0 & 1 & 0 \end{pmatrix} \quad \text{and} \quad \hat{J}_y = \frac{\hbar}{\sqrt{2}} \begin{pmatrix} 0 & -i & 0 \\ i & 0 & -i \\ 0 & i & 0 \end{pmatrix}.$$

It should now be easy (although tedious) to get representations of the states and operators for any value of j.

8·6. DIFFERENTIAL OPERATORS FOR ORBITAL ANGULAR MOMENTUM

We return at last specifically to *orbital* angular momentum. To find the angular part of the wave functions—the actual functions $Y(\theta, \phi)$ of Sec. 1—we need differential recipes in spherical coordinates for $\hat{L}_x$, $\hat{L}_y$, $\hat{L}_z$, $\hat{L}_+$, and $\hat{L}_-$. We found $\hat{L}^2$ in Sec. 1.

In *rectangular* coordinates, the momentum operator $\hat{\mathbf{p}}$, operating on ψ, is

$$\hat{\mathbf{p}}\psi = -i\hbar \left(\hat{\mathbf{x}} \frac{\partial \psi}{\partial x} + \hat{\mathbf{y}} \frac{\partial \psi}{\partial y} + \hat{\mathbf{z}} \frac{\partial \psi}{\partial z} \right) = -i\hbar \nabla \psi ,$$

where ∇ is the gradient operator. In spherical coordinates, the gradient of ψ is

$$\nabla \psi = \hat{\mathbf{r}} \frac{\partial \psi}{\partial r} + \hat{\boldsymbol{\theta}} \frac{1}{r} \frac{\partial \psi}{\partial \theta} + \hat{\boldsymbol{\phi}} \frac{1}{r \sin \theta} \frac{\partial \psi}{\partial \phi} .$$

The denominators of the partial derivatives here are one-to-one with the components of the infinitesimal displacement vector $d\mathbf{r} = \hat{\mathbf{r}}\, dr + \hat{\boldsymbol{\theta}}\, r d\theta + \hat{\boldsymbol{\phi}}\, r \sin \theta\, d\phi$, just as the denominators in the rectangular case are one-to-one with the components of $d\mathbf{r} = \hat{\mathbf{x}}\, dx + \hat{\mathbf{y}}\, dy + \hat{\mathbf{z}}\, dz$. Furthermore, the products of those components in the two coordinate sytems are the infinitesimal volume elements, $dxdydz$ and $r^2 \sin \theta\, drd\theta d\phi$.

The unit vectors $\hat{\mathbf{r}}$, $\hat{\boldsymbol{\theta}}$, and $\hat{\boldsymbol{\phi}}$ are orthogonal (see Fig. 6):

$$\hat{\mathbf{r}} \times \hat{\mathbf{r}} = 0, \quad \hat{\mathbf{r}} \times \hat{\boldsymbol{\theta}} = \hat{\boldsymbol{\phi}}, \quad \text{and} \quad \hat{\mathbf{r}} \times \hat{\boldsymbol{\phi}} = -\hat{\boldsymbol{\theta}} .$$

Then, since $\mathbf{r} = r\hat{\mathbf{r}}$, we get

$$\mathbf{L} = \mathbf{r} \times \mathbf{p} \rightarrow -i\hbar\, r\hat{\mathbf{r}} \times \nabla \psi = -i\hbar \left(\hat{\boldsymbol{\phi}} \frac{\partial \psi}{\partial \theta} - \hat{\boldsymbol{\theta}} \frac{1}{\sin \theta} \frac{\partial \psi}{\partial \phi} \right),$$

which is entirely independent of r.

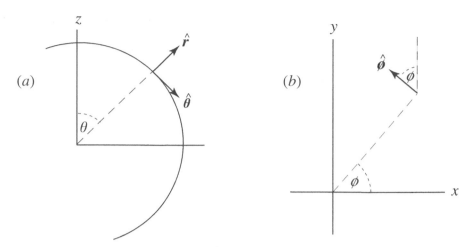

Figure 6. (a) The unit vectors $\hat{\mathbf{r}}$ and $\hat{\boldsymbol{\theta}}$ in the plane they define; $\hat{\boldsymbol{\phi}}$ is perpendicular to this plane and has no z component. The z and xy components of $\hat{\boldsymbol{\theta}}$ are $-\sin\theta$ and $\cos\theta$. (b) The $\hat{\boldsymbol{\phi}}$ vector projected onto the xy plane. Its x and y components are $-\sin\phi$ and $\cos\phi$.

What we want are the *rectangular* orbital-angular-momentum operators, $\hat{L}_x$, $\hat{L}_y$, and $\hat{L}_z$, expressed in terms of *spherical* coordinates. For this we need the unit vectors $\hat{\boldsymbol{\theta}}$ and $\hat{\boldsymbol{\phi}}$ in terms of $\hat{\mathbf{x}}$, $\hat{\mathbf{y}}$, and $\hat{\mathbf{z}}$. A short exercise in spatial geometry (read off from Fig. 6) gives

$$\hat{\boldsymbol{\theta}} = \hat{\mathbf{x}}\cos\theta\cos\phi + \hat{\mathbf{y}}\cos\theta\sin\phi - \hat{\mathbf{z}}\sin\theta$$

$$\hat{\boldsymbol{\phi}} = -\hat{\mathbf{x}}\sin\phi + \hat{\mathbf{y}}\cos\phi .$$

We use these to replace $\hat{\boldsymbol{\theta}}$ and $\hat{\boldsymbol{\phi}}$ in the equation for $\mathbf{L}$, and then collect terms multiplying $\hat{\mathbf{x}}$, $\hat{\mathbf{y}}$, and $\hat{\mathbf{z}}$:

$$\hat{\mathbf{L}}\psi = i\hbar\left[\hat{\mathbf{x}}\left(\sin\phi\frac{\partial\psi}{\partial\theta} + \cos\phi\cot\theta\frac{\partial\psi}{\partial\phi}\right) + \hat{\mathbf{y}}\left(-\cos\phi\frac{\partial\psi}{\partial\theta} + \sin\phi\cot\theta\frac{\partial\psi}{\partial\phi}\right) - \hat{\mathbf{z}}\frac{\partial\psi}{\partial\phi}\right].$$

The components here are $\hat{L}_x\psi$, $\hat{L}_y\psi$, and $\hat{L}_z\psi$, in spherical coordinates. In particular, $\hat{L}_z\psi$ is simply $-i\hbar$ times $\partial\psi/\partial\phi$,

$$\boxed{\hat{L}_z\psi = -i\hbar\frac{\partial\psi}{\partial\phi} .}$$

The ladder operators $\hat{L}_\pm$ are

$$(\hat{L}_x \pm i\hat{L}_y)\psi = i\hbar\left[(\sin\phi \mp i\cos\phi)\frac{\partial\psi}{\partial\theta} + (\cos\phi \pm i\sin\phi)\cot\theta\frac{\partial\psi}{\partial\phi}\right]$$

$$\hat{L}_\pm = \hbar e^{\pm i\phi}\left(\pm\frac{\partial\psi}{\partial\theta} + i\cot\theta\frac{\partial\psi}{\partial\phi}\right) .$$

It is a messy exercise in taking partial derivatives to show that

$$\boxed{\hat{L}^2\psi = (\hat{L}_x^2 + \hat{L}_y^2 + \hat{L}_z^2)\psi = -\hbar^2\left[\frac{1}{\sin\theta}\frac{\partial}{\partial\theta}\left(\sin\theta\frac{\partial\psi}{\partial\theta}\right) + \frac{1}{\sin^2\theta}\frac{\partial^2\psi}{\partial\phi^2}\right].}$$

The simpler argument made in Sec. 1 for $\hat{L}^2\psi$ will suffice here.

8-7. SPHERICAL HARMONICS

With the differential operators for orbital angular momentum in hand, we look for the functions $Y(\theta, \phi)$ that are simultaneously eigenfunctions of $\hat{L}^2$ and $\hat{L}_z$. Notationally, the orbital quantum numbers are ℓ and m instead of j and m. We already know from the general formalism of Sec. 3 that the $Y(\theta, \phi)$ functions will satisfy

$$\hat{L}^2 Y_\ell^m(\theta, \phi) = \ell(\ell+1)\hbar^2 \, Y_\ell^m(\theta, \phi) \qquad \text{and} \qquad \hat{L}_z Y_\ell^m(\theta, \phi) = m\hbar \, Y_\ell^m(\theta, \phi) \ .$$

We just don't yet know what those functions *are*.

We work with the factored $Y(\theta, \phi) = P(\theta) \, \Phi(\phi)$ of Sec. 1. The first step is

$$\hat{L}_z \, Y(\theta, \phi) = \hat{L}_z \big(P(\theta) \, \Phi(\phi) \big) = -i\hbar P(\theta) \frac{d\Phi}{d\phi} = m\hbar P(\theta) \, \Phi(\phi) \ .$$

Cancelling $\hbar P(\theta)$, we get

$$\frac{d\Phi}{d\phi} = im\Phi \ ,$$

and the solution is $\Phi(\phi) = e^{im\phi}$. We know from Sec. 3 that m here can only be integral or half-integral. But wave functions have to be single-valued functions of position, and advancing ϕ by 2π brings us back to the same point in space. Thus we require that

$$e^{im(\phi+2\pi)} = e^{im\phi}, \qquad \text{or} \qquad e^{i2\pi m} = 1 \ .$$

This is only true if m is an *integer*, $m = \ldots, -2, -1, 0, +1, \ldots$ *Half*-integral values of m (and therefore of ℓ) are ruled out for *orbital* angular momenta. The functions $\Phi(\phi) = e^{im\phi}$ solve the separated equation $d^2\Phi/d\phi^2 + \mu\Phi = 0$ that we got near the end of Sec. 1. The value of that θ/ϕ separation constant μ is m^2, where m is an *integer*. So $\mu = 0, 1, 4, 9, \ldots$

We find the top-of-the-ladders $m = +\ell$ states, from the requirement that $\hat{L}_+ Y_\ell^{+\ell} = 0$,

$$\hat{L}_+ Y_\ell^{+\ell}(\theta, \phi) = \hbar e^{+i\phi} \left(\frac{\partial}{\partial\theta} + i\cot\theta \, \frac{\partial}{\partial\phi} \right) P_\ell^{+\ell}(\theta) \, e^{+i\ell\phi} = 0 \ .$$

This reduces to

$$\frac{dP_\ell^{+\ell}(\theta)}{d\theta} = \ell \cot\theta \, P_\ell^{+\ell}(\theta) \ .$$

The (unnormalized) solution is

$$P_\ell^{+\ell}(\theta) = \sin^\ell \theta \ ,$$

because taking the derivative of $\sin^\ell \theta$ amounts to multiplying it by $\ell \cos\theta / \sin\theta = \ell \cot\theta$. Here then is an infinite set of eigenfunctions of $\hat{L}^2$ and $\hat{L}_z$:

$$Y_\ell^{+\ell}(\theta, \phi) = C_\ell^{+\ell} P_\ell^{+\ell}(\theta) \, e^{+i\ell\phi} = C_\ell^{+\ell} \sin^\ell \theta \, e^{+i\ell\phi} \ ,$$

where ℓ is a non-negative integer. The values of the $C_\ell^{+\ell}$'s come from normalization:

$$\langle \ell, m | \ell, m \rangle = \int_{\theta=0}^{\pi} \int_{\phi=0}^{2\pi} |Y_\ell^m(\theta, \phi)|^2 \sin\theta \, d\theta \, d\phi = 1 \ .$$

The normalized $Y_\ell^m(\theta, \phi)$ functions are called spherical harmonics. The normalized Y_0^0 (no θ or ϕ dependence) is $1/\sqrt{4\pi}$. The normalized $m = +\ell$ harmonics with $\ell > 0$ are

$$\boxed{ Y_\ell^{+\ell}(\theta, \phi) = (-1)^\ell \sqrt{\frac{2\ell+1}{4\pi \, (2\ell)!}} \, (2\ell-1)!! \, \sin^\ell \theta \, e^{+i\ell\phi} \ , }$$

161

where $(2\ell - 1)!! \equiv 1 \times 3 \times \cdots \times (2\ell - 1)$. To get another infinite set, $Y_\ell^{-\ell}(\theta, \phi)$, omit the multiplicative factor $(-1)^\ell$ and change $e^{i\ell\phi}$ to $e^{-i\ell\phi}$. See Prob. 6 for the normalization calculation and phase convention.

To get the other members belonging to a family of ℓ harmonics, we use the lowering operator $\hat{L}_-$ from the box near the end of Sec. 3,

$$\hat{L}_- Y_\ell^m = \sqrt{(\ell + m)(\ell - m + 1)}\, \hbar Y_\ell^{m-1} \,,$$

and the differential operator for $\hat{L}_-$. For example, the calculation to get Y_1^0 from the (normalized) Y_1^{+1} in the table below is

$$\hat{L}_- Y_1^{+1} = \sqrt{2}\, \hbar\, Y_1^0 = \hbar e^{-i\phi} \left(-\frac{\partial}{\partial\theta} + i\cot\theta \frac{\partial}{\partial\phi} \right) \left(-\sqrt{\frac{3}{8\pi}} \sin\theta\, e^{i\phi} \right) = 2\hbar \sqrt{\frac{3}{8\pi}} \cos\theta \,,$$

or $Y_1^0(\theta, \phi) = \sqrt{3/4\pi} \cos\theta$. Another lowering gives $Y_1^{-1}(\theta, \phi) = +\sqrt{3/8\pi} \sin\theta\, e^{-i\phi}$. The first few spherical harmonics are

$$Y_0^0 = \sqrt{\frac{1}{4\pi}} \qquad Y_1^{\pm 1} = \mp \sqrt{\frac{3}{8\pi}} \sin\theta\, e^{\pm i\phi} \qquad Y_2^{\pm 2} = +\sqrt{\frac{15}{32\pi}} \sin^2\theta\, e^{\pm 2i\phi}$$

$$Y_1^0 = +\sqrt{\frac{3}{4\pi}} \cos\theta \qquad Y_2^{\pm 1} = \mp \sqrt{\frac{15}{8\pi}} \sin\theta \cos\theta\, e^{\pm i\phi}$$

$$Y_2^0 = +\sqrt{\frac{5}{16\pi}} \left(3\cos^2\theta - 1 \right)$$

The Y_ℓ^m and Y_ℓ^{-m} states are related by $Y_\ell^{-m} = (-1)^m (Y_\ell^m)^*$. The θ-dependent parts of the spherical harmonics are proportional to associated Legendre polynomials.

To summarize, the spherical harmonics are simultaneously eigenstates of $\hat{L}^2$ and $\hat{L}_z$:

$$\hat{L}^2 Y_\ell^m(\theta, \phi) = \ell(\ell + 1)\hbar^2\, Y_\ell^m(\theta, \phi), \qquad \hat{L}_z Y_\ell^m(\theta, \phi) = m\hbar\, Y_\ell^m(\theta, \phi) \,.$$

Here ℓ can be any nonnegative integer ($\ell = 0, 1, 2, \ldots$); and for a given value of ℓ, m can be any of the $(2\ell + 1)$ integers from $+\ell$ to $-\ell$. The spherical harmonics are orthogonal and normalized,

$$\langle \ell'm' | \ell m \rangle = \int_{\theta=0}^{\pi} \int_{\phi=0}^{2\pi} \left[Y_{\ell'}^{m'}(\theta, \phi) \right]^* Y_\ell^m(\theta, \phi) \sin\theta\, d\theta\, d\phi = \delta_{\ell'\ell} \delta_{m'm} \,.$$

They are the solutions of the orbital angular part of the Schrödinger equation for any central-force problem. The separation constant λ in the separated equation for $R(r)$ in Sec. 1 is $\lambda = \ell(\ell + 1)$, and that equation is independent of the quantum number m. We are now set up to solve, in the next chapter, the r-dependent equations for hydrogen and the isotropic harmonic oscillator.

There is another property of the spherical harmonics that will be important later. Reflection through the origin takes one from the point (r, θ, ϕ) to the point $(r, \pi - \theta, \phi + \pi)$. When ℓ is even, Y_ℓ^m is the same at both points; when ℓ is odd, Y_ℓ^m changes sign: The *parity* of a Y_ℓ^m is $(-1)^\ell$. See Prob. 8.

8·8. ANGULAR MOMENTUM AND THE OSCILLATOR

Although the angular-momentum problem involves more operators and is slightly more complicated than the oscillator problem, the same game was played in solving them with operators. Attempted factorization of the quadratic oscillator Hamiltonian and the quadratic $\hat{J}^2 - \hat{J}_z^2 = \hat{J}_x^2 + \hat{J}_y^2$ led to the raising and lowering operators $\hat{a}^\dagger$ and $\hat{a}$ and $\hat{J}_+$ and $\hat{J}_-$. Those operators "created" eigen-ladders with equally spaced rungs. The possible values of the quantum numbers n and j are related by $n = 2j$. The uncertainty relations for the two systems are "saturated" at the extreme rungs. Here are some of the parallels:

$$[\hat{x}, \hat{p}] = i\hbar \qquad\qquad [\hat{J}_x, \hat{J}_y] = i\hbar \hat{J}_z \ \& \text{ cyclicals}$$

$$\hat{H} = \tfrac{1}{2}m\omega^2 \hat{x}^2 + \hat{p}^2/2m \qquad\qquad \hat{J}^2 - \hat{J}_z^2 = \hat{J}_x^2 + \hat{J}_y^2$$

$$[\hat{H}, \hat{a}^\dagger] = +\hbar\omega \hat{a}^\dagger, \ [\hat{H}, \hat{a}] = -\hbar\omega \hat{a} \qquad\qquad [\hat{J}_z, \hat{J}_\pm] = \pm\hbar \hat{J}_\pm$$

$$\hat{H}|n\rangle = (n + \tfrac{1}{2})\hbar\omega |n\rangle, \quad n = 0, 1, 2, \ldots \qquad \hat{J}^2 |j, m\rangle = j(j+1)\hbar^2 |j, m\rangle, \quad j = 0, \tfrac{1}{2}, 1, \tfrac{3}{2}, \ldots$$

$$\hat{J}_z |j, m\rangle = m\hbar |j, m\rangle, \quad m = j, j-1, \ldots, -j$$

$$\hat{a}^\dagger |n\rangle = \sqrt{n+1}\,|n+1\rangle \qquad\qquad \hat{J}_+ |j, m\rangle = \sqrt{(j-m)(j+m+1)}\,\hbar |j, m+1\rangle$$

$$\hat{a}|n\rangle = \sqrt{n}\,|n-1\rangle \qquad\qquad \hat{J}_- |j, m\rangle = \sqrt{(j+m)(j-m+1)}\,\hbar |j, m-1\rangle$$

$$\Delta x\,\Delta p = \tfrac{1}{2}\hbar \ \text{ for } n = 0 \qquad\qquad \Delta J_x\,\Delta J_y = \tfrac{1}{2}|m|\hbar^2 \ \text{ for } m = \pm j$$

$$> \tfrac{1}{2}\hbar \ \text{ for } n > 0 \qquad\qquad > \tfrac{1}{2}|m|\hbar^2 \ \text{ for } |m| < j$$

The physics is different, but the mathematics is almost the same. And systems such as molecules, which both vibrate and rotate, require the physics of both.

Julian Schwinger constructed a model with two independent oscillators (the operators of one commute with those of the other) that duplicates the angular-momentum commutation relations, and therefore can duplicate the angular-momentum eigenstates. Given the parallels, this is perhaps not surprising. See, for example, J.J. Sakurai, *Modern Quantum Mechanics*, p. 217.

PROBLEMS

1. *Some basic commutation relations.*

Use the basic commutator relations $[\hat{x}, \hat{p}_x] = i\hbar$, etc. to show that

$$[\hat{L}_x, \hat{x}] = 0, \quad [\hat{L}_x, \hat{y}] = i\hbar \hat{z}, \quad [\hat{L}_x, \hat{z}] = -i\hbar \hat{y},$$

$$[\hat{L}_x, \hat{p}_x] = 0, \quad [\hat{L}_x, \hat{p}_y] = i\hbar \hat{p}_z, \quad [\hat{L}_x, \hat{p}_z] = -i\hbar \hat{p}_y.$$

Then write down the commutators of $\hat{L}_z$ with $\hat{x}$, $\hat{y}$, $\hat{z}$, $\hat{p}_x$, $\hat{p}_y$, and $\hat{p}_z$.

2. *Getting the equal sign in* $\Delta J_x\,\Delta J_y \geq \tfrac{1}{2}\hbar\,|\langle\psi|\hat{J}_z|\psi\rangle|$.

In deriving the *general* uncertainty relation in Sec. 7·3, we found that there are two conditions for the equal sign to hold. In the present context, these conditions are

$$\langle\psi|(\hat{J}_x - \langle\hat{J}_x\rangle)(\hat{J}_y - \langle\hat{J}_y\rangle) + (\hat{J}_y - \langle\hat{J}_y\rangle)(\hat{J}_x - \langle\hat{J}_x\rangle)|\psi\rangle = 0 \tag{1}$$

$$(\hat{J}_x - \langle\hat{J}_x\rangle)|\psi\rangle = c\,(\hat{J}_y - \langle\hat{J}_y\rangle)|\psi\rangle . \tag{2}$$

(a) Show that (1) is satisfied for any $|j, m\rangle$ state.

(b) Show that (2) is satisfied for $|j, \pm j\rangle$ states but not for other $|j, m\rangle$ states.

(c) As an example, write $\hat{J}_x |1, +1\rangle$, $\hat{J}_y |1, +1\rangle$, $\hat{J}_x |1, 0\rangle$, and $\hat{J}_y |1, 0\rangle$ as 3-component vectors and show that the first two are linearly dependent but the last two are not.

3. *Matrices for $j = 3/2$.*

Find the matrix representations for $\hat{J}^2$, $\hat{J}_z$, $\hat{J}_+$, $\hat{J}_-$, $\hat{J}_x$, and $\hat{J}_y$ when $j = 3/2$.

4. *Eigenstates of $\hat{J}_y$ when $j = 1$.*

Solve the matrix eigenvalue problem for $\hat{J}_y$ when $j = 1$. Normalize the states so that the top component of each is real and positive. Show that the eigenvectors belonging to the different eigenvalues are orthogonal.

5. *The operator $\hat{L}^2$.*

Using the operators $\hat{L}_x$, $\hat{L}_y$, and $\hat{L}_z$ found in Sec. 6, prove that

$$\hat{L}^2 \psi = (\hat{L}_x^2 + \hat{L}_y^2 + \hat{L}_z^2)\psi = -\hbar^2 \left[\frac{1}{\sin\theta} \frac{\partial}{\partial\theta} \left(\sin\theta \frac{\partial\psi}{\partial\theta} \right) + \frac{1}{\sin^2\theta} \frac{\partial^2\psi}{\partial\phi^2} \right].$$

6. *An infinite set of orbital angular-momentum eigenfunctions.*

(a) Show directly using the operators $\hat{L}^2$ and $\hat{L}_z$ that $\sin^\ell\theta\, e^{\pm i\ell\phi}$, where ℓ is any nonnegative integer, are eigenstates of $\hat{L}^2$ and $\hat{L}_z$ with eigenvalues $\ell(\ell+1)\,\hbar^2$ and $\pm\ell\hbar$.

(b) Derive the numerical factors given in Sec. 7 that normalize these $Y_\ell^{\pm\ell}(\theta, \phi)$ functions. By convention, the sign of $Y_\ell^m(\theta, \phi)$ is $(-1)^m$ for $m > 0$ and is $+1$ otherwise.

7. *Flat angular distributions.*

(a) Show that $|Y_1^{+1}|^2 + |Y_1^0|^2 + |Y_1^{-1}|^2$ is independent of θ and ϕ.

(b) Show that $\sum_{m=-2}^{+2} |Y_2^m|^2$ is independent of the angles.

This generalizes: The sum of equally weighted *probability* distributions over all m values for a given ℓ is a uniform distribution over the sphere. We make use of this observation in Sec. 11·6.

8. *Parity and the spherical harmonics.*

Let (r, θ, ϕ) be the spherical coordinates of a point in space. Reflecting this point through the origin takes you to $(r, \pi - \theta, \phi + \pi)$. This reflection (or inversion) is the *parity operation*.

(a) Show that under these changes, $\sin\theta \to \sin\theta$, $\cos\theta \to -\cos\theta$, and $e^{i\phi} \to -e^{i\phi}$: The sine function has *even* (+) *parity*, the other two functions have *odd* (−) *parity*.

(b) Show that the parities of the spherical harmonics $Y_\ell^{+\ell}(\theta, \phi) \propto \sin^\ell\theta\, e^{i\ell\phi}$ are $(-1)^\ell$.

(c) Find the parities of each of the spherical harmonics in the table at the end of Sec. 7. The parities of all the spherical harmonics are $(-1)^\ell$.

The next chapter is about the hydrogen atom, but it is followed by three chapters that are almost entirely about angular momenta.

9. HYDROGEN. THE ISOTROPIC OSCILLATOR

9.1. *The Effective Potential Energy*

9.2. *The Hydrogen Bound-State Energies*

9.3. *The Hydrogen Eigenfunctions*

9.4. *The Isotropic-Oscillator Energies*

9.5. *The Oscillator Eigenfunctions*

9.6. *Hydrogen and the Oscillator*

   Problems

With the angular part of the central-force problem solved, we return to the radial part. In Chapter 8, we found that the constant λ that occurred in getting separate equations for $R(r)$ and $Y(\theta, \phi)$ has the values $\lambda = \ell(\ell + 1)$, where $\ell = 0, 1, 2, \ldots$ Putting this into the equation for $R(r)$, we get

$$\frac{d^2(rR)}{dr^2} + \frac{2m}{\hbar^2}\left(E - V(r) - \frac{\ell(\ell+1)\hbar^2}{2mr^2}\right)rR = 0 \ .$$

It remains to find the values of the last separation constant, E, which of course will depend on the potential energy $V(r)$. In this chapter, we find the energy eigenvalues and eigenfunctions for the bound states of hydrogen, where $V(r) = -e^2/4\pi\epsilon_0 r$, and for the isotropic oscillator, where $V(r) = \frac{1}{2}m\omega^2 r^2$.

Much the more important physical system for the rest of this text is hydrogen, and it would be fine to skip the sections on the isotropic oscillator unless you have an inordinate fondness for series solutions. But do spend a moment with the energy-level diagrams in Figure 2; not every system is hydrogen-like, not all energies vary as $1/n^2$. And perhaps glance at Section 6.

9·1. THE EFFECTIVE POTENTIAL ENERGY

In Sec. 8·1, we separated the time-independent Schrödinger equation for a central potential energy $V(r)$, factoring $\psi(r, \theta, \phi)$ as $R(r)\, Y(\theta, \phi)$. The separated equation for $R(r)$ was

$$\frac{d^2(rR)}{dr^2} + \frac{2m}{\hbar^2}\left(E - V(r) - \frac{\lambda \hbar^2}{2mr^2}\right) rR = 0 \,,$$

where λ was the $r/\theta\phi$ separation constant. Then in solving the angular part, we found that $\lambda = \ell(\ell + 1)$, where $\ell = 0, 1, 2, \ldots$ So with

$$\boxed{V_{eff}(r) \equiv V(r) + \frac{\ell(\ell+1)\hbar^2}{2mr^2} \,,}$$

and $u(r) \equiv rR(r)$, we may write the radial equation as

$$\boxed{\frac{d^2 u}{dr^2} + \frac{2m}{\hbar^2}\big[E - V_{eff}(r)\big]\, u = 0 \,.}$$

Although the equation is for $u(r)$, not $R(r)$, it has the same form as the Schrödinger equation in one dimension for $\psi(x)$. Because of this parallel, $V_{eff}(r)$ is called the *effective* potential energy. When $\ell = 0$, $V_{eff}(r)$ reduces to $V(r)$. Figure 1 shows the (rescaled) effective potential energies for hydrogen and for the isotropic harmonic oscillator when $\ell = 1$. When $\ell > 0$, the ℓ-dependent term in $V_{eff}(r)$ rises to infinity at the origin and is called the *angular-momentum barrier*.

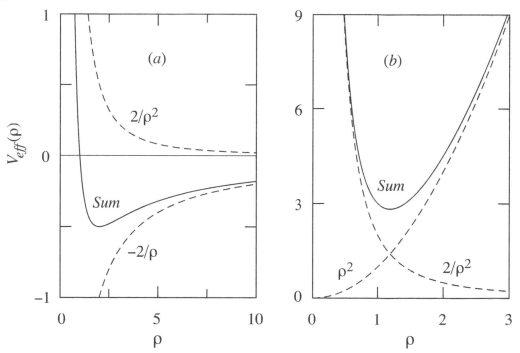

Figure 1. The rescaled effective potential energies $V_{eff}(\rho)$ when $\ell = 1$, (a) for hydrogen and (b) for the isotropic oscillator. See Secs. 2 and 4 for the rescaling; ρ and $V_{eff}(\rho)$ are both dimensionless. The $2/\rho^2$ curves are the angular-momentum terms in $V_{eff}(\rho)$ when $\ell = 1$.

The range of r is 0 to $+\infty$; and since $R(r)$ is everywhere finite, $u(r) = rR(r)$ vanishes at the origin: $u(0) = 0$. This is a boundary condition on $u(r)$, not on $R(r)$. For example, the hydrogen ground-state wave function is $\psi_{100}(r) \propto e^{-r/a_0}$, which is not zero at $r = 0$. However, when $\ell > 0$, the angular-momentum barrier does force $R(r)$ to be 0 at the origin.

The oscillator V_{eff} in Fig. 1(b) looks quite oscillator-like itself, and fitting a parabola to its minimum gives a good approximation to the energy eigenvalues (Prob. 1). However, fitting a parabola to the shallow, highly asymmetric hydrogen V_{eff} in Fig. 1(a) gives just *one* level with $E < 0$. This is a poor approximation indeed, because even though the hydrogen $V(r) \to 0$ as $r \to \infty$, there are an *infinite* number of hydrogen levels with $E < 0$ for *every* value of ℓ.

9·2. THE HYDROGEN BOUND-STATE ENERGIES

For hydrogen, $V(r) = -ke^2/r$ where $k \equiv 1/4\pi\epsilon_0$, and the radial equation for $u(r) = rR(r)$ is

$$\frac{d^2u}{dr^2} + \frac{2m}{\hbar^2}\left(E + \frac{ke^2}{r} - \frac{\ell(\ell+1)\hbar^2}{2mr^2}\right)u = 0 \ .$$

In Sec. 2·8(c), we found that the ground-state energy and wave function are

$$E_1 = -\frac{mk^2e^4}{2\hbar^2} \quad \text{and} \quad \psi_1(r) = \frac{1}{\sqrt{\pi a_0^3}}\,e^{-r/a_0} \ ,$$

where $a_0 \equiv \hbar^2/mke^2$ is the Bohr radius. We now get all the bound-state ($E < 0$) eigenvalues and eigenfunctions with the series-solution method used in Sec. 6·2 to solve the one-dimensional oscillator. Again, the steps are:

(a) Rescale the equation to a dimensionless form.

(b) Investigate the behavior of the solutions as $r \to \infty$ and $r \to 0$.

(c) Factor the "asymptotic" $r \to \infty$ behavior out of the rescaled equation to get a reduced equation.

(d) Try a series solution for the reduced equation that has the $r \to 0$ behavior built in.

(e) Find the constraints imposed on the coefficients of the series by the requirement that the wave function be normalizable.

Since this is much the same path we trod for the oscillator, some of the algebra will be left for the reader.

(a) Rescaling—The natural way to rescale the differential equation for $u(r)$ is to multiply it through by a_0^2 and let $\rho \equiv r/a_0$. We need E_1 in terms of a_0:

$$E_1 = -\frac{mk^2e^4}{2\hbar^2} = -\frac{\hbar^2}{2m}\left(\frac{mke^2}{\hbar^2}\right)^2 = -\frac{\hbar^2}{2m}\frac{1}{a_0^2} \ .$$

Replacing r with ρ, $u(r)$ with $u(\rho)$, and setting $\epsilon \equiv E/E_1$, the radial equation becomes (show this!)

$$\boxed{\frac{d^2u}{d\rho^2} + \left(-\epsilon + \frac{2}{\rho} - \frac{\ell(\ell+1)}{\rho^2}\right)u = 0 \ .}$$

For bound states, ϵ is positive because both E and E_1 are negative. The effective potential energy, shown in Fig. 1(a) for $\ell = 1$, is

$$V_{eff}(\rho) = -\frac{2}{\rho} + \frac{\ell(\ell+1)}{\rho^2} \; .$$

(b) Behavior at limits—As $\rho \to \infty$, the $1/\rho$ and $1/\rho^2$ terms in the equation for $u(\rho)$ become negligible, and the equation reduces to $u'' - \epsilon u \approx 0$. The solution in this limit is

$$u(\rho) = Ae^{-\sqrt{\epsilon}\rho} + A'e^{+\sqrt{\epsilon}\rho} \; ,$$

and we discard the second term (why?). Moreover, $u(\rho) = \rho^n e^{-\sqrt{\epsilon}\rho}$ (with $n > 0$) would still be an *asymptotic* solution (a solution to the highest power in ρ). This is so because, in calculating u'' from this $u(\rho)$, each derivative of ρ^n lowers the power of ρ, whereas each derivative of $e^{-\sqrt{\epsilon}\rho}$ keeps the power constant. Thus, to the *highest* power of ρ, $u'' - \epsilon u$ would still equal 0. For the same reason, $u(\rho) = g_n(\rho)e^{-\sqrt{\epsilon}\rho}$, where $g_n(\rho)$ is a polynomial whose highest power is ρ^n, would be an asymptotic solution.

As $\rho \to 0$, the ϵ and $2/\rho$ terms in the equation for $u(\rho)$ become negligible compared to the $1/\rho^2$ term, and as long as $\ell > 0$ the equation reduces to $\rho^2 u'' - \ell(\ell+1)u \approx 0$. The form of the solution is $u(\rho) = \rho^q$, because then $\rho^2 u''$ is, as required, proportional to u. With $u(\rho) = \rho^q$, we get

$$\rho^2 u'' - \ell(\ell+1)u = \big[q(q-1) - \ell(\ell+1)\big]u = \big[(q+\ell)(q-\ell-1)\big]u = 0 \; .$$

Thus $q = -\ell$ or $q = \ell + 1$, so that when ρ is small (and $\ell > 0$) the mathematical solution is

$$u(\rho) = C\rho^{-\ell} + D\rho^{\ell+1} \; .$$

We discard the first term because it does not satisfy the boundary condition $u(0) = 0$.

When $\ell = 0$, the angular-momentum term is absent, and in the $\rho \to 0$ limit the equation for u reduces to $u'' + (2/\rho)u \approx 0$. Now $u(\rho) \propto \rho^{\ell+1}$ with $\ell = 0$ is not a solution (try it). However, if we back up to the full problem, which is to find solutions over the whole range of ρ, the factor ρ is correct. Of course, we already know an $\ell = 0$ solution: The product of ρ with the large-ρ behavior, $e^{-\sqrt{\epsilon}\rho}$, and with $\epsilon = 1$, is

$$u(\rho) \propto \rho e^{-\rho} \qquad \text{or} \qquad \psi(r) \propto \frac{u(r)}{r} = R(r) \propto e^{-r/a_0} \; .$$

This is of course the actual ground-state wave function for the entire range of r.

(c) A new equation—To cover the whole range $0 \leq \rho < \infty$, we write $u(\rho) = g(\rho)e^{-\sqrt{\epsilon}\rho}$. Below, we will try a power series for $g(\rho)$. As long as $g(\rho)$ is a finite series, the factor $e^{-\sqrt{\epsilon}\rho}$ will overpower $g(\rho)$ as $\rho \to \infty$, and thereby insure normalizability. Putting $u(\rho) = g(\rho)e^{-\sqrt{\epsilon}\rho}$ into the radial equation and canceling the common factor $e^{-\sqrt{\epsilon}\rho}$, we get

$$g'' - 2\sqrt{\epsilon}\,g' + \left(\frac{2}{\rho} - \frac{\ell(\ell+1)}{\rho^2}\right)g = 0 \; .$$

The boundary condition $u(0) = 0$ becomes a boundary condition $g(0) = 0$. Try (in your head, if you can) simplest solutions:

- Does $g = 1$ work? No—it doesn't solve the equation, and $g(0) \neq 0$.
- Does $g = \rho$ work? Yes, with $\ell = 0$ and $\epsilon = 1$.
- Does $g = \rho^2$ work? Yes, with $\ell = 1$ and $\epsilon = 1/4$.
- Does $g = \rho^n$ work for any positive integer n? Yes, with $\ell = n - 1$ and $\epsilon = 1/n^2$.

(Stop now and show this—scribbling allowed.) We get an infinity of bound-state eigenvalues $\epsilon = 1/n^2$ (and so $E_n = E_1/n^2$), each with $\ell = n - 1$. In fact, these are *all* of the bound-state eigen-*values*. But they are (as will appear) only a subset of all the radial parts of the bound-state eigen-*functions*. Aside from normalization, this subset of radial functions, which in general are labeled $R_{n\ell}$, is

$$R_{n,n-1}(\rho) = \frac{u_{n,n-1}(\rho)}{\rho} = \frac{g_{n,n-1}(\rho)}{\rho} e^{-\sqrt{\epsilon}\rho} = \rho^{n-1} e^{-\rho/n} \propto r^{n-1} e^{-r/na_0} .$$

For $n = 1$, we get once again the ground state $R_{10}(r) \propto e^{-r/a_0}$. To make complete $\psi_{n\ell m}(r, \theta, \phi)$ eigenfunctions, each $R_{n,n-1}(r)$ here could be matched with any of the $2\ell + 1$ spherical harmonics $Y_\ell^m(\theta, \phi)$ with $\ell = n - 1$.

(d) **Series solution**—To get all the radial functions $R_{n\ell}(r)$, we try a power series for $g(\rho)$ with the $\rho \to 0$ behavior, $u(\rho) \propto \rho^{\ell+1}$, built in:

$$g(\rho) = \rho^{\ell+1} \sum_{i=0} b_i \rho^i = \sum_{i=0} b_i \rho^{i+\ell+1} \longrightarrow b_0 \rho^{\ell+1} \quad \text{as} \quad \rho \to 0 .$$

Putting this into the differential equation for $g(\rho)$, we get (after some algebra),

$$\sum_{i=0} i(i + 2\ell + 1) b_i \rho^{i+\ell-1} + 2 \sum_{i=0} \left[1 - \sqrt{\epsilon}(i + \ell + 1)\right] b_i \rho^{i+\ell} = 0 .$$

The $i = 0$ term of the first series is zero, so we shift its index, letting $i \to i + 1$. Now the corresponding powers of ρ in the two series line up, and

$$\sum_{i=0} \left\{ (i+1)(i + 2\ell + 2) b_{i+1} + 2\left[1 - \sqrt{\epsilon}(i + \ell + 1)\right] b_i \right\} \rho^{i+\ell} = 0 .$$

(Perhaps write out a few terms of the first series before and after $i \to i+1$.) Since the powers of ρ are linearly independent, the sum can only equal zero if the coefficient of each power of ρ is separately equal to zero. Thus we get a recurrence relation for the coefficients b_i,

$$b_{i+1} = 2b_i \frac{\sqrt{\epsilon}(i + \ell + 1) - 1}{(i+1)(i + 2\ell + 2)} .$$

(e) **Normalizability**—For large i, the ratio of successive coefficients is $b_{i+1}/b_i \approx 2\sqrt{\epsilon}/i$. This is the behavior of consecutive high-order coefficients of the expansion of $e^{2\sqrt{\epsilon}\rho}$ (show this). Allowing an infinite number of terms in $g(\rho)$ would give a series that at large ρ would overpower the factor $e^{-\sqrt{\epsilon}\rho}$ in $u(\rho) = g(\rho)e^{-\sqrt{\epsilon}\rho}$, and $u(\rho)$ would blow up. So the series for $g(\rho)$ has to terminate at some $i = i_{max}$. This requires that $\sqrt{\epsilon}(i_{max} + \ell + 1) = 1$, or that $\sqrt{\epsilon} = 1/n$, where $n \equiv i_{max} + \ell + 1$. Since i_{max} and ℓ are non-negative integers, n is a positive integer, $n = 1, 2, 3, \ldots$ Then the recurrence relation is

$$b_{i+1} = 2b_i \frac{(i + \ell + 1) - (i_{max} + \ell + 1)}{n(i+1)(i + 2\ell + 2)} = \frac{2(i - i_{max})}{n(i+1)(i + 2\ell + 2)} b_i .$$

This makes $b_i = 0$ for $i > i_{max}$, where $i_{max} = n - \ell - 1$. And the highest power in the series for $g(\rho)$ is $\rho^{i_{max}+\ell+1} = \rho^n$. We return to the eigenfunctions in the next section.

(f) The energy eigenvalues—Since $\sqrt{\epsilon} = 1/n$, the energy eigenvalues are

$$\boxed{\epsilon = \frac{1}{n^2}\ , \quad \text{or} \quad E_n = \frac{1}{n^2}E_1 = -\frac{1}{n^2}\frac{mk^2e^4}{2\hbar^2}\ .}$$

The energies depend only on n, called the *principal* quantum number:

$$\boxed{n = i_{max} + \ell + 1 = n_r + \ell + 1\ .}$$

Below, in labeling states, as opposed to working with the series for $g(\rho)$, we let $n_r \equiv i_{max}$; n_r is the *radial* quantum number. And again, n_r and ℓ are nonnegative integers. The number of (n_r, ℓ) pairs for a given value of n is n. Here are the possibilities for the first few values of n:

n	(n_r, ℓ)	E_n/E_1	degeneracy	
1	$(0,0)$	1	1	$n = n_r + \ell + 1$
2	$(1,0),(0,1)$	$1/4$	4	$E_n = E_1/n^2$
3	$(2,0),(1,1),(0,2)$	$1/9$	9	degeneracy $= n^2$
4	$(3,0),(2,1),(1,2),(0,3)$	$1/16$	16	$(n_r \equiv i_{max})$
5	$(4,0),(3,1),(2,2),(1,3),(0,4)$	$1/25$	25	

For a given value of n, the maximum value of ℓ is $n-1$, in which case $n_r = 0$. The degeneracy includes the $2\ell + 1$ values of m for each (n_r, ℓ) pair.

Figure 2(a) shows the bound-state energy levels of hydrogen, with the $\ell = 0, 1, 2, 3, \ldots$ levels placed, as is customary, in separate columns. There is an infinite number of bound states in each column. States with ℓ values 0, 1, 2, 3, ... are also called s, p, d, f ... states, after early spectroscopic terminology based on the characteristics of related series of spectral lines (*sharp*, *principal*, *diffuse*, *fundamental*, ...). The right-most state at each energy belongs to the set of solutions with $\ell = n - 1$ found in Sec. 2(c).

Figure 2(b) shows the entirely different pattern of levels for the isotropic oscillator (see Sec. 4).

9-3. THE HYDROGEN EIGENFUNCTIONS

In Sec. 1, we wrote $u(r) = rR(r)$. In Sec. 2(a), we rescaled this as $u(\rho) = \rho R(\rho)$, where $\rho = r/a_0$ and a_0 is the Bohr radius. In Sec. 2(c), we factored $u(\rho)$ into $g(\rho)\, e^{-\sqrt{\epsilon}\rho}$, and in Sec. 2(d) we tried $g(\rho)$ as a series, $g(\rho) = \rho^{\ell+1} \sum b_i\, \rho^i$. In Sec. 2(e), we got a recurrence relation for the b_i,

$$\boxed{b_{i+1} = \frac{2(i - i_{max})}{n(i+1)(i + 2\ell + 2)}\, b_i\ .}$$

This stops the series at $i = i_{max}$, which makes the eigenfunctions normalizable. At the same time, we found that $\sqrt{\epsilon} = 1/n$, where $n = 1, 2, 3, \ldots$, gives all the bound-state energies, and that $i_{max} = n - \ell - 1$. Reassembling the pieces, we get

$$R_{n\ell}(\rho) = \frac{u_{n\ell}(\rho)}{\rho} = \frac{g_{n\ell}(\rho)}{\rho}\, e^{-\sqrt{\epsilon}\rho} = e^{-\rho/n}\rho^\ell \sum_{i=0}^{n-\ell-1} b_i\, \rho^i\ .$$

The quantum numbers n and ℓ suffice to specify the radial wave functions.

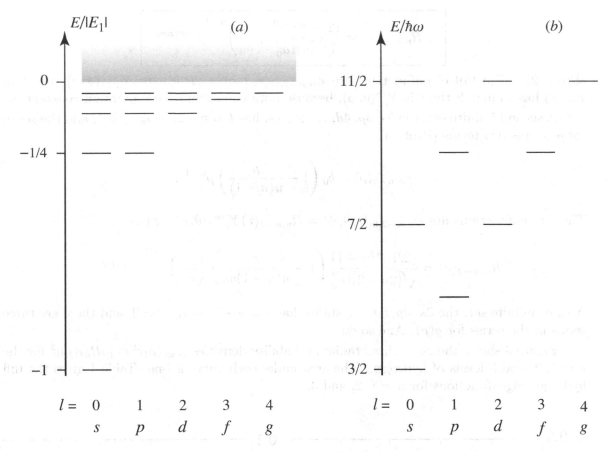

Figure 2. The energy-level diagrams (a) for hydrogen, and (b) for the isotropic oscillator. States with $\ell = 0, 1, 2, 3, \ldots$, called $s, p, d, f, \ldots$ states, are placed in separate columns. Each level here is $(2\ell + 1)$-fold m-state degenerate.

The normalization integral is

$$\int_{r=0}^{\infty} \int_{\theta=0}^{\pi} \int_{\phi=0}^{2\pi} |R_{n\ell}(r)\, Y_\ell^m(\theta, \phi)|^2\, r^2 \sin\theta\, dr\, d\theta\, d\phi = 1 \ .$$

This factors into radial and angular integrals. The spherical harmonics are normalized separately (Sec. 8·7). The radial normalization is

$$\boxed{\int_0^{\infty} |R_{n\ell}(r)|^2\, r^2\, dr = \int_0^{\infty} |u_{n\ell}(r)|^2\, dr = 1 \ .}$$

The probability of finding the particle in the spherical shell of thickness dr at radius r is $|u_{n\ell}(r)|^2\, dr$, not $|R_{n\ell}(r)|^2\, dr$, because the volume of a shell grows as r^2.

Two infinite sets—In Fig. 2(a), the right-most states at each energy, the $n\ell = 1s, 2p, 3d, \ldots$ states, have $\ell = n - 1$. Then $i_{max} = n - \ell - 1 = 0$, and for this *infinite* set of states (already found in Sec. 2(c)), the series for $g(\rho)$ has only the b_0 term. The normalized radial functions are

171

$$R_{n,n-1}(r) = \frac{(2/n)^{n+1/2}}{\sqrt{(2n)!\,a_0^3}} \left(\frac{r}{a_0}\right)^{n-1} e^{-r/na_0}$$

(Prob. 2). The full eigenfunctions are $\psi_{n,n-1,m}(r,\theta,\phi) = R_{n,n-1}(r)\,Y_{n-1}^m(\theta,\phi)$; the ℓ in $R_{n\ell}(r)$ has to match the ℓ in $Y_\ell^m(\theta,\phi)$, because both come from the separation constant λ.

A second infinite set, the $2s$, $3p$, $4d$, ... states, has $\ell = n-2$, so i_{max} is 1, and the series for $g(\rho)$ has two terms (Prob. 3),

$$g_{n,n-2}(\rho) = b_0 \left(1 - \frac{\rho}{n(n-1)}\right) \rho^{n-1}\ .$$

These eigenfunctions are $\psi_{n,n-2,m}(r,\theta,\phi) = R_{n,n-2}(r)\,Y_{n-2}^m(\theta,\phi)$, where

$$R_{n,n-2}(r) = \frac{(2/n)^n (n-1)}{\sqrt{(2n-2)!\,a_0^3}} \left(1 - \frac{r}{n(n-1)a_0}\right) \left(\frac{r}{a_0}\right)^{n-2} e^{-r/na_0}\ .$$

A third infinite set, the $3s$, $4p$, $5d$, ... states, has $\ell = n-3$, so $i_{max} = 2$, and there are three terms in the series for $g(\rho)$. And so on.

Figure 3 shows the normalized radial probability densities $|u_{n\ell}(\rho)|^2 = |\rho R_{n\ell}(\rho)|^2$ for the $n = 1$, 2, and 3 levels of hydrogen. The area under each curve is one. Table 1 gives the full hydrogen eigenfunctions for $n = 1$, 2, and 3.

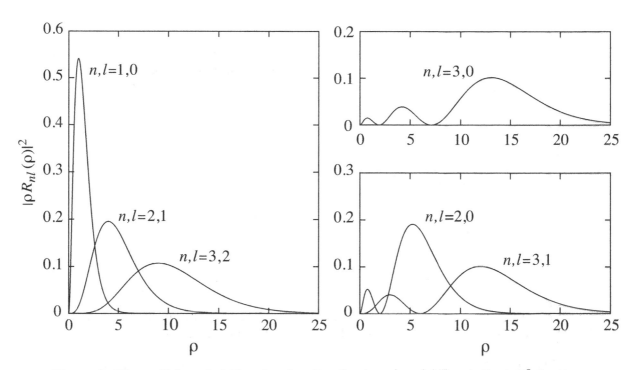

Figure 3. The radial probability-density distributions $|u_{n\ell}(\rho)|^2 = |\rho R_{n\ell}(\rho)|^2$ for the $n = 1$, 2, and 3 levels of hydrogen as functions of $\rho \equiv r/a_0$.

172

Table 1. The hydrogen wave functions $\psi_{n\ell m}(r,\theta,\phi)$ for $n = 1$, 2, and 3, written here as products of separately normalized radial and angular functions: $\psi_{n\ell m}(r,\theta,\phi) = R_{n\ell}(r) \times Y_\ell^m(\theta,\phi)$:

$$\int_0^\infty |R_{n\ell}(r)|^2 \, r^2 \, dr = \int_0^\infty |u_{n\ell}(r)|^2 \, dr = 1, \qquad \int_{\theta=0}^\pi \int_{\phi=0}^{2\pi} |Y_\ell^m(\theta,\phi)|^2 \sin\theta \, d\theta \, d\phi = 1 \, .$$

Factoring the wave functions in this way can simplify calculations. For example, the expectation value of r^2 reduces at once to an integral over just r: $\langle \psi_{n\ell m}|r^2|\psi_{n\ell m}\rangle = \langle R_{n\ell}|r^2|R_{n\ell}\rangle$.

When there is a $\mp$ sign in front of a wave function, it belongs to the $Y_\ell^m(\theta,\phi)$.

$$\psi_{100}(r,\theta,\phi) = R_{10}(r)\, Y_0^0(\theta,\phi) = \frac{2}{\sqrt{a_0^3}}\, e^{-r/a_0} \times \frac{1}{\sqrt{4\pi}}$$

$$\psi_{200}(r,\theta,\phi) = R_{20}(r)\, Y_0^0(\theta,\phi) = \frac{1}{\sqrt{2\,a_0^3}}\left(1 - \frac{1}{2}\frac{r}{a_0}\right) e^{-r/2a_0} \times \frac{1}{\sqrt{4\pi}}$$

$$\psi_{210}(r,\theta,\phi) = R_{21}(r)\, Y_1^0(\theta,\phi) = \frac{1}{\sqrt{24\,a_0^3}}\left(\frac{r}{a_0}\right) e^{-r/2a_0} \times \sqrt{\frac{3}{4\pi}}\cos\theta$$

$$\psi_{21\pm1}(r,\theta,\phi) = R_{21}(r)\, Y_1^{\pm1}(\theta,\phi) = \mp\frac{1}{\sqrt{24\,a_0^3}}\left(\frac{r}{a_0}\right) e^{-r/2a_0} \times \sqrt{\frac{3}{8\pi}}\sin\theta\, e^{\pm i\phi}$$

$$\psi_{300}(r,\theta,\phi) = R_{30}(r)\, Y_0^0(\theta,\phi) = \frac{2}{\sqrt{27\,a_0^3}}\left[1 - \frac{2}{3}\frac{r}{a_0} + \frac{2}{27}\left(\frac{r}{a_0}\right)^2\right] e^{-r/3a_0} \times \frac{1}{\sqrt{4\pi}}$$

$$\psi_{310}(r,\theta,\phi) = R_{31}(r)\, Y_1^0(\theta,\phi) = \frac{8}{27\sqrt{6\,a_0^3}}\left(1 - \frac{1}{6}\frac{r}{a_0}\right)\left(\frac{r}{a_0}\right) e^{-r/3a_0} \times \sqrt{\frac{3}{4\pi}}\cos\theta$$

$$\psi_{31\pm1}(r,\theta,\phi) = R_{31}(r)\, Y_1^{\pm1}(\theta,\phi) = \mp\frac{8}{27\sqrt{6\,a_0^3}}\left(1 - \frac{1}{6}\frac{r}{a_0}\right)\left(\frac{r}{a_0}\right) e^{-r/3a_0} \times \sqrt{\frac{3}{8\pi}}\sin\theta\, e^{\pm i\phi}$$

$$\psi_{320}(r,\theta,\phi) = R_{32}(r)\, Y_2^0(\theta,\phi) = \frac{4}{81\sqrt{30\,a_0^3}}\left(\frac{r}{a_0}\right)^2 e^{-r/3a_0} \times \sqrt{\frac{5}{16\pi}}\,(3\cos^2\theta - 1)$$

$$\psi_{32\pm1}(r,\theta,\phi) = R_{32}(r)\, Y_2^{\pm1}(\theta,\phi) = \mp\frac{4}{81\sqrt{30\,a_0^3}}\left(\frac{r}{a_0}\right)^2 e^{-r/3a_0} \times \sqrt{\frac{15}{8\pi}}\sin\theta\cos\theta\, e^{\pm i\phi}$$

$$\psi_{32\pm2}(r,\theta,\phi) = R_{32}(r)\, Y_2^{\pm2}(\theta,\phi) = \frac{4}{81\sqrt{30\,a_0^3}}\left(\frac{r}{a_0}\right)^2 e^{-r/3a_0} \times \sqrt{\frac{15}{32\pi}}\sin^2\theta\, e^{\pm 2i\phi}$$

For all $n \geq 1$, $\quad R_{n,n-1}(r) = \frac{(2/n)^{n+1/2}}{\sqrt{(2n)!\,a_0^3}}\left(\frac{r}{a_0}\right)^{n-1} e^{-r/na_0}$

For all $\ell \geq 0$, $\quad Y_\ell^{+\ell}(\theta,\phi) = (-1)^\ell\sqrt{\frac{2\ell+1}{4\pi\,(2\ell)!}}\,(2\ell-1)!!\,\sin^\ell\theta\, e^{+i\ell\phi}$

9·4. THE ISOTROPIC-OSCILLATOR ENERGIES

The potential energy for the isotropic oscillator in rectangular and in spherical coordinates is

$$V(x, y, z) = \tfrac{1}{2}m\omega^2(x^2 + y^2 + z^2) \qquad \text{and} \qquad V(r) = \tfrac{1}{2}m\omega^2 r^2 \ .$$

Because $V(x, y, z)$ is a simple sum of x-, y-, and z-dependent pieces, the Schrödinger equation separates in rectangular as well as in spherical coordinates. In rectangular coordinates, the eigenvalues are the sums of one-dimensional eigenvalues (Sec. 2·7(a)),

$$E_{n_x n_y n_z} = E_{n_x} + E_{n_y} + E_{n_z} = (n_x + n_y + n_z + \tfrac{3}{2})\hbar\omega = (n + \tfrac{3}{2})\hbar\omega \ .$$

Here $n \equiv n_x + n_y + n_z$ is the *principal* quantum number, and n_x, n_y, and n_z (and thus also n) can take on the values $0, 1, 2, 3, \ldots$ The eigenfunctions are products of one-dimensional eigenfunctions, which involve Hermite polynomials. The number of ways n_x, n_y, and n_z can sum to n—the degeneracy of the level—is $N = \tfrac{1}{2}(n + 1)(n + 2) = 1, 3, 6, 10, 15, \ldots$ (Prob. 6·2). Therefore we already know, in rectangular coordinates, the solutions of the isotropic oscillator. But sometimes it is useful to know the solutions in terms of the angular-momentum eigenstates $Y_\ell^m(\theta, \phi)$, so we now solve the problem in spherical coordinates. As earlier, some of the algebra will be left to the reader; and the phrase "as in Sec. 6·2" (the series solution of the one-dimensional oscillator) will recur.

(a) **Rescaling**—The radial equation for $u(r) = rR(r)$ for the isotropic oscillator is

$$\frac{d^2 u}{dr^2} + \frac{2m}{\hbar^2}\left(E - \frac{1}{2}m\omega^2 r^2 - \frac{\ell(\ell+1)\hbar^2}{2mr^2}\right) u = 0 \ .$$

We rescale using the one-dimensional ground-state energy $E_0 = \tfrac{1}{2}\hbar\omega$ and the turning-point distance in this state, $\sigma_0 = \sqrt{\hbar/m\omega}$. With $\epsilon \equiv E/\tfrac{1}{2}\hbar\omega$ and $\rho \equiv r/\sigma_0$, we get

$$\boxed{\frac{d^2 u}{d\rho^2} + \left(\epsilon - \rho^2 - \frac{\ell(\ell+1)}{\rho^2}\right) u = 0 \ ,}$$

where now $u = u(\rho)$. The effective potential energy, shown in Fig. 1(b) for $\ell = 1$, is

$$V_{eff}(\rho) = \rho^2 + \frac{\ell(\ell+1)}{\rho^2} \ .$$

When $\ell = 0$, the rescaled equation is $u'' + (\epsilon - \rho^2)u = 0$, which is the same as the rescaled one-dimensional oscillator equation of Sec. 6·2(a), $\psi'' + (\epsilon - y^2)\psi = 0$, where $\psi = \psi(y)$. However, there y ran from $-\infty$ to $+\infty$, whereas here ρ runs from 0 to $+\infty$, with $u(0) = 0$. Therefore, the solutions $u_{\ell=0}(\rho)$ have (aside from normalization) the same form as the $y \geq 0$ parts of the *odd-parity* solutions of the one-dimensional oscillator (see Prob. 2·3, on "half-wells"). Those solutions had these eigenvalues,

$$E_{\ell=0} = \tfrac{3}{2}\hbar\omega, \tfrac{7}{2}\hbar\omega, \tfrac{11}{2}\hbar\omega, \ \ldots$$

See Fig. 2, and note that these levels are separated by $2\hbar\omega$, not $\hbar\omega$.

174

(b) Behavior at limits—To find the solutions for any ℓ, we begin as usual by investigating the behavior at large and small values of ρ. As $\rho \to \infty$, the radial equation becomes $u'' - \rho^2 u \approx 0$, and exactly as in Sec. 6·2(b) the acceptable asymptotic solution is $u(\rho) = e^{-\rho^2/2}$, or any polynomial $f(\rho)$ times $e^{-\rho^2/2}$.

As $\rho \to 0$ and $\ell = 0$, the radial equation becomes $u'' + \epsilon u \approx 0$, and just as in Sec. 6·2(b) the solution in this limit is of the form $u_{\ell=0}(\rho) = A + B\rho$. However, here the boundary condition $u(0) = 0$ makes $A = 0$, and so $u_{\ell=0}(\rho) \approx B\rho$. As $\rho \to 0$ and $\ell \neq 0$, the rescaled equation becomes $\rho^2 u'' - \ell(\ell+1)u \approx 0$. This is the same equation we had in this limit for hydrogen (Sec. 2(b)), because the same angular-momentum term in $V_{eff}(\rho)$ dominates. Thus here as for hydrogen, $u_\ell(\rho) \propto \rho^{\ell+1}$ as $\rho \to 0$. This includes the $\ell = 0$ case, $u_0(\rho) \propto \rho$.

(c) New equation; series solution—To cover the whole range $0 \le \rho < \infty$, we write $u(\rho) = f(\rho)\, e^{-\rho^2/2}$; $f(\rho)$ will be a polynomial, and $e^{-\rho^2/2}$ will kill it off as $\rho \to \infty$ and insure normalizability. Putting this into the Schrödinger equation, we get

$$f'' - 2\rho f' + \left(\epsilon - 1 - \frac{\ell(\ell+1)}{\rho^2} \right) f = 0 \; .$$

Except for the term with ℓ here, this is the same as the equation in Sec. 6·2(c) for $h(y)$, where we wrote $\psi(y) = h(y)\, e^{-y^2/2}$.

We try a power series for $f(\rho)$ with the behavior $u(\rho) \propto \rho^{\ell+1}$ near $\rho = 0$ built in,

$$f(\rho) = \rho^{\ell+1} \sum_{i=0} a_i \, \rho^{i} = \sum_{i=0} a_i \, \rho^{i+\ell+1} \; .$$

Putting this in the equation for $f(\rho)$, we get (after a cancellation)

$$\sum_{i=0} i(i+2\ell+1)\, a_i \, \rho^{i+\ell-1} + \sum_{i=0} (\epsilon - 2i - 2\ell - 3)\, a_i \, \rho^{i+\ell+1} = 0 \; .$$

The factor i in the first series means that the series really starts with $i = 1$. But in order to line up the powers of ρ in the two series, we shift the index in the first series by *two*, $i \to i+2$; and then the first series starts at $i = -1$. We get

$$\sum_{i=-1} (i+2)(i+2\ell+3)\, a_{i+2}\, \rho^{i+\ell+1} + \sum_{i=0} (\epsilon - 2i - 2\ell - 3)\, a_i\, \rho^{i+\ell+1} = 0 \; .$$

(Again, check this by writing out a few terms of the first series before and after $i \to i+2$.) Keeping the $i = -1$ term separate, we get

$$(2\ell+2)a_1\, \rho^\ell + \sum_{i=0} \left[(i+2)(i+2\ell+3)\, a_{i+2} + (\epsilon - 2i - 2\ell - 3)\, a_i \right] \rho^{i+\ell+1} = 0 \; .$$

To satisfy this equation, the coefficient of each power of ρ has separately to be equal to zero. Thus $a_1 = 0$, and there is a recurrence relation for the coefficients,

$$a_{i+2} = \frac{(2i + 2\ell + 3 - \epsilon)}{(i+2)(i+2\ell+3)}\, a_i \; .$$

Since $a_1 = 0$, all the a_i with i odd are zero.

(d) Killing an infinity; the energies—When i is very large, the ratio of coefficients a_{i+2} and a_i is approximately $2/i$. The same ratio occurred with $h(y)$ in Sec. 6·2(e). It meant there (and here) that $f(\rho)$ would behave at large ρ like e^{ρ^2}, which would overwhelm the other factor $e^{-\rho^2/2}$ in $u(\rho)$ and make $R(\rho)$ unnormalizable. Unless, of course, the series for $f(\rho)$ terminates, which will happen if

$$\epsilon = 2i_{\max} + 2\ell + 3 .$$

Then the recurrence relation for the coefficients can be written

$$a_{i+2} = \frac{2(i - i_{max})}{(i+2)(i+2\ell+3)}\, a_i .$$

This makes $a_i = 0$ for all $i > i_{max}$, and the highest power in $f(\rho)$ is $\rho^{i_{max}+\ell+1}$. Since i is even ($a_i = 0$ when i is odd), $i_{\max}$ is any of $0, 2, 4, \ldots$

We return to this series in the next section, but first the energies. When working with the series for $f(\rho)$, i_{max} is useful. However, to honor the convention that quantum numbers advance in unit steps, we (especially when enumerating states) use $n_r \equiv \frac{1}{2}i_{max}$ here, so that n_r, the *radial* quantum number, is $0, 1, 2, \ldots$ And then

$$\epsilon_{i_{max},\ell} = 2i_{\max} + 2\ell + 3 \qquad \text{or} \qquad \epsilon_{n_r,\ell} = 4n_r + 2\ell + 3 ,$$

and

$$\boxed{E_{n_r,\ell} = \tfrac{1}{2}\hbar\omega\, \epsilon_{n_r,\ell} = (2n_r + \ell + \tfrac{3}{2})\hbar\omega, \qquad \text{or} \qquad E_n = (n + \tfrac{3}{2})\hbar\omega ,}$$

where $n \equiv 2n_r + \ell$. For use below, we note that $i_{max} = 2n_r = n - \ell$. Figure 2(b) shows the energy levels.

In rectangular coordinates, we had $E_{n_x n_y n_z} = (n_x + n_y + n_z + \tfrac{3}{2})\hbar\omega$. Comparing equations for E, we get

$$\boxed{n \equiv n_x + n_y + n_z = 2n_r + \ell .}$$

Each of the quantum numbers here can take on any of the values $0, 1, 2, \ldots$ Do we need one more quantum number in rectangular than in spherical coordinates? No, because for each value of ℓ there are $2\ell + 1$ values of m.

Here is a summary of the main results for the isotropic oscillator:

n	(n_r, ℓ)	$E_n/\hbar\omega$	Degeneracy	
0	$(0,0)$	$3/2$	1	$n = 2n_r + \ell$
1	$(0,1)$	$5/2$	3	$E_n = (n + \tfrac{3}{2})\hbar\omega$
2	$(1,0), (0,2)$	$7/2$	6	degeneracy $= \frac{1}{2}(n+1)(n+2)$
3	$(1,1), (0,3)$	$9/2$	10	$i_{max} = 2n_r = n - \ell$
4	$(2,0), (1,2), (0,4)$	$11/2$	15	

For a given value of n, the maximum value of ℓ is n, in which case $n_r = 0$. The degeneracy includes the $2\ell + 1$ values of m for each (n_r, ℓ) pair.

176

9·5 THE OSCILLATOR EIGENFUNCTIONS

In Sec. 1, we wrote $u(r) = rR(r)$. In the previous section, we rescaled this as $u(\rho) = \rho R(\rho)$, where $\rho = r/\sigma_0$ and $\sigma_0 = \sqrt{\hbar/m\omega}$. Then we factored $u(\rho)$ into $f(\rho)\,e^{-\rho^2/2}$, and then tried $f(\rho)$ as a series, $f(\rho) = \rho^{\ell+1} \sum a_i \rho^i$. Then we found that $a_i = 0$ when i is odd, and got a recurrence relation for the a_i when i is even,

$$a_{i+2} = \frac{2(i - i_{max})}{(i+2)(i+2\ell+3)}\, a_i \;.$$

This stops the series at $i = i_{max}$ and makes the wave functions normalizable. In Sec. 4(d), we also found that $i_{max} = 2n_r = n - \ell = 0, 2, 4, \ldots$ Reassembling the pieces, we get

$$R_{n\ell}(\rho) = \frac{u_{n\ell}(\rho)}{\rho} = \frac{f_{n\ell}(\rho)}{\rho}\, e^{-\rho^2/2} = e^{-\rho^2/2}\, \rho^\ell \sum_{i=0}^{n-\ell} a_i \rho^i \;.$$

The quantum numbers n and ℓ suffice to specify the radial wave functions.

In Fig. 2(b), the right-most states at each energy, the $n\ell = 0s, 1p, 2d, \ldots$ states, have $\ell = n$. Then $i_{max} = n - \ell = 0$, and for this set, the series for $f(\rho)$ has only the a_0 term. The normalized radial functions $R_{n\ell}(r)$ when $\ell = n$ are

$$R_{nn}(r) = \sqrt{\frac{2^{n+2}}{(2n+1)!!\sqrt{\pi}\,\sigma_0^3}} \left(\frac{r}{\sigma_0}\right)^n e^{-r^2/2\sigma_0^2}$$

(Prob. 5). Here $(2n+1)!!$ means $1 \times 3 \times 5 \times \cdots \times (2n+1)$. The $n = \ell$ eigenfunctions are $\psi_{nnm}(r, \theta, \phi) = R_{nn}(r)\, Y_n^m(\theta, \phi)$.

A second infinite set, the $n\ell = 2s, 3p, 4d, \ldots$ states, has $i_{max} = n - \ell = 2$, and the series for $R_{n,n-2}(r)$ will have two terms. The $4s, 5p, 6d, \ldots$ states have $i_{max} = n - \ell = 4$, and the series for $R_{n,n-4}(r)$ will have three terms; and so on. Table 2 gives the eigenfunctions for $n = 0, 1,$ and 2. Again, the exponential factors e^{-r/na_0} in hydrogen eigenfunctions depended on n; the exponential factors for the oscillator are all the same $e^{-r^2/2\sigma_0^2}$.

9·6. HYDROGEN AND THE OSCILLATOR

(a) Simple things—The energy levels for the bound states of hydrogen and the isotropic oscillator are

$$E_n = -\frac{1}{n^2}\frac{mk^2e^4}{2\hbar^2}, \quad n = 1, 2, 3, \ldots \qquad \text{and} \qquad E_n = \left(n + \tfrac{3}{2}\right)\hbar\omega, \quad n = 0, 1, 2, \ldots$$

In Sec. 3·1, we remarked on two properties these equations would have:

(1) The combinations $mk^2e^4/\hbar^2$ and $\hbar\omega$ are the *only* ways to make quantities having the dimensions of energy from the parameters in the Schrödinger equations, and therefore would *have* to occur that way.

(2) To leading order, the quantum number n goes linearly with each factor of $\hbar$.

The work it took in this chapter to go farther can make one appreciate dimensional arguments.

Table 2. The isotropic oscillator wave functions $\psi_{n\ell m}(r, \theta, \phi)$ for $n = 0$, 1, and 2, written here as products of separately normalized radial and angular functions: $\psi_{n\ell m}(r, \theta, \phi) = R_{n\ell}(r) \times Y_\ell^m(\theta, \phi)$. See also the caption of Table 1. The $\mp$ signs in front of some of the functions belong to the $Y_\ell^m(\theta, \phi)$'s.

$$\psi_{000}(r, \theta, \phi) = R_{00}(r)\, Y_0^0(\theta, \phi) = \frac{2}{\sqrt{\sqrt{\pi}\,\sigma_0^3}}\, e^{-r^2/2\sigma_0^2} \times \frac{1}{\sqrt{4\pi}}$$

$$\psi_{110}(r, \theta, \phi) = R_{11}(r)\, Y_1^0(\theta, \phi) = \sqrt{\frac{8}{3\sqrt{\pi}\,\sigma_0^3}} \left(\frac{r}{\sigma_0}\right) e^{-r^2/2\sigma_0^2} \times \sqrt{\frac{3}{4\pi}} \cos\theta$$

$$\psi_{11\pm1}(r, \theta, \phi) = R_{11}(r)\, Y_1^{\pm1}(\theta, \phi) = \mp\sqrt{\frac{8}{3\sqrt{\pi}\,\sigma_0^3}} \left(\frac{r}{\sigma_0}\right) e^{-r^2/2\sigma_0^2} \times \sqrt{\frac{3}{8\pi}} \sin\theta\, e^{\pm i\phi}$$

$$\psi_{200}(r, \theta, \phi) = R_{20}(r)\, Y_0^0(\theta, \phi) = \frac{\sqrt{6}}{\sqrt{\sqrt{\pi}\,\sigma_0^3}} \left[1 - \frac{2}{3}\left(\frac{r}{\sigma_0}\right)^2\right] e^{-r^2/2\sigma_0^2} \times \frac{1}{\sqrt{4\pi}}$$

$$\psi_{220}(r, \theta, \phi) = R_{22}(r)\, Y_2^0(\theta, \phi) = \frac{4}{\sqrt{15\sqrt{\pi}\,\sigma_0^3}} \left(\frac{r}{\sigma_0}\right)^2 e^{-r^2/2\sigma_0^2} \times \sqrt{\frac{5}{16\pi}} \left(3\cos^2\theta - 1\right)$$

$$\psi_{22\pm1}(r, \theta, \phi) = R_{22}(r)\, Y_2^{\pm1}(\theta, \phi) = \mp\frac{4}{\sqrt{15\sqrt{\pi}\,\sigma_0^3}} \left(\frac{r}{\sigma_0}\right)^2 e^{-r^2/2\sigma_0^2} \times \sqrt{\frac{15}{8\pi}} \sin\theta \cos\theta\, e^{\pm i\phi}$$

$$\psi_{22\pm2}(r, \theta, \phi) = R_{22}(r)\, Y_2^{\pm2}(\theta, \phi) = \frac{4}{\sqrt{15\sqrt{\pi}\,\sigma_0^3}} \left(\frac{r}{\sigma_0}\right)^2 e^{-r^2/2\sigma_0^2} \times \sqrt{\frac{15}{32\pi}} \sin^2\theta\, e^{\pm 2i\phi}$$

In Secs. 3 and 5, we found two simple, infinite sets of solutions for $R_{n\ell}(r)$:

$$R_{n,n-1}(r) \propto (r/a_0)^{n-1} e^{-r/na_0} \quad \text{for hydrogen levels with } \ell = n - 1$$
$$R_{nn}(r) \propto (r/\sigma_0)^n e^{-r^2/2\sigma_0^2} \quad \text{for oscillator levels with } \ell = n .$$

Since the numbering starts with $n = 1$ for hydrogen and $n = 0$ for the oscillator, the factors r^{n-1} for hydrogen and r^n for the oscillator are the same set: 1, r, r^2, r^3, ... The exponential for hydrogen, e^{-r/na_0}, varies with n; the exponential for the oscillator, $e^{-r^2/2\sigma_0^2}$, does not. These radial functions pair up nicely with the doubly infinite set of spherical harmonics, $Y_\ell^{\pm\ell} \propto \sin^\ell\theta\, e^{\pm i\ell\phi}$ (Sec. 8·7).

(b) "Accidental" degeneracies—Figure 2 showed the energy levels. Each level in the $\ell = 1$ columns is 3-fold degenerate ($m = +1, 0, -1$); each level in the $\ell = 2$ columns is 5-fold degenerate; and so on. Those "directional" degeneracies simply result from the spherical symmetry of central forces. However, as Fig. 2 shows, there are additional (sometimes called accidental) degeneracies of levels having different values of ℓ. The surface reason is that different combinations of the radial and angular-momentum quantum numbers n_r and ℓ can give the same value of n, and the energies depend only on n. The tables at the end of Secs. 2

and 4, repeated here in summary, gave the (n_r, ℓ) choices for the lowest few values of n:

n	(n_r, ℓ)	E_n/E_1	Degeneracy	Hydrogen
1	$(0,0)$	1	1	
2	$(1,0),(0,1)$	$1/4$	4	$n = n_r + \ell + 1$
3	$(2,0),(1,1),(0,2)$	$1/9$	9	$E_n = E_1/n^2$
4	$(3,0),(2,1),(1,2),(0,3)$	$1/16$	16	degeneracy $= n^2$
5	$(4,0),(3,1),(2,2),(1,3),(0,4)$	$1/25$	25	

n	(n_r, ℓ)	$E_n/\hbar\omega$	Degeneracy	Oscillator
0	$(0,0)$	$3/2$	1	
1	$(0,1)$	$5/2$	3	$n = 2n_r + \ell$
2	$(1,0),(0,2)$	$7/2$	6	$E_n = (n + \tfrac{3}{2})\hbar\omega$
3	$(1,1),(0,3)$	$9/2$	10	degeneracy $= \tfrac{1}{2}(n+1)(n+2)$
4	$(2,0),(1,2),(0,4)$	$11/2$	15	

A deeper reason for the degeneracy of levels with different values of ℓ is that the Coulomb and Hooke's-law forces are the only two central forces that (classically) have *closed* elliptical orbits. For these cases, there is an extra conserved quantity; see H. Goldstein, C. Poole, and J. Safko, *Classical Mechanics*, 3rd ed., Sec. 3·9. Figure 4 shows classical orbits for an attractive Coulomb force and for an isotropic oscillator. The concentric circles indicate minimum and maximum values of r as measured from the centers of force, which lie at the centers of the circles. Try to draw a closed ellipse that encircles the center of force and touches both circles in any *other* way. In one complete angular cycle the Coulomb orbit makes only *one* radial oscillation, whereas the oscillator orbit makes *two* complete radial oscillations. This perhaps hints at why n_r counts equally with ℓ in the hydrogen equation ($n = n_r + \ell + 1$), but counts doubly in the oscillator equation ($n = 2n_r + \ell$). If the gravitational force were Hooke's law, we would (ignoring circular orbits and axis tilt) have two summers and two winters per cycle around the Sun.

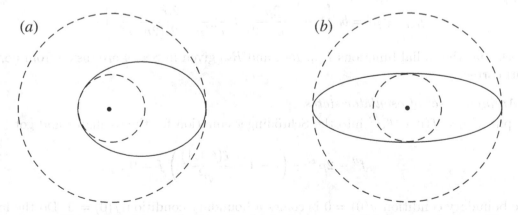

Figure 4. The classical bound orbits for (a) hydrogen and (b) the isotropic oscillator are elliptical.

PROBLEMS

1. *An oscillator approximation for the oscillator $V_{eff}(r)$ when $\ell > 0$.*

(a) When $\ell > 0$, find the value of r (call it r_0) at which $V_{eff}(r)$ for the isotropic oscillator is a minimum, and find $V_{eff}(r_0)$.

(b) Use a Taylor series to approximate $V_{eff}(r)$ near r_0 with an oscillator potential (see Sec. 6·1). Show that the the angular frequency for *this* oscillator (call it ω_0) is 2ω, where ω is the frequency in $V(r) = \frac{1}{2}m\omega^2 r^2$.

(c) Thus write down approximate energy levels for the oscillator, and compare with the exact answers, $E_{n_r,\ell} = (2n_r + \ell + 3/2)\,\hbar\omega$, where n_r and $\ell = 0, 1, 2, \ldots$ Note that for each value of ℓ the spacing between levels is correct, and that the absolute energies get better as ℓ increases.

2. *The 1s, 2p, 3d, ... ($\ell = n - 1$) hydrogen states.*
The normalized eigenfunctions for the hydrogen states with $\ell = n - 1$ are

$$\psi_{n,n-1,m}(r,\theta,\phi) = \frac{(2/n)^{n+1/2}}{\sqrt{(2n)!\,a_0^3}} \left(\frac{r}{a_0}\right)^{n-1} e^{-r/na_0}\, Y_{n-1}^m(\theta,\phi)\ .$$

For these states:

(a) Show that the normalization constant is correct.

(b) Find the mean radius $\langle r \rangle$.

(c) Show that the most probable radius r_P is $n^2 a_0$ (the radius of the nth orbit in the Bohr model).

d) Which states are these in Figs. 2(a) and 3?

3. *The $\ell = n - 2$ and $\ell = n - 3$ hydrogen states.*
Show that the radial series $g_{n\ell}(\rho)$ for hydrogen when $\ell = n - 2$ and $\ell = n - 3$ are

$$g_{n,n-2}(\rho) = b_0 \left(1 - \frac{\rho}{n(n-1)}\right)\rho^{n-1}$$

and

$$g_{n,n-3}(\rho) = b_0 \left[1 - \frac{2\rho}{n(n-2)} + \frac{2\rho^2}{n^2(n-2)(2n-3)}\right]\rho^{n-2}\ .$$

Check that the radial functions R_{20}, R_{31}, and R_{30} given in Sec. 4 are (aside from normalization) correct.

4. *An infinite set of oscillator states.*
We put $u(\rho) = f(\rho)\,e^{-\rho^2/2}$ into the Schrödinger equation for the oscillator and got

$$f'' - 2\rho f' + \left(\epsilon - 1 - \frac{\ell(\ell+1)}{\rho^2}\right)f = 0\ .$$

The boundary condition $u(0) = 0$ becomes a boundary condition $f(0) = 0$. Do the following functions $f(\rho)$ satisfy the equation and the boundary condition? If not, why not? If so, what is the relation between n and ℓ, and what are ϵ and E?

180

(a) Does $f(\rho) = 1$ work?

(b) Does $f(\rho) = \rho$ work?

(c) Does $f(\rho) = \rho^2$ work?

(d) Does $f(\rho) = \rho^N$ work for any positive integer N? (Since the oscillator quantum numbering begins with 0, the principle quantum number n is $n = N - 1$.)

See the next problem.

5. *The 0s, 1p, 2d, ... ($\ell = n$) oscillator states.*

The eigenfunctions for the oscillator states with $\ell = n$ are

$$\psi_{nnm}(r, \theta, \phi) = \sqrt{\frac{2^{n+2}}{(2n+1)!!\sqrt{\pi}\,\sigma_0^3}} \left(\frac{r}{\sigma_0}\right)^n e^{-r^2/2\sigma_0^2}\, Y_n^m(\theta, \phi) ,$$

where $\sigma_0 \equiv \sqrt{\hbar/m\omega}$ and $(2n+1)!! \equiv 1 \times 3 \times 5 \times \cdots \times (2n+1)$. For these states:

(a) Show that the normalization constant is correct.

(b) Show that the most probable radius r_P is $\sqrt{n+1}\,\sigma_0$.

(c) Which states are these in Fig. 2(b)?

6. *The virial theorem for a central potential energy $V(r)$.*

In Sec. 5·7(c), we proved the virial theorem in one dimension,

$$\left\langle \frac{p_x^2}{2m} \right\rangle_n = \langle K \rangle_n = E_n - \langle V(x) \rangle_n = \frac{1}{2}\left\langle x\frac{\partial V}{\partial x} \right\rangle_n .$$

This relates the expectation values of the kinetic and potential energies in an energy eigenstate $\psi_n(x)$.

(a) Prove the three-dimensional version of the theorem for a central force,

$$\langle K \rangle_{n\ell m} = E_{n\ell m} - \langle V(r) \rangle_{n\ell m} = \frac{1}{2}\left\langle r\frac{\partial V}{\partial r} \right\rangle_{n\ell m} .$$

The simplest way is to sum three one-dimensional versions of the theorem. The subscripts indicate the $\psi_{n\ell m}$ eigenstate.

(b) Thus show that $\langle K \rangle = \langle V \rangle = \frac{1}{2}E_n$ in an eigenstate $\psi_{n\ell m}$ of the isotropic oscillator, and that $\langle r^2 \rangle = (n + 3/2)\,\sigma_0^2$.

(c) Show that $\langle K \rangle = -\frac{1}{2}\langle V \rangle = -E_n$ in an an eigenstate $\psi_{n\ell m}$ of hydrogen, and that $\langle 1/r \rangle = 1/(n^2 a_0)$.

7. *A family of states and potential energies.*

Consider an (unnormalized) wave function of the form $\psi_\lambda(\rho) = e^{-\rho - \lambda\rho^2}$. For $\lambda = 0$, this is $\psi_0(\rho) = e^{-\rho}$, the (unnormalized) hydrogen ground state; and $\psi_\lambda(\rho)$ will be normalizable for $\lambda \geq 0$. Calculate $\nabla^2\psi_\lambda$, and arrange the result in the form of a Schrödinger equation for $\psi_\lambda(y)$,

$$-\nabla^2\psi_\lambda + \left[(4\rho - 6)\lambda + 4\lambda^2\rho^2 - 2\rho^{-1}\right]\psi_\lambda = -1\psi_\lambda .$$

The function in the square brackets is a continuous infinity of potential energies $V_\lambda(\rho)$, all with the energy eigenvalue $\epsilon = -1$ (or $E_1 = -13.6$ eV). From J.P. Killingbeck; see Prob. 6·10 for the reference.

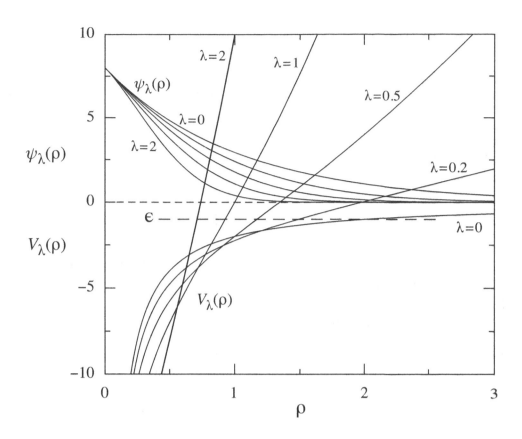

10. SPIN-1/2 PARTICLES

10.1. *Spinors. Eigenvalues and Eigenstates*

10.2. *The Polarization Vector*

10.3. *Magnetic Interactions and Zeeman Splitting*

10.4. *Time Dependence and Larmor Precession*

10.5. *Time Dependence and Magnetic Resonance*

10.6. *Stern-Gerlach Experiments*

10.7. *Polarization and Light*

    Problems

The simplest system with a nonzero angular momentum is a spin-1/2 particle. Quarks, the building blocks of protons, neutrons, and the other baryons, as well as of the mesons, are spin-1/2 particles. Protons and neutrons, the building blocks of nuclei, are spin-1/2 particles. Electrons, which with nuclei are the building blocks of atoms, are spin-1/2 particles. And so are muons and neutrinos. Thus all the matter we know about is made of spin-1/2 particles. The spin of an electron is as intrinsic to it—as much a part of what an electron is—as is its mass and charge: The electron always has spin 1/2. Of course, a particle will have kinetic energy, momentum, and orbital angular momentum, but here we are only concerned with the spin angular momentum of a spin-1/2 particle.

Not all particles have spin 1/2. The "carriers" of the fundamental forces—the photon, gluons, W and Z bosons, and graviton—have spin 1 or (the graviton) spin 2. Composite systems—nuclei and atoms, for example—can have higher angular momenta, depending on how the various spin and orbital angular momenta of its constituents couple together. For example, in the ground states of carbon, nitrogen, oxygen, iron, silver, and iridium, the six, seven, eight, 26, 47, and 77 electrons conspire to have total angular momenta of 0, 3/2, 2, 4, 1/2, and 9/2. In Chapter 14, we find how to get any of those values in a moment or two.

We start with the two-component vectors and 2-by-2 operators for $j=1/2$ that were introduced in Chapters 7 and 8. We find how to write any spin state in terms of the "up" and "down" states with respect to any direction, and how to associate a polarization vector with the spin. A particle with spin (except perhaps a neutrino) will also have a magnetic moment, and these moments interact with magnetic fields. This has several consequences:

- In a magnetic field, the spin-up and -down states with respect to the field direction have different energies (Zeeman splitting).

- In a static magnetic field, the spin precesses about the field direction (Larmor precession).

- In an arranged time-dependent field, the spin spirals around the field direction and back and forth between the spin-up and -down states (magnetic resonance).

- In an inhomogeneous field, a beam of particles with randomly oriented spins can be separated into beams of up and down states (Stern-Gerlach apparatus).

In one form or another, these properties and arrangements underlie a vast range of scientific experiments and technological applications.

10·1. SPINORS. EIGENVALUES AND EIGENSTATES

(a) Spin states as vectors—We use the symbol s instead of j for *spin* angular momentum, but the commutation relations, $[\hat{s}_x, \hat{s}_y] = i\hbar\hat{s}_z$, etc., and the relations that follow from them, such as $\hat{s}^2|s,m\rangle = s(s+1)\hbar^2|s,m\rangle$, are of course the same. For $s = 1/2$, the only simultaneous eigenstates of the operators $\hat{s}^2$ and $\hat{s}_z$ are $|s,m\rangle = |\frac{1}{2}, +\frac{1}{2}\rangle$ and $|\frac{1}{2}, -\frac{1}{2}\rangle$. These two states, introduced in Secs. 7·6 and 8·5, are represented in the literature in several short-hand ways, among them ($s = 1/2$ being implicit)

$$|+z\rangle = |+\rangle = |\alpha\rangle = |1\rangle = \uparrow \qquad \text{and} \qquad |-z\rangle = |-\rangle = |\beta\rangle = |2\rangle = \downarrow .$$

Here we shall favor $|+z\rangle$ and $|-z\rangle$, in order later to distinguish them from $|-x\rangle$, $|+y\rangle$, and other such states. We often call $|+z\rangle$ and $|-z\rangle$ the "spin up" and "spin down" states (with respect to the $+z$ direction), even though in the $|+z\rangle$ state, for example, the vector $\mathbf{s}$ is at a large angle (55°) to the $+z$ direction (see Fig. 8·5). In the $|+z\rangle$ state, $\mathbf{s}$ is as up as it can get, consistent with the uncertainty principle. That is, the expectation value of s_z is as large as it can get, $+\frac{1}{2}\hbar$.

In Sec. 8·5, we wrote the $|+z\rangle$ and $|-z\rangle$ states as two-component orthonormal vectors, also called *spinors*,

$$\boxed{|+z\rangle = \begin{pmatrix} 1 \\ 0 \end{pmatrix} \quad \text{and} \quad |-z\rangle = \begin{pmatrix} 0 \\ 1 \end{pmatrix}.}$$

They are orthogonal because they are eigenstates of $\hat{s}_z$ with different eigenvalues. They can be used as a basis for writing any spin state of a spin-1/2 particle,

$$|\chi\rangle = c_+|+z\rangle + c_-|-z\rangle = c_+\begin{pmatrix} 1 \\ 0 \end{pmatrix} + c_-\begin{pmatrix} 0 \\ 1 \end{pmatrix} = \begin{pmatrix} c_+ \\ c_- \end{pmatrix}.$$

Then $c_+ = \langle +z|\chi\rangle$ and $c_- = \langle -z|\chi\rangle$ are probability amplitudes, and $|c_+|^2$ and $|c_-|^2$ are the probabilities we would get $+\frac{1}{2}\hbar$ or $-\frac{1}{2}\hbar$ on measuring the z component of the spin. Since these are the only possible outcomes of this measurement, the normalization constraint is $\langle \chi|\chi\rangle = |c_+|^2 + |c_-|^2 = 1$.

We call a spin-1/2 system a two-state system, by which we mean there are two *basis* states. There are actually of course an infinite number of states, depending on the values of c_+ and c_-. Spin-1/2 is a two-state system in the sense that a plane is a two-dimensional space: Any vector in the plane can be written in terms of two unit vectors, $\hat{\mathbf{x}}$ and $\hat{\mathbf{y}}$. These two in fact are often represented by the same two-component vectors we use for $|+z\rangle$ and $|-z\rangle$. However, the spinor components c_+ and c_- are in general complex numbers. Spin angular momentum exists *physically* in ordinary three-dimensional space (where else?), even though in non-relativistic quantum mechanics its mathematical description requires the 2-component spinor add-on to $\psi(r, \theta, \phi)$. Dirac found (to his amazement) that joining relativity to quantum mechanics joined the (r, θ, ϕ) and spin spaces in states with *four* components (Chap. 17). In this regard, see again the Wigner quote that begins Chap. 5.

(b) Spin operators as matrices—In Sec. 8·5, we found that if we represent the $|s,m\rangle$ states as two-component vectors, we can represent the operators $\hat{s}_x$, $\hat{s}_y$, and $\hat{s}_z$, as 2-by-2 matrices,

$$\hat{s}_x = \frac{\hbar}{2}\begin{pmatrix} 0 & 1 \\ 1 & 0 \end{pmatrix}, \quad \hat{s}_y = \frac{\hbar}{2}\begin{pmatrix} 0 & -i \\ i & 0 \end{pmatrix}, \quad \hat{s}_z = \frac{\hbar}{2}\begin{pmatrix} 1 & 0 \\ 0 & -1 \end{pmatrix}.$$

These matrices, which are Hermitian, satisfy all the operational relations, such as $[\hat{s}_x, \hat{s}_y] = i\hbar\hat{s}_z$ (show this). The eigenvalues of $\hat{s}_z$ are obviously $\pm\frac{1}{2}\hbar$, and by symmetry (and by calculation) the eigenvalues of $\hat{s}_x$ and of $\hat{s}_y$ are also $\pm\frac{1}{2}\hbar$. The matrices themselves, without the factors $\frac{1}{2}\hbar$, are the *Pauli spin matrices*,

$$\hat{\sigma}_x = \begin{pmatrix} 0 & 1 \\ 1 & 0 \end{pmatrix}, \quad \hat{\sigma}_y = \begin{pmatrix} 0 & -i \\ i & 0 \end{pmatrix}, \quad \hat{\sigma}_z = \begin{pmatrix} 1 & 0 \\ 0 & -1 \end{pmatrix}.$$

In short, $\hat{\mathbf{s}} = \frac{1}{2}\hbar\hat{\boldsymbol{\sigma}}$, where $\hat{\boldsymbol{\sigma}}$ is a vector of operators, $\hat{\boldsymbol{\sigma}} = \hat{\sigma}_x\hat{\mathbf{x}} + \hat{\sigma}_y\hat{\mathbf{y}} + \hat{\sigma}_z\hat{\mathbf{z}}$. The eigenvalues of each Pauli matrix are ± 1.

(c) General eigenvalues—We shall want the eigenvalues and eigenstates of the general 2-by-2 Hermitian operator, not just of $\hat{s}_x$, $\hat{s}_y$, and $\hat{s}_z$. The most general such operator is

$$\begin{pmatrix} a + b_z & b_x - ib_y \\ b_x + ib_y & a - b_z \end{pmatrix} = a\hat{I} + b_x\hat{\sigma}_x + b_y\hat{\sigma}_y + b_z\hat{\sigma}_z = a\hat{I} + \mathbf{b}\cdot\hat{\boldsymbol{\sigma}}.$$

Here $\hat{I}$ is the 2-by-2 identity matrix, and the parameters a, b_x, b_y, and b_z are real (else the matrix would not be Hermitian) and independent (else the matrix would not be general). The sum of the diagonal elements (the trace) is $2a$; the difference of the diagonal elements is $2b_z$; the real and imaginary parts of the lower left element are b_x and b_y.

The b_i define a vector, $\mathbf{b} = (b_x, b_y, b_z) = b\hat{\mathbf{n}}$, where $\hat{\mathbf{n}}$ is the unit vector in the direction of $\mathbf{b}$. The eigenvalues of the above matrix are (see Prob. 1)

$$\lambda_{\pm} = a \pm (b_x^2 + b_y^2 + b_z^2)^{1/2} = a \pm b.$$

The eigenvalues are just a plus or minus the length of the vector $\mathbf{b}$. Given *any* 2-by-2 Hermitian matrix, we can simply identify a and b and write down the eigenvalues. The sum of the two eigenvalues is the trace $2a$, the difference is $2b$, and if $b = 0$ the eigenvalues are degenerate.

Example—In

$$\begin{pmatrix} 2 & -i \\ i & 0 \end{pmatrix},$$

$a = 1$, $b_z = 1$, $b_x = 0$, and $b_y = 1$. Thus $b = \sqrt{2}$ and the eigenvalues are $\lambda_{\pm} = 1 \pm \sqrt{2}$. And $\mathbf{b}$ is a vector lying in the yz plane, with polar and azimuthal angles $\theta = 45°$ and $\phi = 90°$.

(d) General eigenstates—The eigenstates of $a\hat{I} + \mathbf{b}\cdot\hat{\boldsymbol{\sigma}}$ belonging to the eigenvalues $a \pm b$ are associated with the direction of $\mathbf{b} = b\hat{\mathbf{n}}$, just as the eigenstates of $\hat{s}_z$ or $\hat{\sigma}_z$ are associated with the direction $\hat{\mathbf{z}}$. It could hardly be otherwise because $\mathbf{b}$ is the only direction in the problem. Let θ and ϕ be the polar and azimuthal angles of $\mathbf{b}$. Then

$$(b_x, b_y, b_z) = (b\sin\theta\cos\phi, b\sin\theta\sin\phi, b\cos\theta),$$

and the general Hermitian matrix becomes

$$a\hat{I} + \mathbf{b}\cdot\hat{\boldsymbol{\sigma}} = \begin{pmatrix} a + b\cos\theta & b\,e^{-i\phi}\sin\theta \\ b\,e^{+i\phi}\sin\theta & a - b\cos\theta \end{pmatrix}.$$

185

The normalized eigenstates belonging to the eigenvalues $\lambda_\pm = a \pm b$ are (more Prob. 1)

$$|+n\rangle = \begin{pmatrix} \cos \frac{1}{2}\theta \\ e^{+i\phi} \sin \frac{1}{2}\theta \end{pmatrix} \quad \text{and} \quad |-n\rangle = \begin{pmatrix} \sin \frac{1}{2}\theta \\ -e^{+i\phi} \cos \frac{1}{2}\theta \end{pmatrix}.$$

These are the states that have spin up and spin down with respect to the direction of $\mathbf{b} = b\hat{\mathbf{n}}$. They are normalized and orthogonal; they depend on the direction of $\mathbf{b}$ but not on its magnitude.

The spinors $|\pm n\rangle$ would still be the eigenstates if they were multiplied by an overall phase factor $e^{i\delta}$; that is, the eigenstates are not uniquely determined. The choice of phase used here makes the upper component of both spinors real and nonnegative, but other choices are often used. A different choice, consistently observed, would not affect any experimental results. Probability *amplitudes* might be different, but *probabilities* would not.

Examples—If the matrix is either $\hat{\sigma}_x$ or $\hat{\sigma}_y$, then $a = 0$ and $b = 1$, and the eigenvalues are ± 1. The corresponding eigenstates of $\hat{\sigma}_x$ and $\hat{\sigma}_y$ are

$$|+x\rangle = \begin{pmatrix} 1/\sqrt{2} \\ 1/\sqrt{2} \end{pmatrix}, \quad |-x\rangle = \begin{pmatrix} 1/\sqrt{2} \\ -1/\sqrt{2} \end{pmatrix}, \quad |+y\rangle = \begin{pmatrix} 1/\sqrt{2} \\ i/\sqrt{2} \end{pmatrix}, \quad |-y\rangle = \begin{pmatrix} 1/\sqrt{2} \\ -i/\sqrt{2} \end{pmatrix}.$$

You can read these off from the above general $|\pm n\rangle$ states by inserting the appropriate values of θ and ϕ. Thus for $\hat{\sigma}_x$, $\mathbf{b} = \hat{\mathbf{x}}$ and the angles are $\theta = 90°$ and $\phi = 0°$.

The eigenvalues of $\hat{s}_x$ and $\hat{s}_y$ are $\pm\frac{1}{2}\hbar$, and the eigenstates are the same as for $\hat{\sigma}_x$ and $\hat{\sigma}_y$ (why?). In the $|+y\rangle$ state, the spin is up along the y direction, and so on. (Operate on $|+y\rangle$ with $\hat{s}_y$ and see what you get.)

(e) Amplitudes for any direction—The spinors $|\pm n\rangle$ given above as functions of θ and ϕ satisfy

$$|+n\rangle \langle +n| + |-n\rangle \langle -n| = \hat{I}$$

(still more Prob. 1). Therefore, for any spinor $|\chi\rangle$,

$$|\chi\rangle = \hat{I}|\chi\rangle = |+n\rangle \langle +n|\chi\rangle + |-n\rangle \langle -n|\chi\rangle.$$

This is an expansion of $|\chi\rangle$ in terms of the basis states $|+n\rangle$ and $|-n\rangle$, where the amplitudes are the (generally complex) scalar products $\langle +n|\chi\rangle$ and $\langle -n|\chi\rangle$. And then $|\langle +n|\chi\rangle|^2$ and $|\langle -n|\chi\rangle|^2$ are the probabilities of getting $+\frac{1}{2}\hbar$ or $-\frac{1}{2}\hbar$ on measuring the component of the spin along the direction $\hat{\mathbf{n}} = \hat{\mathbf{n}}(\theta, \phi)$. Of course, *after* such a measurement, the spin state is either $|+n\rangle$ or $|-n\rangle$.

Example—Suppose the initial state is $|-y\rangle$ and a measurement is made of s_x. Then the probability amplitude for getting $+\frac{1}{2}\hbar$ is

$$\langle \text{final state}|\text{initial state}\rangle = \langle +x|-y\rangle = \begin{pmatrix} 1/\sqrt{2} & 1/\sqrt{2} \end{pmatrix} \begin{pmatrix} 1/\sqrt{2} \\ -i/\sqrt{2} \end{pmatrix} = \frac{1-i}{2}.$$

The probability then is $|\langle +x|-y\rangle|^2 = (1+i)(1-i)/4 = 1/2$.

(f) Remarks—It bears repeating: The only possible results of a good measurement of a physical quantity *are the eigenvalues of the corresponding operator*. The only possible results of a measurement of the component of spin of a spin-1/2 particle are $+\frac{1}{2}\hbar$ and $-\frac{1}{2}\hbar$, *no matter what state the particle is initially in or what direction we measure along*. The *probabilities* of getting the two possible results depend on that state, but the *possible results* do not. And *after* the measurement, the particle is in the eigenstate that belongs to the result of the measurement. In classical physics, where one imagines that measurements can be made that do not affect the state of a system, this is incomprehensible.

We now know everything about the eigen-solutions of 2-by-2 Hermitian matrices (symmetric 2-by-2 matrices being a subset). After doing Prob. 1, you will never again have to solve a 2-by-2 Hermitian-matrix eigenvalue problem—you will have already solved them all. While the present application is to spin-1/2, the mathematics is of course independent of the physical context (if there be any such).

10·2. THE POLARIZATION VECTOR

(a) What it is—The two complex components of a spinor $|\chi\rangle$ completely specify a spin state, but it is sometimes easier and clearer to work with an ordinary vector that has three real components. The *polarization vector* **P** is a unit vector that points in the direction along which, were a measurement made of the component of **s**, the result would be $+\frac{1}{2}\hbar$ with certainty: **P** points in the direction in which the spin is "up." Thus for the states $|+z\rangle$, $|-x\rangle$, $|-y\rangle$, and $|+n\rangle$, the **P** vectors are $+\hat{\mathbf{z}}$, $-\hat{\mathbf{x}}$, $-\hat{\mathbf{y}}$, and $+\hat{\mathbf{n}}$.

The tip of **P** lies on a unit sphere in ordinary 3-dimensional space. In terms of the polar and azimuthal angles of **P**, its components are the real quantities

$$(P_x, P_y, P_z) = (\sin\theta\cos\phi, \sin\theta\sin\phi, \cos\theta) ,$$

and $P_x^2 + P_y^2 + P_z^2 = 1$. This is one-to-one with the spinor

$$|+n\rangle = \begin{pmatrix} \cos\frac{1}{2}\theta \\ e^{+i\phi}\sin\frac{1}{2}\theta \end{pmatrix}$$

of Sec. 1(d), which is spin up along the direction (θ, ϕ). Given a direction (θ, ϕ), we can write down either or both of **P** and $|+n\rangle$.

The three (real) components of **P** can also be got directly from the two (complex) components $c_+ = \langle+z|\chi\rangle$ and $c_- = \langle-z|\chi\rangle$ of a spinor $|\chi\rangle$. The components are simply the expectation values of $\hat{\sigma}_x$, $\hat{\sigma}_y$, and $\hat{\sigma}_z$ in the state $|\chi\rangle$,

$$P_x = \langle\chi|\hat{\sigma}_x|\chi\rangle, \quad P_y = \langle\chi|\hat{\sigma}_y|\chi\rangle, \quad P_z = \langle\chi|\hat{\sigma}_z|\chi\rangle .$$

These work out to be (Prob. 2)

$$P_x = 2\,\mathrm{Re}\,(c_+^* c_-), \quad P_y = 2\,\mathrm{Im}\,(c_+^* c_-), \quad P_z = |c_+|^2 - |c_-|^2 .$$

For example, if $c_+ = 1$ and $c_- = 0$ (the $|+z\rangle$ state), then $P_x = P_y = 0$ and $P_z = 1$. Or if $c_+ = c_- = 1/\sqrt{2}$ (the $|+x\rangle$ state), then $P_x = 1$ and $P_y = P_z = 0$. Thus we can find **P** either from the angles θ and ϕ or from the spinor components c_+ and c_-,

$$\mathbf{P} = \begin{pmatrix} \sin\theta\cos\phi \\ \sin\theta\sin\phi \\ \cos\theta \end{pmatrix} = \begin{pmatrix} 2\,\mathrm{Re}\,(c_+^* c_-) \\ 2\,\mathrm{Im}\,(c_+^* c_-) \\ |c_+|^2 - |c_-|^2 \end{pmatrix}.$$

For example, if you put the components $c_+ = \cos\frac{1}{2}\theta$ and $c_- = e^{+i\phi}\sin\frac{1}{2}\theta$ of the spinor $|+n\rangle$ into the second column, you get the first column. (Do it.)

(b) Why it is useful—Since $P_z = |c_+|^2 - |c_-|^2$ and also $|c_+|^2 + |c_-|^2 = 1$, we get

$$P_z = |c_+|^2 - (1 - |c_+|^2) = 2|c_+|^2 - 1 \ .$$

Therefore if, before measuring, we know the component of $\mathbf{P}$ along z and then measure s_z, the probability we will get $+\frac{1}{2}\hbar$ is

$$|c_+|^2 = \tfrac{1}{2}(1 + P_z) = \tfrac{1}{2}(1 + \cos\theta) = \cos^2\tfrac{1}{2}\theta \ ,$$

and the probability we would get $-\frac{1}{2}\hbar$ is $\sin^2\frac{1}{2}\theta$. After a measurement that gets $s_z = +\frac{1}{2}\hbar$, $\mathbf{P} = \hat{\mathbf{z}}$.

Suppose, knowing $\mathbf{P}$, we want the probability of getting $+\frac{1}{2}\hbar$ on measuring along some arbitrary direction $\mathbf{d}$, as in Fig. 1. The *physical* relation between this probability and the angle between $\mathbf{P}$ and $\mathbf{d}$ has to be the same as it is between $\mathbf{P}$ and the z axis. There, P_z was $\cos\theta$, here P_d is $\cos\beta$, and so

$$\text{the probability of getting spin up along } \mathbf{d} = \tfrac{1}{2}(1 + \cos\beta) = \cos^2\tfrac{1}{2}\beta \ .$$

For example, the probability of getting spin up for a measurement along any direction in the plane perpendicular to $\mathbf{P}$ is $1/2$ ($\beta = 90°$). In such calculations, it is usually easier to use the frame-independent angle β than to find the spinors $|d\rangle$ and $|P\rangle$ associated with the directions of $\mathbf{d}$ and $\mathbf{P}$ and then calculate $|\langle d|P\rangle|^2$.

Again: $\mathbf{P}$ is an unit vector in ordinary 3-dimensional space that points in the direction along which the spin is up. In the time-dependent examples to follow, we shall examine the motion of the tip of $\mathbf{P}$ on the surface of the unit sphere.

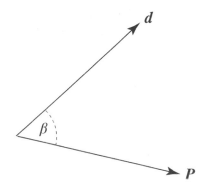

Figure 1. If a measurement of the component of the spin is made along the direction $\mathbf{d}$, the probabilities of getting $+\frac{1}{2}\hbar$ or $-\frac{1}{2}\hbar$ are $\cos^2\frac{1}{2}\beta$ and $\sin^2\frac{1}{2}\beta$.

10·3. MAGNETIC INTERACTIONS AND ZEEMAN SPLITTING

(a) Magnetic moments—So far, we have only introduced ways to describe the *state* of a spin-1/2 particle. Figure 2 shows a (classical) particle with charge q and mass m moving with speed v in a circular orbit of radius r. The circling frequency is $f = v/2\pi r$, so there is effectively a current $I = qf = qv/2\pi r$ past any point on the orbit. The magnetic moment μ of a flat current loop enclosing area A is IA (see any introductory text). Thus here,

$$\mu = IA = qfA = \frac{qv}{2\pi r}\pi r^2 = \frac{1}{2}qvr = \frac{q}{2m}L\,,$$

where $L = mvr$ is the orbital angular momentum of the particle. The direction associated with μ is related to the direction of the current by the usual right-hand rule, so that

$$\boxed{\boldsymbol{\mu} = \frac{q}{2m}\mathbf{L}\,.}$$

Thus $\boldsymbol{\mu}$ and $\mathbf{L}$ are parallel or antiparallel as q is positive or negative. The quantum-mechanical operator for $\boldsymbol{\mu}$ comes from using the quantum-mechanical operator for $\mathbf{L}$ in this relation. For example, $\hat{\mu}_z = (q/2m)\hat{L}_z$.

For a spin-1/2 particle, where $\hat{\mathbf{s}} = \frac{1}{2}\hbar\hat{\boldsymbol{\sigma}}$, the magnetic moment is

$$\boxed{\hat{\boldsymbol{\mu}} = g\frac{q}{2m}\hat{\mathbf{s}} = g\frac{q\hbar}{4m}\hat{\boldsymbol{\sigma}} = \mu\hat{\boldsymbol{\sigma}}\,.}$$

Here $\mu \equiv g(q\hbar/4m)$, and g is a factor that in nonrelativistic quantum mechanics simply has to be put in by hand to get agreement with experiment. For the electron, $q = -e$ and $g_e = 2$, so that

$$\boxed{\hat{\boldsymbol{\mu}}_e = \mu_e\hat{\boldsymbol{\sigma}} = -\frac{e\hbar}{2m}\hat{\boldsymbol{\sigma}} = -\mu_B\hat{\boldsymbol{\sigma}}\,,}$$

where $\mu_B \equiv e\hbar/2m$ is the *Bohr magneton*: $\mu_B = 9.274 \times 10^{-24}$ J/T $= 5.788 \times 10^{-5}$ eV/T (T for tesla). The value $g_e = 2$ for the electron appears miraculously in Dirac's relativistic quantum mechanics (as does spin itself; see the quotations that begin Chap. 17). However, in the even deeper theory of quantum electrodynamics (QED), g is not exactly 2.

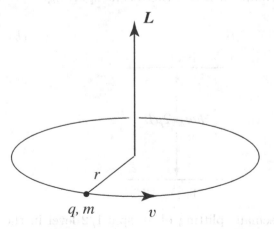

Figure 2. A (classical) particle with charge q in a circular orbit.

189

For the proton, g is *not* 2; and the neutron has a nonzero moment even though its charge is zero. This is because the proton and neutron, unlike the electron, are not elementary but are made of quarks (which have charge).

For the general spin-1/2 case, we shall write $\hat{\boldsymbol{\mu}} = \mu\hat{\boldsymbol{\sigma}}$, where (again) μ can be either positive or negative. The eigenvalues of $\hat{\mu}_z = \mu\hat{\sigma}_z$ are $\pm\mu$:

$$\hat{\mu}_z = \mu\hat{\sigma}_z = \begin{pmatrix} \mu & 0 \\ 0 & -\mu \end{pmatrix}.$$

Sometimes $\hat{\boldsymbol{\mu}}$ is written $\hat{\boldsymbol{\mu}} = \gamma\hat{\mathbf{s}}$, where γ is the *gyromagnetic ratio*, the ratio of the magnetic moment to the angular momentum. Thus μ and γ are related by $\mu = \gamma\hbar/2$. For the electron, $\gamma = -e/m$.

(b) Interaction with magnetic fields—Classically, the interaction energy of a magnetic moment $\boldsymbol{\mu}$, such as the moment of a compass needle, with a magnetic field $\mathbf{B}$ is $V = -\boldsymbol{\mu} \cdot \mathbf{B}$, where $V = 0$ when $\boldsymbol{\mu}$ is at $90°$ to $\mathbf{B}$ (again, see any introductory text). For a spin-1/2 particle, the quantum-mechanical Hamiltonian is then

$$\hat{H} = -\boldsymbol{\mu} \cdot \mathbf{B} = -\mu\hat{\boldsymbol{\sigma}} \cdot \mathbf{B} = -\mu(\hat{\sigma}_x B_x + \hat{\sigma}_y B_y + \hat{\sigma}_z B_z)$$
$$= -\mu \begin{pmatrix} B_z & B_x - iB_y \\ B_x + iB_y & -B_z \end{pmatrix}.$$

The eigenvalues of $\hat{H}$ are $-\mu B$ and $+\mu B$. When $\mathbf{B} = B\hat{\mathbf{z}}$, $\hat{H}$ is diagonal

$$\hat{H} = -\mu B\hat{\sigma}_z = -\mu B \begin{pmatrix} 1 & 0 \\ 0 & -1 \end{pmatrix} = \begin{pmatrix} -\mu B & 0 \\ 0 & +\mu B \end{pmatrix}.$$

and the eigenstates are $|+z\rangle$ and $|-z\rangle$; see Fig. 3. If μ is positive, the lower-energy state is $|+z\rangle$; if μ is negative, as for the electron, the lower state is $|-z\rangle$. For $\mathbf{B} = B\hat{\mathbf{n}}$, where the direction of $\hat{\mathbf{n}}$ is given by angles θ and ϕ, the eigenvalues are still $\mp\mu B$, but the eigenstates are the spinors $|\pm n\rangle$ of Sec. 1(d).

The splitting of the energies of angular-momentum states in a magnetic field is called *Zeeman splitting*. Of course, if $B = 0$ there is no splitting.

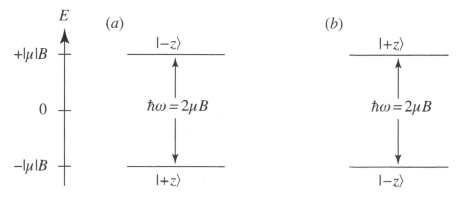

Figure 3. The Zeeman splitting of an spin-1/2 level in the field $\mathbf{B} = B\hat{\mathbf{z}}$: (a) for $\mu > 0$, (b) for $\mu < 0$. The splitting is $\hbar\omega \equiv 2\mu B$, where ω is the Larmor frequency.

190

10·4. TIME DEPENDENCE AND LARMOR PRECESSION

(a) **Time dependence**—We investigate the time-dependent behavior of a spinor and its associated polarization vector in the presence of a constant magnetic field $\mathbf{B}_0 = B_0\,\hat{\mathbf{z}}$ (the subscript on $\mathbf{B}$ anticipates the next section). The time-dependent Schrödinger equation is

$$i\hbar\,\frac{\partial \Psi}{\partial t} = \hat{H}\Psi .$$

This equation (as experiment confirms!) also tells how a spinor $\chi(t)$ evolves in time when the Hamiltonian is $\hat{H} = -\mu\hat{\boldsymbol{\sigma}}\cdot\mathbf{B}_0$. Since we have, for mathematical simplicity, taken $\mathbf{B}_0$ to define the z direction, $\hat{H}$ is $-\mu B_0\hat{\sigma}_z$ and the Schrödinger equation is

$$i\hbar\frac{\partial\chi}{\partial t} = i\hbar\begin{pmatrix} \dot{c}_+ \\ \dot{c}_- \end{pmatrix} = \begin{pmatrix} -\mu B_0 & 0 \\ 0 & +\mu B_0 \end{pmatrix}\begin{pmatrix} c_+ \\ c_- \end{pmatrix}.$$

Here $c_\pm = c_\pm(t)$, and a dot indicates differentiation with respect to time. Since $\hat{H}$ is diagonal, the equations for c_+ and c_- are not coupled:

$$\frac{dc_+}{dt} = i\,\frac{\mu B_0}{\hbar}\,c_+ = i\,\frac{\omega_0}{2}\,c_+ \qquad \text{and} \qquad \frac{dc_-}{dt} = -i\,\frac{\mu B_0}{\hbar}\,c_- = -i\,\frac{\omega_0}{2}\,c_- .$$

The ω_0 is the $\omega = 2\mu B/\hbar$ of the Zeeman splitting (Fig. 3). If μ is negative, then so is ω_0. The solutions of the equations are

$$c_+(t) = c_+(0)\,e^{+i\omega_0 t/2} \qquad \text{and} \qquad c_-(t) = c_-(0)\,e^{-i\omega_0 t/2} .$$

The time dependences here are just the usual factors $e^{-iEt/\hbar}$, with $E_\pm = \mp\mu B_0 = \mp\frac{1}{2}\hbar\omega_0$. The time-dependent spinor is

$$\boxed{|\chi(t)\rangle = \begin{pmatrix} c_+(t) \\ c_-(t) \end{pmatrix} = \begin{pmatrix} c_+(0)\,e^{+i\omega_0 t/2} \\ c_-(0)\,e^{-i\omega_0 t/2} \end{pmatrix}.}$$

(b) **Larmor precession**—Let $\mathbf{P}(t)$ at $t = 0$ lie at an angle θ to $\mathbf{B}_0$, and let $\mathbf{P}(0)$ and $\mathbf{B}_0$ define the xz plane. The azimuthal angle ϕ of $\mathbf{P}(0)$ is then zero, and $|\chi(0)\rangle$ and $\mathbf{P}(0)$ are

$$|\chi(0)\rangle = \begin{pmatrix} c_+(0) \\ c_-(0) \end{pmatrix} = \begin{pmatrix} \cos\frac{1}{2}\theta \\ \sin\frac{1}{2}\theta \end{pmatrix} \qquad \text{and} \qquad \mathbf{P}(0) = \begin{pmatrix} \sin\theta \\ 0 \\ \cos\theta \end{pmatrix}.$$

The time-dependence of $\mathbf{P}(t)$ comes from those $e^{\pm i\omega_0 t/2}$ factors in $\chi(t)$ in the above box:

$$\mathbf{P}(t) = \begin{pmatrix} 2\,\mathrm{Re}\left[c_+^*(t)c_-(t)\right] \\ 2\,\mathrm{Im}\left[c_+^*(t)c_-(t)\right] \\ |c_+(t)|^2 - |c_-(t)|^2 \end{pmatrix} = \begin{pmatrix} 2\,\mathrm{Re}\left[c_+^*(0)c_-(0)e^{-i\omega_0 t}\right] \\ 2\,\mathrm{Im}\left[c_+^*(0)c_-(0)e^{-i\omega_0 t}\right] \\ |c_+(0)|^2 - |c_-(0)|^2 \end{pmatrix} = \begin{pmatrix} +\sin\theta\cos\omega_0 t \\ -\sin\theta\sin\omega_0 t \\ \cos\theta \end{pmatrix}.$$

Figure 4 shows the precessional motion of $\mathbf{P}$ around $\mathbf{B}_0$; the polar angle θ stays constant and the azimuthal angle ϕ varies as $-\omega_0 t$. The angular frequency ω_0 is the *Larmor frequency*. Looked in from out along the $+z$ axis, $\mathbf{P}$ precesses in the clockwise or counterclockwise direction as μ (and thus ω_0) is positive or negative. If $\theta = 90°$, the tip of $\mathbf{P}$ circles around the equator of the polarization sphere.

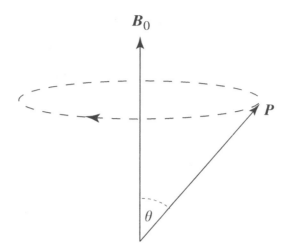

Figure 4. The precession of $\mathbf{P}$ around $\mathbf{B}_0$. The angular frequency of the precession is $\omega_0 = 2\mu B_0/\hbar$, independent of θ. The arrow on the circle shows the direction of precession when μ (and thus ω_0) is positive.

If $\theta = 0$ or $180°$, $\mathbf{P}$ is straight up or down and there is no precession. Straight up and down are the pure energy eigenstates $E = \mp\mu B_0$. Precession occurs when there is a superposition of different energy states, here the up and down states, because they have different time dependences. The precession angular frequency ω_0 is proportional to the energy difference $2\mu B_0$, independent of the values (as long as neither is zero) of $c_+(0)$ and $c_-(0)$. The circular precession is the superposition of $90°$ out-of-phase sinusoidal sloshing of the x and y components of $\mathbf{P}$. In Prob. 2·2, there was sloshing of the expectation value $\langle x \rangle$ of the position of a particle in an infinite square well when two energy eigenstates of the well were superimposed. There too the frequency was proportional to the energy difference.

The equations for c_+ and c_- were uncoupled because we took z along $\mathbf{B}_0$, which made the Hamiltonian diagonal. But the *physical* relation between $\mathbf{P}$ and $\mathbf{B}_0$ is the same no matter what direction $\mathbf{B}_0$ points in: $\mathbf{P}$ precesses around $\mathbf{B}_0$ with an angular frequency $\omega_0 = 2\mu B_0/\hbar$, and the angle between $\mathbf{P}$ and $\mathbf{B}_0$ stays constant.

10·5. TIME DEPENDENCE AND MAGNETIC RESONANCE

(a) A time-dependent Hamiltonian—We now add to the static $\mathbf{B}_0$ a time-dependent field $\mathbf{B}_1(t)$ that rotates in the xy plane; see Fig. 5. The full field is

$$\mathbf{B} = \mathbf{B}_0 + \mathbf{B}_1 = B_0\,\hat{\mathbf{z}} + B_1(\hat{\mathbf{x}}\cos\omega t - \hat{\mathbf{y}}\sin\omega t)\,.$$

For a given kind of particle, the magnetic moment μ is fixed, but B_0, B_1, and ω are independent, experimentally adjustable quantities—we could turn knobs. In applications, B_1 is usually much smaller than B_0, and ω can be of either sign. The reason for the minus sign with the y component of $\mathbf{B}_1$ is this: When $\omega_0 \equiv 2\mu B_0/\hbar$ and ω have the *same* sign, $\mathbf{B}_1$ circles about the z axis in the *same* sense as $\mathbf{P}$ precesses about $\mathbf{B}_0$ when $\mathbf{B}_1 = 0$ (see Fig. 4). Of special interest will be the case $\omega = \omega_0$ (resonance). Without that minus sign, this would occur at $\omega = -\omega_0$.

192

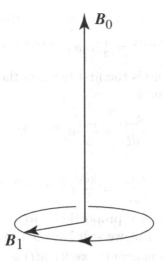

Figure 5. A fixed field $\mathbf{B}_0$ and a rotating field $\mathbf{B}_1$. The arrow on the circle shows the direction of rotation when ω is positive.

The Hamiltonian is now

$$\hat{H} = -\boldsymbol{\mu} \cdot (\mathbf{B}_0 + \mathbf{B}_1) = -\mu B_0 \hat{\sigma}_z - \mu B_1 (\hat{\sigma}_x \cos \omega t - \hat{\sigma}_y \sin \omega t)$$

$$= -\mu \begin{pmatrix} B_0 & B_1(\cos \omega t + i \sin \omega t) \\ B_1(\cos \omega t - i \sin \omega t) & -B_0 \end{pmatrix}$$

$$= -\mu \begin{pmatrix} B_0 & B_1 e^{+i\omega t} \\ B_1 e^{-i\omega t} & -B_0 \end{pmatrix}.$$

This is our first Hamiltonian with *explicit time dependence*. It is one of the few such that can be solved in exact, closed form.

The time-dependent Schrödinger equation is

$$\boxed{i\hbar \begin{pmatrix} \dot{c}_+ \\ \dot{c}_- \end{pmatrix} = -\mu \begin{pmatrix} B_0 & B_1 e^{+i\omega t} \\ B_1 e^{-i\omega t} & -B_0 \end{pmatrix} \begin{pmatrix} c_+ \\ c_- \end{pmatrix},}$$

or

$$\frac{dc_+}{dt} = \frac{i\mu}{\hbar}(B_0 c_+ + B_1 e^{+i\omega t} c_-) = \frac{i}{2}(\omega_0 c_+ + \omega_1 e^{+i\omega t} c_-)$$

$$\frac{dc_-}{dt} = \frac{i\mu}{\hbar}(B_1 e^{-i\omega t} c_+ - B_0 c_-) = \frac{i}{2}(\omega_1 e^{-i\omega t} c_+ - \omega_0 c_-) .$$

Here $\omega_0 \equiv 2\mu B_0/\hbar$ as before, $\omega_1 \equiv 2\mu B_1/\hbar$, and ω is the angular rotation frequency of $\mathbf{B}_1$. The field in the xy plane *couples* the equations for c_+ and c_-. The rate of change of *each* of c_+ and c_- now depends on the current values of *both* of them. The first step in finding $c_+(t)$ and $c_-(t)$ is to uncouple the equations. Divide and conquer.

For use below, we note that if $c_+(0) = 1$, then dc_-/dt at $t = 0$ is $\dot{c}_-(0) = i\omega_1/2$. These will be the initial conditions for an example.

(b) Uncoupling the equations—To uncouple them, we first factor out of $c_+(t)$ and $c_-(t)$ their time dependences in the *absence* of B_1. In Sec. 4(a), we found that these dependences were $e^{\pm i\omega_0 t/2}$. We write $c_\pm(t) = a_\pm(t) e^{\pm i\omega_0 t/2}$, and then dc_+/dt is

$$\dot{c}_+ = \left(\dot{a}_+ + \tfrac{1}{2} i\omega_0 a_+\right) e^{+i\omega_0 t/2} .$$

193

Putting this into the equation that relates dc_+/dt to the current values of c_+ and c_-, we get

$$\left(\dot{a}_+ + \tfrac{1}{2}i\omega_0 a_+\right)e^{+i\omega_0 t/2} = \tfrac{1}{2}i\omega_0 a_+\, e^{+i\omega_0 t/2} + \tfrac{1}{2}i\omega_1\, e^{+i\omega t}a_-e^{-i\omega_0 t/2}\ .$$

The second term on the left cancels the first term on the right (which was the reason for the factorization), and what remains is

$$\frac{da_+}{dt} = \frac{i\omega_1}{2}e^{+i(\omega-\omega_0)t}\,a_-\ .$$

Similarly,

$$\frac{da_-}{dt} = \frac{i\omega_1}{2}e^{-i(\omega-\omega_0)t}\,a_+\ .$$

The rate of change of a_+ is directly proportional to a_-, and vice versa—the two (reduced) amplitudes "feed" one another. But we still have not uncoupled the equations.

For brevity (and just for a moment), we let $u(t) \equiv \tfrac{1}{2}i\omega_1\, e^{i(\omega-\omega_0)t}$. With this shorthand, the equations for $da_\pm/dt$ are

$$\dot{a}_+(t) = u(t)\,a_-(t) \qquad \text{and} \qquad \dot{a}_-(t) = -u^*(t)\,a_+(t)\ .$$

Next, we use these two equations to replace a_- and $\dot{a}_-$ in the time derivative of $\dot{a}_+$:

$$\ddot{a}_+ = \dot{u}\,a_- + u\,\dot{a}_- = (\dot{u}/u)\dot{a}_+ - uu^*a_+\ .$$

But $\dot{u}/u = i(\omega - \omega_0)$ and $uu^* = \tfrac{1}{4}\omega_1^2$, so that

$$\ddot{a}_+ - i(\omega - \omega_0)\dot{a}_+ + \tfrac{1}{4}\omega_1^2 a_+ = 0\ .$$

This is a linear, homogeneous, second-order, constant-coefficient differential equation for $a_+(t)$. Uncoupled! At the price of becoming second order.

(c) **Solving the equations**—The usual way to attack such equations is to try a trial function of the form $a_+(t) = e^{\beta t}$, and then cancel the common factor $e^{\beta t}$. We get

$$\beta^2 - i(\omega - \omega_0)\beta + \tfrac{1}{4}\omega_1^2 = 0\ .$$

The two roots are

$$\beta_\pm = \tfrac{1}{2}i(\omega - \omega_0) \pm \tfrac{1}{2}\sqrt{-(\omega-\omega_0)^2 - \omega_1^2} = \tfrac{1}{2}i(\omega - \omega_0) \pm \tfrac{1}{2}i\Omega\ ,$$

where $\Omega \equiv \sqrt{(\omega-\omega_0)^2 + \omega_1^2}$ is real and positive. When $\omega = \omega_0$, $\Omega = |\omega_1|$.

The general solution for $a_+(t)$ is a superposition of the two particular solutions:

$$a_+(t) = Ce^{\beta_+ t} + De^{\beta_- t} = e^{i(\omega-\omega_0)t/2}\left(Ce^{+i\Omega t/2} + De^{-i\Omega t/2}\right)\ .$$

And then

$$c_+(t) = a_+(t)\,e^{+i\omega_0 t/2} = e^{i\omega t/2}\left(Ce^{+i\Omega t/2} + De^{-i\Omega t/2}\right)\ .$$

Similarly,

$$c_-(t) = c^{-i\omega t/2}\left(Fe^{+i\Omega t/2} + Ge^{-i\Omega t/2}\right)\ .$$

Here C and D and F and G are arbitrary complex constants, to be determined by initial ($t = 0$) conditions.

(d) Spin up at t = 0—As an example, we take the spin to be up at $t = 0$: $c_+(0) = 1$ and $c_-(0) = 0$. From $c_-(0) = 0$ and the equation for $c_-(t)$ just obtained, we get $G = -F$, and so

$$c_-(t) = 2iFe^{-i\omega t/2}\sin(\tfrac{1}{2}\Omega t) .$$

In that case,

$$\dot{c}_-(t) = iFe^{-i\omega t/2}\left[-i\omega\sin(\tfrac{1}{2}\Omega t) + \Omega\cos(\tfrac{1}{2}\Omega t)\right] ,$$

which at $t = 0$ is $\dot{c}_-(0) = iF\Omega$. In the last lines of Sec. 5(a), we noted that if $c_+(0) = 1$ then $\dot{c}_-(0) = i\omega_1/2$. Therefore $F = \omega_1/2\Omega$, and

$$c_-(t) = \frac{i\omega_1}{\Omega}\, e^{-i\omega t/2}\sin(\tfrac{1}{2}\Omega t) .$$

Then the time-dependent probability the spin is down is

$$\boxed{|c_-(t)|^2 = \frac{\omega_1^2}{\Omega^2}\sin^2(\tfrac{1}{2}\Omega t) = A(\omega)\sin^2(\tfrac{1}{2}\Omega t) ,}$$

where

$$\boxed{\Omega^2 \equiv (\omega - \omega_0)^2 + \omega_1^2 \qquad \text{and} \qquad A(\omega) \equiv \frac{\omega_1^2}{\Omega^2} = \frac{1}{1 + (\omega - \omega_0)^2/\omega_1^2} .}$$

Of course, $A(\omega)$ is a function of all three ω's. But a common experimental arrangement is to fix ω_0 and ω_1 (by fixing B_0 and B_1), and then search for where $A(\omega) = 1$ (when $\omega = \omega_0$) by varying ω. The ω knob is what you explore with.

(e) Resonance—We started with the spin up at $t = 0$. The maximum probability the spin is ever down is then $A(\omega)$. Figure 6 shows $A(\omega)$ versus $(\omega - \omega_0)/\omega_1$. It is largest—it *resonates*—at $\omega = \omega_0$. The value of ω_0 is determined by B_0, since $\hbar\omega_0 = 2\mu B_0$. The full width at half maximum (FWHM) of the $A(\omega)$ curve is $2|\omega_1|$ (see the problems). Thus $A(\omega)$ is relatively narrow when B_1/B_0 is small: the "tuning" is sharper when B_1/B_0 is small.

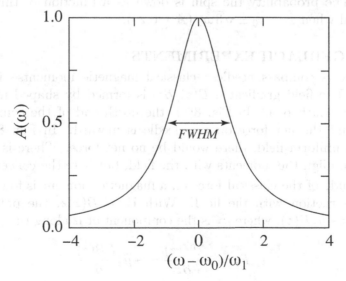

Figure 6. The function $A(\omega)$ versus $(\omega - \omega_0)/\omega_1$. The full width at half maximum (FWHM) is $2|\omega_1|$.

195

If ω is far from ω_0, then $A(\omega)$ is small and the tip of the polarization vector never wanders far from the north pole. It simply spirals out a little and then back, over and over. But if $\omega = \omega_0$ then $A(\omega_0) = 1$ and $\mathbf{P}$ spirals all the way down to the south pole and then back to the north pole, and repeats this over and over (Prob. 7). Figure 7 shows the probability the spin is down as a function of time for two cases. At resonance ($\omega = \omega_0$), $\Omega = |\omega_1|$, and the spin is down at times

$$t_- = n\,\frac{\pi}{|\omega_1|}\ ,$$

where $n = 1, 3, 5, \ldots$ When $\omega = \omega_0 \pm \sqrt{3}\,|\omega_1|$, $A(\omega) = 0.25$ and $\Omega = 2|\omega_1|$. The oscillation frequency Ω is smallest at resonance.

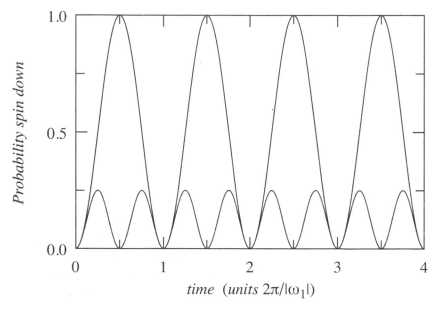

Figure 7. The probability the spin is down as a function of time when $\omega = \omega_0$ ($A = 1$), and when $\omega = \omega_0 \pm \sqrt{3}\,\omega_1$ ($A = 0.25$).

10·6. STERN-GERLACH EXPERIMENTS

Figure 8 shows compass needles—classical magnetic moments—in an inhomogeneous magnetic field. The field gradient, $\partial B(z)/\partial z$, is formed by shaped magnets—a flat north pole and a sharp south pole. In Fig. 8(a), the north end of the compass needle is in the stronger field, and the net force on the needle is upward. In Fig. 8(b), the net force is downward. In a uniform field, there would be no net force. There is also a torque on the needles, acting to align the moments with the field, but here the concern is the net force.

The magnitude of the classical force on a magnetic moment is found from the potential energy of its interaction with the field. With $\mathbf{B} = B(z)\hat{\mathbf{z}}$, the potential energy is now $V(z) = -\boldsymbol{\mu} \cdot \mathbf{B} = -\mu_z B(z)$, where μ_z is the component of $\boldsymbol{\mu}$ along the z axis. Then the force is

$$F_z = -\frac{\partial V(z)}{\partial z} = +\mu_z\frac{\partial B(z)}{\partial z}\ .$$

Classically, μ_z can be anything between $+\mu$ and $-\mu$, where $\mu = |\boldsymbol{\mu}|$. Therefore, classically, the force along the z direction can be anything from $+\mu\,\partial B(z)/\partial z$ to $-\mu\,\partial B(z)/\partial z$.

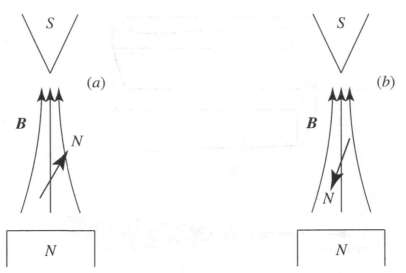

Figure 8. Compass needles in an inhomogeneous magnetic field created by shaped magnetic poles (we are looking end-on to the magnets; see Fig. 9). The net force is up in (a), down in (b).

In Fig. 9(a), a narrow beam of spin-1/2 particles of well-defined momentum enters a channel of shaped magnets—a *Stern-Gerlach apparatus*. The shaping makes a field gradient perpendicular to the beam throughout the length. A given particle will exit the far end deflected up or down by an amount proportional to the component of μ along the z direction. If the moments of the incoming particles are initially oriented randomly in direction, classically one expects that the particles will be spread over a continuous range on the screen, as in Fig. 9(b).

Figure 9(c) shows what actually happens. The interaction of a moment with the field constitutes a measurement of the component of μ (and thus of the component of the spin) along the field direction. For spin 1/2, μ_z can only be $\pm\mu$. Therefore, the *force* F_z is quantized. If the particles in the incident beam are randomly polarized, statistically equal numbers will be bent up and down. If the incident beam is polarized with the moments aligned with the field, so that $\mu_z = |\mu|$, all the particles will be bent upward. (What happens if the incident beam is polarized along a direction perpendicular to the magnetic field?)

The first experiments of this kind were carried out by Otto Stern and Walther Gerlach in 1922. They gave the first *direct* demonstration of angular-momentum quantization. The particles were neutral silver atoms boiled out of an oven and collimated. Silver atoms in their ground state have spin 1/2 (Chap. 14). The atoms were splashed onto a glass plate, sticking there in two clumps. For a simple discussion, with a reproduction of a postcard sent to Niels Bohr showing a photo of the plate, see A.P. French and E.F. Taylor, *An Introduction to Quantum Physics*, Chap. 10.

Figure 10 shows schematically a three-stage Stern-Gerlach apparatus, in which the field direction is reversed in the center half, as indicated by the fields beneath the three sections. If the spin-down beam is not blocked, the two beams are brought back together. Blocking one of the beams as in Fig. 10 produces a purely polarized output beam. And the exiting beam is going in the same direction as the incident beam was. This *ideal SG* apparatus is used to brilliant pedagogical effect in Chaps. 5 and 6 of *The Feynman Lectures*, Vol. III.

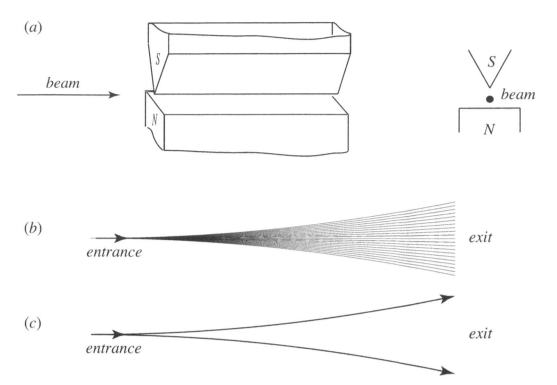

Figure 9. (a) A Stern-Gerlach apparatus, in perspective and end-on, with shaped poles to make an inhomogeneous magnetic field. (b) Classically, the beam in passing through the apparatus would be spread out over a range of angles. This does not happen. (c) For a spin-1/2 beam, there are just two beams at the exit.

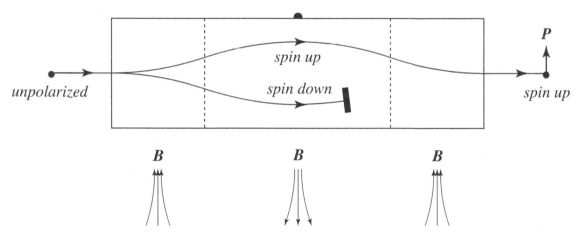

Figure 10. A schematic of an initially unpolarized beam passing through a three-stage Stern-Gerlach apparatus. The field configuration in each stage, as seen from the direction of the incoming beam, is shown below. Because the lower beam is blocked, the beam that emerges is entirely in the spin-up state (assuming $\mu > 0$).

A Stern-Gerlach apparatus splits a spin-1 beam into $m = +1$, 0, and -1 beams. If two of those beams are blocked, the output is a pure m state. And so on for higher spins.

10·7. POLARIZATION AND LIGHT

(a) **Classical light**—Polarized light is another two-state system. The oscillating electric and magnetic fields (they *are* the light) oscillate in sync at right angles to the beam direction and to each other. Two axes orthogonal to the beam direction and to each other are labeled H and V, for horizontal and vertical. *Linearly* polarized light is light in which the electric field $\mathcal{E}(t)$ oscillates back-and-forth along a fixed direction. Linear polarizations along the H or V axes are represented by the same two orthonormal vectors used for spin-1/2 particles,

$$|H\rangle = \begin{pmatrix} 1 \\ 0 \end{pmatrix}, \qquad |V\rangle = \begin{pmatrix} 0 \\ 1 \end{pmatrix}.$$

For light, these are called Jones vectors instead of spinors. As with spin-1/2 particles, any superposition of these basis states is a possible polarization state.

However, the parallels in mathematics can obscure differences in physics. The foregoing physics of spin-1/2 was essentially about a particle at rest, and in the absence of an external magnetic field all directions were equal. In contrast, light is by its very nature kinematic, in two ways: (1) Light is always *going* in some direction, and (2) its fields are *oscillating*. And whereas the spin-1/2 $|+z\rangle$ and $|-z\rangle$ states are at 180° to one another ($\mathbf{P}$ is up *or* down along z), the $|H\rangle$ and $|V\rangle$ states are at 90° to one another ($\mathcal{E}$ goes back *and* forth along H, or up *and* down along V). The table shows correspondences for the basic polarization states.

spinor/ Jones vector	$\begin{pmatrix} 1 \\ 0 \end{pmatrix}$	$\begin{pmatrix} 0 \\ 1 \end{pmatrix}$	$\begin{pmatrix} 1/\sqrt{2} \\ 1/\sqrt{2} \end{pmatrix}$	$\begin{pmatrix} 1/\sqrt{2} \\ -1/\sqrt{2} \end{pmatrix}$	$\begin{pmatrix} 1/\sqrt{2} \\ i/\sqrt{2} \end{pmatrix}$	$\begin{pmatrix} 1/\sqrt{2} \\ -i/\sqrt{2} \end{pmatrix}$
spin-1/2	$\uparrow\,|+z\rangle$	$\downarrow\,|-z\rangle$	$\overrightarrow{\;}\;|+x\rangle$	$\overleftarrow{\;}\;|-x\rangle$	$\otimes\,|+y\rangle$	$\odot\,|-y\rangle$
light	$\leftrightarrow\;|H\rangle$	$\updownarrow\;|V\rangle$	↗	↘	$\circlearrowleft\,|L\rangle$	$\circlearrowright\,|R\rangle$

The polarization vectors of the $|+y\rangle$ and $|-y\rangle$ states point in and out of the page. For all six cases here, the light is coming out of the page ($\mathbf{H} \times \mathbf{V}$), and the electric fields oscillate parallel to the plane of the page. The $\pm i$ factors in the last two Jones vectors make the H and V oscillations $\pm 90°$ out of phase with one another, and the electric field spirals about the direction the light is going in.

(b) **Making polarized light**—Much of light that has been reflected or scattered before it reaches the eye is partially polarized. This can be seen by rotating a piece of Polaroid while looking through it at light reflected obliquely from a lake, streets, a table top, ..., or at light scattered by a clear sky at 90° to the Sun's rays. Polaroid is made of long-chain hydrocarbon molecules. In manufacturing, the material is heated and stretched to make the molecules parallel. The molecules are then "doped" to make them conducting—like tiny wires. The *transmission axis* (TA) lies in the plane of the sheet and is perpendicular to those "wires."

The intensity I of light is the rate at which energy is transported across a surface of unit area. It is proportional to $\mathcal{E}_0^2$, where $\mathcal{E}_0$ is the amplitude of the electric field. In Fig. 11(a), linearly polarized light is incident upon a sheet of *ideal* linear polarizer. (Polaroid is not ideal.) The axis of polarization of the incident light is at an angle θ to the TA. The energy associated with the component parallel to the wires, $\mathcal{E}_0 \sin\theta$, drives current along them and is absorbed as heat. The intensity of light that is transmitted is proportional to the square of the component of the electric field $\mathcal{E}$ along the TA,

$$\frac{I(\text{transmitted})}{I(\text{incident})} = \frac{(\mathcal{E}_0 \cos\theta)^2}{\mathcal{E}_0^2} = \cos^2\theta \; ,$$

where $0 \leq \theta \leq 90°$. There is is a nice lecture demonstration with, in sequence, a source of unpolarized microwaves, a grid of parallel wires or metal rods, and a linear receiving antenna that can be rotated about the beam axis.

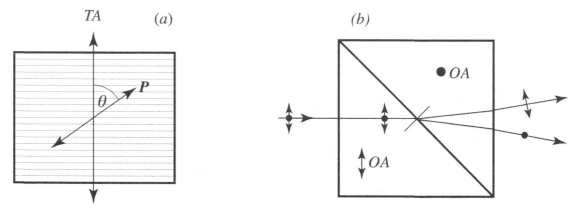

Figure 11. (a) A sheet of polarizer with linearly polarized light incident normally upon it. The polarization axis of the incident light is at an angle θ to the transmission axis (TA) of the polarizer. (The horizontal lines are the "wires" that absorb the energy of the horizontal component of the incident wave.) (b) A double prism of crystal calcite, seen from the side, with the optical axes (OA) oriented as shown. It splits unpolarized light into H and V polarized beams.

A sheet of linear polarizer does for light what a Stern-Gerlach apparatus with one channel blocked does for spin-1/2 particles: What gets through is polarized. If the incident beam is unpolarized, then (ideally) half in number or intensity gets through. The quantities analogous to $\cos^2\theta$ and $\sin^2\theta$ with $0 \leq \theta \leq 90°$ for light were $\cos^2\frac{1}{2}\beta$ and $\sin^2\frac{1}{2}\beta$ with $0 \leq \beta \leq 180°$ for spin-1/2. Two sheets of polarizer with transmission axes at $90°$ totally block light, just as two SG apparatuses with the "down" channels blocked and the open channels at $180°$ (like the first and last of the SG's in the figure for Prob. 8) totally block spin-1/2 particles.

Polarized light can be produced in several other ways. Figure 11(b) shows a device made of two prisms of calcite crystal. It splits a beam of unpolarized light into $|H\rangle$ and $|V\rangle$ beams, just as an SG apparatus with both channels open splits a beam of spin-1/2 particles into $|+\rangle$ and $|-\rangle$ beams. The constituents of calcite are $CaCO_3$ molecules. The three oxygen atoms form an equilateral triangle, and in a pure crystal of calcite the planes of all the O_3 triangles are parallel. The direction perpendicular to all those triangles is the *optic axis OA*. The electrons in the crystal react with different strengths to forcing by $\mathcal{E}$-field oscillations parallel and perpendicular to that axis. This makes the indices of refraction different for light polarized parallel and perpendicular to the axis. The indices for visible light in calcite are $n_\parallel = 1.4864$ and $n_\perp = 1.6584$.

In Fig. 11(b), the OA in the first prism is vertical in the plane of the page, and the OA in the second prism is horizontal and perpendicular to the page. Unpolarized light is incident normally from the left. The vertically polarized component ($\updownarrow$) of the light "sees" indices $n_{\parallel}$ in the first prism and $n_{\perp}$ in the second; and $n_{\parallel} < n_{\perp}$. By the law of refraction (Snell's law), this component is bent, as shown, toward the normal to the interface (as is light entering water or glass from air). The horizontally polarized component ($\bullet$) of the light sees $n_{\perp}$ first and $n_{\parallel}$ second and is bent away from the normal. If the interface between the prisms is at $45°$, the opening angle between the rays is $12.8°$ (which is further increased when the rays exit the second prism).

(c) Quantum light—Everything said so far about polarized light is classical, and was known long before quantum mechanics. It becomes quantum mechanical when applied to the polarization states of individual "particles" of light—photons. An incident photon polarized at an angle θ to the *TA* of the polarizer in Fig. 11(a) will be transmitted or absorbed with probabilities $\cos^2\theta$ and $\sin^2\theta$. At low enough intensities, photon counters at the two outputs of the prism arrangement of Fig. 11(b) will count at one of the other in some random fashion, but not simultaneously. For an unpolarized incident beam, statistically half of the photons will strike each counter. We saw something of (unpolarized) photons in Chap. 1, and will learn more about them later on.

PROBLEMS

1. *Eigenvalues and eigenstates.*

(a) Show that the eigenvalues of the most general 2-by-2 Hermitian matrix,

$$a\hat{I} + b_x\hat{\sigma}_x + b_y\hat{\sigma}_y + b_z\hat{\sigma}_z = a\hat{I} + \mathbf{b}\cdot\hat{\boldsymbol{\sigma}} = \begin{pmatrix} a+b_z & b_x - ib_y \\ b_x + ib_y & a - b_z \end{pmatrix},$$

are $\lambda_{\pm} = a \pm (b_x^2 + b_y^2 + b_z^2)^{1/2} = a \pm b$. Here all of a, b_x, b_y, and b_z are real and independent, and $\mathbf{b} = (b_x, b_y, b_z) = b\,\hat{\mathbf{n}}$ is an ordinary vector of magnitude b in the direction $\hat{\mathbf{n}}$.

(b) Show that the normalized eigenstates belonging to the eigenvalues $a \pm b$ are, aside from an arbitrary overall phase factor,

$$|{+}n\rangle = \begin{pmatrix} \cos\frac{1}{2}\theta \\ e^{+i\phi}\sin\frac{1}{2}\theta \end{pmatrix} \quad \text{and} \quad |{-}n\rangle = \begin{pmatrix} \sin\frac{1}{2}\theta \\ -e^{+i\phi}\cos\frac{1}{2}\theta \end{pmatrix},$$

where θ and ϕ are the polar and azimuthal angles of $\mathbf{b}$.

(c) Show that $|{+}n\rangle$ and $|{-}n\rangle$ are orthogonal and that $|{+}n\rangle\langle{+}n| + |{-}n\rangle\langle{-}n| = \hat{I}$.

2. *The polarization vector.*

(a) Show that the components of $\mathbf{P}$, where $P_x \equiv \langle\chi|\hat{\sigma}_x|\chi\rangle$, etc., are in terms of the components c_+ and c_- of the spinor $|\chi\rangle$,

$$P_x = 2\,\mathrm{Re}\,(c_+^* c_-), \quad P_y = 2\,\mathrm{Im}\,(c_+^* c_-), \quad P_z = |c_+|^2 - |c_-|^2.$$

(b) Thus show from the spinors $|{+}z\rangle$, $|{-}x\rangle$, $|{-}y\rangle$, and $|{+}n\rangle$ given in Sec. 1(d) that the corresponding polarization vectors are $+\hat{\mathbf{z}}$, $-\hat{\mathbf{x}}$, $-\hat{\mathbf{y}}$, and $+\hat{\mathbf{n}}$.

(c) Find (if you can, by inspection) the eigenvalues and the polar and azimuthal angles θ and ϕ of the polarization vector for each of these matrices:

$$\begin{pmatrix} 0 & 1 \\ 1 & 2 \end{pmatrix}, \quad \begin{pmatrix} 1 & -i \\ i & 1 \end{pmatrix}, \quad \begin{pmatrix} 0 & 1+i \\ 1-i & 0 \end{pmatrix}.$$

(d) Find the probabilities of getting $+\hbar/2$ on making a measurement along the $+x$, $-x$, $+y$, $-y$, $+z$, or $-z$ directions when $\mathbf{P} = (1/\sqrt{3}, 1/\sqrt{6}, 1/\sqrt{2})$.

(e) Find the normalized spinor that corresponds to the $\mathbf{P}$ vector in (d).

3. *Sequential measurements.*

A spin-1/2 particle is initially spin-up along the $+z$ direction. A measurement is made of the component of the spin along the direction $(\theta, \phi) = (73°, 205°)$, and then a measurement is made of the component along the $-z$ direction.

Find the probabilities of getting the four possible sequences of measurement results:

$+\hbar/2$ along $(73°, 205°)$, then $+\hbar/2$ along $-z$;

$+\hbar/2$ along $(73°, 205°)$, then $-\hbar/2$ along $-z$;

$-\hbar/2$ along $(73°, 205°)$, then $+\hbar/2$ along $-z$;

$-\hbar/2$ along $(73°, 205°)$, then $-\hbar/2$ along $-z$.

The sum of the four probabilities should add up to one.

4. *Spin precession again.*

(a) Show that $[\hat{\sigma}_x, \hat{\sigma}_y] = 2i\hat{\sigma}_z$, and then write down $[\hat{\sigma}_y, \hat{\sigma}_z]$ and $[\hat{\sigma}_z, \hat{\sigma}_x]$.

(b) Get simple expressions for $[\hat{H}, \hat{\sigma}_x]$, $[\hat{H}, \hat{\sigma}_y]$, and $[\hat{H}, \hat{\sigma}_z]$ when $\hat{H} = -\boldsymbol{\mu} \cdot \mathbf{B} = -\mu\hat{\sigma}_z B$.

(c) In Sec. 5·7, it was shown that the time rate of change of the expectation value $\langle A \rangle$ of a physical quantity A when there is no explicit time dependence in the operator $\hat{A}$ is

$$\frac{d\langle A \rangle}{dt} = \frac{i}{\hbar} \langle [\hat{H}, \hat{A}] \rangle .$$

Use this to find the rates of change of P_x, P_y, and P_z in terms of P_x, P_y, and P_z.

(d) Take the time derivative of the equation for dP_x/dt, and get an equation that only involves P_x. Then show that P_x varies sinusoidally with time, and that so does P_y, but $90°$ out of phase with P_x. The combined motion is precession.

5. *Spin rotations in beams.*

A beam of neutrons is traveling in the x direction. Initially, all the spins are up along the z direction. The neutrons pass through a region in which there is a uniform 1-tesla magnetic field in the y direction. What is the minimum time the neutrons would have to remain in the field to exit it with their spins along $-z$? The magnetic moment of the neutron is $\mu_n = -0.966 \times 10^{-26}$ J/T (joules/tesla).

6. *Resonance curves.*

The resonance form $A(\omega)$ of Sec. 5(d) occurs in mechanics, electromagnetism, optics, and atomic, nuclear, and particle physics. (See also Fig. 7·3.)

(a) Show that the full width at half maximum of $A(\omega)$ is $2\omega_1$. For "sharp tuning," the ratio $\omega_1/\omega_0 = B_1/B_0$ should be small.

(b) Graph $A(\omega)$ versus ω carefully for ω between $0.8\,\omega_0$ and $1.2\,\omega_0$ for the cases $B_1/B_0 = 0.1$ and 0.03. Scale the range of $A(\omega_0)$ (from 0 to 1) to 8 cm, and the range of ω to 15 cm.

7. *Spin flipping at resonance.*

In the situation of Sec. 5, with spin up at $t = 0$, we found that

$$c_-(t) = i\,\frac{\omega_1}{\Omega}\,e^{-i\omega t/2}\,\sin(\tfrac{1}{2}\Omega t)\,.$$

(a) Show that when $\omega = \omega_0$ (resonance), $c_-(t)$ and $c_+(t)$ are

$$c_-(t) = ie^{-i\omega_0 t/2}\sin(\tfrac{1}{2}\omega_1 t) \quad\text{and}\quad c_+(t) = e^{+i\omega_0 t/2}\cos(\tfrac{1}{2}\omega_1 t)\,.$$

Don't start from scratch. Get $c_+(t)$ from $c_-(t)$; or get $a_+(t)$ from $a_-(t)$.

(b) Thus show that at resonance the components of the polarization vector are

$$P_x(t) = \sin\omega_1 t\,\sin\omega_0 t\,,\quad P_y(t) = \sin\omega_1 t\,\cos\omega_0 t\,,\quad P_z(t) = \cos\omega_1 t\,.$$

(c) How then do the polar and azimuthal angles θ and ϕ vary with time?

(d) If $B_1/B_0 = 10^{-4}$, how many times does **P** spiral about the z axis as the spin goes from up to down?

8. *Stern-Gerlachs or polarizers in series.*

(a) A beam of spin-1/2 particles is incident upon a sequence of N ideal Stern-Gerlach apparatuses, SG_1, SG_2, ..., SG_N. The magnetic axis of each apparatus is perpendicular to the beam direction, and each apparatus allows only its up beam to pass. The magnetic axis of SG_{i+1} is at an angle $180°/(N-1)$ to that of SG_i, so that the N axes form a fan with SG_N at $180°$ to SG_1. The figure shows the arrangement for $N = 3$. Make a table showing what fraction of particles that *exit* from SG_1 exit from SG_N, for $N = 2, 3, 5, 10, 20, 50, 100$.

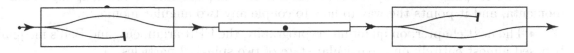

(b) A beam of photons is incident normally upon a sequence of N ideal linear polarizers, P_1, P_2, ..., P_N. The transmission axis of P_{i+1} is at an angle $90°/(N-1)$ to that of P_i, so that the N axes form a fan with P_N at $90°$ to P_1. Make a table showing what fraction of photons that *exit* from P_1 also exit from P_N, for $N = 2, 3, 5, 10, 20, 50, 100$.

9. *Projection operators.*

Make a table with the operators down the side expressed as matrices, and the states across the top expressed as spinors. Then *in* the table, show the result of what each operator does to each spinor.

	$\lvert+z\rangle$	$\lvert-z\rangle$	$\lvert+x\rangle$	$\lvert-x\rangle$	$\lvert+y\rangle$	$\lvert-y\rangle$
$\lvert+z\rangle\langle+z\rvert$						
$\lvert-z\rangle\langle-z\rvert$						
$\lvert+x\rangle\langle+x\rvert$						
$\lvert-x\rangle\langle-x\rvert$						
$\lvert+y\rangle\langle+y\rvert$						
$\lvert-y\rangle\langle-y\rvert$						

What (in a few words) does the $\lvert+z\rangle\langle+z\rvert$ operator do to every spinor it operates on? What experimental arrangement is this analogous to?

11. HYPERFINE SPLITTING. TWO ANGULAR MOMENTA. ISOSPIN

11.1. *Hyperfine Structure of the Hydrogen Ground State*

11.2. *The 21-cm Line and Astronomy*

11.3. *Coupling Two Spin-1/2 Particles*

11.4. *Coupling Any Two Angular Momenta*

11.5. *Clebsch-Gordan Coefficients*

11.6. *Particle Multiplets and Isospin*

      Problems

This chapter is about the angular-momentum states of the combined system when two angular momenta are involved. For example, in most of the excited states of hydrogen the electron has a non-zero orbital angular momentum as well as spin (and there is also the spin of the proton). In a later chapter, we consider the possible angular-momentum states of atoms with many electrons. For the present, two constituent angular momenta will be enough.

We begin (somewhat indirectly) with a simple example, the "hyperfine splitting" of the ground state of the hydrogen atom: a tiny 5.9×10^{-6} eV splitting of the -13.6 eV level. The splitting is due to the interaction of the magnetic moments of the spin-1/2 electron and proton. The example is important for several reasons:

• The wavelength of the transition between the hyperfine-split levels plays a major role in radio astronomy.

• The coupling of two spin-1/2 particles is the simplest case of the coupling of angular momenta, and it points the way to how to couple any two angular momenta.

• The next chapter, on quantum cryptography, the EPR argument, and Bell's inequality, is based almost entirely on a particular state of two spin-1/2 particles.

• The hyperfine calculation is an example of (degenerate) perturbation theory, and gives a glimpse of the subject of Chapter 13.

Section 1 follows early sections of Chapter 12 of *The Feynman Lectures*, Vol. III. In Sec. 3, we find a quicker way to calculate the hyperfine structure, but it will be instructive to follow Feynman for a while.

11·1. HYPERFINE STRUCTURE OF THE HYDROGEN GROUND STATE

(a) Two spins—The hydrogen atom is a bound state of two spin-1/2 particles, an electron and a proton. It takes two states, $|+z\rangle$ and $|-z\rangle$, to span the spin space of each particle. The two spaces are independent, and so it takes $2 \times 2 = 4$ states to span the spin space of the two particles. We shall use

$$|++\rangle, \quad |+-\rangle, \quad |-+\rangle, \quad \text{and} \quad |--\rangle$$

as the basis states. The first symbol will indicate the spin state of the electron and the second that of the proton. Thus in the $|+-\rangle$ state the electron spin is up and the proton spin is down. Up and down are with respect to the direction chosen as z. In the absence of an external field or some otherwise preferred direction, any direction will do.

Implicit in our abbreviated notation is that the two particles have spin 1/2. A full characterization of, say, the $|+-\rangle$ state would be

$$|+-\rangle \rightarrow |s_e, s_p; m_e, m_p\rangle = |\tfrac{1}{2}, \tfrac{1}{2}; +\tfrac{1}{2}, -\tfrac{1}{2}\rangle .$$

Again, a sign is given on m values even when m is positive, but is never given on an s (or ℓ or j) value, which is never negative. A sign indicates that a value *is* an m value.

We often represent the four spin states with these four orthonormal column vectors,

$$|++\rangle = \begin{pmatrix} 1 \\ 0 \\ 0 \\ 0 \end{pmatrix}, \quad |+-\rangle = \begin{pmatrix} 0 \\ 1 \\ 0 \\ 0 \end{pmatrix}, \quad |-+\rangle = \begin{pmatrix} 0 \\ 0 \\ 1 \\ 0 \end{pmatrix}, \quad |--\rangle = \begin{pmatrix} 0 \\ 0 \\ 0 \\ 1 \end{pmatrix}.$$

We could instead use electron and proton spinors side by side, as for example,

$$|+-\rangle = \begin{pmatrix} 1 \\ 0 \end{pmatrix}_e \begin{pmatrix} 0 \\ 1 \end{pmatrix}_p .$$

However, the 4-component vectors give exactly the same information and are usually easier to work with. The most general *spin* state of the hydrogen atom is

$$|\chi\rangle = c_{++}|++\rangle + c_{+-}|+-\rangle + c_{-+}|-+\rangle + c_{--}|--\rangle = \begin{pmatrix} c_{++} \\ c_{+-} \\ c_{-+} \\ c_{--} \end{pmatrix},$$

where $\langle \chi|\chi\rangle = |c_{++}|^2 + |c_{+-}|^2 + |c_{-+}|^2 + |c_{--}|^2 = 1$. The probability that both spins are up is $|c_{++}|^2$; the probability that the proton spin is down is $|c_{+-}|^2 + |c_{--}|^2$. The full specification of the ground state of hydrogen would be

$$\psi_{100}(r)\,|\chi\rangle = \frac{1}{\sqrt{\pi a_0^3}}\, e^{-r/a_0} \begin{pmatrix} c_{++} \\ c_{+-} \\ c_{-+} \\ c_{--} \end{pmatrix}.$$

This is the only state we are concerned with here.

(b) Two moments, an interaction—Associated with the spins of the electron and proton are magnetic moments, $\hat{\boldsymbol{\mu}}_e = -\mu_B \hat{\boldsymbol{\sigma}}_e$ and $\hat{\boldsymbol{\mu}}_p = \mu_p \hat{\boldsymbol{\sigma}}_p$, where $\mu_p \ll \mu_B$ because $m_p \gg m_e$. (See Sec. 10·3(a); the ratio μ_p/μ_e is -1.521×10^{-3}, which is not equal to m_e/m_p.) These moments, like tiny compass needles, interact with each other. When $\ell \neq 0$, they also interact with the magnetic moment associated with the orbital angular momentum; but $\ell = 0$ in the hydrogen ground state. The magnetic interaction is a tiny "perturbation," an add-on to the ground-state energy of -13.6 eV. The latter is of course due to the *electrical* interaction of the electron and proton.

What is $\hat{H}_{hf}$, the hyperfine Hamiltonian? The classical interaction energy of the two magnetic moments is a complicated function of $\boldsymbol{\mu}_e$, $\boldsymbol{\mu}_p$, and the radius vector from the proton to the electron; see, for example, D.J. Griffiths, *Introduction to Quantum Mechanics*, 2nd ed., pp. 283–285. The quantum-mechanical Hamiltonian comes from integrating that interaction function, weighted by the spatial probability distribution $|\psi_{100}(r)|^2$ of the electron with respect to the proton, over the coordinates r, θ, and ϕ. After that integration, $\hat{H}_{hf}$ still depends on the two magnetic moments, and these have to enter in a way that makes no preference for any direction in space—because there *is* no special direction. The only way to do this with two *vector* operators is to take their *scalar* product,

$$\hat{H}_{hf} \propto \hat{\boldsymbol{\mu}}_e \cdot \hat{\boldsymbol{\mu}}_p, \qquad \text{or} \qquad \hat{H}_{hf} = A\, \hat{\boldsymbol{\sigma}}_e \cdot \hat{\boldsymbol{\sigma}}_p \;,$$

where we absorb μ_B and μ_p into the proportionality constant A. To calculate A would take introducing that (very likely unfamiliar) classical interaction energy and doing the integral. But the *form* of the result has to be that given. Below, we take the value of A from experiment. Dimensionally, A is an energy, and it proves to be positive.

As in Chap. 10, each symbol $\hat{\boldsymbol{\sigma}}$ represents a triad of matrix operators, and so

$$\boxed{\hat{H}_{hf} = A\, \hat{\boldsymbol{\sigma}}_e \cdot \hat{\boldsymbol{\sigma}}_p = A(\hat{\sigma}_x^e \hat{\sigma}_x^p + \hat{\sigma}_y^e \hat{\sigma}_y^p + \hat{\sigma}_z^e \hat{\sigma}_z^p) \;.}$$

The electron operators commute with the proton operators because they operate on independent degrees of freedom; just as, say, $[\hat{p}_x, \hat{p}_y] = 0$ because $\hat{p}_x$ and $\hat{p}_y$ act on x- and y-dependent functions.

(c) Four states—To find the energy eigenvalues and eigenstates of $\hat{H}_{hf}$, we first find what $\hat{H}_{hf}$ does to each of the basis states, $|++\rangle$, $|+-\rangle$, $|-+\rangle$, and $|--\rangle$. The $\hat{\sigma}_i^e$'s in $\hat{H}_{hf}$ will affect only the first (electron) symbol in the kets, the $\hat{\sigma}_i^p$'s only the second (proton) symbol. When we get the eigenvalues of $\hat{H}_{hf}$, we will add the tiny perturbations to the ground-state energy, -13.6 eV.

The Pauli spin operators, from Sec. 10·1(b), are

$$\hat{\sigma}_x = \begin{pmatrix} 0 & 1 \\ 1 & 0 \end{pmatrix}, \quad \hat{\sigma}_y = \begin{pmatrix} 0 & -i \\ i & 0 \end{pmatrix}, \quad \hat{\sigma}_z = \begin{pmatrix} 1 & 0 \\ 0 & -1 \end{pmatrix}.$$

The matrices provide at once a look-up table, because the first and second columns of $\hat{\sigma}_x$, say, are the spinors $\hat{\sigma}_x |+\rangle$ and $\hat{\sigma}_x |-\rangle$ (the picket-fence recipe again, Sec. 5·5). Thus

$$\hat{\sigma}_x|+\rangle = +|-\rangle \qquad \hat{\sigma}_y|+\rangle = +i\,|-\rangle \qquad \hat{\sigma}_z|+\rangle = +|+\rangle$$
$$\hat{\sigma}_x|-\rangle = +|+\rangle \qquad \hat{\sigma}_y|-\rangle = -i\,|+\rangle \qquad \hat{\sigma}_z|-\rangle = -|-\rangle \;.$$

(If there is any doubt here, multiply a couple of cases out.) Then

$$\hat{\sigma}_x^e \hat{\sigma}_x^p \left|++\right\rangle = +\left|--\right\rangle$$
$$\hat{\sigma}_y^e \hat{\sigma}_y^p \left|++\right\rangle = i^2 \left|--\right\rangle = -\left|--\right\rangle$$
$$\hat{\sigma}_z^e \hat{\sigma}_z^p \left|++\right\rangle = +\left|++\right\rangle .$$

The sum of these is simply $\left|++\right\rangle$, and so

$$\hat{H}_{hf} \left|++\right\rangle = A\, \hat{\boldsymbol{\sigma}}_e \cdot \hat{\boldsymbol{\sigma}}_p \left|++\right\rangle = +A \left|++\right\rangle .$$

Thus $\left|++\right\rangle$ is already an eigenstate of $\hat{H}_{hf}$, with eigenvalue $+A$. Similarly, $\hat{H}_{hf}\left|--\right\rangle = +A\left|--\right\rangle$, and so $\left|--\right\rangle$ is a second eigenstate with the same eigenvalue $+A$. We already have two (of four) eigenstates and eigenvalues of $\hat{H}_{hf}$.

The $\left|+-\right\rangle$ and $\left|-+\right\rangle$ states are not eigenstates of $\hat{H}_{hf}$:

$$\hat{\sigma}_x^e \hat{\sigma}_x^p \left|+-\right\rangle = +\left|-+\right\rangle$$
$$\hat{\sigma}_y^e \hat{\sigma}_y^p \left|+-\right\rangle = i(-i)\left|-+\right\rangle = +\left|-+\right\rangle$$
$$\hat{\sigma}_z^e \hat{\sigma}_z^p \left|+-\right\rangle = -\left|+-\right\rangle .$$

The sum is $2\left|-+\right\rangle - \left|+-\right\rangle$, so that

$$\hat{H}_{hf} \left|+-\right\rangle = A\left(2\left|-+\right\rangle - \left|+-\right\rangle\right) .$$

Similarly,

$$\hat{H}_{hf} \left|-+\right\rangle = A\left(2\left|+-\right\rangle - \left|-+\right\rangle\right) .$$

The last two equations are mirror images of one another—they differ only in an interchange of the $+$ and $-$ signs in all the kets. A symmetry of this sort cries out, "Add and subtract the two equations!" We get

$$\hat{H}_{hf}(\left|+-\right\rangle + \left|-+\right\rangle) = +A(\left|+-\right\rangle + \left|-+\right\rangle)$$
$$\hat{H}_{hf}(\left|+-\right\rangle - \left|-+\right\rangle) = -3A(\left|+-\right\rangle - \left|-+\right\rangle) .$$

The sum of the $\left|+-\right\rangle$ and $\left|-+\right\rangle$ states is an eigenstate of $\hat{H}_{hf}$ with eigenvalue $+A$; the difference is an eigenstate with eigenvalue $-3A$. The sum of all four eigenvalues is zero.

We now have three orthogonal states with the same eigenvalue $+A$. Normalized, these are

$$\left|++\right\rangle, \quad \frac{1}{\sqrt{2}}(\left|+-\right\rangle + \left|-+\right\rangle), \quad \left|--\right\rangle .$$

There is an infinity of different orthonormal mixes of these three states, but it is most useful to choose those for which the sums of the z components of the spins of the electron and proton are, as here, $+\hbar$, 0, and $-\hbar$. In Sec. 3(b), we show that these three states form an angular-momentum triplet, $\left|S, M\right\rangle$, with $S = 1$ and $M = +1$, 0, and -1. The fourth normalized state, with eigenvalue $-3A$, is

$$\frac{1}{\sqrt{2}}(\left|+-\right\rangle - \left|-+\right\rangle) .$$

The total z component of the spin of this state is zero, and it is (we shall find) an angular-momentum singlet, $\left|S, M\right\rangle = \left|0, 0\right\rangle$. Two *constituents* with spin $1/2$ can combine to make a *system* spin of either 1 or 0. In one *basis* there are $2 \times 2 = 4$ states; in the other there are $3 + 1 = 4$ states. Either basis is a complete set for describing the spin state of hydrogen.

The columns of the full matrix operator for the hyperfine Hamiltonian are the vectors that $\hat{H}_{hf}$ produces operating on the the $|m_e, m_p\rangle$ states $|++\rangle$, $|+-\rangle$, $|-+\rangle$, and $|--\rangle$. That is, $\hat{H}_{hf}|++\rangle = +A|++\rangle$ is the first column, etc., the picket-fence recipe yet again. We have already calculated those columns:

$$\hat{H}_{hf} = \begin{pmatrix} A & 0 & 0 & 0 \\ 0 & -A & 2A & 0 \\ 0 & 2A & -A & 0 \\ 0 & 0 & 0 & A \end{pmatrix} \qquad \text{in the } |m_e, m_p\rangle \text{ basis .}$$

We could have set up the eigenvalue problem as a matrix equation, but it was easy to solve the coupled equations directly. If you absorbed Sec. 10·1(c), you can *see* that the eigenvalues of the center 2-by-2 sub-matrix are $-A \pm 2A = +A$ and $-3A$.

11·2. THE 21-CM LINE AND ASTRONOMY

Figure 1 shows the energy-level diagram of the hyperfine structure of the ground state of hydrogen. The dashed line is at -13.6 eV, the zero from which the splitting is measured. The energy difference between the $+A$ and $-3A$ levels is $4A$. The frequency of the transition, $\nu = 4A/h$, has been measured with incredible precision,

$$\nu = 1,420,405,751.7667 \pm 0.0007 \text{ Hz .}$$

The corresponding wavelength, $\lambda = c/\nu$, is about 21.1 cm. The splitting is about 5.9×10^{-6} eV, more than a million times smaller than the 10.2 eV between the $n = 1$ and 2 levels of hydrogen. *Hyper*fine indeed. (We come to *fine* structure in Chap. 13.)

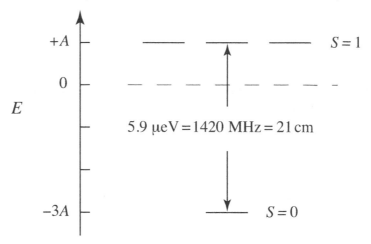

Figure 1. The hyperfine splitting $4A$ of the ground state of hydrogen. The zero is at -13.6 eV.

The Earth's atmosphere is opaque to many wavelengths outside the visible range. At such wavelengths, the Earth from space is just a cloudy ball, the oceans and continents invisible. However, there are other "windows" in the atmosphere besides the visible, and one of these is open to 21-cm radiation. Much of the matter of galaxies, particularly of spiral galaxies like the Milky Way, is enormous clouds of hydrogen gas. In the denser regions of these clouds, stars are born. Even in the otherwise almost empty regions of deep space, the $T = 2.7$ K cosmic background radiation keeps the four hydrogen hyperfine levels equally populated, because the thermal energy, $kT = 2.3 \times 10^{-4}$ eV, is much greater than the splitting, $4A = 5.9 \times 10^{-6}$ eV. Radio telescopes tuned to 1,420 MHz record the emissions due to the eternal slow shuffling of atoms between the $+A$ and $-3A$ states. From the strength of the signal, one can estimate the "column mass" of hydrogen in any direction. Using the Doppler shift of frequencies, one can measure the velocity distribution of the hydrogen along the line of sight. One can even measure the magnetic field strength in the hydrogen clouds using the Zeeman effect (see Prob. 1). All this information reveals much about the structure of our Galaxy.

And more. Light of visible wavelengths is attenuated by dust in the plane of the Galaxy, and optical telescopes cannot see all the way to the Galaxy's center. However, radiation of wavelengths far larger (such as 21 cm) than the size of dust particles is scarcely attenuated, and so with radio telescopes one can see into the otherwise veiled inner regions of the Galaxy. And finally, the search for extraterrestrial intelligence (SETI) is often carried out close to the 21-cm line, a universal marker.

11·3. COUPLING TWO SPIN-1/2 PARTICLES

(a) **Commutation relations**—We define spin operators for the hydrogen atom as a whole in terms of those for its constituents, the electron and proton. For example,

$$\hat{S}_x \equiv \hat{s}_x^e + \hat{s}_x^p, \quad \hat{\mathbf{S}} \equiv \hat{\mathbf{s}}^e + \hat{\mathbf{s}}^p, \quad \hat{S}^2 \equiv (\hat{\mathbf{s}}^e + \hat{\mathbf{s}}^p)^2 = (\hat{\mathbf{s}}^e)^2 + (\hat{\mathbf{s}}^p)^2 + 2\hat{\mathbf{s}}^e \cdot \hat{\mathbf{s}}^p .$$

Again, the electron operators act only on the electron spinor, the proton operators act only on the proton spinor, and so the electron and proton operators commute with one another. Thus, for example, using $[\hat{A} + \hat{B}, \hat{C}] = [\hat{A}, \hat{C}] + [\hat{B}, \hat{C}]$, etc., we get

$$[\hat{S}_x, \hat{S}_y] = [\hat{s}_x^e + \hat{s}_x^p, \hat{s}_y^e + \hat{s}_y^p] = [\hat{s}_x^e, \hat{s}_y^e] + [\hat{s}_x^e, \hat{s}_y^p] + [\hat{s}_x^p, \hat{s}_y^e] + [\hat{s}_x^p, \hat{s}_y^p]$$
$$= i\hbar \hat{s}_z^e + i\hbar \hat{s}_z^p = i\hbar \hat{S}_z ,$$

because $[\hat{s}_x^e, \hat{s}_y^p]$ and $[\hat{s}_x^p, \hat{s}_y^e]$ are zero. Thus the spin operators for the composite system—the hydrogen atom—obey exactly the same commutation relations as do the spin operators for its elementary constituents—the electron and proton.

The same is true for the operators belonging to any two distinct angular momenta, such as the spin and orbital angular momenta of an electron; the operators for the spin commute with those for the orbital motion. Furthermore, it is true for *any number* of operators, $\mathbf{j}_1, \mathbf{j}_2,$ $\ldots, \mathbf{j}_N$, having to do with N distinct angular momenta: the operators for any one commute with the operators for all the others. Then

$$[\hat{J}_x, \hat{J}_y] = [\hat{j}_{1x}, \hat{j}_{1y}] + \cdots + [\hat{j}_{Nx}, \hat{j}_{Ny}] = i\hbar (\hat{j}_{1z} + \hat{j}_{2z} + \cdots + \hat{j}_{Nz}) = i\hbar \hat{J}_z ,$$

where $\hat{J}_x = \hat{j}_{1x} + \cdots + \hat{j}_{Nx}$, etc. All the cross terms such as $[\hat{j}_{1x}, \hat{j}_{2y}]$ are zero; the commutation relations for the components of $\hat{\mathbf{J}}$ are exactly the same as for a single angular momentum, and everything in Chap. 8 that followed from the angular-momentum commutation relations holds true for *all* angular momenta. Here again are the basics:

$$\hat{J}^2 \,|j,m\rangle = j(j+1)\hbar^2\,|j,m\rangle, \qquad \text{where } j = 0, \tfrac{1}{2}, 1, \tfrac{3}{2}, \ldots$$

$$\hat{J}_z\,|j,m\rangle = m\hbar\,|j,m\rangle, \qquad \text{where } m = j, j-1, \ldots, -j$$

$$\hat{J}_+\,|j,m\rangle = \sqrt{(j-m)(j+m+1)}\,\hbar\,|j,m+1\rangle,$$

$$\hat{J}_-\,|j,m\rangle = \sqrt{(j+m)(j-m+1)}\,\hbar\,|j,m-1\rangle\ .$$

The magnitudes of *all* angular momenta are of the form $\sqrt{j(j+1)}\,\hbar$, where $j = 0,\ 1/2,\ 1,\ 3/2,\ \ldots$ The component of *any* angular momentum, when measured along *any* direction, is either an integral or a half-integral multiple of $\hbar$. The rules apply to everything from a single electron to the nucleus of a uranium atom with its 92 protons + 146 neutrons = 238 nucleons, to the whole atom with its 238 nucleons + 92 electrons = 330 constituents, to a molecule of water, and so on. This is why angular momentum is the most used topic in quantum mechanics.

(b) System spin states—The three hyperfine states with $E = +A$ have M values of $+1$, 0, and -1, where $M \equiv m_1 + m_2$ is the total z component of the spin (in units of $\hbar$). We show that these three make up a triplet of spin $S = 1$ states. First we operate on the $|{++}\rangle$ state with $\hat{S}_z = \hat{s}_z^e + \hat{s}_z^p$,

$$\hat{S}_z\,|{++}\rangle = (\hat{s}_z^e + \hat{s}_z^p)|{++}\rangle = \left(\tfrac{1}{2}\hbar + \tfrac{1}{2}\hbar\right)|{++}\rangle = +\hbar\,|{++}\rangle\ .$$

This is what we expect of an $|S,M\rangle = |1,+1\rangle$ state. Next, we operate with the $\hat{S}^2$ operator, which is

$$\hat{S}^2 = (\hat{\mathbf{s}}_e + \hat{\mathbf{s}}_p)^2 = \hat{s}_e^2 + \hat{s}_p^2 + 2\,\hat{\mathbf{s}}_e \cdot \hat{\mathbf{s}}_p\ .$$

In Sec. 1, we found that $(\hat{\boldsymbol{\sigma}}_e \cdot \hat{\boldsymbol{\sigma}}_p)|{++}\rangle = +|{++}\rangle$. Therefore $(\hat{\mathbf{s}}_e \cdot \hat{\mathbf{s}}_p)|{++}\rangle = \tfrac{1}{4}\hbar^2|{++}\rangle$, and

$$\left(\hat{s}_e^2 + \hat{s}_p^2 + 2\hat{\mathbf{s}}_e \cdot \hat{\mathbf{s}}_p\right)|{++}\rangle = \left[\tfrac{1}{2}\left(\tfrac{1}{2}+1\right) + \tfrac{1}{2}\left(\tfrac{1}{2}+1\right) + \tfrac{2}{4}\right]\hbar^2\,|{++}\rangle = 2\hbar^2|{++}\rangle\ .$$

This too is what we expect of an $|S,M\rangle = |1,+1\rangle$ state, where $S(S+1) = 2$. Thus we identify the *constituent* state $|m_1, m_2\rangle = |{++}\rangle$ with the *system* state $|S,M\rangle = |1,+1\rangle$.

Using the lowering operation $\hat{J}_-\,|j,m\rangle = \sqrt{(j+m)(j-m+1)}\,\hbar\,|j,m-1\rangle$, we get

$$\hat{S}_-\,|1,+1\rangle = \left(\hat{s}_-^e + \hat{s}_-^p\right)|{++}\rangle, \qquad \text{or} \qquad \sqrt{2}\,|1,0\rangle = |{-+}\rangle + |{+-}\rangle\ ,$$

so that

$$|S,M\rangle = |1,0\rangle = \frac{1}{\sqrt{2}}\bigl(|{+-}\rangle + |{-+}\rangle\bigr)\ .$$

A further lowering (do it!) gives $|1,-1\rangle = |{--}\rangle$. It remains to find a second combination of the $|{+-}\rangle$ and $|{-+}\rangle$ states that is orthogonal to the $|S,M\rangle = |1,0\rangle$ state. The state that is orthogonal to $|I\rangle = a|1\rangle + b|2\rangle$, where $|1\rangle$ and $|2\rangle$ are orthonormal, is $|II\rangle = b|1\rangle - a|2\rangle$. So,

$$|S,M\rangle = |0,0\rangle = \frac{1}{\sqrt{2}}\bigl(|{+-}\rangle - |{-+}\rangle\bigr)\ .$$

Some things to notice:

• The $|m_e, m_p\rangle$ basis uses the $2 \times 2 = 4$ spin states of the *constituents*. The $|S, M\rangle$ basis uses the $3 + 1 = 4$ spin states of the whole *system*. The number of constituent and system basis states is the same because they are different ways to span the same space.

• Just as the $|m_e, m_p\rangle$ kets are shorthand for $|s_e, s_p; m_e, m_p\rangle$ kets, the $|S, M\rangle$ kets are shorthand for $|s_e, s_p; S, M\rangle$ kets. It is implicit in both the $|m_e, m_p\rangle$ and $|S, M\rangle$ notations that the constituents here are spin-1/2 particles.

• Although the electron and proton are different in nearly every way, they both have spin-1/2. Nothing about the $|S, M\rangle$ states is specific to the particles being an electron and a proton. Those four $|S, M\rangle$ states are the system spin states for *any* two spin-1/2 particles.

• The three $S = 1$ states are symmetric (do not change) under interchange of the m_e and m_p values. The $S = 0$ state is antisymmetric (changes sign) under this interchange:

$$|0,0\rangle = \frac{1}{\sqrt{2}}(|+-\rangle - |-+\rangle) \longrightarrow \frac{1}{\sqrt{2}}(|-+\rangle - |+-\rangle) = -|0,0\rangle .$$

This difference between the $S = 1$ and $S = 0$ states will be of fundamental importance when we come to consider identical particles.

• The numerical coefficients in, say, the expansion of the $|0,0\rangle$ state in terms of the $|+-\rangle$ and $|-+\rangle$ states are *probability amplitudes*. If we measure the component of the spin of the first particle along the z axis, the probabilities of getting $+\frac{1}{2}\hbar$ or $-\frac{1}{2}\hbar$ (the only possible results) are each one-half. And if we get $-\frac{1}{2}\hbar$, the particles are no longer in the $|0,0\rangle$ state; they are in the $|-+\rangle$ state.

(c) Clebsch-Gordan coefficients—We can write the mapping or transformation from the constituent $|m_1, m_2\rangle$ basis to the system $|S, M\rangle$ basis as a matrix equation:

$$\begin{pmatrix} |1,+1\rangle \\ |1,0\rangle \\ |0,0\rangle \\ |1,-1\rangle \end{pmatrix} = \begin{pmatrix} 1 & 0 & 0 & 0 \\ 0 & 1/\sqrt{2} & 1/\sqrt{2} & 0 \\ 0 & 1/\sqrt{2} & -1/\sqrt{2} & 0 \\ 0 & 0 & 0 & 1 \end{pmatrix} \begin{pmatrix} |++\rangle \\ |+-\rangle \\ |-+\rangle \\ |--\rangle \end{pmatrix} .$$

(This expresses a transformation, not an eigenvalue problem.) The elements of the matrix are called *Clebsch-Gordan coefficients*. The elements are all real. The columns of the matrix, considered as vectors, are normalized, and they are orthogonal to one another. The rows are an orthonormal set too. It follows that $QQ^T = Q^TQ = I$, where Q is the above matrix, Q^T is its transpose and I is the 4-by-4 identity matrix. (Multiply Q by Q^T if this is not clear.) Thus Q is an *orthogonal* matrix, and $Q^T = Q^{-1}$ is the matrix for transforming the other way—from the $|S, M\rangle$ to the $|m_1, m_2\rangle$ basis. All these things are true for Clebsch-Gordan tables for any j_1 and j_2. In the present example, $Q^T = Q$, but this is not usually so, even when $j_1 = j_2$. Section 5 has more about Clebsch-Gordan coefficients.

(d) Hyperfine splitting again—Knowing now that the total spin of a system of two spin-1/2 particles can only be $S = 1$ or 0, we can see how to find the eigenvalues of $\hat{H}_{hf}$ and similar Hamiltonians more quickly than we did in Sec. 1. Starting with

$$\hat{S}^2 \equiv (\hat{\mathbf{s}}_e + \hat{\mathbf{s}}_p)^2 = \hat{s}_e^2 + \hat{s}_p^2 + 2\hat{\mathbf{s}}_e \cdot \hat{\mathbf{s}}_p ,$$

211

we get

$$\hat{H}_{hf} = A\,\hat{\boldsymbol{\sigma}}_e \cdot \hat{\boldsymbol{\sigma}}_p = \frac{4A}{\hbar^2}\,\hat{\mathbf{s}}_e \cdot \hat{\mathbf{s}}_p = \frac{2A}{\hbar^2}\left(\hat{S}^2 - \hat{s}_e^2 - \hat{s}_p^2\right).$$

Then, using $\hat{J}^2|j,m\rangle = j(j+1)\hbar^2|j,m\rangle$, we get

$$\hat{H}_{hf}|S,M\rangle = 2A\big[S(S+1) - s_e(s_e+1) - s_p(s_p+1)\big]|S,M\rangle,$$

which is independent of M. Evaluating this for $s_e = s_p = 1/2$ and, in turn, for $S = 1$ and $S = 0$, we get as before,

$$\hat{H}_{hf}|1,M\rangle = +A\,|1,M\rangle \qquad \text{and} \qquad \hat{H}_{hf}|0,0\rangle = -3A\,|0,0\rangle.$$

The three $S = 1$ states all have the same energy because the Hamiltonian is independent of which direction was chosen as z: $\hat{H}_{hf} = A\,\hat{\boldsymbol{\sigma}}_e \cdot \hat{\boldsymbol{\sigma}}_p$, a scalar product, is rotationally invariant.

It is the basis states of the *system* that have definite energies. With $|S,M\rangle$ states instead of $|m_e,m_p\rangle$ states as the unit vectors,

$$|1,+1\rangle = \begin{pmatrix} 1 \\ 0 \\ 0 \\ 0 \end{pmatrix}, \quad |1,0\rangle = \begin{pmatrix} 0 \\ 1 \\ 0 \\ 0 \end{pmatrix}, \quad |0,0\rangle = \begin{pmatrix} 0 \\ 0 \\ 1 \\ 0 \end{pmatrix}, \quad |1,-1\rangle = \begin{pmatrix} 0 \\ 0 \\ 0 \\ 1 \end{pmatrix},$$

the Hamiltonian is diagonal with the eigenvalues along the diagonal:

$$\hat{H}_{hf} = \begin{pmatrix} +A & 0 & 0 & 0 \\ 0 & +A & 0 & 0 \\ 0 & 0 & -3A & 0 \\ 0 & 0 & 0 & +A \end{pmatrix} \qquad \text{in the } |S,M\rangle \text{ basis}.$$

The ordering is by the values of M, not S, the same as in tables of Clebsch-Gordan coefficients.

The relations

$$\hat{\mathbf{s}}_1 \cdot \hat{\mathbf{s}}_2 = \frac{1}{2}\left(\hat{S}^2 - \hat{s}_1^2 - \hat{s}_2^2\right) \qquad \text{or} \qquad \hat{\mathbf{j}}_1 \cdot \hat{\mathbf{j}}_2 = \frac{1}{2}\left(\hat{J}^2 - \hat{j}_1^2 - \hat{j}_2^2\right)$$

will be useful again. For example, suppose we need the value of $\hat{\mathbf{L}} \cdot \hat{\mathbf{S}}$ for an atom in the state $|L,S;J,M\rangle$, where the orbital, spin, and total angular-momentum quantum numbers are $L = 2$, $S = 1$, and $J = 3$. It is

$$\hat{\mathbf{L}} \cdot \hat{\mathbf{S}} = \frac{1}{2}\left(3 \times 4 - 2 \times 3 - 1 \times 2\right)\hbar^2 = 2\hbar^2.$$

(Angular-momentum quantum numbers are often written lower case (ℓ, s, and j) for single electrons, but upper case for states with more than one electron.)

11·4. COUPLING ANY TWO ANGULAR MOMENTA

(a) Coupling rules—The coupling of two spin-1/2 particles to make system spins of 1 and 0 has most of the ingredients of coupling any two angular momenta. Let the two be j_1 and j_2, with $j_1 \geq j_2$. Each m_1 state can combine with each m_2 state, so there are $(2j_1+1)(2j_2+1)$ different $|m_1,m_2\rangle$ kets. The values J of total angular momentum that occur in coupling j_1 and j_2 are

$$\boxed{J - j_1 + j_2, \quad j_1 + j_2 - 1, \quad \dots, \quad j_1 - j_2 + 1, \quad j_1 - j_2.}$$

There are $2j_2 + 1$ different values of J here. For $s_1 = s_2 = 1/2$, there are two values of S (1 and 0). For $j_1 = 2$ and $j_2 = 1$, there are three values of J (3, 2, and 1—but no $J = 0$!).

We justify the above rule by counting states. Each value of J gives $2J + 1$ values of M. Thus, with $j_1 \geq j_2$, the total number of $|J, M\rangle$ states is

$$
\begin{array}{ll}
2(j_1 + j_2) & +1 \\
2(j_1 + j_2 - 1) & +1 \\
\quad\vdots & \quad\vdots \\
2(j_1 - j_2 + 1) & +1 \\
2(j_1 - j_2) & +1 \\
\hline
2j_1(2j_2 + 1) \quad + \quad (2j_2 + 1) & = \quad (2j_1 + 1)(2j_2 + 1) \text{ states .}
\end{array}
$$

$(2j_2 + 1 \text{ rows})$

There are $2j_2 + 1$ rows above the summation line (for the $2j_2 + 1$ different values of J). The sum at the bottom is equal to the number of states in the $|m_1, m_2\rangle$ basis. For $s_1 = s_2 = 1/2$, there are $2 \times 2 = 3 + 1 = 4$ basis states. For $j_1 = 2$ and $j_2 = 1$, there are $5 \times 3 = 7 + 5 + 3 = 15$ basis states. The $|m_1, m_2\rangle$ and $|J, M\rangle$ bases have the same number of states because they are two different ways of spanning the same $(2j_1 + 1)(2j_2 + 1)$-dimensional space.

(b) Example—Here in outline is how to couple $j_1 = 1$ with $j_2 = 1/2$. These might be $\ell = 1$ and $s = 1/2$ orbital- and spin-angular momenta of an electron in hydrogen, a spherical harmonic $Y_1^m(\theta, \phi)$ and a spinor $|\chi\rangle$. The values of J are 3/2 and 1/2. We arrange the $3 \times 2 = 6$ $|m_1, m_2\rangle$ states by the values of $M = m_1 + m_2$:

$$
\begin{array}{ll}
|m_1, m_2\rangle = |+1, +\tfrac{1}{2}\rangle & m_1 + m_2 = M = +3/2 \\
= |+1, -\tfrac{1}{2}\rangle, \ |0, +\tfrac{1}{2}\rangle & = +1/2 \\
= |0, -\tfrac{1}{2}\rangle, \ |-1, +\tfrac{1}{2}\rangle & = -1/2 \\
= |-1, -\tfrac{1}{2}\rangle & = -3/2
\end{array}
$$

The first line maps the constituent $|m_1, m_2\rangle = |+1, +\tfrac{1}{2}\rangle$ to the $|J, M\rangle = |\tfrac{3}{2}, +\tfrac{3}{2}\rangle$ state. From there, we work down to the $|J, M\rangle = |\tfrac{3}{2}, -\tfrac{3}{2}\rangle$ state in three steps, using $\hat{J}_-$ on the $|J, M\rangle$ side and $\hat{j}_{1-} + \hat{j}_{2-}$ on the $|m_1, m_2\rangle$ side, with $\hat{j}_{1-}$ operating on the m_1 symbol, etc. We would get

$$
\begin{aligned}
|J, M\rangle = |\tfrac{3}{2}, +\tfrac{3}{2}\rangle &= |+1, +\tfrac{1}{2}\rangle \\
= |\tfrac{3}{2}, +\tfrac{1}{2}\rangle &= \sqrt{\tfrac{1}{3}} \, |+1, -\tfrac{1}{2}\rangle + \sqrt{\tfrac{2}{3}} \, |0, +\tfrac{1}{2}\rangle \\
= |\tfrac{3}{2}, -\tfrac{1}{2}\rangle &= \sqrt{\tfrac{2}{3}} \, |0, -\tfrac{1}{2}\rangle + \sqrt{\tfrac{1}{3}} \, |-1, +\tfrac{1}{2}\rangle \\
= |\tfrac{3}{2}, -\tfrac{3}{2}\rangle &= |-1, -\tfrac{1}{2}\rangle .
\end{aligned}
$$

Here, using $\hat{J}_- |j, m\rangle = \sqrt{(j + m)(j - m + 1)}\, \hbar \, |j, m - 1\rangle$, is how the first step goes:

$$
\hat{J}_- |\tfrac{3}{2}, +\tfrac{3}{2}\rangle = (\hat{j}_{1-} + \hat{j}_{2-}) \, |+1, +\tfrac{1}{2}\rangle \quad \text{or} \quad \sqrt{3} |\tfrac{3}{2}, +\tfrac{1}{2}\rangle = \sqrt{2} \, |0, +\tfrac{1}{2}\rangle + |+1, -\tfrac{1}{2}\rangle .
$$

Division of the latter equation by the factor, here $\sqrt{3}$, given by the action of $\hat{J}_-$ on $|\tfrac{3}{2}, +\tfrac{3}{2}\rangle$, gives the normalized $|J, M\rangle = |\tfrac{3}{2}, +\tfrac{1}{2}\rangle$ state, as above.

The numerical coefficients, $1, \sqrt{2/3}, \sqrt{1/3}$, etc., are the Clebsch-Gordan coefficients. When J has its maximum value, $J = j_1 + j_2$, the coefficients are all positive, because the ladder operators do not create minus signs. Again, it helps to distinguish $|J, M\rangle$ states from

213

$|m_1, m_2\rangle$ states by always putting a sign on a (nonzero) M or m value, but never on a J value (because J is never negative).

There are four $J = 3/2$ states, but we need altogether $3 \times 2 = 6$ states. The other two are of course the $|J, M\rangle = |\frac{1}{2}, \pm\frac{1}{2}\rangle$ states, which have to be orthogonal to the corresponding $|J, M\rangle = |\frac{3}{2}, \pm\frac{1}{2}\rangle$ states. To get them, simply interchange the Clebsch-Gordan coefficients on those $|\frac{3}{2}, \pm\frac{1}{2}\rangle$ states and put a minus sign on the latter of the two $|m_1, m_2\rangle$ states:

$$|J, M\rangle = |\tfrac{1}{2}, +\tfrac{1}{2}\rangle = \sqrt{\tfrac{2}{3}} \, |+1, -\tfrac{1}{2}\rangle - \sqrt{\tfrac{1}{3}} \, |0, +\tfrac{1}{2}\rangle$$

$$= |\tfrac{1}{2}, -\tfrac{1}{2}\rangle = \sqrt{\tfrac{1}{3}} \, |0, -\tfrac{1}{2}\rangle - \sqrt{\tfrac{2}{3}} \, |-1, +\tfrac{1}{2}\rangle \, .$$

(Why does this work?) Another way to get the top rung of the $J = 1/2$ ladder is to require that $\hat{J}_+ \equiv \hat{j}_{1+} + \hat{j}_{2+}$, operating on

$$a \, |+1, -\tfrac{1}{2}\rangle + b \, |0, +\tfrac{1}{2}\rangle \, ,$$

gives 0. This makes $b = -a/\sqrt{2}$, and then normalization (and a choice of overall phase) does the rest.

We now know how to couple any two constituent angular momenta to get the possible system angular-momentum states. To couple three angular momenta, first couple j_1 and j_2, then couple each of the J values so obtained with j_3. Of course, this would be a tedious process, and there are sophisticated group-theoretical methods for complicated cases.

11·5. CLEBSCH-GORDAN COEFFICIENTS

Once we know what Clebsch-Gordan coefficients are good for, we look them up in tables when we need them. Table 1 gives the coefficients for several j_1-j_2 pairs. The schematic at the upper right shows the layout. To get more coefficients on a page, the arrays are partly collapsed. One convention is to suppress all the square-root signs on coefficients: Where the table gives a coefficient as $-3/5$, read $-\sqrt{3/5}$. Another convention is to omit the coefficients around the periphery that are automatically zero because $M \neq m_1 + m_2$. For example, the array of coefficients given in Sec. 3(c) for $s_1 = s_2 = 1/2$ is four-by-four, but there is no reason to include the zeros along the top that show that the system $|1, +1\rangle$ state contains nothing of the constituent $|+-\rangle$, $|-+\rangle$, or $|--\rangle$ states.

A Clebsch-Gordan coefficient is a bracket telling how much of a constituent state contributes to a system state—and vice versa. In full detail,

$$\langle j_1, j_2; J, M | j_1, j_2; m_1, m_2 \rangle = \langle j_1, j_2; m_1, m_2 | j_1, j_2; J, M \rangle \, .$$

Clebsch-Gordan coefficients are real numbers—no complex conjugation is needed here. Here is an example of reading the $j_1 \times j_2 = 2 \times 1$ table. The system $|J, M\rangle = |3, -1\rangle$ state is, in terms of constituent $|m_1, m_2\rangle$ states,

$$|J, M\rangle = |3, -1\rangle = \sqrt{\tfrac{2}{5}} \, |0, -1\rangle + \sqrt{\tfrac{8}{15}} \, |-1, 0\rangle + \sqrt{\tfrac{1}{15}} \, |-2, +1\rangle \, .$$

The coefficients are probability amplitudes. In this $|J, M\rangle$ state ($j_1 = 2$ and $j_2 = 1$ being understood), the probabilities for getting 0, -1, or -2 on measuring m_1 are 2/5, 8/15, and 1/15 (which sum to one).

Table 1. Clebsch-Gordan coefficients for several j_1-j_2 pairs. The coefficients themselves are shaded. The schematic at the upper right shows the arrangement. A square-root sign is to be understood over each coefficient, except for a minus sign: for $-2/3$, read $-\sqrt{2/3}$, etc. The tables are from the Review of Particle Physics, Particle Data Group (any edition).

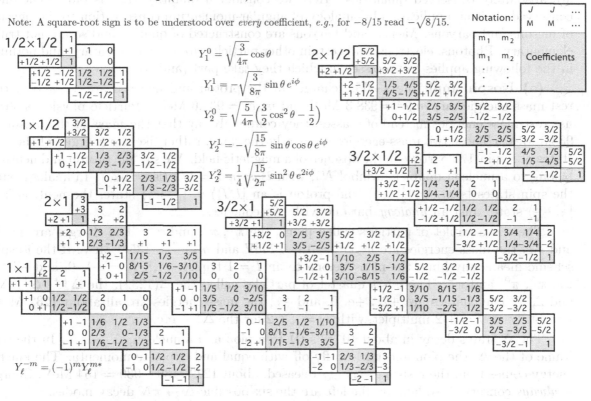

The mean or *expectation value* of m_1, on making repeated measurements of m_1 on the $|J, M\rangle = |3, -1\rangle$ state, is

$$\langle m_1 \rangle = \langle 2, 1; 3, -1 | m_1 | 2, 1; 3, -1 \rangle = \tfrac{2}{5} \times 0 + \tfrac{8}{15} \times (-1) + \tfrac{1}{15} \times (-2) = -\tfrac{2}{3}.$$

After measuring m_1 on a $|J, M\rangle = |3, -1\rangle$ state and getting, say, $m_1 = -2$, the state is the constituent $|m_1, m_2\rangle = |-2, +1\rangle$ state, not the system $|J, M\rangle = |3, -1\rangle$ state we started with. We look up, again in the $j_1 \times j_2 = 2 \times 1$ table, the constituent $|m_1, m_2\rangle = |-2, +1\rangle$ state in terms of system $|J, M\rangle$ states:

$$|m_1, m_2\rangle = |-2, +1\rangle = \sqrt{\tfrac{1}{15}}\, |3, -1\rangle - \sqrt{\tfrac{1}{3}}\, |2, -1\rangle + \sqrt{\tfrac{3}{5}}\, |1, -1\rangle.$$

The probabilities of getting 3, 2, or 1 on measuring J are now 1/15, 1/3, and 3/5. If we made many measurements of J, each one on a $|j_1, j_2; m_1, m_2\rangle = |2, 1; -2, +1\rangle$ state, the expectation value of J would be

$$\langle J \rangle = \langle 2, 1; -2, +1 | J | 2, 1; -2, +1 \rangle = 3 \times \tfrac{1}{15} + 2 \times \tfrac{1}{3} + 1 \times \tfrac{3}{5} = \tfrac{22}{15}.$$

215

11·6. PARTICLE MULTIPLETS AND ISOSPIN

In the interactions and decays of atoms, nuclei, or elementary particles, certain quantities are always conserved—their totals are the same before and after, no matter what happens. The total energy, the total momentum, the total angular momentum, and the total charge are absolutely conserved quantities. Here we consider a quantity that is only sometimes conserved. It is an application-by-analogy of angular-momentum conservation to the physics of mesons and baryons. Mesons and baryons are constructed of quarks, and so are not truly elementary. Photons, electrons, and certain other particles *are* truly elementary, and nothing in the following applies to processes in which they take part (and see below).

(a) Isospin multiplets—The charges of the proton and neutron are $+e$ and 0, the rest-mass energies are $m_p c^2 = 938.3$ MeV and $m_n c^2 = 939.6$ MeV. (Particle physicists often use rest-mass energies instead of masses; they commonly say that the *mass* of the proton is 938.3 MeV.) Those two mass-energies are nearly the same, rather like the energies of the two states of a spin-1/2 particle in the absence of a magnetic field. We say the proton and neutron belong to a nucleon doublet (symbol N), with *isospin* $I = 1/2$ (iso = like). In analogy with the spin states $|s, m\rangle = |\frac{1}{2}, \pm\frac{1}{2}\rangle$, the proton is an $|I, I_z\rangle = |\frac{1}{2}, +\frac{1}{2}\rangle$ state, the neutron is a $|\frac{1}{2}, -\frac{1}{2}\rangle$ state. Just an *analogy*, but one with consequences.

There is a triplet of particles called pions, the π^+, π^0, and π^-. The charges are $+e$, 0, and $-e$, the mass-energies are $m_{\pi^\pm} c^2 = 139.6$ MeV and $m_{\pi^0} c^2 = 135.0$ MeV. If the isospin scheme means anything, the pions belong to an $I = 1$ triplet, with $I_z = +1$, 0, and -1 for the π^+, π^0, and π^-. There is a quartet of particles called the $\Delta(1232)$, the Δ^{++}, Δ^+, Δ^0, and Δ^-. The charges are $+2e$, $+e$, 0, and $-e$, the mass-energies are all about 1232 MeV. Obviously an $I = 3/2$ multiplet, with $I_z = +3/2$ for the Δ^{++}, etc.

The Δ states decay in about 10^{-23} s (!) to a pion and a nucleon, $\Delta \to \pi N$. In the rest frame of the Δ, the pion and nucleon fly off with equal and opposite momenta. The kinetic energy comes from the rest-mass energy released, about $1232 - 139 - 939 = 154$ MeV. Charge is *always* conserved, so here on the left are the six possible $\Delta \to \pi N$ decay modes:

$$\Delta^{++} \to \pi^+ p \qquad |\tfrac{3}{2}, +\tfrac{3}{2}\rangle = |1, +1\rangle\, |\tfrac{1}{2}, +\tfrac{1}{2}\rangle$$

$$\Delta^+ \to \pi^+ n \qquad |\tfrac{3}{2}, +\tfrac{1}{2}\rangle = \sqrt{\tfrac{1}{3}}\, |1, +1\rangle\, |\tfrac{1}{2}, -\tfrac{1}{2}\rangle$$

$$\to \pi^0 p \qquad\qquad\quad + \sqrt{\tfrac{2}{3}}\, |1, 0\rangle\, |\tfrac{1}{2}, +\tfrac{1}{2}\rangle$$

$$\Delta^0 \to \pi^0 n \qquad |\tfrac{3}{2}, -\tfrac{1}{2}\rangle = \sqrt{\tfrac{2}{3}}\, |1, 0\rangle\, |\tfrac{1}{2}, -\tfrac{1}{2}\rangle$$

$$\to \pi^- p \qquad\qquad\quad + \sqrt{\tfrac{1}{3}}\, |1, -1\rangle\, |\tfrac{1}{2}, +\tfrac{1}{2}\rangle$$

$$\Delta^- \to \pi^- n \qquad |\tfrac{3}{2}, -\tfrac{3}{2}\rangle = |1, -1\rangle\, |\tfrac{1}{2}, -\tfrac{1}{2}\rangle\,.$$

(b) Isospin conservation—As it happens, Nature in most of the reactions and decays of mesons and baryons (but not in all of them; see below) honors the spin-isospin analogy by *conserving isospin*, just as it (always) conserves angular momentum. Since the Δ's have isospin 3/2, the pions isospin 1, and the nucleons isospin 1/2, the coupling equations for $\Delta \to \pi N$ use the Clebsch-Gordan coefficients for $j_1 = 1$ and $j_2 = 1/2$ coupling to $J = 3/2$. The results are at the right of the decays.

What does this tell us? Well, the coefficients are as usual probability amplitudes, and when squared they give probabilities. So all of Δ^{++} decays to πN are to $\pi^+ p$, and all of Δ^-

decays are to $\pi^- n$. But the Δ^+ *branches* to $\pi^+ n$ and $\pi^0 p$ in the ratio 1/3 and 2/3. And the *branching fractions* of the Δ^0 to $\pi^0 n$ and $\pi^- p$ are 2/3 and 1/3. And this is how Nature actually works.

Say your experiment found 600 $\Delta^0 \to \pi^- p$ decays, but (as is often the case) it could only detect charged decay particles. How many Δ^0's were actually produced? Altogether about 1800, because $\Delta^0 \to \pi^0 n$ decays occur at twice the rate of $\pi^- p$ decays. Isospin conservation is also used to reduce the number of independent amplitudes in reactions, such as the scattering of pions by nucleons.

(c) **Branching fractions by counting**—There is a quicker way to get branching fractions due to isospin conservation. We start with equal numbers of the charge states of the decaying multiplet. For example, take 75 of each of the four Δ charge states, for 300 in all:

$$\begin{array}{llll}
75\ \Delta^{++} & \to \pi^+ p & & 75\ \pi^+,\ 75\ p \\
75\ \Delta^+ & \to \pi^+ n & & 25\ \pi^+,\ 25\ n \\
& \to \pi^0 p & & 50\ \pi^0,\ 50\ p \\
75\ \Delta^0 & \to \pi^0 n & & 50\ \pi^0,\ 50\ n \\
& \to \pi^- p & & 25\ \pi^-,\ 25\ p \\
75\ \Delta^- & \to \pi^- n & & 75\ \pi^-, 75\ n
\end{array}$$

The 300 decays produce 300 pions and 300 nucleons. A consequence of isospin conservation (not proven here, but certainly the symmetrical outcome; see also Prob. 8·7) is this: The decays of the uniform initial population of Δ charge states produce equal numbers of the pion charge states and equal numbers of the nucleon charge states. Here that means 100 each of π^+'s, π^0's, and π^-'s, and 150 each of the protons and neutrons. Since the 75 $\Delta^{++} \to \pi^+ p$ decays produce 75 π^+s, only 25 of the 75 Δ^+s can decay to $\pi^+ n$, leaving 50 to decay to $\pi^0 p$. Similarly, 25 of the 75 Δ^0s have to decay to $\pi^- p$. These branchings also produce 150 protons and 150 neutrons. These are the same branching *fractions* (1, 1/3, 2/3, 2/3, 1/3, and 1) we found using Clebsch-Gordan coefficients. The counting method calculates the squares of Clebsch-Gordan coefficients directly.

(d) **Charge symmetry**—The charge-symmetry operation turns a meson or a baryon into the particle that is the same distance from the other end of the isospin multiplet it belongs to. For example,

$$p \longleftrightarrow n, \qquad \Delta^{++} \longleftrightarrow \Delta^-, \qquad \pi^+ \longleftrightarrow \pi^-, \qquad \pi^0 \longleftrightarrow \pi^0 .$$

The π^0, being the middle member of a multiplet with an odd number of particles, is its own partner. Charge symmetry is not the same as an "antiparticle" operation, which would turn a proton into an antiproton, not a neutron.

When isospin is conserved, the branching fractions for modes related by charge symmetry are equal. The $\Delta \to \pi N$ decay modes in the above example are arranged so that turning all the particles into their charge-symmetric partners reflects the four modes about the middle of the list—and the branching fractions for paired modes are the same.

Now although the charge-symmetry operation changes a π^+ to a π^-, and vice-versa, it still leaves a charged pion charged; and it leaves the π^0 a π^0. The over-all rule for the uniform set of Δ decays was that they produce equal numbers of π^+'s, π^0's, and π^-'s. Therefore the rule for either charge-symmetric *half* of the set is that they produce twice as many charged pions as π^0's. This can be quite useful (see the problems).

(e) Limitations—Unlike angular momentum, isospin is not always conserved. The basic forces of Nature fall into four classes: strong, electromagnetic, weak, and gravitational. Isospin is only conserved in the strong interactions, which involve only the particles made of quarks—the mesons and baryons such as the pion and nucleon. Any interaction or decay involving a photon, electron, neutrino, or certain other particles, is not strong; and there are decays of some mesons and baryons that involve the weak or electromagnetic forces. Even when isospin is conserved, it is not conserved exactly; the small differences in the masses in a multiplet (which themselves indicate the inexactness of isospin conservation) mean small differences in energy release, and this slightly alters the relative decay rates. Nevertheless, isospin conservation greatly reduces the numbers of independent rates of strong decays and interactions of mesons and baryons. For more on mesons, baryons, the other species of particles, and their various interactions, see any elementary text on particle physics.

PROBLEMS

1. *From constituent states and operators to system states and operators.*

(a) Formally, the *tensor product* of two two-component vectors is a four-component vector defined by

$$\begin{pmatrix} a_1 \\ b_1 \end{pmatrix} \otimes \begin{pmatrix} a_2 \\ b_2 \end{pmatrix} = \begin{pmatrix} a_1 a_2 \\ a_1 b_2 \\ b_1 a_2 \\ b_1 b_2 \end{pmatrix}.$$

Show using the usual $|+\rangle = |+z\rangle$ and $|-\rangle = |-z\rangle$ vectors that this recipe gives the four-component vectors $|++\rangle$, $|+-\rangle$, $|-+\rangle$, and $|--\rangle$ given in Sec. 2(a).

(b) The tensor product of two 2-by-2 matrices is a 4-by-4 matrix,

$$\begin{pmatrix} a_1 & c_1 \\ b_1 & d_1 \end{pmatrix} \otimes \begin{pmatrix} a_2 & c_2 \\ b_2 & d_2 \end{pmatrix} = \begin{pmatrix} a_1 a_2 & a_1 c_2 & c_1 a_2 & c_1 c_2 \\ a_1 b_2 & a_1 d_2 & c_1 b_2 & c_1 d_2 \\ b_1 a_2 & b_1 c_2 & d_1 a_2 & d_1 c_2 \\ b_1 b_2 & b_1 d_2 & d_1 b_2 & d_1 d_2 \end{pmatrix}.$$

The first column on the right is the tensor product of the first columns of the 2-by-2 matrices; the second column is the tensor product of the first column of the first 2-by-2 matrix and the second column of the second one; etc. Calculate the tensor product of the 2-by-2 $\hat{\sigma}_x^e$ and $\hat{\sigma}_x^p$ matrices. Does the result do what it is supposed to do when it operates on each of the four-component basis states of part (a)? (What *is* it supposed to do?)

(c) Calculate the tensor products $\hat{\sigma}_y^e \otimes \hat{\sigma}_y^p$ and $\hat{\sigma}_z^e \otimes \hat{\sigma}_z^p$. Add them to $\hat{\sigma}_x^e \otimes \hat{\sigma}_x^p$ and show that, aside from the factor A, you get the 4-by-4 Hamiltonian at the end of Sec. 1.

(d) What 2-by-2 matrices would you use to get a 4-by-4 matrix that flips the spin state of just the second particle?

2. *Zeeman splitting of the hydrogen ground state.*

In the presence of an external magnetic field $\mathbf{B} = B\hat{\mathbf{z}}$, two terms are added to the hyperfine Hamiltonian for the hydrogen ground state,

$$\hat{H} = A\,\hat{\boldsymbol{\sigma}}^e \cdot \hat{\boldsymbol{\sigma}}^p - \mu_e\hat{\sigma}^e_z B - \mu_p\hat{\sigma}^p_z B \ ,$$

where $\mu_e = -\mu_B$, the Bohr magneton.

(a) Set up the Hamiltonian as a 4×4 matrix using the $|m_e, m_p\rangle$ states as the basis (the $A\,\hat{\boldsymbol{\sigma}}^e \cdot \hat{\boldsymbol{\sigma}}^p$ matrix is in Sec. 1). Let $\mu_B + \mu_p \simeq \mu_B - \mu_p = \mu_B$ (because $\mu_p \ll \mu_B$). Solve for the four energy eigenvalues as functions of $\mu_B B/A$.

(b) The figure shows the eigenvalues for $0 \leq \mu_B B/A \leq 4$. Check that your answers in (a) give the correct values at the far right.

(c) What is the value of B, in tesla, when $\mu_B B/A = 1$, where $\mu_B = 5.788 \times 10^{-5}$ eV/T?

(d) What frequencies would you observe from a hydrogen cloud in an arm of the Galaxy for this value of B? Assume that transitions between any two of the four energy levels are allowed. One of the frequencies is $\frac{1}{2} \times 1420$ MHz. Can you see that in the figure?

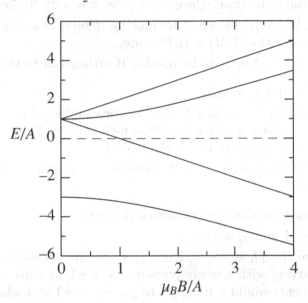

3. *Hyperfine splitting of deuterium.*

Deuterium is like ordinary hydrogen except its nucleus contains a neutron as well as a proton. The nuclear charge is still $+e$, and so the only changes to the basic energy levels, $E_n = -13.6/n^2$ eV, and wave functions come from a 0.027% change of the reduced mass.

(a) But the hyperfine *structure* of deuterium is different because the spin of the deuteron is 1. What then are the spin states of the ground state of deuterium? Sketch the hyperfine structure, analogous to Fig. 1 for ordinary hydrogen, with no further calculation.

(b) In parallel with $\hat{\mathbf{s}}_p = \frac{1}{2}\hbar\hat{\boldsymbol{\sigma}}_p$ and $\boldsymbol{\mu}_p = \mu_p\hat{\boldsymbol{\sigma}}_p$, we write $\hat{\mathbf{S}}_d = \hbar\hat{\boldsymbol{\Sigma}}_d$ and $\boldsymbol{\mu}_d = \mu_d\hat{\boldsymbol{\Sigma}}_d$. The hyperfine Hamiltonian has the form $\hat{H}_{hf} = C\,\hat{\boldsymbol{\sigma}}^e \cdot \hat{\boldsymbol{\Sigma}}^d$. Use the $\hat{\mathbf{j}}_1 \cdot \hat{\mathbf{j}}_2$ identity to find the energy levels in terms of C.

(c) The ratio of the deuteron and proton magnetic moments is $\mu_d/\mu_p = 0.307$. Show that the frequency of the hyperfine transition in deuterium is $\frac{3}{4}\mu_d/\mu_p \times 1420 = 327$ MHz. (How is C related to A?)

4. *Drill.*

Fill in the table:

j_1	j_2	J values	# $|m_1, m_2\rangle$ states	# $|J, M\rangle$ states
3/2	1			
2	1			
3/2	3/2			
7/2	2			
7/2	5/2			

5. *Coupling $j_1 = 1$ and $j_2 = 1$.*

(a) Find the Clebsch-Gordan coefficients for the coupling of $j_1 = j_2 = 1$ states to make $|J.M\rangle = |2, M\rangle$ states. Since $j_1 = j_2$, you ought to be able to write down without calculation the $|J, M\rangle = |2, \pm 2\rangle$ and $|2, \pm 1\rangle$ states, using m_1-m_2 symmetries. Then use the lowering operator $\hat{J}_-$ to get the $|2, 0\rangle$ state. Check along the way with Table 1.

(b) Get the $|J, M\rangle = |1, +1\rangle$ state by making it be orthogonal to the $|J, M\rangle = |2, +1\rangle$ state. Then lower to get the $|J, M\rangle = |1, 0\rangle$ state.

(c) Get the $|J, M\rangle = |0, 0\rangle$ state by making it orthogonal to the $|2, 0\rangle$ and $|1, 0\rangle$ states.

6. *Coupling $j_1 = 3/2$ and $j_2 = 1/2$.*

(a) Write down the $|J, M\rangle = |2, +2\rangle$ state for the coupling of $j_1 = 3/2$ and $j_2 = 1/2$ states. Then use the lowering operator to get the Clebsch-Gordan coefficients for the $|J.M\rangle = |2, +1\rangle$ and $|2, 0\rangle$ states. Check along the way with Table 1.

(b) Get the $|J, M\rangle = |1, +1\rangle$ and $|1, 0\rangle$ states by making them orthogonal to the $|J, M\rangle = |2, +1\rangle$ and $|2, 0\rangle$ states.

7. *Expectation values from Clebsch-Gordan coefficients.*

In the following, $j_1 = 1$ and $j_2 = 1/2$.

(a) You prepare the $|J, M\rangle = |3/2, +1/2\rangle$ state and measure m_1. You repeat this 99 times, each time starting with a newly prepared $|3/2, +1/2\rangle$ state. For about what fraction of the 100 measurements would you expect to get $m_1 = +1$ and what fraction $m_1 = 0$?

(b) What is the mean or expectation value of $\langle m_1 \rangle \equiv \langle 3/2, +1/2|m_1|3/2, +1/2\rangle$ for your 100 measurements?

(c) Repeat (a) and (b) for 100 measurements of m_2 instead of m_1.

(d) What is the sum of $\langle m_1 \rangle$ from (b) and $\langle m_2 \rangle$ from (c)?

(e) Say you get $m_1 = +1$ on your first measurement in part (a). And you make your next 99 measurements *without* restoring the $|J, M\rangle$ state to $|3/2, +1/2\rangle$. What then is your value of $\langle m_1 \rangle$?

8. *More angular-momentum operators.*

The operator relation $\hat{\mathbf{L}} \cdot \hat{\mathbf{S}} = \frac{1}{2}(\hat{J}^2 - \hat{L}^2 - \hat{S}^2)$, where $\hat{\mathbf{J}} = \hat{\mathbf{L}} + \hat{\mathbf{S}}$, was used at the end of Sec. 3. Operating on a state $|L, S; J, M\rangle$, it gives

$$\hat{\mathbf{L}} \cdot \hat{\mathbf{S}} |L, S; J, M\rangle = \frac{1}{2}\big[J(J+1) - L(L+1) - S(S+1)\big]\hbar^2 |L, S; J, M\rangle .$$

220

Although the operators are quantum mechanical and the vectors are directional, the relation is little more than the law of cosines for triangles with sides a, b, and c: $\mathbf{a} \cdot \mathbf{b} = ab \cos \theta = \frac{1}{2}(c^2 - a^2 - b^2)$, where c is the side opposite θ.

(a) Prove the operator relations $\hat{\mathbf{L}} \cdot \hat{\mathbf{J}} = \frac{1}{2}(\hat{J}^2 + \hat{L}^2 - \hat{S}^2)$ and $\hat{\mathbf{S}} \cdot \hat{\mathbf{J}} = \frac{1}{2}(\hat{J}^2 + \hat{S}^2 - \hat{L}^2)$.

(b) In Chap. 13, the *Landé g factor* will be important, because $-gJ$ is the magnetic moment μ of an atom in units of the Bohr magneton μ_B. The calculation of g involves $\hat{\mathbf{S}} \cdot \hat{\mathbf{J}}$, and in the state $|L, S; J, M\rangle$ the value of g is

$$g = 1 + \frac{J(J+1) + S(S+1) - L(L+1)}{2J(J+1)}.$$

The ground state of an iron atom is $|L, S; J, M\rangle = |2, 2; 4, M\rangle$ (Chap. 14). Show that $g = 3/2$ and that the magnetic moment μ of iron is $-6\,\mu_B$.

(c) Find g and μ for the states $|L, S; J, M\rangle = |1, \frac{1}{2}; \frac{3}{2}, M\rangle$ and $|1, \frac{1}{2}; \frac{1}{2}, M\rangle$.

9. *Particle decay branching fractions.*

Following are the isospins and charge states of some particle multiplets. Numbers in parentheses are mass-energies mc^2 in MeV. The masses are usually omitted from symbols for the nucleon (p and n) and the pion (π^+, π^0, and π^-). Particles in the first two rows are mesons, those in the last two are baryons.

Particle	Isospin	Charges
$\pi(139)$, $\rho(770)$	1	$+e$, 0, $-e$
$\omega(782)$, $f(980)$	0	0
$N(939)$, $N^*(1675)$	1/2	$+e$, 0
$\Delta(1232)$	3/2	$+2e$, $+e$, 0, $-e$

(a) Write down all the charge modes for the following decays, and use Clebsch-Gordan coefficients to get their branching fractions:

$$N^*(1675) \to \pi N \qquad\qquad f(980) \to \pi\pi$$
$$N^*(1675) \to \Delta(1232)\pi \qquad\qquad \rho(770) \to \pi\pi.$$

For the $f \to \pi\pi$ and $\rho \to \pi\pi$ decays, treat the $\pi^+\pi^-$ and $\pi^-\pi^+$ decays as independent, and then add the squares of the two Clebsch-Gordan coefficients to get the total $\pi^+\pi^- + \pi^-\pi^+$ branching fraction.

(b) Find the branching fractions again using the counting method. Here you do not have to treat $\pi^+\pi^-$ and $\pi^-\pi^+$ decays as independent.

(c) Use the counting method to find the branching fractions for $\omega \to 3\pi$ decays.
See, perhaps, C.G. Wohl, "Isospin Relations by Counting," Am. J. Phys. **50** (1982) 748.

10. *A more complicated decay.*

If its mass is large enough, an N^* baryon (there are a number of them) can decay to $\pi\pi N$ as well as to πN. The possible $N^{*+} \to \pi\pi N$ charge modes are

$$N^{*+} \to \pi^+\pi^- p \qquad n_{+-}$$
$$\to \pi^+\pi^0 n \qquad n_{+0}$$
$$\to \pi^0\pi^0 p \qquad n_{00}.$$

Let n_{+-}, n_{+0}, and n_{00} be the numbers of decays to the three modes, and let $n_{tot} = n_{+-} + n_{+0} + n_{00}$. Things are now more complicated: A pion and a nucleon can couple to make $I = 1/2$ and $3/2$, and the second pion can then couple to each of these to match the $I = 1/2$ of the N^{*+}. Thus there are two independent amplitudes for $N^* \to \pi\pi N$ decays, and the ratios of n_{+-} to n_{+0} to n_{00} are not unique. There *is* a relation between the three numbers, but working it out with Clebsch-Gordan coefficients would take some time.

(a) Use the counting method to show in two lines that

$$2n_{+-} = n_{+0} + 4n_{00} \ .$$

Changing n_{+0} to n_{-0} would give the relation for the three $N^{*0} \to \pi\pi N$ modes.

(b) Thus only two of the three numbers have to be measured to get the third. Show then that the total is

$$n_{tot} = 3(n_{+-} - n_{00})$$
$$= 3(n_{+0} + 2n_{00})/2$$
$$= 3(n_{+0} + 2n_{+-})/4 \ ,$$

and that if only n_{+-} is measured then n_{tot} is bound by

$$\tfrac{3}{2}n_{+-} \le n_{tot} \le 3n_{+-} \ .$$

11. *Pion production in the atmosphere by cosmic rays.*

The Earth's (dry) atmosphere is about 78% nitrogen, 21% oxygen, and 1% argon. The nuclei are almost entirely ^{14}N, ^{16}O, and ^{40}Ar, all of which have isospin 0. Thus the atmosphere is an almost perfect isospin-0 target.

High-energy cosmic rays are about 74% protons, 18% alpha particles (^{4}H nuclei), and 8% heavier nuclei. The alpha and nearly all the heavier nuclei have $I = 0$. The cosmic-ray energies can be enormous, far beyond anything we can produce at an accelerator. Striking the atmosphere, the rays can produce enormous sprays of particles. The interactions of the rays with the atomic electrons in the atmosphere are in comparison negligible.

Ratchet up the argument of Prob. 10 to show that cosmic rays striking the atmosphere produce twice as many charged pions as π^0's.

222

12. CRYPTOGRAPHY. THE EPR ARGUMENT. BELL'S INEQUALITY

12.1. *Quantum Cryptography*

12.2. *The EPR Argument*

12.3. *Bell's Inequality*

Problems

A short chapter, based entirely on the $|S, M\rangle = |0, 0\rangle$ state of two spin-1/2 particles,

$$|0, 0\rangle = \frac{1}{\sqrt{2}} \left(|+-\rangle - |-+\rangle \right).$$

The kets on the right side give the $|m_1, m_2\rangle$ states of the first and second particles. In early sections of Chapter 11, they were electron and proton. Plus and minus, spin up or down, are with respect to the direction taken to be the z axis: $|+-\rangle$ is short for $|+z, -z\rangle$, and so on. Suppose, starting in the $|0, 0\rangle$ state, a measurement of the z component of the spin of one of the particles gets $-$. Then a measurement of the z component of the other particle would get $+$, with certainty; the results are strictly anti-correlated. This in itself is not mysterious. But according to quantum mechanics, neither result, $+$ or $-$, *exists* until someone makes a measurement. And this, as Einstein insisted, makes the anti-correlation a very big mystery.

In general, a rotation of axes from xyz to $x'y'z'$ would change what in the unprimed frame was, say, an $|S, M\rangle = |1, +1\rangle$ state into a mix of all three $|1, M'\rangle$ states. But when $S = 0$, there is only $M = 0$, so there is no mixing: The $|0, 0\rangle$ state "looks" the same from any direction. The strict anti-correlation of $+$ and $-$ states is true, no matter which direction the two measurements are made along (see Problem 2).

We shall need an equation from Section 10·2. Let $P_d = \cos \beta$ be the component of the polarization vector of a spin-1/2 particle along a direction $\mathbf{d}$. Then on making a measurement of the spin along $\mathbf{d}$, the probabilities of getting spin up or down are $\cos^2(\frac{1}{2}\beta)$ and $\sin^2(\frac{1}{2}\beta)$.

This chapter is only an introduction to its subjects. There are some references for going farther at the end of the chapter.

223

12·1. QUANTUM CRYPTOGRAPHY

(a) Cryptography ("secret writing")—In business, government, warfare, crime, affairs of the heart, and other human activities, secrecy of communications is vital. Encryption of messages goes back millennia, but beginning in the 1930s with a world war looming and the first primitive computers available—"war is the mother of invention"—enormous efforts have been put into the making and breaking of ever more sophisticated codes. In World War II, shipping of food and munitions from the United States to Britain and the Soviet Union was absolutely critical. The breaking of the German "Enigma" code by Polish and British workers played a vital role in, among other things, fighting off the "wolf packs" of German submarines that preyed upon the convoys.

In modern communications, most messages, whether encrypted or not, are sent as streams of bits, like 1001011000... Given the stream, one has to know how to read the message from it—to know the "key." Encrypted or not, the bits are usually read in groups of eight, using a standard 8-bits-per-symbol code called ASCII (for American Standard Code for Information Interchange). For example, in ASCII the letter A is 01000001, the letter B is 01000010, the number 1 is 10110001, and so on.

In most encryption schemes, the message is encoded directly in the stream of bits, encrypted at the sender's end and decrypted at the receiver's end. This happens every time you send your credit-card information to buy something using a computer. It wouldn't help to know the stream, because you wouldn't know how to read it. But in the following scheme, there is no message in the stream of bits. How to read the bits comes later, after checking for a spy. It is the stream itself—the values of its bits—that has to be kept secret.

The idea of quantum cryptography is simple. The hard part is to physically implement it, but there are now several small networks and several companies that sell working systems. They use a state of two polarized photons instead of the $|0,0\rangle$ state of two spin-1/2 particles. The photon state is much easier to create, and the photons can be sent down optical fibers.

(b) Alice and Bob—Following custom, we name the communicating parties Alice and Bob. Alice has a device that creates $|S, M\rangle = |0,0\rangle$ states of two spin-1/2 particles:

$$|0,0\rangle = \frac{1}{\sqrt{2}}\left(|+-\rangle - |-+\rangle\right) .$$

Bob would have such a device too, but we discuss messages sent by Alice to Bob. Alice uses her device to generate a long, numbered sequence of $|0,0\rangle$ states, each of which breaks up into its two spin-1/2 constituents. One of these stays with Alice, while the other is sent off to Bob. Alice and Bob each measure, along an agreed-upon axis z, the component of the spin of each particle they receive (see Fig. 1).

Alice's SG   *Bob's SG*

Figure 1. Alice and Bob, using Stern-Gerlach apparatuses, each measure a component of the spin of the spin-1/2 particles that fly apart from a $|0,0\rangle$ source. In this example, Alice gets spin up along z and Bob gets spin down.

224

Spin-up and spin-down are the only possible results of a good measurement of the component of the spin of a spin-1/2 particle. For any particular $|0,0\rangle$ state, Alice gets, *equally probably* and *entirely* randomly, $+$ or $-$. If she gets $+$, the state is no longer $|0,0\rangle$, it is $|+-\rangle$. (Alice's result is the first symbol in the ket.) And then Bob, later, being farther off, will get $-$. Whichever she gets, Bob gets the opposite. A sequence might begin

$$\text{Alice}: \quad + + - + - - - + - + \cdots$$
$$\text{Bob}: \quad - - + - + + + - + - \cdots$$

By the rules of quantum mechanics—which are supposedly Fundamental Laws of Nature—there is absolutely no pattern in this sequence. It is truly random, with about half of the results $+$ and half $-$. Getting *truly* random sequences from number generators is not easy.

Now the *message*, in which $+$ and $-$ will stand in for the bits 1 and 0. To send the letter A ($= 01000001$) in ASCII, Alice would tell Bob to read bits for which *she* got $+-+++++-$. For example, she might say, "Read your 1st, 3rd, 4th, 8th, 10th, ... bits." In this way, perhaps in groups of eight instructions, using no bit twice, she can send the letters, etc. of her message. The message is indecipherable without knowing what each bit *is*, $+$ or $-$. In fact, Alice could send Bob the ordered list of the measurements he is to read publicly. It is just a list of numbers, and in itself it would reveal nothing.

(c) Eve—But suppose a spy, the lovely Eve, is lurking somewhere between Alice and Bob, and knows the z axis they are using. If Eve measures the z component of the particles going to Bob, she does not change the results, $+$ or $-$, that Bob gets (why not?). And if she can also learn the instructions for how Bob is to read the message, she can read it too.

There is, however, a foolproof way that Alice can know if a spy is listening. Alice and Bob choose *two* axes, an orthogonal z and x. For each particle they receive, they decide independently and randomly which axis to measure along. About half the time they will decide differently—Alice x and Bob z or vice versa. For these cases, Bob's result will be the opposite of Alice's only about half the time (why?). But when they measure along the same axis, they should, as before, get opposite results. Here is an example (with no spy):

$$\text{Alice} \left\{ \begin{array}{cccccccccccccc} x & x & z & z & x & z & z & x & x & x & z & x & z & z \\ + & + & + & - & + & - & - & + & + & - & + & + & - & - \end{array} \right\}$$

$$\text{Bob} \left\{ \begin{array}{cccccccccccccc} x & z & x & x & x & z & z & z & x & x & x & z & x & z \\ - & - & - & - & - & + & + & - & - & + & + & - & + & + \end{array} \right\}$$

Now, even if Eve knows the axes x and z, she has to guess which axis to use as each particle heading for Bob goes by. Suppose, for example, for particle-pair #1143, Alice and Bob both measure along the x axis, and Eve does too. Then Eve gets the same result that Bob does, and does not disturb the Alice-Bob anticorrelation. But if for #1143 Eve measures along the z axis, the probability that Bob's result is the opposite of Alice's is only 50% (why?).

(d) The protocol (quantum key distribution)—So here is what Alice and Bob do with their long strings of N numbered measurements:

(1) Bob sends Alice a report of a subset of n ($1 \ll n \ll N$) of his measurements scattered along the length of his string: the sequence number, the axis used, and the result. For example, #3/x/$-$, #4/x/$-$, #6/z/$+$, ... He can send this to Alice publicly.

(2) Alice looks over this subset, marks those (about $n/2$ of them) for which they both chose the same axis, and checks to see if these results are always opposite. If so, the way is clear. If not, Eve is lurking, and Bond or Bourne will have to deal with the problem before a message can be sent.

(3) If the way is clear, Alice tells Bob to send the remaining $(N - n)$ of his axis settings, but not the results. For example, #1/x, #2/z, #5/x, ... Then she sends the message using the roughly half of these for which she and Bob used the same axis and so got opposite results.

The scheme may seem complicated, and for a lengthy message N would need to be large. But of course it would be automated, as are nearly all the apparently simple but in fact often extremely complex things we do using computers.

Quantum cryptography is a good warm-up for a deeper subject ...

12·2. THE EPR ARGUMENT

(a) "Spooky action at a distance"—Consider again a source of $|0,0\rangle$ states that each break into two spin-1/2 particles that fly apart to Alice and Bob, who are now each light years from the source (see Fig. 2), but Bob is just one meter farther away (and the particles travel at the same speed).

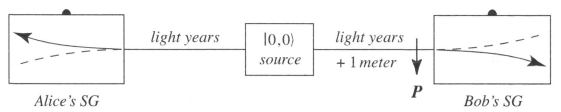

Figure 2. Alice and Bob are now each light years from the $|0,0\rangle$ source. For this particle pair, Alice gets spin up and Bob gets spin down. According to quantum mechanics, that polarization vector **P** pops into existence only in the last extra meter to Bob.

Let Alice measure along some axis the component of the spin of each particle that reaches her. After each measurement, she will know instantly that Bob, were he to measure along the same axis, would get the opposite of her result. She gets, say, $+$, and an instant later he gets, with *certainty*, $-$. Can Alice's measurement, at the instant it occurs, somehow directly affect Bob's particle? No, because the cause would have to travel far faster than the speed of light to reach him in time, and this is forbidden by special relativity. Doesn't that mean that the polarization vector for this particular particle about to reach Bob has to be as it is shown in Fig. 2, even before anyone makes a measurement? How else, with *certainty*, is Bob going to get $-$?

Thus it would seem that Bob's result actually preexists Alice's measurement. Bob's particle, it seems, must (speaking loosely) *know in advance* the answer it would give, whether or not anyone measures anything. And if Bob's result pre-exists measurement, then conversely so also must Alice's result. This is in accord with classical thinking, in which the dynamical properties of particles have actual values, whether or not those properties are measured. But this is not in accord with quantum mechanics. Quantum mechanics insists that whether the spin is up or down is not determined—the answer *does not exist*—unless or until Alice (or Bob) makes a measurement. Instead, the potentialities $|+-\rangle$ and $|-+\rangle$ coexist in the

$|0, 0\rangle$ state—they are *entangled until a measurement is made*. The polarization vector of the particle about to reach Bob pops into existence at the instant Alice, light years away, makes her measurement. Until then, it does not *exist*.

"Spooky action at a distance" indeed. That is Einstein's phrase, and he did not believe that Nature works that way.

(b) The EPR argument—A 1935 paper by A. Einstein, B. Podolsky, and N. Rosen, "Can [the] Quantum-Mechanical Description of Reality be Considered Complete?", sharpened the argument and deepened the conundrum. Note that the title questions the *completeness* of quantum mechanics, not the *correctness* of its predictions about the results of measurements. The "EPR" paper argues from the following position (italics in original):

> *If, without in any way disturbing the system, we can predict with certainty ... the value of a physical quantity, then there exists an element of reality corresponding to this physical quantity.*

The original EPR argument was stated in terms of measurements of position and momentum. It was restated in the simpler, binary terms of the break-up of a $|0, 0\rangle$ state by David Bohm. Here it would read:

> *Since, without in any way disturbing Bob's particle, Alice can predict with certainty the value of a component of its spin, there exists an element of reality corresponding to that component.*

It seems (as we have already said) that the component must be "real," and have a definite value, $+$ or $-$, prior to anyone's measurement.

But the thrust of the EPR argument goes much deeper. Alice can wait till the last moment to decide *which direction* to measure along. If she chooses the z direction, then as she gets $+$ or $-$, Bob's polarization vector will point along the $-z$ or $+z$ direction. If instead she chooses x, then as she gets $+$ or $-$, Bob's polarization vector will point along $-x$ or $+x$; and so on for any direction she chooses. Thus according to the EPR argument, *each and every* component of Bob's spin is "real"—*each* component has a definite value, $+$ or $-$. But according to the uncertainty relation $\Delta s_x \Delta s_y \geq \frac{1}{2}\hbar|\langle s_z\rangle|$, if you know the z component of a spin, its x and y components do not have definite values. The notion that all three components can simultaneously have exact values violates quantum mechanics.

The view that dynamical particle properties have actual values, whether measured or not (realism), and that causes cannot travel faster than light (locality), is called "local realism." A hypothetical deeper local-realistic theory is called a "hidden-variables" theory: There must be extra variables that, presently anyway, lie beyond our ken, and they would determine the particle's answers to all questions. This view claims that quantum mechanics only gives us probabilities because it doesn't see deeply enough. For nearly 30 years, the EPR argument was thought to be without experimental consequence—or at least beyond experimental exploration—and it was largely ignored. How could you even tell if all three components of a spin are really real? You can only *measure* one component at a time.

12·3. BELL'S INEQUALITY

In 1964, John Bell discovered a way to distinguish experimentally between quantum mechanics and local-realistic theories. He derived an inequality for certain spin-correlation measurements, assuming that the values of the spin components actually exist prior to measurement. The inequality conflicts with some of the predictions of quantum mechanics. It took years for the full implications of Bell's work to be appreciated, in part because there did not yet exist the experimental tools with which to do the implied experiments. But since about 1970 many experiments have been done—and the results *violate Bell's inequality and agree with quantum mechanics.*

The simplest proof of Bell's inequality for our Alice-and-Bob experiments with $|0,0\rangle$ states is due to Eugene Wigner, Am. J. Phys. **38** (1970) 1005. We consider correlations of results when Alice and Bob measure spin components along *different* axes. Figure 3 shows the arrangement. In Fig. 3(b), the 1 and 3 axes are at right angles, and the 2 axis is half way in between. They are coplanar in the plane perpendicular to the beam axis. We derive Bell's inequality for this experimental arrangement, and compare it with the predictions of quantum mechanics.

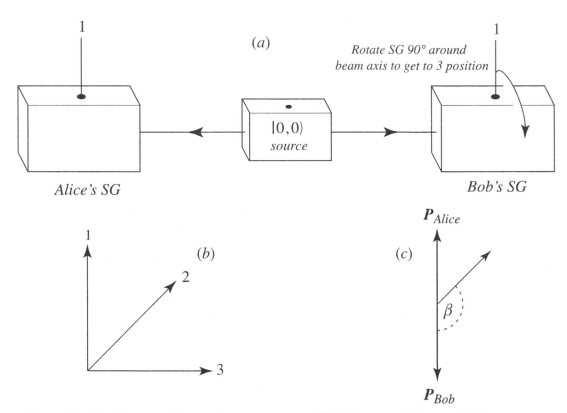

Figure 3. (a) The experimental arrangement. (b) The axes Alice and Bob measure along. The axes all lie in the plane perpendicular to the beam axis. (c) The polarization vectors when Alice gets + measuring along the 1 direction.

(a) Consequences of being "real"—Assume that all three components of the spin of a particle really have a value, + or −, whether or not anyone measures anything. This is, of course, not in accord with quantum mechanics. Let Alice and Bob make measurements with some large number N of $|0,0\rangle$ states. Let $f(+++,---)$ be the fraction of the N particles that come to Alice for which it is predetermined that she will get +, no matter which of the axes, 1, 2, or 3, she measures along. The first three signs in f tell what Alice would get for the axes 1, 2, and 3. The next three signs tell (redundantly) Bob's necessarily opposite results for those axes.

228

Now there are altogether $2 \times 2 \times 2 = 8$ such fractions:

$$f_1 = f(+++, ---)$$
$$f_2 = f(++-, --+)$$
$$f_3 = f(+-+, -+-)$$
$$f_4 = f(-++, +--)$$
$$f_5 = f(+--, -++)$$
$$f_6 = f(-+-, +-+)$$
$$f_7 = f(--+, ++-)$$
$$f_8 = f(---, +++) .$$

The fractions are real nonnegative numbers, and they sum to one. In each f, the first and fourth signs ($+$ or $-$), the second and fifth, and the third and sixth are opposite, because of the equal-and-opposite constraint when Alice and Bob measure along the same axis. This is a constraint *from experiment* that any theory must obey. Without it, there would be $2^6 = 64$ fractions instead of only eight. Again, the fractions are consequences of assuming that properties of particles have actual values, whether they are measured or not.

Let Alice measure along the 1 axis and Bob along the 3 axis. For what fraction of the total of N events will Alice and Bob *both* get $+$? We look down the f's for a $+$ in the first place (Alice's result along the 1 axis) *and* a $+$ in the sixth place (Bob's result along the 3 axis). These are f_2 and f_5. Thus, with an obvious notation, the joint fraction is

$$f_{A1+}^{B3+} = f_2 + f_5 .$$

This is the sum of the two fractions in which the values along the 1 and 3 axes are determined by measurement, but the value along the 2 axis can be either $+$ or $-$, because no measurement is being made along that axis.

Similarly, the joint fractions for both getting $+$ when Alice measures along the 1 axis and Bob measures along the 2 axis, or when Alice measures along the 2 axis and Bob measures along the 3 axis, are

$$f_{A1+}^{B2+} = f_3 + f_5 \qquad \text{and} \qquad f_{A2+}^{B3+} = f_2 + f_6 .$$

Adding these two joint fractions, we get

$$f_{A1+}^{B2+} + f_{A2+}^{B3+} = f_3 + f_5 + f_2 + f_6 \geq f_2 + f_5 = f_{A1+}^{B3+} .$$

This, for this set of measurements, is Bell's inequality. It says that the sum of the two joint probabilities for the axes at $45°$ to one another is an upper limit on the joint probability for the axes at $90°$. In fact, it says that $f_{A1+}^{B2+} = f_{A2+}^{B3+} \geq \frac{1}{2} f_{A1+}^{B3+}$, because f_{A1+}^{B2+} and f_{A2+}^{B3+} are both measurements at $45°$, and are equal by symmetry.

(b) Quantum mechanics—Now for the predictions of quantum mechanics. We remember (Sec. 10·2) that if $P_d = \cos \beta$ is the component of the polarization vector of a spin-1/2 particle along a direction **d**, the probability of finding the spin to be up along **d** is

$$\text{Prob}(+ \text{ along } \mathbf{d}) = \tfrac{1}{2}(1 + \cos \beta) = \cos^2(\tfrac{1}{2}\beta) .$$

229

Again, let Alice measure along the 1 axis and Bob along the 3 axis. Alice gets + half the time, and then Bob's polarization vector is along the negative 1 axis, and $\beta = 90°$; see Fig. 3. Thus, according to quantum mechanics, the joint probability of both Alice and Bob getting + is

$$QM_{A1+}^{B3+} = \tfrac{1}{2}\cos^2(\tfrac{1}{2} \times 90°) = 0.25 \; .$$

Similarly,

$$QM_{A1+}^{B2+} = \tfrac{1}{2}\cos^2(\tfrac{1}{2} \times 135°) = 0.073 \; ,$$

and QM_{A2+}^{B3+} is the same. Whereas "local realism" gives

$$f_{A1+}^{B2+} + f_{A2+}^{B3+} \geq f_{A1+}^{B3+} \; ,$$

quantum mechanics gives

$$QM_{A1+}^{B2+} + QM_{A2+}^{B3+} = 2 \times 0.073 = 0.146 < 0.25 = QM_{A1+}^{B3+} \; .$$

The sum of quantum-mechanical probabilities for measurements at 45° is *much less than* the probability for measurements at 90°. Many experiments have been done, and they violate the consequences of local realism and agree with the predictions of quantum mechanics. As Feynman said about the double-slit experiment, "Nobody knows how it can be like that."

The derivation of Bell's inequality did not refer to the specific angles used by Alice and Bob (although the values of the f fractions would of course depend on those angles); the quantum-mechanical calculation did specify angles. Figure 4 shows the joint quantum-mechanical probabilities as functions of θ_{13}, for $0 \leq \theta_{13} \leq 180°$. The calculations above were for $\theta_{13} = 90°$.

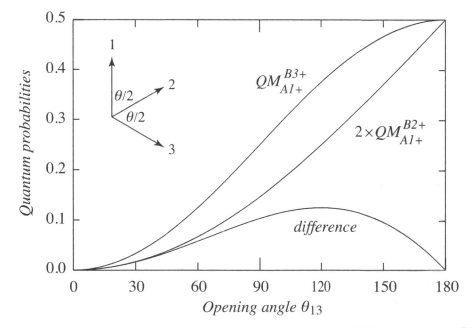

Figure 4. The quantum-mechanical joint probabilities QM_{A1+}^{B3+} and $2 \times QM_{A1+}^{B2+}$ as functions of the angle $\theta = \theta_{13}$ shown in the insert. The three axes are coplanar. The difference between the joint probabilities is positive. According to local realism, it cannot be positive.

(c) References—There is a large literature on the subjects of this chapter. Twenty two articles by Bell on "quantum philosophy" are collected in

> J.S. Bell, *Speakable and Unspeakable in Quantum Mechanics*, (Cambridge University Press, Cambridge, 1987).

A wide range of subjects related to quantum mysteries is discussed in

> George Greenstein and Arthur G. Zajonc, *The Quantum Challenge: Modern Research on the Foundations of Quantum Mechanics*, (Jones & Bartlett Publishers, Sudbury MA, 2006).

There are some old but "elementary" articles on Bell's inequality. Old is good, because near beginnings people try hard to get at the fundamentals:

> Eugene P. Wigner, "On Hidden Variables and Quantum Mechanical Probabilities," American Journal of Physics **38** (1970) 1005.

> Bernard d'Espagnat, "The Quantum Theory and Reality," Scientific American **241** (November 1979) 158.

> N. David Mermin, "Is the Moon There When Nobody Looks? Reality and the Quantum Theory," Physics Today **38** (April 1985) 38.

PROBLEMS

1. *A spy?*

(a) Bob e-mails Alice his axis settings and results for the 1st, 3rd, 5th, ... measurements in the set below. Does Alice have evidence that a spy is lurking? If she does, circle the evidence and skip to part (b). If she does not, she tells Bob to send the rest of his settings, but not the results. How does she then send Bob the message "+−"? Circle the measurement pairs you use.

$$
\text{Alice} \left\{
\begin{matrix}
x & z & z & z & x & x & z & x & x & z & x & z & z & x \\
+ & + & - & + & + & - & + & + & - & - & - & + & - & +
\end{matrix}
\right\}
$$

$$
\text{Bob} \left\{
\begin{matrix}
z & x & z & x & x & x & z & x & z & z & z & z & x & x \\
+ & - & + & + & - & - & + & - & - & - & - & - & - & +
\end{matrix}
\right\}
$$

(b) Repeat (a) for the set:

$$
\text{Alice} \left\{
\begin{matrix}
x & x & z & x & z & z & x & x & z & z & x & z & x & z \\
+ & - & + & + & - & - & + & + & + & - & + & - & - & +
\end{matrix}
\right\}
$$

$$
\text{Bob} \left\{
\begin{matrix}
x & x & x & z & x & z & x & z & x & x & x & z & z & z \\
- & + & + & - & + & + & - & - & - & - & - & + & - & -
\end{matrix}
\right\}
$$

2. *The $|S, M\rangle = |0,0\rangle$ and $|1,0\rangle$ states of two spin-1/2 particles.*

This and the next problem use the tensor product introduced in Prob. 11·3,

$$
\begin{pmatrix} a_1 \\ b_1 \end{pmatrix} \otimes \begin{pmatrix} a_2 \\ b_2 \end{pmatrix} = \begin{pmatrix} a_1 a_2 \\ a_1 b_2 \\ b_1 a_2 \\ b_1 b_2 \end{pmatrix}.
$$

Find the two-component vectors $|+z\rangle$, $|-x\rangle$, etc. in Sec. 10·1.

(a) Write the $|S, M\rangle = |0, 0\rangle$ and $|1, 0\rangle$ states

$$\frac{1}{\sqrt{2}} \left(|+z, -z\rangle - |-z, +z\rangle \right) \qquad \text{and} \qquad \frac{1}{\sqrt{2}} \left(|+z, -z\rangle + |-z, +z\rangle \right)$$

as four-component vectors. The answers are just linear superpositions of the four-component vectors in Sec. 11·1(a).

(b) Write

$$\frac{1}{\sqrt{2}} \left(|+x, -x\rangle - |-x, +x\rangle \right) \qquad \text{and} \qquad \frac{1}{\sqrt{2}} \left(|+x, -x\rangle + |-x, +x\rangle \right)$$

as four-component vectors. Which of these is, aside from an overall factor, still the same four-component vector as in (a)?

(c) Write

$$\frac{1}{\sqrt{2}} \left(|+y, -y\rangle - |-y, +y\rangle \right) \qquad \text{and} \qquad \frac{1}{\sqrt{2}} \left(|+y, -y\rangle + |-y, +y\rangle \right)$$

as four-component vectors. Which of these is, aside from an overall factor, still the same four-component vector as in (a)?

3. *Quantum correlations.*

(a) Explain why the probability curves in Fig. 4 are at zero at $\theta \equiv \theta_{13} = 0$, and are at 0.5 at $\theta = 180°$.

(b) Find the equations for the QM curves in Fig. 4 as functions of θ_{13}, and show that the equation for the difference curve is

$$\text{difference} = 0.5 \, \cos(\tfrac{1}{2}\theta) \left[1 - \cos(\tfrac{1}{2}\theta) \right] .$$

(c) Find the angle at which the difference curve is a maximum.

4. *Symmetry relations between the f fractions.*

(a) The eight f fractions are not all independent. Use a symmetry argument to show that

$$f_1 = f_8, \qquad f_2 = f_7, \qquad f_3 = f_6, \quad \text{and} \quad f_4 = f_5 .$$

(What change of axes would simply switch all the + signs to − signs, and vice versa, but would be physically equivalent? Or, even simpler, what change of labels would switch all those signs but leave the axes unchanged?)

(b) In Sec. 3(b), we concluded that

$$f_{A1+}^{B2+} = f_3 + f_5 = f_{A2+}^{B3+} = f_2 + f_6 .$$

Thus show that

$$f_2 = f_4 = f_5 = f_7 .$$

So all the relations of f fractions could be written in terms of just f_1, f_2, and f_3. And the normalization constraint would reduce the number of independent fractions to two.

13. TIME-INDEPENDENT PERTURBATION THEORY.

13.1. *The Nondegenerate Recipes*

13.2. *Examples of Nondegenerate Theory*

13.3. *The Nondegenerate Derivations*

13.4. *The Degenerate Recipe*

13.5. *Two Selection Rules. A Useful Relation*

13.6. *The Stark Effect in Hydrogen (Strong-Field Case)*

13.7. *Hydrogen Fine Structure: Experiment*

13.8. *Hydrogen Fine Structure: Theory*

13.9. *Atomic Magnetic Moments*

13.10. *The Zeeman Effect in Hydrogen (Weak-Field Case)*

   Problems

We assume we already know the eigen-solutions E_n^0 and ψ_n^0 of the Hamiltonian $\hat{H}^0$: $\hat{H}^0\psi_n^0 = E_n^0\psi_n^0$. But what we really want are the solutions of $\hat{H}\psi_n = E_n\psi_n$, where $\hat{H} = \hat{H}^0 + \hat{H}'$, and $\hat{H}'$ is small compared to $\hat{H}^0$. Roughly, "small" means that $\hat{H}'$ only shifts an E_n^0 by an amount that is small compared to the differences between it and the energies of the nearest states for which $E_{n'}^0 \neq E_n^0$. The small effects are called perturbations. The effects of massive Jupiter on the orbits of the other planets, the effects of the $-\theta^3/3!$ term in the expansion of $\sin\theta$ in small oscillations of a pendulum, are classical perturbations.

We use nondegenerate perturbation theory when a ψ_n^0 is nondegenerate, and degenerate theory when it is degenerate (although sometimes we can use what amounts to nondegenerate theory on degenerate levels). Perturbation theory usually only gives approximate answers, in the form of a series of (usually) ever smaller corrections: "Great fleas have little fleas upon their backs to bite 'em, and little fleas have lesser fleas, and so ad infinitum." (Augustus de Morgan). Here we shall usually stop at the first nonzero corrections.

Examples abound. Even for the hydrogen atom there is an $\hat{H}'$ for the neglect of relativity in the Hamiltonian $\hat{H}^0$; for the interaction of the magnetic moments associated with the spin and orbital angular momenta of the electron; for the interaction of the magnetic moment of the proton with those of the electron (solved for the ground state in Chapter 11); for the effects of external electric and magnetic fields, when present. The investigation of ever finer details of atomic spectra, and especially of hydrogen spectra, drove much of early quantum mechanics, with experiment usually ahead of theory.

Some things are easier to learn when you see *how* they work before trying to see *why* they work. For nondegenerate perturbation theory, we give the recipes first, examples second, and derivations last. But (first-order) degenerate perturbation theory is just standard simultaneous-equation matrix eigenvalue theory, already used in Chapters 10 and 11 on problems with one or two spins.

13·1. THE NONDEGENERATE RECIPES

Perturbation theory assumes that we already know the eigenvalues E_n^0 and normalized eigenfunctions ψ_n^0 of a Hamiltonian $\hat{H}^0$: $\hat{H}^0 \psi_n^0 = E_n^0 \psi_n^0$. However, there is a small additional piece, $\hat{H}'$, of the full Hamiltonian $\hat{H}$,

$$\hat{H} = \hat{H}^0 + \hat{H}',$$

and we want the eigen-solutions of $\hat{H}\psi_n = E_n\psi_n$. We can usually only solve this equation approximately. *Nondegenerate* perturbation theory is used for eigenfunctions of $\hat{H}^0$ that are nondegenerate, such as those of the one-dimensional harmonic oscillator. The solutions are given by "perturbation series" of (usually) ever smaller correction terms,

$$E_n \approx E_n^0 + E_n' + E_n'' + \cdots \qquad \text{and} \qquad \psi_n \approx \psi_n^0 + \psi_n' + \psi_n'' + \cdots$$

Primes here denote terms in the series, *not* differentiation. If $\hat{H}'$ is indeed small, then E_n', E_n'', ... , should be small compared to the distances $(E_n^0 - E_{n-1}^0)$ and $(E_{n+1}^0 - E_n^0)$ of E_n^0 from its nearest neighbors.

(a) **Recipes**—The first-order correction to E_n^0 (to be derived in Sec. 3) is

$$\boxed{E_n' = \langle \psi_n^0 | \hat{H}' | \psi_n^0 \rangle \equiv H_{nn}' \; .}$$

This is a much used and very important equation. The calculation of, say, E_5' involves only the unperturbed wave function ψ_5^0 (and $\hat{H}'$), not any of the functions ψ_m^0 with $m \neq 5$. In fact, E_5' is simply the expectation value of $\hat{H}'$ in the state ψ_5^0.

The second-order correction to E_n^0 (to be derived in Sec. 3) is

$$\boxed{E_n'' = \sum_{m \neq n} \frac{\left| \langle \psi_m^0 | \hat{H}' | \psi_n^0 \rangle \right|^2}{E_n^0 - E_m^0} = \sum_{m \neq n} \frac{|H_{mn}'|^2}{E_n^0 - E_m^0} \; ,}$$

where $H_{mn}' \equiv \langle \psi_m^0 | \hat{H}' | \psi_n^0 \rangle$. The sum is over all $m \neq n$, with n fixed; the denominators are never zero because E_n^0 is nondegenerate. We often neglect E_n'' if E_n' is nonzero; E_n'' is, we hope, small compared to E_n'. We can check this if the calculations are not too difficult.

The first-order correction to ψ_n^0 is a sum over the other unperturbed wave functions,

$$\boxed{\psi_n' = \sum_{m \neq n} \frac{\langle \psi_m^0 | \hat{H}' | \psi_n^0 \rangle}{E_n^0 - E_m^0} \, \psi_m^0 = \sum_{m \neq n} \frac{H_{mn}'}{E_n^0 - E_m^0} \, \psi_m^0 \; .}$$

The terms in this series tell how much of each ψ_m^0 is to be added to ψ_n^0 to get the approximate ψ_n. Because we are adding small amounts of $m \neq n$ wave functions to the normalized ψ_n^0, the total, $\psi_n \approx \psi_n^0 + \psi_n'$, will not be quite normalized. However, because ψ_n^0 is nondegenerate, every ψ_m^0 with $m \neq n$ is orthogonal to ψ_n^0, and so $\langle \psi_n' | \psi_n^0 \rangle = 0$. Therefore ψ_n' spoils the normalization only quadratically:

$$\langle \psi_n | \psi_n \rangle = \langle \psi_n^0 + \psi_n' | \psi_n^0 + \psi_n' \rangle = \langle \psi_n^0 | \psi_n^0 \rangle + \langle \psi_n' | \psi_n' \rangle = 1 + \langle \psi_n' | \psi_n' \rangle \; .$$

Having found ψ_n', we can of course renormalize ψ_n, but this is a second-order effect and it is often neglected. The second-order corrections ψ_n'' to the ψ_n^0 are also often neglected.

234

In any particular application, the work is in getting those matrix elements, the H'_{mn}. The calculations use the known unperturbed wave functions ψ_n^0 and ψ_m^0 and the known perturbation Hamiltonian $\hat{H}'$. The first-order corrections, E_n' and ψ_n', are linear in the matrix elements; the second-order corrections, E_n'', are quadratic in them. Higher-order corrections are still more complicated and will not concern us.

When a number of calculations with the same $\hat{H}'$ are to be made, it helps to display the structure of the arrays of matrix elements of $\hat{H}^0$ and $\hat{H}'$,

$$\hat{H}^0 = \begin{pmatrix} E_1^0 & 0 & 0 & \cdots \\ 0 & E_2^0 & 0 & \cdots \\ 0 & 0 & E_3^0 & \cdots \\ \vdots & \vdots & \vdots & \ddots \end{pmatrix} \quad \text{and} \quad \hat{H}' = \begin{pmatrix} H_{11}' & H_{12}' & H_{13}' & \cdots \\ H_{21}' & H_{22}' & H_{23}' & \cdots \\ H_{31}' & H_{32}' & H_{33}' & \cdots \\ \vdots & \vdots & \vdots & \ddots \end{pmatrix}.$$

The matrices are of course Hermitian: $H_{mn}' = (H_{nm}')^*$. The diagonal elements H_{nn}' of the H_{mn}' array are the first-order corrections E_n'. The only matrix elements you need to get E_1'' and ψ_1' are in column 1 of the H_{mn}' array; the only elements you need to get E_2'' and ψ_2' are in column 2; and so on. The H_{mn}' matrix is often "sparse," meaning that most of its elements are zero. For the oscillator, the numbering of states begins with 0, not (as above) 1.

(b) An example—Equations with summation signs can take some getting used to, so here is a simple numerical example. Let the matrices for a 3-state system be (in appropriate energy units)

$$\hat{H}^0 = \begin{pmatrix} 1 & 0 & 0 \\ 0 & 4 & 0 \\ 0 & 0 & 9 \end{pmatrix} \quad \text{and} \quad \hat{H}' = \begin{pmatrix} 0.10 & 0.05 & 0.01 \\ 0.05 & 0.07 & 0.02 \\ 0.01 & 0.02 & 0.04 \end{pmatrix},$$

where

$$\psi_1^0 = \begin{pmatrix} 1 \\ 0 \\ 0 \end{pmatrix}, \text{ etc.}$$

Since the unperturbed energy eigenvalues, 1, 4, and 9, are all different, the three states are nondegenerate. If these eigenvalues were 1, 4, and 4, we could only use nondegenerate theory for the ground state.

The first-order perturbation corrections, E_n', $n = 1, 2, 3$, are the diagonal elements of $\hat{H}'$, 0.10, 0.07, and 0.04. The second-order correction to the *second* level is

$$E_2'' = \frac{(0.05)^2}{4-1} + \frac{(0.02)^2}{4-9} = (+8.33 - 0.80) \times 10^{-4} = 0.000753 .$$

This is negligible compared to $E_2' = 0.07$, and so $E_2 \simeq E_2^0 + E_2' = 4.07$. The state vector ψ_2, corrected to *first* order is

$$\psi_2 = \psi_2^0 + \psi_2' = \begin{pmatrix} 0 \\ 1 \\ 0 \end{pmatrix} + \frac{0.05}{4-1} \begin{pmatrix} 1 \\ 0 \\ 0 \end{pmatrix} + \frac{0.02}{4-9} \begin{pmatrix} 0 \\ 0 \\ 1 \end{pmatrix} = \begin{pmatrix} +0.017 \\ 1 \\ -0.004 \end{pmatrix}.$$

The sum of the squares of the components is 1.00029, so renormalizing here is of little importance. Take a minute or two and calculate E'' and ψ' for the first and third states.

As this shows, once you have the elements H_{mn}', calculating E_n', E_n'', and ψ_n' is easy.

13·2. EXAMPLES OF NONDEGENERATE THEORY

(a) **Example: oscillator Stark effect**—A particle is in a one-dimensional oscillator potential well $V^0(x) = \frac{1}{2}m\omega^2 x^2$. The particle has a charge q and is acted upon by an additional force $F = q\mathcal{E}_0$ due to an electric field $\mathcal{E}_0$ in the x direction. A perturbation caused by an external electric field is called a Stark effect. The perturbation Hamiltonian is

$$\hat{H}' = V'(x) = -\int_0^x q\mathcal{E}_0\, dx = -q\mathcal{E}_0 x \;.$$

We found the array of oscillator matrix elements $\langle \psi_m^0 | \hat{x} | \psi_n^0 \rangle$ of x in Sec. 6·4(b). Multiplied by $-q\mathcal{E}_0$, it is

$$\hat{H}' = -q\mathcal{E}_0 \sqrt{\frac{\hbar}{2m\omega}} \begin{pmatrix} 0 & \sqrt{1} & 0 & 0 & \cdots \\ \sqrt{1} & 0 & \sqrt{2} & 0 & \cdots \\ 0 & \sqrt{2} & 0 & \sqrt{3} & \cdots \\ 0 & 0 & \sqrt{3} & 0 & \cdots \\ \vdots & \vdots & \vdots & \vdots & \ddots \end{pmatrix},$$

which is indeed sparse. The diagonal elements are zero because the integrands $-q\mathcal{E}_0 x\, |\psi_n^0(x)|^2$ of the integrals $\langle n|\hat{H}'|n \rangle$ are odd functions of x. Thus $E_n' = 0$ for all n.

Since the array is sparse, there are only two terms in each second-order correction (or one for $n = 0$):

$$E_n'' = \sum_{m \neq n} \frac{|H_{mn}'|^2}{E_n^0 - E_m^0} = q^2\mathcal{E}_0^2 \frac{\hbar}{2m\omega} \left(\frac{n}{\hbar\omega} + \frac{n+1}{-\hbar\omega} \right) = -\frac{q^2\mathcal{E}_0^2}{2m\omega^2} \;.$$

The result is independent of n,

$$E_n = E_n^0 + E_n'' = \left(n + \tfrac{1}{2}\right)\hbar\omega - q^2\mathcal{E}_0^2/2m\omega^2 \;,$$

and this is in fact exact (see Prob. 1). Figure 1 shows how the linear-in-x $\hat{H}'$ simply shifts the parabola $\frac{1}{2}m\omega^2 x^2$ over and down, without changing the curvature of $V(x)$. Thus all the eigenvalues are simply lowered by that downward shift, $q^2\mathcal{E}_0^2/2m\omega^2$.

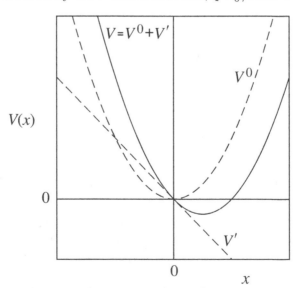

Figure 1. Adding the linear term $V'(x) = -q\mathcal{E}_0 x$ to the quadratic $V^0(x) = \frac{1}{2}m\omega^2 x^2$ simply shifts the parabola over and down.

(b) Example: finite size of the proton—The radius of the proton is very much smaller than the Bohr radius a_0, but it is not zero. (The proton is a finite construct of three quarks.) We shall model the charge of the proton to be spread uniformly throughout a sphere of radius R, where $R \ll a_0$. Figure 2 then shows how the potential energy $V(r)$ for hydrogen differs from $V^0(r) = -ke^2/r$ for a point proton. The perturbation Hamiltonian is (see Prob. 2),

$$\hat{H}' = V'(r) = V(r) - V^0(r) = -\frac{ke^2}{2R}\left(3 - \frac{r^2}{R^2}\right) + \frac{ke^2}{r} \qquad \text{for } r < R .$$

For $r > R$, $V'(r) = 0$. The perturbation $V'(r)$ is what you add to the unperturbed $V^0(r)$ to get the actual $V(r)$. The first-order correction to the ground-state energy, $E_1^0 = -13.6$ eV, is

$$E_1' = \langle \psi_{100}|V'|\psi_{100}\rangle = 4\pi \int_0^R |\psi_{100}^0(r)|^2\, V'(r)\, r^2\, dr .$$

Because R is tiny (about 10^{-15} m) compared to a_0 (about 5×10^{-11} m), the probability distribution $|\psi_{100}(r)|^2 = (\pi a_0^3)^{-1} e^{-2r/a_0}$ is very nearly constant over the range of integration and can be set equal to its value at the origin, $1/\pi a_0^3$, and taken outside the integral. The calculation gives E_1' of order 10^{-9} eV (Prob. 2).

We used nondegenerate perturbation theory here even though the ground state of hydrogen is (counting spin states) four-fold degenerate. This is because the perturbation here is independent of the spin state, and so all four states are shifted by the same amount. We shall see other examples where degeneracy can be ignored.

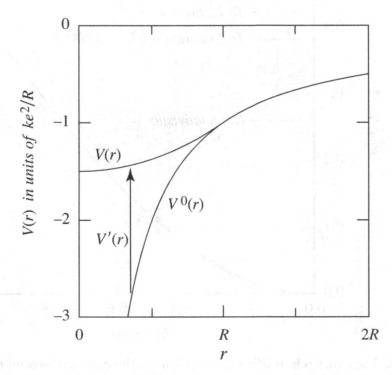

Figure 2. The potential energy $V(r)$ of the hydrogen atom, assuming that the charge of the proton is spread uniformly throughout a sphere of radius R.

(c) Example: relativistic corrections—The relativistic kinetic energy of a particle is

$$K_{rel} = (m^2 c^4 + p^2 c^2)^{1/2} - mc^2 = \left[\left(1 + \frac{p^2}{m^2 c^2} \right)^{1/2} - 1 \right] mc^2 \ .$$

Here p is the *relativistic* momentum, $p = \gamma m v$, where $\gamma \equiv (1 - \beta^2)^{-1/2}$ and $\beta \equiv v/c$. We expand K_{rel} as a power series in the ratio $u \equiv p/mc = \gamma \beta$:

$$K_{rel} = mc^2 \left(\frac{u^2}{2} - \frac{u^4}{8} + \frac{u^6}{16} + \cdots \right) = \frac{p^2}{2m} - \frac{p^4}{8m^3 c^2} + \frac{p^6}{16 m^5 c^4} + \cdots$$

The $p^2/2m$ term has the *form* of the nonrelativistic kinetic energy, but it only slides into $\frac{1}{2} mv^2$ for $v \ll c$. Figure 3 shows, for $u \leq 1$, the exact relativistic kinetic energy $K(u)$, in units of mc^2, and the p^2- and p^4-order approximations to it. When $u = p/mc$ is small, the perturbation Hamiltonian $\hat{H}'_{rel}$ is, to a good approximation,

$$\hat{H}'_{rel} \approx - \frac{\hat{p}^4}{8m^3 c^2} = - \frac{1}{2mc^2} \left(\frac{\hat{p}^2}{2m} \right)^2 \ .$$

This is negative (the p^4 term is negative), and it slightly overcorrects (the p^6 term is positive).

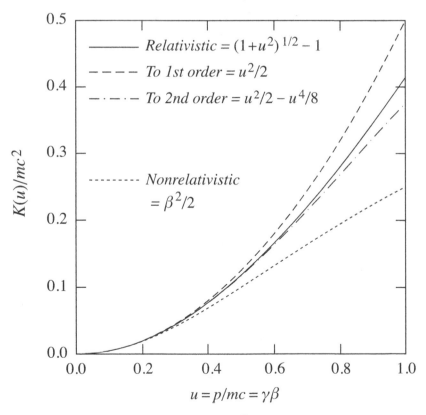

Figure 3. The exact relativistic ratio $K(u)/mc^2$, the first- and second-order approximations to it, and the nonrelativistic $K/mc^2 = \frac{1}{2}\beta^2$, versus $u \equiv p/mc = \gamma \beta$, for $u \leq 1$. The momentum p here is the relativistic $\gamma m v$, not mv.

238

Instead of calculating the matrix elements of $\hat{H}'_{rel}$ directly, we use a trick. In the non-relativistic limit, $\hat{H}^0$ is the nonrelativistic Hamiltonian,

$$\hat{H}^0 = \frac{\hat{p}^2}{2m} + \hat{V}, \qquad \text{or} \qquad \frac{\hat{p}^2}{2m} = \hat{H}^0 - \hat{V} \ .$$

Then

$$\hat{H}'_{rel} \approx -\frac{1}{2mc^2}\left(\frac{\hat{p}^2}{2m}\right)^2 = -\frac{1}{2mc^2}\left[(\hat{H}^0)^2 - \hat{H}^0\hat{V} - \hat{V}\hat{H}^0 + \hat{V}^2\right] \ ,$$

so that

$$\boxed{E'_n = \langle\psi_n^0|\hat{H}'_{rel}|\psi_n^0\rangle = -\frac{1}{2mc^2}\left[(E_n^0)^2 - 2E_n^0\langle\psi_n^0|\hat{V}|\psi_n^0\rangle + \langle\psi_n^0|\hat{V}^2|\psi_n^0\rangle\right] \ .}$$

Problem 3 is to find E'_n for the one-dimensional oscillator. For that problem, we already know from the virial theorem that $\langle\psi_n^0|\hat{V}|\psi_n^0\rangle = \frac{1}{2}E_n^0$ (Sec. 5·7(c)). In Sec. 8, we find the corrections for hydrogen (where the energy levels are degenerate). In the Bohr model of hydrogen, the fastest speed of the electron is $v = \alpha c = c/137$ (Sec. 1·4(b)). This makes $u = p/mc$ very close to $v/c = 1/137$, and the relevant part of Fig. 3 is way down in the lower left-hand corner.

13·3. THE NONDEGENERATE DERIVATIONS

We know the solutions of $\hat{H}^0\psi_n^0 = E_n^0\psi_n^0$, but want the solutions of $\hat{H}\psi_n = E_n\psi_n$, where $\hat{H} = \hat{H}^0 + \hat{H}'$. We write $\hat{H}$ like this,

$$\hat{H} = \hat{H}^0 + \lambda\hat{H}' \ .$$

Here λ is a "knob" (not an eigenvalue!) that we can turn from 0 (the perturbation is off) up to 1 (the perturbation is all the way on). Sometimes there is an actual knob, as when we turn on an external electric or magnetic field. Other times, as for the small effects in the internal workings of an atom, the knob can only be imagined. Mathematically, the λ knob is, as we shall see, a bookkeeping device.

We try writing the eigenvalues and eigenfunctions as series in λ,

$$E_n = E_n^0 + \lambda E'_n + \lambda^2 E''_n + \cdots \qquad \text{and} \qquad \psi_n = \psi_n^0 + \lambda\psi'_n + \lambda^2\psi''_n + \cdots$$

We are assuming that, as the λ-knob turns from 0 to 1, a given eigenvalue will vary smoothly from E_n^0 to the full E_n; see Fig. 4. And the eigenfunction ψ_n^0 will be slightly distorted by the addition to it of small amounts of the ψ_m^0 with $m \neq n$.

In terms of these series, the Schrödinger equation is

$$(\hat{H}^0 + \lambda\hat{H}')(\psi_n^0 + \lambda\psi'_n + \lambda^2\psi''_n + \cdots) = (E_n^0 + \lambda E'_n + \lambda^2 E''_n + \cdots)(\psi_n^0 + \lambda\psi'_n + \lambda^2\psi''_n + \cdots) \ .$$

Collecting terms on each side according to the powers of λ, we get

$$\hat{H}^0\psi_n^0 + \lambda(\hat{H}^0\psi'_n + \hat{H}'\psi_n^0) + \lambda^2(\hat{H}^0\psi''_n + \hat{H}'\psi'_n) + \cdots$$

$$= E_n^0\psi_n^0 + \lambda(E_n^0\psi'_n + E'_n\psi_n^0) + \lambda^2(E_n^0\psi''_n + E'_n\psi'_n + E''_n\psi_n^0) + \cdots$$

Every term linear in λ has one prime, every term quadratic in λ has two primes, etc.

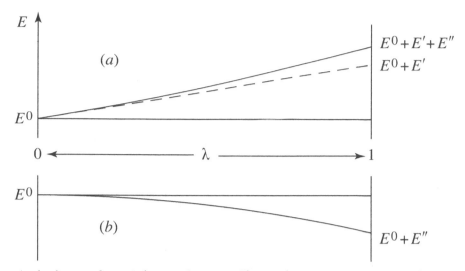

Figure 4. As λ runs from 0 (perturbation off) to 1 (perturbation fully on), the energy runs from E^0 to $E^0 + E' + E''$ (ignoring higher contributions). As a function of λ, $\lambda E'$ is a straight line with slope E', and $\lambda^2 E''$ is a parabola with slope 0 at $\lambda = 0$ and curvature $2E''$. Examples: (a) E' and E'' are both positive, with $E'' = E'/3$. (b) $E' = 0$ and $E'' < 0$, as for the Stark effect for the oscillator in Sec. 2(a).

Putting all the terms on one side of the equation, we get a power series in λ that is equal to zero: $a_0 + a_1\lambda + a_2\lambda^2 + \cdots = 0$. Because the powers of λ are linearly independent functions of λ, this equation is only satisfied for all values of λ between 0 and 1 if each and every coefficient a_n is zero. The resulting equations, rearranged to better show the structure, are

$$\lambda^0 \text{ terms}: \qquad (\hat{H}^0 - E_n^0)\psi_n^0 = 0$$

$$\lambda^1 \text{ terms}: \qquad (\hat{H}^0 - E_n^0)\psi_n' = (E_n' - \hat{H}')\psi_n^0$$

$$\lambda^2 \text{ terms}: \qquad (\hat{H}^0 - E_n^0)\psi_n'' = (E_n' - \hat{H}')\psi_n' + E_n''\psi_n^0$$

The λ^0 equation is simply the unperturbed eigenvalue equation, whose solutions E_n^0 and ψ_n^0 we already know.

Now take the scalar product of ψ_n^0 with the λ^1 equation,

$$\langle\psi_n^0|(\hat{H}^0 - E_n^0)|\psi_n'\rangle = \langle\psi_n^0|(E_n' - \hat{H}')|\psi_n^0\rangle .$$

The left side is zero because $\hat{H}^0$ simply rolls over onto the bra and produces E_n^0, which cancels the E_n^0 term. The right side gives

$$E_n' = \langle\psi_n^0|\hat{H}'|\psi_n^0\rangle \equiv H_{nn}' ,$$

as claimed in Sec. 1. This is our most important result.

To get ψ_n', take the scalar product of ψ_m^0, where $m \neq n$, with the λ^1 equation,

$$\langle\psi_m^0|(\hat{H}^0 - E_n^0)|\psi_n'\rangle - E_n'\langle\psi_m^0|\psi_n^0\rangle - \langle\psi_m^0|\hat{H}'|\psi_n^0\rangle .$$

Here $\langle\psi_m^0|\psi_n^0\rangle = 0$ because $m \neq n$ and ψ_n^0 is nondegenerate. Thus

$$(E_m^0 - E_n^0)\langle\psi_m^0|\psi_n'\rangle = -\langle\psi_m^0|\hat{H}'|\psi_n^0\rangle \equiv -H_{mn}' ,$$

or

$$\langle \psi_m^0 | \psi_n' \rangle = \frac{H_{mn}'}{E_n^0 - E_m^0}, \qquad \text{and so} \qquad |\psi_m^0\rangle \langle \psi_m^0 | \psi_n' \rangle = \left(\frac{H_{mn}'}{E_n^0 - E_m^0} \right) \psi_m^0 .$$

But $|\psi_m^0\rangle \langle \psi_m^0 | \psi_n' \rangle$ is just the projection of ψ_n' onto ψ_m^0, and the sum of all the projections for $m \neq n$ is ψ_n',

$$\psi_n' = \sum_{m \neq n} |\psi_m^0\rangle \langle \psi_m^0 | \psi_n' \rangle = \sum_{m \neq n} \frac{H_{mn}'}{E_n^0 - E_m^0} \, \psi_m^0 ,$$

as claimed in Sec. 1. If the ratios $H_{mn}'/(E_n^0 - E_m^0)$ are not small, the approximations are not likely to be good.

As was remarked in Sec. 1, the reduction of the amplitude of ψ_n^0 due to renormalizing $\psi_n = \psi_n^0 + \psi_n'$ is usually insignificant and ignored. What we really find from ψ_n' is how adding it to ψ_n^0 changes the *shape* of the wave function.

To get E_n'', take the scalar product of ψ_n^0 with the λ^2 equation. The left side is zero for the same reason it was zero in getting E_n'; and the term $E_n' \langle \psi_n^0 | \psi_n' \rangle$ is zero because ψ_n^0 is orthogonal to ψ_n'. What remains of the equation gives

$$E_n'' = \langle \psi_n^0 | \hat{H}' | \psi_n' \rangle = \sum_{m \neq n} \frac{H_{mn}'}{E_n^0 - E_m^0} \langle \psi_n^0 | \hat{H}' | \psi_m^0 \rangle = \sum_{m \neq n} \frac{|H_{mn}'|^2}{E_n^0 - E_m^0} ,$$

because $\langle \psi_n^0 | \hat{H}' | \psi_m^0 \rangle = H_{nm}' = (H_{mn}')^*$.

13·4. THE DEGENERATE RECIPE

Now suppose that some or all of the eigenstates of $\hat{H}^0$ are degenerate. Examples: When the spin states of the electron and proton are counted, the hydrogen ground state is four-fold degenerate (Sec. 11·1). Even when the spins are neglected, the hydrogen excited states are n^2-fold degenerate (Sec. 9·2). The problem is to find how a perturbation $\hat{H}'$ shifts and splits N initially degenerate states, all of which start with the same energy E^0.

(a) **The recipe**—Choose an orthonormal set of the N degenerate states that spans the N-dimensional subspace, and calculate the elements $H_{ij}' = \langle \psi_i^0 | \hat{H}' | \psi_j^0 \rangle$ for $i, j = 1, 2, \ldots, N$. Then the eigenvalues of $\hat{H}^0 + \hat{H}'$ are, to first order,

$$E_k = E^0 + E_k' ,$$

where the E_k', $k = 1, 2, \ldots, N$, are the eigenvalues of the N-by-N determinant

$$|\hat{H}' - E'\hat{I}| = \begin{vmatrix} H_{11}' - E' & H_{12}' & \cdots & H_{1N}' \\ H_{21}' & H_{22}' - E' & \cdots & H_{2N}' \\ \cdots & \cdots & \cdots & \cdots \\ H_{N1}' & H_{N2}' & \cdots & H_{NN}' - E' \end{vmatrix} = 0 .$$

The corresponding energy states are the N eigenstates belonging to the N eigenvalues. Often, however, $\hat{H}'$ does not remove all the degeneracy.

In Sec. 1, on nondegenerate states, the H'_{mn} array included *all* the energy states. Here the H'_{ij} array (and the *numbering* of its elements) only includes the N initially degenerate states having energy E^0. Second-order theory would bring in the states having other energies, but here we stop with first-order corrections.

When a state ψ^0 is actually nondegenerate ($N = 1$), the above determinant reduces to

$$|H'_{11} - E'| = 0 \, ,$$

or $E' = H'_{11}$, as in Sec. 1. The first-order correction E' for a *non*degenerate state uses only its ψ^0. The first-order corrections E'_k, $k = 1, 2, \ldots, N$, for *degenerate* states use only the N degenerate ψ_i^0's for that level. First-order degenerate theory sorts out which combinations of the initially degenerate states are eigenstates of $\hat{H}'$. Since any linear superposition of the initially degenerate states is still an eigenstate of $\hat{H}^0$, the eigenstates of $\hat{H}'$ are eigenstates of $\hat{H} = \hat{H}^0 + \hat{H}'$.

(b) **Example: hyperfine splitting**—We have already, in Sec. 11·1, had an example of degenerate theory—the hyperfine interaction of the magnetic moments of the electron and proton in the ground state of hydrogen. There, E^0 was -13.6 eV, and $\hat{H}'_{hf}$ was $A\hat{\boldsymbol{\sigma}}_e \cdot \hat{\boldsymbol{\sigma}}_p$. We chose as the basis the four orthonormal constitutent spin states $|m_1, m_2\rangle = |++\rangle$, $|+-\rangle$, $|-+\rangle$, and $|--\rangle$, and found that

$$\hat{H}'|++\rangle = +A|++\rangle \qquad \hat{H}'|+-\rangle = A\left(2|-+\rangle - |+-\rangle\right)$$
$$\hat{H}'|--\rangle = +A|--\rangle \qquad \hat{H}'|-+\rangle = A\left(2|+-\rangle - |-+\rangle\right) .$$

Those equations were simple enough to solve directly, but the eigenvalue problem as a determinant has columns $\hat{H}'|++\rangle$, $\hat{H}'|+-\rangle$, $\hat{H}'|-+\rangle$, and $\hat{H}'|--\rangle$,

$$|\hat{H}' - E'\hat{I}| = \begin{vmatrix} A - E' & 0 & 0 & 0 \\ 0 & -A - E' & 2A & 0 \\ 0 & 2A & -A - E' & 0 \\ 0 & 0 & 0 & A - E' \end{vmatrix} = 0 \, .$$

The roots are $E' = +A, +A, -3A$, and $+A$ (three of the states are still degenerate). The corresponding eigenstates are the orthonormal *system* spin states, $|S, M\rangle = |1, +1\rangle$, $|1, 0\rangle$, $|0, 0\rangle$, and $|1, -1\rangle$. With these states as the basis, the $|\hat{H}' - E'\hat{I}| = 0$ determinant was

$$\begin{vmatrix} A - E' & 0 & 0 & 0 \\ 0 & A - E' & 0 & 0 \\ 0 & 0 & -3A - E' & 0 \\ 0 & 0 & 0 & A - E' \end{vmatrix} = 0 \, .$$

Now $\hat{H}'$ is diagonal, and its diagonal elements E'_i are the perturbative corrections. It obviously simplifies matters to choose a basis that makes the perturbation matrix H' most nearly diagonal. If we know enough to choose a basis that makes it entirely diagonal, then there is no need to even set up a matrix; the basis states are the eigenstates and the diagonal elements $H'_{ii} = \langle \psi_i^0 | \hat{H}' | \psi_i^0 \rangle$ for $i = 1, 2, \ldots, N$ are the corrections. This looks very like the recipe for the nondegenerate E'_n.

(c) Example: two-fold degeneracy—Two-by-two matrices often occur by themselves or as sub-matrices of larger problems (as in middle part of the above hyperfine matrix in the constituent basis). Whatever the physical context (if any), the *mathematics* is identical to that for spin-1/2 problems. The array of matrix elements of the 2-by-2 $\hat{H}'$ is

$$\begin{pmatrix} H'_{11} & H'_{12} \\ H'_{21} & H'_{22} \end{pmatrix} .$$

Here H'_{11} and H'_{22} are real and $H'_{12} = (H'_{21})^*$, because $\hat{H}'$ is Hermitian. In Sec. 10·1, we found that the eigenvalues of

$$\begin{pmatrix} a + b_z & b_x - ib_y \\ b_x + ib_y & a - b_z \end{pmatrix} ,$$

where a, b_x, b_y, and b_z are all real, are

$$\lambda_\pm = a \pm (b_x^2 + b_y^2 + b_z^2)^{1/2} = a \pm b .$$

Comparing the two arrays, we see that the eigenvalues E'_1 and E'_2 of $\hat{H}'$ are

$$E'_{1,2} = \tfrac{1}{2}\left(H'_{11} + H'_{22}\right) \pm \tfrac{1}{2}\left[\left(H'_{11} - H'_{22}\right)^2 + 4|H'_{12}|^2\right]^{1/2} .$$

It is worth remembering that when $H'_{11} = H'_{22}$, the eigenvalues are simply $E'_{1,2} = H'_{11} \pm |H'_{12}|$.

13·5. TWO SELECTION RULES. A USEFUL RELATION

(a) Two rules—Most of the rest of the chapter is about hydrogen. The calculation of matrix elements between hydrogen states often involves integrals whose angular part is

$$\boxed{ \langle \ell'm'| \cos\theta |\ell m\rangle = \int_{\theta=0}^{\pi} \int_{\phi=0}^{2\pi} (Y_{\ell'}^{m'})^* \cos\theta \, Y_\ell^m \, d\Omega . }$$

The $Y_\ell^m = Y_\ell^m(\theta, \phi)$ are the spherical harmonics, and $d\Omega = \sin\theta \, d\theta \, d\phi$. It is *extremely useful* to know that these integrals are zero unless

$$\boxed{ m' = m \quad (\Delta m = 0) \qquad \text{and} \qquad \ell' = \ell \pm 1 \quad (\Delta\ell = \pm 1) . }$$

The $\cos\theta$ in the middle of $\langle \ell'm'|\cos\theta|\ell m\rangle$ only "connects" spherical harmonics when $m' = m$ and $\ell' = \ell \pm 1$. These same angular integrals also occur in other applications, such as getting decay rates of excited states (Chap. 15). Rules like these that say matrix elements are "zero unless" are called *selection rules*. When, as here, the matrix elements are of x, y, or z between angular-momentum states, they are called *dipole* selection rules (see the next section).

Proof of the $\Delta m = 0$ rule is simple. The ϕ-dependent factor in each $Y_\ell^m(\theta, \phi)$ is $e^{im\phi}$, where m is an integer. Thus the ϕ integrations, for $m' = m$ and for $m' \neq m$, are

$$\int_0^{2\pi} d\phi = 2\pi \qquad \text{and} \qquad \int_0^{2\pi} e^{i(m-m')\phi} d\phi = \left.\frac{e^{i(m-m')\phi}}{i(m-m')}\right|_0^{2\pi} = 0 .$$

For the $\Delta\ell = \pm 1$ rule, we use a *recurrence relation* for the spherical harmonics,

$$\boxed{ \cos\theta \, Y_\ell^m = \sqrt{\frac{(\ell+1)^2 - m^2}{4(\ell+1)^2 - 1}} \, Y_{\ell+1}^m + \sqrt{\frac{\ell^2 - m^2}{4\ell^2 - 1}} \, Y_{\ell-1}^m . }$$

The $\Delta\ell$ and Δm rules for the matrix elements $\langle \ell'm'|\cos\theta|\ell m\rangle$ both follow at once because the spherical harmonics are orthonormal.

(b) Uses—The recurrence relation and the selection rules are very useful!

• The example in the next section involves a four-by-four matrix. The $\Delta m = 0$ and $\Delta \ell = \pm 1$ rules will tell us at once that 14 of the 16 matrix elements are zero.

• The relation gives the *values* of integrals when $\Delta \ell = \pm 1$ and $\Delta m = 0$. For example,

$$\langle 7, -3 | \cos \theta | 8, -3 \rangle = \int_{\theta=0}^{\pi} \int_{\phi=0}^{2\pi} \left(Y_7^{-3} \right)^* \cos \theta \, Y_8^{-3} \, d\Omega = \sqrt{\frac{8^2 - (-3)^2}{4 \times 8^2 - 1}} = \sqrt{\frac{11}{51}} \ .$$

You don't have to *do* the integral—or even to know the formulas for the spherical harmonics.

• With the ladder operators $\hat{L}_\pm$, one can get $Y_\ell^{m\pm 1}$ from Y_ℓ^m (ℓ stays the same). With the recurrence relation, one can get $Y_{\ell+1}^m$ from Y_ℓ^m and $Y_{\ell-1}^m$ (m stays the same). For example,

$$\cos \theta \, Y_0^0 = \frac{1}{\sqrt{3}} Y_1^0, \qquad \text{or} \qquad Y_1^0 = \sqrt{3} Y_0^0 = \sqrt{\frac{3}{4\pi}} \cos \theta \ .$$

Then from Y_0^0 and Y_1^0 one can calculate Y_2^0, and so on. Why does the recurrence relation here produce only one term on the right? (See below.)

(c) Remarks—The spherical harmonics are a complete set for expansions of angular distributions in three dimensions. Therefore any product of spherical harmonics can be written as a *linear* superposition of the harmonics. The simplest nontrivial products are $Y_1^{m'} Y_\ell^m$, and the product $\cos \theta \, Y_\ell^m$ is just $Y_1^0 Y_\ell^m$ without the factor $\sqrt{3/4\pi}$ that normalizes Y_1^0. In

$$\cos \theta \, Y_\ell^m = \sqrt{\frac{(\ell+1)^2 - m^2}{4(\ell+1)^2 - 1}} \, Y_{\ell+1}^m + \sqrt{\frac{\ell^2 - m^2}{4\ell^2 - 1}} \, Y_{\ell-1}^m \ ,$$

the coefficients of $Y_{\ell+1}^m$ and $Y_{\ell-1}^m$ have the same form, with $\ell+1 \to \ell$. The coefficient of $Y_{\ell+1}^m$ is found by matching its leading term with the leading term of $\cos \theta \, Y_\ell^m$. [The leading term in, say, $Y_2^0 = \sqrt{5/16\pi} \, (3\cos^2 \theta - 1)$ is the $\cos^2 \theta$ term.] The factor $\ell^2 - m^2$ in the coefficient of $Y_{\ell-1}^m$ is zero when $m = \pm \ell$; otherwise it would multiply a non-existant $Y_{\ell-1}^{\pm\ell}$. This is like the end-of-ladder zeros of $\hat{a}|0\rangle = 0$ or $\hat{L}_+|\ell, +\ell\rangle = 0$ with oscillator and angular-momentum operators. We omit a derivation of the recurrence relation, but see Prob. 6 for a simpler recurrence for the $Y_\ell^{\pm\ell}$ harmonics.

In applications, the $\cos \theta$ term will usually come from $z = r\cos \theta$ sandwiched between wave functions in a matrix element. In spherical coordinates,

$$x = r\sin \theta \cos \phi, \ y = r\sin \theta \sin \phi, \quad \text{or} \quad x \pm iy = r\sin \theta \, e^{\pm i\phi} \ ,$$

and there is a recurrence relation for, say, $\sin \theta \, e^{+i\phi} \propto Y_1^{+1}$ of the form

$$\sin \theta \, e^{i\phi} Y_\ell^m = F Y_{\ell+1}^{m+1} + G Y_{\ell-1}^{m+1} \ .$$

Thus $\Delta \ell = \pm 1$ for matrix elements of x or y too, but now $\Delta m = \pm 1$.

Problem 8·8 was to show that the parity of a Y_ℓ^m is $(-1)^\ell$. Thus the parities of $\cos \theta$ and $\sin \theta \, e^{i\phi}$ are -1, which makes the parities of the $Y_{\ell\pm 1}$ terms on the right of each recurrence relation the opposite of the parity of the Y_ℓ^m on the left.

There are three-term recurrence relations for *every* set of orthogonal polynomials, and tables of mathematical functions usually list some of them. They are often given in a form like $A Y_{\ell-1}^m + \cos \theta \, Y_\ell^m + B Y_{\ell+1}^m = 0$.

13·6. THE STARK EFFECT IN HYDROGEN (STRONG-FIELD CASE)

(a) The Stark effect for n=2—When $n = 2$, there are four orbital-angular-momentum states of hydrogen, $|\ell m\rangle = |0\,0\rangle$, $|1+1\rangle$, $|1\,0\rangle$, and $|1-1\rangle$. An external electric field $\mathcal{E}_0$ partially splits these states. The electric field does not couple to spin, so the spin states can be ignored here. We shall calculate for the *strong-field* case, in which the splitting due to $\mathcal{E}_0$ is much larger than fine-structure splitting. (Fine structure is the subject of the next section.) Fine structure would then be treated as a second-stage perturbation.

With z taken along the field direction, the potential energy of the electron due to the perturbing force $\mathbf{F} = -e\mathcal{E}_0\hat{\mathbf{z}}$ is

$$V'(z) = -\int_0^z (-e\mathcal{E}_0)\,dz = +e\mathcal{E}_0 z = +e\mathcal{E}_0 r\cos\theta \ .$$

By pushing the proton and electron in opposite directions, the electric field gives the atom an electric dipole moment. The *dipole* matrix elements of $\hat{H}' = \hat{V}'$ for $n = 2$ are

$$e\mathcal{E}_0\,\langle 2\ell'm'|r\cos\theta|2\ell m\rangle = e\mathcal{E}_0\,\langle 2\ell'|r|2\ell\rangle \times \langle\ell'm'|\cos\theta|\ell m\rangle \ .$$

Here $|2\ell m\rangle = \psi_{2\ell m}(r,\theta,\phi)$, $|2\ell\rangle = R_{2\ell}(r)$, and $|\ell m\rangle = Y_\ell^m(\theta,\phi)$. The $n = 2$ wave functions are in Table 1.

Since there are four $|2\ell m\rangle$ states—three with $\ell = 1$ and one with $\ell = 0$—the $\hat{H}'$ matrix will be 4-by-4. Happily, the $\hat{H}'$ array is very sparse—14 of its 16 elements are zero! It looks like this:

$$
\begin{array}{c}
\\
\langle 2\,1+1| \\
\langle 2\,1\,0| \\
\langle 2\,0\,0| \\
\langle 2\,1-1|
\end{array}
\begin{array}{cccc}
|2\,1+1\rangle & |2\,1\,0\rangle & |2\,0\,0\rangle & |2\,1-1\rangle \\
\left(\begin{array}{cccc}
0 & 0 & 0 & 0 \\
0 & 0 & \mathcal{M} & 0 \\
0 & \mathcal{M}^* & 0 & 0 \\
0 & 0 & 0 & 0
\end{array}\right)
\end{array} \ .
$$

- The $\Delta m = 0$ rule makes 10 of the 12 off-diagonal elements zero. Check this!

- The $\Delta\ell = \pm 1$ rule makes the four diagonal elements (and six off-diagonal elements already found to be zero) zero. Check this!

- The remaining two elements are symmetrically placed across the main diagonal and are related by complex conjugation ($\hat{H}'$ is Hermitian). In fact $\mathcal{M}$ is real, so $\mathcal{M}^* = \mathcal{M}$.

Therefore, thanks to the selection rules, there is only one integral to be done—instead of 10! (Why 10 instead of 16?) Arranging the array with the $m=0$ states adjacent, as above, keeps the array more nearly diagonal (why?), and this makes solving $|\hat{H}' - E'\hat{I}| = 0$ simpler. The value of $\mathcal{M}$ is found below. The result is

$$\mathcal{M} = e\mathcal{E}_0\langle 2\,1\,0|r\cos\theta|2\,0\,0\rangle = -3e\mathcal{E}_0 a_0 \ ,$$

where a_0 is the Bohr radius. By now, of course, you can just write down the eigenvalues of the determinant $|\hat{H}' - E'\hat{I}| = 0$. They are,

$$E' = 0,\ 0,\ -3e\mathcal{E}_0 a_0, \text{ and } +3e\mathcal{E}_0 a_0 \ .$$

The $E' = 0$ values go with the $|2\,1\pm 1\rangle$ states, which at least to first order remain degenerate.

245

Table 1 (from Sec. 9·3). The hydrogen wave functions $\psi_{n\ell m}(r,\theta,\phi)$ for $n = 1$, 2, and 3, written here as products of separately normalized radial and angular functions: $\psi_{n\ell m}(r,\theta,\phi) = R_{n\ell}(r) \times Y_\ell^m(\theta,\phi)$:

$$\int_0^\infty |R_{n\ell}(r)|^2\, r^2\, dr = \int_0^\infty |u_{n\ell}(r)|^2\, dr = 1, \qquad \int_{\theta=0}^\pi \int_{\phi=0}^{2\pi} |Y_\ell^m(\theta,\phi)|^2 \sin\theta\, d\theta\, d\phi = 1\ .$$

Factoring the wave functions in this way can simplify calculations. For example, the expectation value of r^2 reduces at once to an integral over just r: $\langle \psi_{n\ell m}|r^2|\psi_{n\ell m}\rangle = \langle R_{n\ell}|r^2|R_{n\ell}\rangle$.

When there is a $\mp$ sign in front of a wave function, it belongs to the $Y_\ell^m(\theta,\phi)$.

$$\psi_{100}(r,\theta,\phi) = R_{10}(r)\, Y_0^0(\theta,\phi) = \frac{2}{\sqrt{a_0^3}}\, e^{-r/a_0} \times \frac{1}{\sqrt{4\pi}}$$

$$\psi_{200}(r,\theta,\phi) = R_{20}(r)\, Y_0^0(\theta,\phi) = \frac{1}{\sqrt{2\,a_0^3}} \left(1 - \frac{1}{2}\frac{r}{a_0}\right) e^{-r/2a_0} \times \frac{1}{\sqrt{4\pi}}$$

$$\psi_{210}(r,\theta,\phi) = R_{21}(r)\, Y_1^0(\theta,\phi) = \frac{1}{\sqrt{24\,a_0^3}} \left(\frac{r}{a_0}\right) e^{-r/2a_0} \times \sqrt{\frac{3}{4\pi}}\,\cos\theta$$

$$\psi_{21\pm1}(r,\theta,\phi) = R_{21}(r)\, Y_1^{\pm1}(\theta,\phi) = \mp\frac{1}{\sqrt{24\,a_0^3}} \left(\frac{r}{a_0}\right) e^{-r/2a_0} \times \sqrt{\frac{3}{8\pi}}\,\sin\theta\, e^{\pm i\phi}$$

$$\psi_{300}(r,\theta,\phi) = R_{30}(r)\, Y_0^0(\theta,\phi) = \frac{2}{\sqrt{27\,a_0^3}} \left[1 - \frac{2}{3}\frac{r}{a_0} + \frac{2}{27}\left(\frac{r}{a_0}\right)^2\right] e^{-r/3a_0} \times \frac{1}{\sqrt{4\pi}}$$

$$\psi_{310}(r,\theta,\phi) = R_{31}(r)\, Y_1^0(\theta,\phi) = \frac{8}{27\sqrt{6\,a_0^3}} \left(1 - \frac{1}{6}\frac{r}{a_0}\right)\left(\frac{r}{a_0}\right) e^{-r/3a_0} \times \sqrt{\frac{3}{4\pi}}\,\cos\theta$$

$$\psi_{31\pm1}(r,\theta,\phi) = R_{31}(r)\, Y_1^{\pm1}(\theta,\phi) = \mp\frac{8}{27\sqrt{6\,a_0^3}} \left(1 - \frac{1}{6}\frac{r}{a_0}\right)\left(\frac{r}{a_0}\right) e^{-r/3a_0} \times \sqrt{\frac{3}{8\pi}}\,\sin\theta\, e^{\pm i\phi}$$

$$\psi_{320}(r,\theta,\phi) = R_{32}(r)\, Y_2^0(\theta,\phi) = \frac{4}{81\sqrt{30\,a_0^3}} \left(\frac{r}{a_0}\right)^2 e^{-r/3a_0} \times \sqrt{\frac{5}{16\pi}}\,(3\cos^2\theta - 1)$$

$$\psi_{32\pm1}(r,\theta,\phi) = R_{32}(r)\, Y_2^{\pm1}(\theta,\phi) = \mp\frac{4}{81\sqrt{30\,a_0^3}} \left(\frac{r}{a_0}\right)^2 e^{-r/3a_0} \times \sqrt{\frac{15}{8\pi}}\,\sin\theta\,\cos\theta\, e^{\pm i\phi}$$

$$\psi_{32\pm2}(r,\theta,\phi) = R_{32}(r)\, Y_2^{\pm2}(\theta,\phi) = \frac{4}{81\sqrt{30\,a_0^3}} \left(\frac{r}{a_0}\right)^2 e^{-r/3a_0} \times \sqrt{\frac{15}{32\pi}}\,\sin^2\theta\, e^{\pm 2i\phi}$$

For all $n \geq 1$, $\quad R_{n,n-1}(r) = \dfrac{(2/n)^{n+1/2}}{\sqrt{(2n)!\,a_0^3}} \left(\dfrac{r}{a_0}\right)^{n-1} e^{-r/na_0}$

For all $\ell \geq 0$, $\quad Y_\ell^{+\ell}(\theta,\phi) = (-1)^\ell \sqrt{\dfrac{2\ell+1}{4\pi\,(2\ell)!}}\,(2\ell-1)!!\,\sin^\ell\theta\, c^{+i\ell\phi}$

246

The $E' = \pm 3e\mathcal{E}_0 a_0$ values go with equal mixes of the two $m = 0$ states,

$$\frac{1}{\sqrt{2}}\big(|2\,1\,0\rangle - |2\,0\,0\rangle\big) \quad \text{for} \quad E' = +3e\mathcal{E}_0 a_0 \ ,$$

$$\frac{1}{\sqrt{2}}\big(|2\,1\,0\rangle + |2\,0\,0\rangle\big) \quad \text{for} \quad E' = -3e\mathcal{E}_0 a_0 \ .$$

Figure 5 shows the splitting.

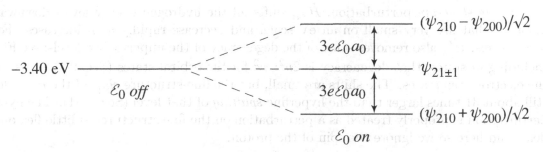

Figure 5. The splitting of the $n = 2$ level of hydrogen by an external electric field $\mathcal{E}_0\hat{\mathbf{z}}$. This is the strong-field case, in which fine structure is ignored.

(b) The matrix element—The element $\mathcal{M} = \langle 2\,1\,0|r\cos\theta|2\,0\,0\rangle$ factors into radial and angular integrals: $\langle 2\,1\,0|r\cos\theta|2\,0\,0\rangle = \langle 2\,1|r|2\,0\rangle \times \langle 1\,0|\cos\theta|0\,0\rangle$. The radial functions are in Table 1, from which we get

$$\langle 2\,1|r|2\,0\rangle = \int_0^\infty \frac{1}{\sqrt{24\,a_0^3}}\left(\frac{r}{a_0}\right)e^{-r/2a_0}\frac{1}{\sqrt{2\,a_0^3}}\left(1 - \frac{1}{2}\frac{r}{a_0}\right)e^{-r/2a_0}\,r^3\,dr \ .$$

With $s \equiv r/a_0$ and some rearrangement, this becomes

$$\frac{a_0}{\sqrt{48}}\int_0^\infty \left(s^4 - \frac{1}{2}s^5\right)e^{-s}\,ds = \frac{a_0}{4\sqrt{3}}\left(4! - \frac{1}{2}5!\right) = -3\sqrt{3}\,a_0 \ ,$$

where we have used

$$\int_0^\infty s^n e^{-s}\,ds = n! \ .$$

The recurrence relation for the spherical harmonics (Sec. 5) gives the angular integral,

$$\langle 1\,0|\cos\theta|0\,0\rangle = \sqrt{\frac{(0+1)^2 - 0^2}{4(0+1)^2 - 1}} \times \langle 1\,0|1\,0\rangle = \frac{1}{\sqrt{3}} \ .$$

The product of the radial and angular integrals is $-3a_0$, and so

$$\mathcal{M} = e\mathcal{E}_0\langle 2\,1\,0|r\cos\theta|2\,0\,0\rangle = -3e\mathcal{E}_0 a_0 \ ,$$

as claimed in (a).

13·7. HYDROGEN FINE STRUCTURE: EXPERIMENT

The full specification of the state of a hydrogen atom requires not just the spatial quantum numbers n, ℓ, and m_ℓ, but also the spin $s = 1/2$ and m_s and the total angular momentum j. When $\ell = 0$, $j = s = 1/2$; when $\ell > 0$, $j = \ell \pm 1/2$. The state of a hydrogen atom is then given by $n^{2s+1}\ell_j = n^2\ell_j$. The ground state is $1\,^2s_{1/2}$ ("one-doublet-ess-one-half"), the $n = 2$ states are $2\,^2s_{1/2}$, $2\,^2p_{1/2}$, and $2\,^2p_{3/2}$. The degenerate m_ℓ and m_s states are usually not specified until particular matrix elements are needed. For all but hydrogen, the angular momenta are (vector) sums over more than one electron, and the labelling will be $n\,^{2S+1}L_J$.

A *fine-structure* perturbation, $\hat{H}'_{fs}$, shifts all the hydrogen energy levels downward by amounts that are very small on an eV scale, and decrease rapidly as n increases. For the $n > 1$ levels, $\hat{H}'_{fs}$ also removes some of the degeneracy of the unperturbed levels; see Fig. 9·2. Including m states, that degeneracy is $2n^2$: n^2 for the orbital states (Sec. 9·4) times two for the electron spin states. The shifts are small, but the fine-structure *shift* of the $n = 1$ level is still about 31 times larger than the hyperfine *splitting* of that level (Sec. 11·1). The hyperfine perturbation is properly treated as a perturbation on the fine structure—a little flea on a big flea—and here so we ignore the spin of the proton.

(a) Splitting of energy levels—The calculation of fine structure is complicated, and it will be helpful to see the results before we derive them. The unperturbed energies are $E_n^0 = E_1^0/n^2$ (Sec. 9·2), where

$$E_1^0 = -\frac{mk^2e^4}{2\hbar^2} = -\frac{1}{2}mc^2\left(\frac{ke^2}{\hbar c}\right)^2 = -\frac{1}{2}\alpha^2 mc^2\,,$$

and $\alpha \equiv ke^2/\hbar c \simeq 1/137$ is the fine-structure constant. The first-order fine-structure *corrections* to the unperturbed energy levels are

$$\boxed{(E'_{fs})_{nj} = \frac{1}{8}\alpha^4 mc^2 \times \frac{1}{n^4}\left(3 - \frac{4n}{j + 1/2}\right).}$$

Whereas the unperturbed energies, E_n^0, are proportional to $\alpha^2 = 5.325 \times 10^{-5}$, the fine-structure corrections, E'_{fs}, are proportional to $\alpha^4 = 2.836 \times 10^{-9}$, a factor of α^2 smaller. The corrections are functions of n and the total angular-momentum quantum number j, but not (explicitly) of the orbital quantum number ℓ.

The fine-structure corrections for the $n = 1$, 2, and 3 levels, in units of $\frac{1}{8}\alpha^4 mc^2 = 1.81 \times 10^{-4}$ eV, which is the magnitude of the shift for the ground state, are

n	j	ℓ	$\frac{1}{n^4}\left(3 - \frac{4n}{j+1/2}\right)$
1	1/2	s	-1
2	1/2	s, p	$-5/16$
	3/2	p	$-1/16$
3	1/2	s, p	$-1/9$
	3/2	p, d	$-1/27$
	5/2	d	$-1/81$

Figure 6 shows this structure. The shifts, all downward, decrease rapidly with n because of the factor $1/n^4$ in E'_{fs}. The number of *different* energies for a given n is equal to the number of different j values, which equals n.

248

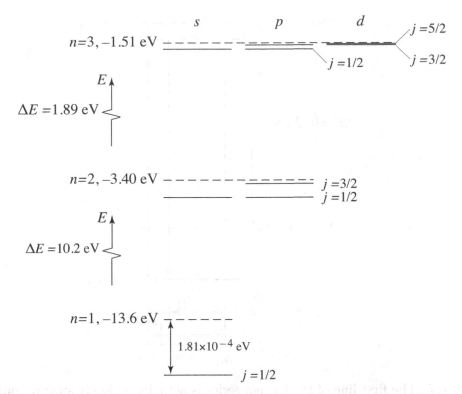

Figure 6. Hydrogen fine structure for the $n = 1$, 2, and 3 levels. The splittings from the unperturbed energy levels (the dashed lines) are all drawn to the same scale, but are greatly magnified compared to the separations between E_1^0, E_2^0, and E_3^0.

(b) **Splitting of spectral lines**—The shifts and splittings of fine structure are seen by measuring shifts and splittings of spectral lines. The transitions between levels involve dipole matrix elements of x, y, and z, and so are (to a high degree) forbidden unless $\Delta \ell = \pm 1$. In Fig. 6 (and Fig. 9·2), the levels with adjacent values of ℓ are placed in adjacent columns. Then transitions only occur between adjacent columns: $s \leftrightarrow p$ and $p \leftrightarrow d$ but not $s \leftrightarrow d$, and so on. We learn how to calculate transition rates from the matrix elements in Chap. 15.

The fine structure splitting of ℓ states with $\ell > 0$ into j states requires another selection rule, one for transitions between j states. The transitions involve the emission or absorption of a photon. The spin J and parity P of a photon are $J^P = 1^-$, and the rule is $\Delta j = 0$ or ± 1. For example, both $2\,{}^2p_{1/2} \to 1\,{}^2s_{1/2}$ and $2\,{}^2p_{3/2} \to 1\,{}^2s_{1/2}$ transitions are allowed.

The *Lyman series* of frequencies involves transitions to or from the ground state, the $np \leftrightarrow 1s$ transitions. Figure 7 shows the $2p \to 1s$ transitions—two frequencies because there are two energy differences, for $2\,{}^2p_{1/2} \to 1\,{}^2s_{1/2}$ and $2\,{}^2p_{3/2} \to 1\,{}^2s_{1/2}$. The frequencies are slightly higher than what the frequency would be in the absence of fine-structure (why?). The shifts $\Delta \nu$ from the unperturbed frequency ν_0 are in the ratio 11-to-15 (show this). Other lines of the Lyman series, involving p states with $n > 2$, are also doublets, but the splitting of the p states decreases rapidly with n and therefore so also does the frequency splitting.

A figure like Fig. 7 for the $n = 3$ to $n = 2$ transitions would be more complicated. For $n = 2$, there are ${}^2s_{1/2}$, ${}^2p_{1/2}$, and ${}^2p_{3/2}$ levels; for $n = 3$, there are ${}^2s_{1/2}$, ${}^2p_{1/2}$, ${}^2p_{3/2}$, ${}^2d_{3/2}$, and ${}^2d_{5/2}$ levels. Altogether, there are seven allowed $n = 3$ to $n = 2$ transitions, but only five distinct frequencies. See the problems.

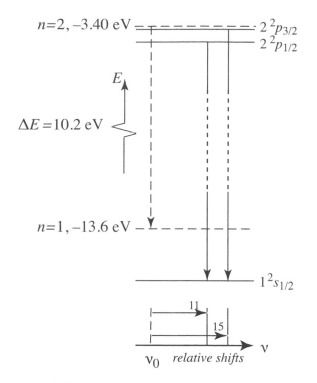

Figure 7. The first line of the Lyman series is actually a closely spaced doublet.

13·8. HYDROGEN FINE STRUCTURE: THEORY

Now to calculate. Two perturbations contribute to the fine structure, one relativistic, the other a spin-orbit interaction (which is really another relativistic effect). The two effects fall out, indivisibly and without approximation, from the fully relativistic Dirac equation for hydrogen. But here we approximate the two perturbations in turn and add the results. For the relativistic perturbation, we shall need the expectation values of $V(r) = -ke^2/r$ and $V^2(r) = k^2e^4/r^2$ in the unperturbed states $\psi_{n\ell m}^0$ of hydrogen, so we find them first. For this, we follow a guided problem at the end of Chap. 17 in R. Shankar, *Principles of Quantum Mechanics*. It is essentially an in-detail exploration of the *Feynman-Hellmann theorem*. The method is of some interest, so we take our time.

(a) **Expectation values of V and V^2**—Section 3 was about a Hamiltonian $\hat{H} = \hat{H}^0 + \lambda\hat{H}'$, where $\hat{H}^0$ has known (nondegenerate) eigenvalues E_n^0, and $\hat{H}'$ is a perturbation. We wrote

$$E_n = E_n^0 + \lambda E_n' + \lambda^2 E_n'' + \cdots,$$

where λ is a mathematical "knob" we can turn from 0 to 1 to turn $\hat{H}'$ from off to fully on. In particular, the recipe for E_n' was $\langle\psi_n^0|\hat{H}'|\psi_n^0\rangle$. Now if we can find the *exact* eigenvalues E_n for $\hat{H} = \hat{H}^0 + \lambda\hat{H}'$ as a function of λ, then

$$E_n^0 = (E_n)_{\lambda=0}, \qquad E_n' = \left(\frac{\partial E_n}{\partial \lambda}\right)_{\lambda=0}, \qquad E_n'' = \frac{1}{2}\left(\frac{\partial^2 E_n}{\partial \lambda^2}\right)_{\lambda=0},$$

and so on. Knowing the *exact* eigenvalues E_n, we can get its *pieces*, E_n', E_n'', and so on.

250

For example, let us nudge the hydrogen potential energy $V(r) = -ke^2/r$ with a perturbation $\lambda V'(r) = \lambda(A/r)$, where $A \ll ke^2$ (but $\dim A = \dim ke^2$). The total potential energy becomes $-(ke^2 - \lambda A)/r$. The unperturbed energies are $E^0_{n\ell m} = E^0_n = -mk^2e^4/2n^2\hbar^2$. So, replacing ke^2 here with $ke^2 - \lambda A$, we get the exact new energies,

$$E_{n\ell m} = -\left(k^2e^4 - 2\lambda Ake^2 + \lambda^2 A^2\right)\frac{m}{2n^2\hbar^2} = E^0_{n\ell m} + \lambda E'_{n\ell m} + \lambda^2 E''_{n\ell m} \; .$$

But then the coefficient of the λ term is

$$E'_{n\ell m} = \langle \psi^0_{n\ell m}|V'(r)|\psi^0_{n\ell m}\rangle = \left\langle \frac{A}{r} \right\rangle_n = \frac{A}{n^2}\frac{mke^2}{\hbar^2}, \qquad \text{and} \qquad \left\langle \frac{1}{r} \right\rangle_n = \frac{1}{n^2}\frac{mke^2}{\hbar^2} \; .$$

This depends on n, but not on ℓ or m. Finally, multiplying the last result by $-ke^2$, we get the desired

$$\langle \psi^0_{n\ell m}|V(r)|\psi^0_{n\ell m}\rangle = -\left\langle \frac{ke^2}{r} \right\rangle_n = -\frac{1}{n^2}\frac{mk^2e^4}{\hbar^2} = 2E^0_n \; .$$

In the ground state of hydrogen, $E = -13.6$ eV and $\langle V \rangle = -27.2$ eV; and then $\langle K \rangle = +13.6$ eV. Problem 6 in Chap. 9 was to get this same result using the virial theorem.

Next we nudge $\hat{H}^0$ with a $1/r^2$ term $\lambda V'(r) = \lambda(C/r^2)$. This in effect changes the orbital-angular-momentum term, just as the A/r perturbation did the $V(r) = ke^2/r$ term:

$$\frac{\ell(\ell+1)\hbar^2}{2mr^2} \longrightarrow \frac{\ell(\ell+1)\hbar^2}{2mr^2} + \frac{\lambda C}{r^2} = \frac{\ell'(\ell'+1)\hbar^2}{2mr^2} \; ,$$

where $\ell'(\ell'+1) = \ell(\ell+1) + 2m\lambda C/\hbar^2$. In Sec. 9·2(f), we found that the principle quantum n and the radial and orbital numbers n_r and ℓ are related by $n = n_r + \ell + 1$. That is,

$$E^0_{n\ell m} = -\frac{mk^2e^4}{2n^2\hbar^2} = -\frac{mk^2e^4}{2(n_r + \ell + 1)^2\hbar^2} \; .$$

Then the exact new energies $E_{n\ell'm}$ are given by replacing ℓ in this equation with ℓ'. The series for $E_{n\ell m}$ in powers of λ is not laid out here simply, as it was for $V' = A/r$. However, using the chain rule for chain dependences, we get

$$E'_{n\ell m} = \left(\frac{\partial E_{n\ell'm}}{\partial \lambda}\right)_{\lambda=0} = \left(\frac{\partial E_{n\ell'm}}{\partial \ell'}\right)_{\ell'=\ell} \times \left(\frac{\partial \ell'}{\partial \lambda}\right)_{\lambda=0} = \frac{mk^2e^4}{n^3\hbar^2} \times \frac{2mC}{(2\ell+1)\hbar^2} \; .$$

But $E'_{n\ell m}$ is $\langle \psi^0_{n\ell m}|V'(r)|\psi^0_{n\ell m}\rangle = \langle C/r^2 \rangle_{n\ell m}$, so replacing C with k^2e^4, we get

$$\langle \psi^0_{n\ell m}|V^2(r)|\psi^0_{n\ell m}\rangle = \left\langle \frac{k^2e^4}{r^2} \right\rangle_{n\ell} = \frac{4n}{(\ell+1/2)}\left(\frac{mk^2e^4}{2n^2\hbar^2}\right)^2 = \frac{4n}{(\ell+1/2)}(E^0_n)^2 \; .$$

Dividing this by k^2e^4, we get

$$\left\langle \frac{1}{r^2} \right\rangle_{n\ell} = \frac{4n}{(\ell+1/2)}\left(\frac{mke^2}{2n^2\hbar^2}\right)^2 = \frac{1}{(\ell+1/2)\,n^3a_0^2} \; .$$

This result depends on ℓ as well as n. We need $\langle V^2 \rangle_{n\ell}$ in (b) below, and $\langle 1/r^2 \rangle_{n\ell}$ in (c).

251

(b) The relativistic perturbation—In Sec. 2(c), we found that the approximate relativistic perturbation Hamiltonian is

$$\hat{H}'_{rel} \approx -\frac{1}{2mc^2}\left[(\hat{H}^0)^2 - \hat{H}^0\hat{V} - \hat{V}\hat{H}^0 + \hat{V}^2\right] .$$

Since $\hat{H}'_{rel}$ is independent of the spin of the electron, the shifts will be the same for $j = \ell + 1/2$ and $j = \ell - 1/2$. Since the number of $Y_\ell^m(\theta, \phi)$ states for a given value of n is n^2, degenerate theory calls for an n-by-n array of elements $\langle\psi^0_{n\ell'm'}|\hat{H}'_{rel}|\psi^0_{n\ell m}\rangle$ for each n. However, $\hat{H}^0$ operating on $\psi^0_{n\ell m}$ simply multiplies it by E^0_n; and $\hat{V} = V(r)$ is independent of θ and ϕ. Therefore nothing in $\hat{H}'_{rel}$ "spoils" the angular integrands, and the orthogonality of the spherical harmonics makes all the off-diagonal elements zero. Then the diagonal elements $\langle\psi^0_{n\ell m}|\hat{H}'_{rel}|\psi^0_{n\ell m}\rangle$ are the first-order perturbation energies directly, just as in nondegenerate theory.

Now all we have to do is read off the three terms in $\hat{H}'_{rel}$. Aside from the factor $-1/2mc^2$, the terms are

$$\langle\psi^0_{n\ell m}|(\hat{H}^0)^2|\psi^0_{n\ell m}\rangle = (E^0_n)^2$$

$$-\langle\psi^0_{n\ell m}|(\hat{H}^0\hat{V} + \hat{V}\hat{H}^0)|\psi^0_{n\ell m}\rangle = -2E^0_n\langle\psi^0_{n\ell m}|\hat{V}|\psi^0_{n\ell m}\rangle = -4(E^0_n)^2$$

$$\langle\psi^0_{n\ell m}|\hat{V}^2|\psi^0_{n\ell m}\rangle = \frac{4n}{\ell + 1/2}(E^0_n)^2 .$$

The sum multiplied by $-1/2mc^2$ is

$$(E'_{rel})_{n\ell} = \langle\psi^0_{n\ell m}|\hat{H}'_{rel}|\psi^0_{n\ell m}\rangle = \frac{(E^0_n)^2}{2mc^2}\left(3 - \frac{4n}{\ell + 1/2}\right).$$

Using $E^0_n = E^0_1/n^2$ and $E^0_1 = -\frac{1}{2}\alpha^2 mc^2$, where $\alpha = ke^2/\hbar c$, we rewrite this as

$$(E'_{rel})_{n\ell} = \frac{\alpha^4 mc^2}{8n^4}\left(3 - \frac{4n}{\ell + 1/2}\right).$$

This finishes the first (relativistic) part of the calculation of fine structure. The second (spin-orbit) part is also complicated, but adding the two parts will just turn that ℓ in this last equation into a j.

(c) The expectation value of $1/r^3$—For the spin-orbit perturbation, we shall need the expectation value of $1/r^3$. But with no term in $\hat{H}^0$ proportional to $1/r^3$ to nudge, we need a new idea. (I again follow Shankar's sketched solution.) The radial part of the kinetic-energy term in the three-dimensional $\hat{H}^0\psi$ is, from Sec. 8·1,

$$-\frac{\hbar^2}{2m}\frac{1}{r}\frac{\partial^2(r\psi)}{\partial r^2} = \frac{\hat{p}_r^2}{2m}\psi ,$$

where $\hat{p}_r$ is the radial-momentum operator. But then $\hat{p}_r$ cannot simply be $-i\hbar\partial/\partial r$, like $\hat{p}_x$ is $-i\hbar\partial/\partial x$; if it were, the $\hat{p}_r^2$ operator would be $-\hbar^2(\partial^2/\partial r^2)$. Spherical coordinates are not rectangular coordinates, and the range of r is 0 to ∞, not $-\infty$ to $+\infty$. The proper radial-momentum operator is

$$\hat{p}_r = -i\hbar\left(\frac{\partial}{\partial r} + \frac{1}{r}\right).$$

This gives the proper radial term in the Hamiltonian (see the problems).

Now at the end of Sec. 5·7(a), we found that $\langle\psi_n|[\hat{H},\hat{A}]|\psi_n\rangle = 0$ if ψ_n is an eigenstate of $\hat{H}$, even when $[\hat{H},\hat{A}]$ is not zero. With $\hat{A} = \hat{p}_r$, we have

$$[\hat{H},\hat{p}_r] = \frac{1}{2m}[\hat{p}_r^2,\hat{p}_r] + \frac{\ell(\ell+1)\hbar^2}{2m}[1/r^2,\hat{p}_r] + [\hat{V}(r),\hat{p}_r] \ .$$

The first term is zero (why?). Evaluation of $\langle\psi_{n\ell m}^0|[\hat{H},\hat{p}_r]|\psi_{n\ell m}^0\rangle = 0$ (again, a problem) gives a relation between $\langle 1/r^2\rangle_{n\ell}$ and $\langle 1/r^3\rangle_{n\ell}$. We know $\langle 1/r^2\rangle_{n\ell}$ from the end of part (a), so the relation gives $\langle 1/r^3\rangle_{n\ell}$. That relation is

$$\left\langle\frac{1}{r^3}\right\rangle_{n\ell} = \frac{1}{\ell(\ell+1)\,a_0}\left\langle\frac{1}{r^2}\right\rangle_{n\ell} = \frac{1}{\ell(\ell+1)(\ell+1/2)\,n^3 a_0^3} \ .$$

This is infinite when $\ell = 0$ (but see below).

(d) The spin-orbit perturbation—Figure 8(a) shows the electron circling the proton. (Once again, we start with a classical picture to get a quantum operator.) The circling frequency is $f = v/2\pi r$, the effective current of the loop is $I = -ef$, and the current creates a magnetic field $B = \mu_0 I/2r$ at the center of the loop (see an introductory text). Here $\mu_0 = 1/\epsilon_0 c^2$ is the permeability constant. Altogether, we get

$$B = \frac{-ev}{4\pi\epsilon_0 c^2 r^2} = \frac{-ke}{mc^2 r^3} L \ ,$$

where $L = mvr$ is the orbital angular momentum of the electron. As the figure shows, **B** and **L** are antiparallel.

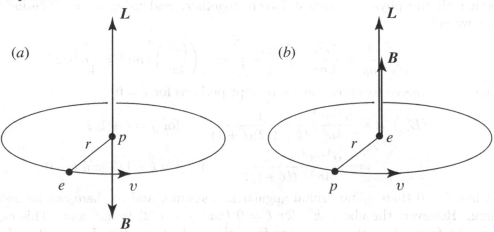

Figure 8. (a) A circling electron makes a magnetic field at the proton. (b) As seen by the electron, the proton is circling, making a magnetic field at the electron.

Figure 8(b) shows the motion of the proton from the point of view of the electron. An "obvious" reciprocity makes the magnetic field the electron sees in (b) the same as the field the proton sees in (a), except now **B** and **L** are parallel (why?). The interaction of the magnetic moment $\boldsymbol{\mu}_e = -\mu_B\boldsymbol{\sigma} = -(e/m)\mathbf{s}$ of the electron with **B** is

$$E'_{so} = -\boldsymbol{\mu}_e\cdot\mathbf{B} = +\frac{e}{m}\,\mathbf{s}\cdot\mathbf{B} = +\frac{ke^2}{m^2 c^2 r^3}\,\mathbf{s}\cdot\mathbf{L} \ .$$

This, however, misses a remarkable relativistic effect discovered in 1926 by L.H. Thomas. Even though the electron is quite non-relativistic ($v/c \le 1/137$), relativity puts a factor 2 in the denominator. How is that possible? It surprised even Einstein.

The derivation, beyond us here, is given in advanced texts on *classical* mechanics and electromagnetism. The subtle derivation follows the circling electron through a continuous sequence of inertial frames, each making an infinitesimal angle with respect to the previous frame. Here we must simply take the factor as given, just as we have the g-factor of 2 for the magnetic moment of the electron.

Using $\mathbf{J} \equiv \mathbf{L} + \mathbf{s}$ and the $\mathbf{j}_1 \cdot \mathbf{j}_2$ identity near the end of Sec. 11·3, we get

$$\hat{H}'_{so} = +\frac{ke^2}{4m^2c^2r^3}\left(\hat{J}^2 - \hat{L}^2 - \hat{s}^2\right).$$

The 4 downstairs is half Thomas, half from the $\mathbf{s} \cdot \mathbf{L}$ identity. The perturbation matrix is diagonal if we use as the basis the system states $|n\ell s; jm\rangle$. We get

$$\langle n\ell s; jm|\hat{H}'_{so}|n\ell s; jm\rangle = \frac{ke^2\hbar^2}{4m^2c^2}\left[j(j+1) - \ell(\ell+1) - s(s+1)\right]\left\langle\frac{1}{r^3}\right\rangle_{n\ell},$$

For hydrogen, j is either $\ell + 1/2$ or (when $\ell > 0$) $\ell - 1/2$. We eliminate j in favor of ℓ,

$$j(j+1) - \ell(\ell+1) - s(s+1) = \begin{cases} +\ell & \text{for } j = \ell + 1/2 \\ -(\ell+1) & \text{for } j = \ell - 1/2. \end{cases}$$

(Check this.) From part (c),

$$\left\langle\frac{1}{r^3}\right\rangle_{n\ell} = \frac{1}{\ell(\ell+1/2)(\ell+1)n^3a_0^3}.$$

Collecting all the physical-constant factors together, and using $a_0 = \hbar^2/mke^2$ and $\alpha = ke^2/\hbar c$, we get

$$\frac{ke^2\hbar^2}{4m^2c^2}\frac{1}{a_0^3} = \frac{ke^2\hbar^2}{4m^2c^2}\left(\frac{mke^2}{\hbar^2}\right)^3 = \frac{1}{4}\left(\frac{ke^2}{\hbar c}\right)^4 mc^2 = \frac{1}{4}\alpha^4mc^2.$$

Putting all the pieces together, we get (except perhaps for $\ell = 0$)

$$(E'_{so})_{n\ell j} = \frac{\alpha^4mc^2}{4n^3}\frac{1}{(\ell+1/2)(\ell+1)} \qquad \text{for } j = \ell + 1/2$$

$$= -\frac{\alpha^4mc^2}{4n^3}\frac{1}{\ell(\ell+1/2)} \qquad \text{for } j = \ell - 1/2 \text{ and } \ell > 0.$$

Now when $\ell = 0$ there *is* no orbital angular momentum, and so there can be no spin-orbit coupling. However, the above E'_{so} for $\ell = 0$ (and $j = 1/2$) is *not* zero. This came about because the factor ℓ in the numerator from the evaluation of $\mathbf{s} \cdot \mathbf{L}$ cancelled the ℓ in the denominator of $\langle 1/r^3\rangle$. Surprisingly, our nonzero E'_{so} for $\ell = 0$, when added to E'_{rel}, gives exactly the same α^4 term as the fully relativistic Dirac equation does.

(e) The sum: fine structure—The sum of E'_{rel} and E'_{so} is the fine-structure shift E'_{fs}. With a little algebra, for $j = \ell + 1/2$ we get,

$$(E'_{fs})_{n\ell} = \frac{\alpha^4mc^2}{8n^4}\left(3 - \frac{4n}{\ell+1/2} + \frac{2n}{(\ell+1/2)(\ell+1)}\right),$$

or

$$\boxed{(E'_{fs})_{nj} = \frac{1}{8}\alpha^4mc^2 \times \frac{1}{n^4}\left(3 - \frac{4n}{j+1/2}\right).}$$

For $j = \ell - 1/2$, we get the same thing—the shifts can be written in terms of just n and j.

13·9. ATOMIC MAGNETIC MOMENTS

In the following, it will be important to distinguish between:

(1) the quantum numbers J, L, and S;

(2) the operators $\hat{\mathbf{J}}$, $\hat{\mathbf{L}}$, and $\hat{\mathbf{S}}$, where $\hat{\mathbf{J}} \cdot \hat{\mathbf{J}} = \hat{J}^2$ and $\hat{J}^2|J,M\rangle = J(J+1)\hbar^2|J,M\rangle$, etc;

(3) the vector representations of $\mathbf{J}$, $\mathbf{L}$, and $\mathbf{S}$, with lengths $\sqrt{J(J+1)}\,\hbar$, etc.

We use capital letters for the quantum numbers because the results will apply for multi-electron atoms as well as for hydrogen. In the next chapter, we find how to get the values of L, S, and J for the ground state of nearly any element in about a minute.

(a) Atomic magnetic moments—In Sec. 10·3(a), we introduced the operators for the magnetic moments associated with orbital and spin angular momenta and for the interactions of those moments with magnetic fields. Let $\hat{\mathbf{L}}$ and $\hat{\mathbf{S}}$ be the operators for the summed orbital and spin angular momenta of the electrons in an atom. The magnetic-moment operators are

$$\hat{\boldsymbol{\mu}}_L = -\frac{e}{2m}\hat{\mathbf{L}} \quad \text{and} \quad \hat{\boldsymbol{\mu}}_S = -2\frac{e}{2m}\hat{\mathbf{S}}\,.$$

Note again that extra factor 2 with $\hat{\boldsymbol{\mu}}_S$. The total magnetic-moment operator is

$$\hat{\boldsymbol{\mu}} = \hat{\boldsymbol{\mu}}_L + \hat{\boldsymbol{\mu}}_S = -\frac{e}{2m}\left(\hat{\mathbf{L}} + 2\hat{\mathbf{S}}\right) = -\frac{\mu_B}{\hbar}\left(\hat{\mathbf{L}} + 2\hat{\mathbf{S}}\right) = -\frac{\mu_B}{\hbar}\left(\hat{\mathbf{J}} + \hat{\mathbf{S}}\right)\,,$$

where $\mu_B \equiv e\hbar/2m = 5.788\times10^{-5}$ eV/T is the Bohr magneton, and $\hat{\mathbf{J}} = \hat{\mathbf{L}} + \hat{\mathbf{S}}$ is the operator for the total angular momentum. The Hamiltonian for the interaction of $\boldsymbol{\mu}$ with $\mathbf{B} = B\hat{\mathbf{z}}$ is

$$\boxed{\hat{H}'_{Zee} = -\hat{\boldsymbol{\mu}} \cdot \mathbf{B} = +\frac{\mu_B}{\hbar}\left(\hat{\mathbf{L}} + 2\hat{\mathbf{S}}\right) \cdot \mathbf{B} = +\frac{\mu_B}{\hbar}\left(\hat{\mathbf{J}} + \hat{\mathbf{S}}\right) \cdot \mathbf{B}\,.}$$

By μ we shall mean the projection of $\boldsymbol{\mu}$ along $\mathbf{B} = B\hat{\mathbf{z}}$ in the state $|J,+J\rangle$:

$$\mu = -\langle J,+J|(M_L + 2M_S)|J,+J\rangle\,\mu_B = -\langle J,+J|(J + M_S)|J,+J\rangle\,\mu_B\,.$$

Without that factor 2 in $\hat{\boldsymbol{\mu}}_S$, μ would simply be $-J\mu_B$.

For given L and S, J can take on any of the values from $L+S$ down (in unit steps) to $|L-S|$ (Sec. 11·4). Now suppose that J is as big as it can be ($J = L+S$), and that $M_J = +J$. Then M_L is $+L$, M_S is $+S$, and the component of $\mathbf{L} + 2\mathbf{S}$ along $\mathbf{B} = B\hat{\mathbf{z}}$ is $(L+2S)\hbar$. In this case, μ is simply

$$\boxed{\mu = -(L+2S)\mu_B = -(J+S)\mu_B \qquad \text{(only when } J = L+S)\,.}$$

Cases (when $J = L + S$):

- If $S = 0$, then $\mu = -L\mu_B$.
- If $L = 0$, then $\mu = -2S\mu_B$ ($= -\mu_B$ for a single electron).
- If $L = 2$, $S = 2$, and $J = L+S = 4$ (the quantum numbers of the ground state of iron), then $\mu = -(2 + 2\times2)\mu_B = -6\mu_B$.
- If $S = 1/2$ and $J = L+S = L+1/2$ (as for the $^2S_{1/2}$, $^2P_{3/2}$, $^2D_{5/2}$, ... states of hydrogen), then $\mu = -(L+1)\,\mu_B = -\mu_B, -2\mu_B, -3\mu_B, \ldots$, for $L = 0, 1, 2, \ldots$

255

Getting μ when $J < L + S$ is more complicated, because then $|J, +J\rangle$ states are a mix of $|M_L, M_S\rangle$ states. For example, in Sec. 11·4(b) we found that when $L = 1$, $S = 1/2$, and $J = 1/2$ (a $^2P_{1/2}$ state), the system $|J, +J\rangle = |\frac{1}{2}, +\frac{1}{2}\rangle$ state in terms of constituent $|M_L, M_S\rangle$ states is

$$|J, +J\rangle = |\tfrac{1}{2}, +\tfrac{1}{2}\rangle = \sqrt{\tfrac{2}{3}}\,|+1, -\tfrac{1}{2}\rangle - \sqrt{\tfrac{1}{3}}\,|0, +\tfrac{1}{2}\rangle\;.$$

The weighted averages (expectation values) of M_L and M_S in this state are

$$\langle M_L\rangle = \tfrac{2}{3} \times (+1) + \tfrac{1}{3} \times 0 = \tfrac{2}{3}, \qquad \langle M_S\rangle = \tfrac{2}{3} \times (-\tfrac{1}{2}) + \tfrac{1}{3} \times (+\tfrac{1}{2}) = -\tfrac{1}{6}\;.$$

(Check: The sum is $M_J = +1/2$.) Then $\mu/\mu_B = -\langle M_L\rangle - 2\langle M_S\rangle = -2/3 + 1/3 = -1/3$ for the $^2P_{1/2}$ state. This is only one-sixth the value $\mu/\mu_B = -2$ for the $^2P_{3/2}$ state.

Since we do not readily have Clebsch-Gordan coefficients for all L-S pairs (and it would be cumbersome to use them if we did), we need a better way to get μ when $J < L + S$. The main problem is to calculate the expectation value of M_S in the state $|J, +J\rangle$. Figure 9(a) shows the geometry. In a given state, the vectors $\mathbf{L}$ and $\mathbf{S}$ have fixed projections on $\mathbf{J}$, but not on the z axis; the dashed circle around $\mathbf{J}$ shows the uncertainty of their orientations in space. We get the expectation value of M_S using two geometrical relations that *are* fixed:

(1) For given L, S, and J, the component of $\mathbf{S}$ along $\mathbf{J}$ is fixed.

(2) The angle a $|J, +J\rangle$ state makes with the z axis is fixed.

To emphasize that the calculation is purely geometrical, we begin with the triangle in Fig. 9(b), where $\mathbf{a} + \mathbf{b} = \mathbf{c}$ and the lengths of the vectors are a, b, and c.

First, to get the component of $\mathbf{b}$ along $\mathbf{c}$, we square the third side, $\mathbf{a} = \mathbf{c} - \mathbf{b}$,

$$a^2 = c^2 - 2\,\mathbf{b}\cdot\mathbf{c} + b^2, \qquad \text{or} \qquad \mathbf{b}\cdot\mathbf{c} = (b\cos\theta)\,c = \frac{1}{2}\left(c^2 + b^2 - a^2\right)\;,$$

where θ is the angle between $\mathbf{b}$ and $\mathbf{c}$. This is just the law of cosines, and the component of $\mathbf{b}$ along $\mathbf{c}$ is $b\cos\theta$. Secondly, the projection of this component on the z axis is

$$\frac{\mathbf{b}\cdot\mathbf{c}}{c} \times \frac{c_z}{c} = \frac{c^2 + b^2 - a^2}{2c^2} \times c_z\;.$$

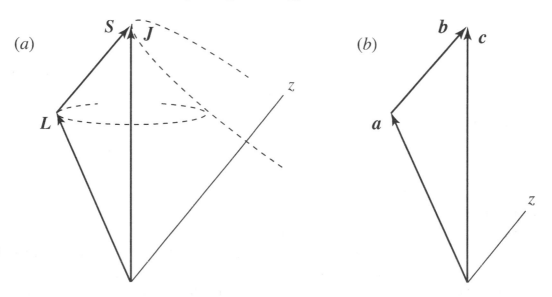

Figure 9. Geometry for the calculation of the expectation value of M_S.

Now in this equation, set $a^2 = L(L+1)\hbar^2$, $b^2 = S(S+1)\hbar^2$, $c^2 = J(J+1)\hbar^2$, and $c_z = +J\hbar$. This projection of $\mathbf{S}$ on the z axis is the expectation value of M_S in the $|J, M\rangle$ state,

$$\langle J, +J|M_S|J, +J\rangle = \frac{J(J+1) + S(S+1) - L(L+1)}{2J(J+1)} \times J \ .$$

And adding the component of $\mathbf{J}$ along z, which here is $+J$, we get

$$\boxed{\langle J, +J|M_J + M_S|J, +J\rangle = \left[1 + \frac{J(J+1) + S(S+1) - L(L+1)}{2J(J+1)}\right] \times J = gJ \ .}$$

The quantity in the square brackets is the *Landé g factor*, and finally $\mu = -gJ\mu_B$.

Check that for $L = 2$, $S = 2$, and $J = L+S = 4$ (iron again), $g = 3/2$ and $\mu = -gJ\mu_B = -6\mu_B$, as earlier. And that for the $^2P_{1/2}$ state, $g = 2/3$ and $\mu = -gJ\mu_B = -(1/3)\mu_B$, as earlier. Figure 10 show the geometry of the g-factors for the $^2P_{3/2}$ and $^2P_{1/2}$ states.

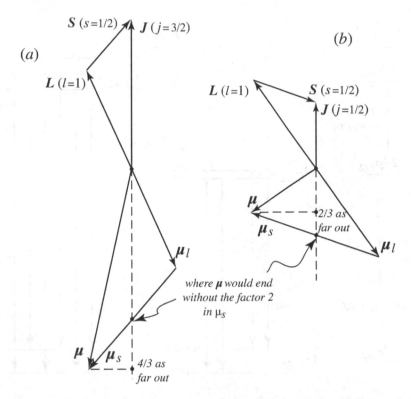

Figure 10. The geometry of angular momenta and magnetic moments (to different scales) for $^2P_{3/2}$ and $^2P_{1/2}$ states. The g-factors are 4/3 and 2/3.

13·10. THE ZEEMAN EFFECT IN HYDROGEN (WEAK-FIELD CASE)

We have shown that when $M_J = +J$ the component of $\boldsymbol{\mu}$ along $\mathbf{B} = B\hat{z}$ is $\mu = -gJ\mu_B$. Now in units of $\hbar$, the component of $\mathbf{J}$ along $\mathbf{B}$ can be any of the $(2J+1)$ values $M_J = +J$, $+J - 1$, ..., $-J$. This quantization of M_J means that the component of $\boldsymbol{\mu}$ along $\mathbf{B}$ is quantized, and the perturbation energies are discrete,

$$\boxed{\mu_z = -gM_J\mu_B, \qquad \text{and} \qquad E'_{Zee} = -\boldsymbol{\mu} \cdot \mathbf{B} = -\mu_z B = +gM_J\mu_B B \ .}$$

The values of g, $\mu/\mu_B = -gJ$, and gm_j, for the $^2s_{1/2}$, $^2p_{3/2}$, and $^2p_{1/2}$ states are

State	g	$\mu/\mu_B = -gJ$	m_j	gm_j
$^2s_{1/2}$	2	-1	$+1/2$	$+1$
			$-1/2$	-1
$^2p_{3/2}$	$4/3$	-2	$+3/2$	$+2$
			$+1/2$	$+2/3$
			$-1/2$	$-2/3$
			$-3/2$	-2
$^2p_{1/2}$	$2/3$	$-1/3$	$+1/2$	$+1/3$
			$-1/2$	$-1/3$

Figure 7 showed the fine-structure splitting of the $n = 1$ and 2 states of hydrogen in the absence of an external field. Figure 11 shows the further splitting of those states in a magnetic field small enough for the Zeeman effect to be treated as a small perturbation on the *separate* $2\,^2L_J$ states.

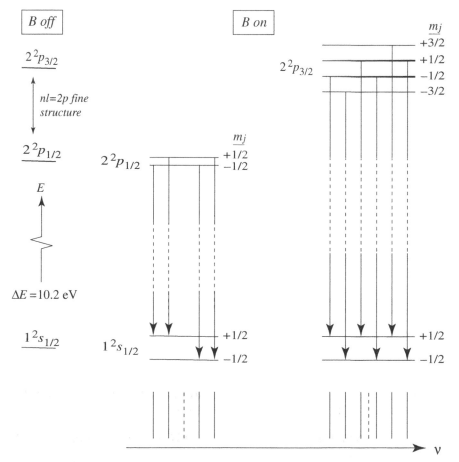

Figure 11. The two dashed lines in the spectrum at the bottom here are the fine-structure splitting of the first spectral line of the hydrogen Lyman series in the absence of an external magnetic field (see Fig. 7). This figure shows the further splitting of those two into four and six components caused by a (weak) magnetic field. A selection rule forbids transitions with $|\Delta m_j| > 1$.

PROBLEMS

1. *Oscillator with an electric field.*

(a) Show that adding the perturbation $V'(x) = -q\mathcal{E}_0 x$ to the oscillator $V^0(x) = \frac{1}{2}m\omega^2 x^2$ simply shifts the parabola over and down; and find out by how much. Write down the exact energy eigenvalues and compare with those in Sec. 2(a).

(b) The unperturbed $n = 0$ and 1 wave functions are, from Sec. 6·2,

$$\psi_0^0(x) = \left(\frac{m\omega}{\pi\hbar}\right)^{1/4} e^{-m\omega x^2/2\hbar} \qquad \text{and} \qquad \psi_1^0(x) = \left(\frac{m\omega}{2\hbar}\right)^{1/2} 2x\, \psi_0^0(x) \ .$$

Find the first-order correction $\psi_0'(x)$ to $\psi_0^0(x)$. Write down the exact perturbed $\psi_0(x)$, and find the first two terms of its expansion about the minimum, x_0, of the perturbed potential energy. Compare with $\psi_0^0(x) + \psi_0'(x)$.

2. *Finite size of the proton.*

(a) Use Gauss's law to show that the potential energy of interaction of a point charge $-e$ with a charge $+e$ that is spread uniformly throughout a sphere of radius R is, for $r < R$,

$$V(r) = -\frac{ke^2}{2R}\left(3 - \frac{r^2}{R^2}\right).$$

For $r > R$, $V(r) = V^0(r) = -ke^2/r$. The two are of course equal at $r = R$. See Fig. 2.

(b) Find the first-order correction E' to the hydrogen ground-state energy due to the perturbation $V'(r) = V(r) - V^0(r)$. Evaluate E', in eV, for $R = 10^{-15}$ m.

3. *Relativity and the oscillator.*

Calculate the first-order corrections E_n' due to relativity for the one-dimensional oscillator. The result of Prob. 6·7, $\langle \psi_n^0 | \hat{x}^4 | \psi_n^0 \rangle = (6n^2 + 6n + 3)(\hbar/2m\omega)^2$, is useful. Note that E_n' will not be small compared to $\hbar\omega$ for large values of n.

4. *A small additional magnetic field.*

A spin-1/2 particle with a magnetic moment $\boldsymbol{\mu} = \mu\hat{\boldsymbol{\sigma}}$ is in a magnetic field $\mathbf{B} = B_0\hat{\mathbf{z}} + B_1\hat{\mathbf{x}}$, where $B_1 \ll B_0$.

(a) Write down the exact eigenvalues $E_\pm$ in terms of the angle between $\mathbf{B}$ and $\mathbf{B}_0$. Expand them as series through the first terms involving B_1.

(b) Now treat $\hat{H}' = -\boldsymbol{\mu} \cdot \mathbf{B}_1$ as a perturbation on $\hat{H}^0 = -\boldsymbol{\mu} \cdot \mathbf{B}_0$. Write down the matrix for $\hat{H}'$, and find the first- and second-order corrections to the unperturbed eigenvalues $E_\pm^0 = \mp\mu B_0$. Compare with (a).

(c) Write down the exact spinor that goes with E_+, using $|+z\rangle$ and $|-z\rangle$ as the basis. Expand this, keeping only the lowest nonzero approximation to each component. Do not bother to renormalize.

(d) Find the first-order perturbation correction to the unperturbed spinor. Compare with (c).

5. *Some $\langle \ell' m | \cos\theta | \ell\, m \rangle$ brackets.*

    (a) Is $\langle \ell', -m | \cos\theta | \ell, -m \rangle$ equal to $\langle \ell' m | \cos\theta | \ell\, m \rangle$?

    (b) Is $\langle \ell' m | \cos\theta | \ell\, m \rangle$ equal to $\langle \ell\, m | \cos\theta | \ell' m \rangle$?

    (c) Use the recurrence relation of Sec. 5 to calculate these $\langle \ell' m | \cos\theta | \ell\, m \rangle$ brackets:

$$\langle 1,0 | \cos\theta | 0,0 \rangle \qquad\qquad \langle 3,+2 | \cos\theta | 2,+2 \rangle$$
$$\langle 2,+1 | \cos\theta | 1,+1 \rangle \qquad\qquad \langle 3,+1 | \cos\theta | 2,+1 \rangle$$
$$\langle 2,0 | \cos\theta | 1,0 \rangle \qquad\qquad \langle 3,0 | \cos\theta | 2,0 \rangle$$

6. *A simpler recurrence relation.*

The angular dependence of the spherical harmonics $Y_\ell^{+\ell}(\theta,\phi)$ is $\sin^\ell\theta\, e^{i\ell\phi}$. Thus it is apparent that $Y_{\ell+1}^{\ell+1} = C_\ell \sin\theta\, e^{i\phi}\, Y_\ell^{+\ell}$. Show that

$$C_\ell = \sqrt{\frac{2\ell+3}{2\ell+2}} = \sqrt{\frac{3}{2}},\ \sqrt{\frac{5}{4}},\ \sqrt{\frac{7}{6}},\ \ldots \quad \text{for } \ell = 0,1,2,\ \ldots$$

The normalized $Y_\ell^{+\ell}(\theta,\phi)$'s are in Table 1 of this chapter.

7. *The $n=3$ Stark effect in hydrogen (ignoring fine structure).*

Consider the Stark effect in hydrogen for $n=3$. Use as a basis the $|\ell, m\rangle$ states, ordered as $|2,+2\rangle$, $|2,+1\rangle$, $|1,+1\rangle$, $|2,0\rangle$, $|1,0\rangle$, $|0,0\rangle$, $|1,-1\rangle$, $|2,-1\rangle$, and $|2,-2\rangle$.

    (a) Only eight of the 81 elements of the 9×9 perturbation matrix are nonzero. Which ones?

    (b) Calculate the four independent nonzero elements of the matrix. The useful integral is (again) $\int_0^\infty x^n e^{-ax} dx = n!/a^{n+1}$. It is often more efficient to calculate separately all the radial integrals you will need and all the angular integrals, and then combine them as needed.

    (c) Show that the energy shifts are 0, $\pm 9e\mathcal{E}_0 a_0/2$, and $\pm 9e\mathcal{E}_0 a_0$. How many times does each root occur?

    (d) Draw the energy-level diagram.

8. *Dirac and hydrogen.*

The energy levels of the Dirac equation (which is fully relativistic) for hydrogen are

$$E_{nj} = mc^2 \left[1 + \left(\frac{\alpha}{n - (j+1/2) + \sqrt{(j+1/2)^2 - \alpha^2}} \right)^2 \right]^{-1/2},$$

where α is the fine-structure constant. Expand E_{nj} in powers of α, keeping terms through α^4. Work from the inside out. Show that the α^0, α^2, and α^4 terms are mc^2, E_n^0, and E_{fs}'.

9. *Hydrogen Lyman-series fine-structure doublet splitting.*

Perhaps you can solve parts (a) and (b) here just by examining the table in Sec. 7(a).

    (a) Show that the fine-structure frequency shifts from the unperturbed frequency ν_0 of the first Lyman line are in the ratio 11-to-15, as Fig. 7 shows.

    (b) Find the ratio of the fine-structure shifts for the second unperturbed Lyman frequency (the $n=3$ to $n=1$ transition).

    (c) Do the same for the $n=4$ to $n=1$ transition.

10. *The $n = 3 \to n = 2$ spectrum for hydrogen.*

Figure 6 shows the fine structure of the $n = 1$, 2, and 3 levels of hydrogen.

(a) Given the restrictions (selection rules) on allowed transitions,

(1) $\Delta \ell = \pm 1$ only, and (2) $\Delta j = 0$ or ± 1 only,

how many different transitions between levels with different n are there? One such is $n^2 L_j = 3^2 P_{3/2} \longrightarrow 2^2 S_{1/2}$. List them.

(b) How many *distinct* frequencies will there be? Calculate the relative frequency shifts from the unperturbed frequency. As in the previous problem, the table in Sec. 7(a) is useful.

(c) Draw to scale the frequency spectrum for transitions between the $n = 2$ and $n = 3$ levels. Make the horizontal distance between the smallest and largest frequencies be 8 cm. Label the frequencies with the associated transitions. Show also the unperturbed frequency.

11. *The radial momentum operator and expectation values of $1/r^3$.*

(a) Show that

$$\frac{1}{2m}\,\hat{p}_r^2 \psi = -\frac{\hbar^2}{2m}\frac{1}{r}\frac{\partial^2 (r\psi)}{\partial r^2}, \qquad \text{where} \quad \hat{p}_r = -i\hbar \left(\frac{\partial}{\partial r} + \frac{1}{r} \right).$$

(b) A plane wave of particles is scattered from a scattering center centered at the origin. The incident wave function is e^{ikz}, where k is the wave number. In Chap. 16, we find that at large distances from the center the form of the scattered wave function is

$$\psi_{scat} = f(\theta, k)\frac{e^{ikr}}{r},$$

where θ is the polar angle. Thus $|\psi_{scat}|^2$ falls off as $1/r^2$, so that the same number of particles cross spheres of different radii. Show that $\hat{p}_r \psi_{scat} = k\hbar\,\psi_{scat}$, as it should.

(c) Calculate $[1/r, \hat{p}_r]$ and $[1/r^2, \hat{p}_r]$. Then use the relation between the $1/r$ and $1/r^2$ terms in the hydrogen Hamiltonian to find the expectation value of $1/r^3$ in $\psi_{n\ell m}$ states.

12. *Expectation values by integrating.*

In Sec. 8 and Prob. 11, we found these expectation values for hydrogen states $\psi_{n\ell m}$,

$$\left\langle \frac{1}{r} \right\rangle_n = \frac{1}{n^2 a_0}, \qquad \left\langle \frac{1}{r^2} \right\rangle_{n\ell} = \frac{1}{(\ell + 1/2)\,n^3 a_0^2}, \qquad \left\langle \frac{1}{r^3} \right\rangle_{n\ell} = \frac{1}{\ell(\ell + 1)(\ell + 1/2)\,n^3 a_0^3}.$$

The last one is only for $\ell > 0$. Calculate these quantities for the $\ell = n - 1$ states by direct integration, and compare. The radial wave functions are at the bottom of Table 1.

13 *The magnetic field is ... ?*

What is the approximate value of the magnetic field that causes the Zeeman splitting shown in Fig. 11?

14. *Some g factors.*

(a) Evaluate the Landé g factor for a hydrogen state with an orbital quantum number of L when $J = L + \frac{1}{2}$ and when $J = L - \frac{1}{2}$. Show that the average of the two values is one, and that both values approach one as L increases. Why?

(b) Evaluate the Landé g factor for helium in terms of the orbital quantum number L for the three values of J when the spin S equals 1. Show that the average of the three is one.

15. *An interval rule.*

The total orbital and spin quantum numbers of the state of an atom (one perhaps having many electrons) are L and S. A spin-orbit perturbation $\hat{H}' \propto \mathbf{L} \cdot \mathbf{S}$ splits the energies of the initially degenerate states having different values of the total angular momentum J. Thus the perturbed energies are

$$E_J = E_0 + A\big[J(J+1) - L(L+1) - S(S+1)\big] ,$$

where E_0 is the unperturbed energy, and A scales the perturbation.

(a) Evaluate $[J(J+1) - L(L+1) - S(S+1)]$ for each of the possible J values when $L = 2$ and $S = 3/2$. Draw to scale the energy-level diagram, showing the splittings from E_0.

(b) Show that the energy differences

$$E_{7/2} - E_{5/2}, \qquad E_{5/2} - E_{3/2}, \qquad \text{and } E_{3/2} - E_{1/2},$$

are in the ratio 7:5:3. This is an example of the "interval rule": The difference between adjacent levels is proportional to the larger of the two J values involved in the difference.

(c) Show that the values obtained in (a), when weighted by the number of M_J states for each J, sum to zero. Thus the "center of mass" of all the $|L, S, J, M_J\rangle$ states is E_0.

14. IDENTICAL PARTICLES

14.1. *Electrons in a Box*

14.2. *Electrons in an Atom: the Periodic Table*

14.3. *Electrons in an Atom: More Pauli*

14.4. *Two-Electron Symmetries*

14.5. *The Helium Ground State*

14.6. *The Electron-Electron Repulsion Integral*

14.7. *Helium Excited States. Exchange Degeneracy*

14.8. *Fermions and Bosons. How to Count. Symmetries*

     Problems

Nearly all of this chapter is about applications of the Pauli exclusion principle. The discovery of the principle was entwined with the discovery of spin 1/2, which doubled the number of quantum states. The first triumph was the explanation of the periodicity of the periodic table of the elements. Physics underlies Chemistry, but the patterns discovered by Chemistry greatly aided discoveries of new Physics.

The exclusion principle states that no two electrons in the same system can share all the same quantum numbers. That being so, the electrons in the ground states of systems with many electrons fill the energy levels up from the bottom. We first calculate the minimum total energy of the sea of conduction electrons in a block of metal, where perhaps 10^{23} electrons stack up. Then we explain the structure of the periodic table, and calculate the angular momenta of the ground states of the elements.

Much of the rest of the chapter is about the energy states of helium, the second element in the periodic table. This is less for the sake of helium than for the physics involved. There are here only the nucleus and two electrons, but few problems with three or more bodies, either classical or quantum, can be solved exactly. We calculate an upper limit on the ground-state energy using the variational method, introduced in Chapter 3. The excited states of helium are treated only qualitatively, but it is in fact the qualitative features that are of most interest.

Nature of course has got more than electrons. Every particle, whether it is elementary (like the electron) or composite (like an atom), belongs to one of two classes: It is either a *fermion* (after Enrico Fermi) or a *boson* (after Satendra Nath Bose). Any particle that has a half-integral angular momentum ($J=1/2$, $3/2$, $5/2$, ...) is a fermion. Fermions all obey the exclusion principle: No two *identical* fermions in the same system can share all the same quantum numbers. Any particle that has an integral angular momentum ($J=0$, 1, 2, ...) is a boson. Bosons behave entirely differently from fermions: Identical bosons *like* to be in the same quantum state. Photons are bosons, and a laser beam is a lot of identical photons, all in the same state.

14·1. ELECTRONS IN A BOX

In a block of metal such as copper, there is about one "conduction electron" per atom. These electrons move about almost freely (and so are also called free electrons), like the molecules of a gas in a box, through the fixed, neutralizing background of the positive ions. They belong to no particular atom, and given their wave nature, each of them is better thought of as belonging to the whole block. If a voltage difference is applied to opposite sides of the block—or to the ends of a wire—the conduction electrons drift in response.

A puzzle before the discovery of quantum mechanics was why the free electrons contribute almost nothing to the heat capacity of a metal. According to the (classical) equipartition theorem, the average kinetic energy per free electron ought to be $\frac{3}{2}kT$, where k is the Boltzmann constant and T is the absolute temperature. Thus the free electrons ought to contribute an amount $\frac{3}{2}N_A kT = \frac{3}{2}RT$ to the molar heat capacity, where N_A is Avogadro's number and $R = N_A k$ is the ideal gas constant.

The quantum answer to the puzzle is that at ordinary temperatures nearly all the free electrons are stacked up in the lowest quantum states consistent with the Pauli exclusion principle. (At ordinary temperatures, few of the molecules of an ordinary gas are in the lowest quantum states.) And then only a small fraction of all the free electrons—those with the highest energies—are close enough to empty states to be able to be excited by a small increase in temperature; for most of the electrons, there are simply no empty states nearby. A rough analogy: Only the molecules of water near the surface of a sea can evaporate.

We shall model a block of metal as a cubical infinite square well. The walls of the block are the walls of the well. The energy levels in one-, two-, and three-dimensional boxes with sides of length L are given by the first one, two, or three terms of

$$E_{n_x n_y n_z} = (n_x^2 + n_y^2 + n_z^2) \frac{\pi^2 \hbar^2}{2mL^2} ,$$

where n_x, n_y, and n_z are (independently) 1, 2, 3, ... (Sec. 2·7). Figure 1 shows again the lowest energy levels for one, two, and three dimensions. There is a level for each value of n_x, and for each set of (n_x, n_y) and of (n_x, n_y, n_z) quantum numbers. The states with, say, $(n_x, n_y) = (2, 3)$ and $(3, 2)$ have the same energy (are two-fold degenerate). The (n_x, n_y, n_z) states $(2, 2, 3)$ and $(1, 3, 4)$ belong to three- and six-fold degenerate levels. In Fig. 1, the densities of levels in the two and three-dimensional cases are rather ragged. On a larger scale, the average density of levels is constant in two dimensions and increases with energy in three (see Prob. 3). In the degenerate case, the free electrons fill up those levels from the bottom.

(a) Two dimensions—It is easier to begin in two dimensions. Since electrons are fermions, no more than two of the free electrons can occupy a given spatial quantum state in the well—*two* because there remains the spin-up or -down degree of freedom, $m_s = \pm 1/2$. Thus in a two-dimensional well, no more than two electrons can be in a given (n_x, n_y) state; or no more than *one* electron can be in a given (n_x, n_y, m_s) state.

Now suppose a very large number, N, of free electrons settle into the *lowest* energy states, consistent with the Pauli principle. (In three dimensions, N will typically be on the order of Avogadro's number.) This is called a completely degenerate electron gas, and is the situation when the absolute temperature T is 0 K. As we noted above, it is also very nearly the situation at room temperatures (see Prob. 2). We want to find both the average energy of a free electron and the highest energy of any of them.

264

Example: In Fig. 1 there are 19 (n_x, n_y) levels with $n_x^2 + n_y^2 \leq 30$, so it takes 38 electrons to completely fill those levels. In units of $\pi^2\hbar^2/2mL^2$, the highest energy of an electron is 29, the total energy of all 38 is 636 (do the sum), and the average energy per electron is 16.7.

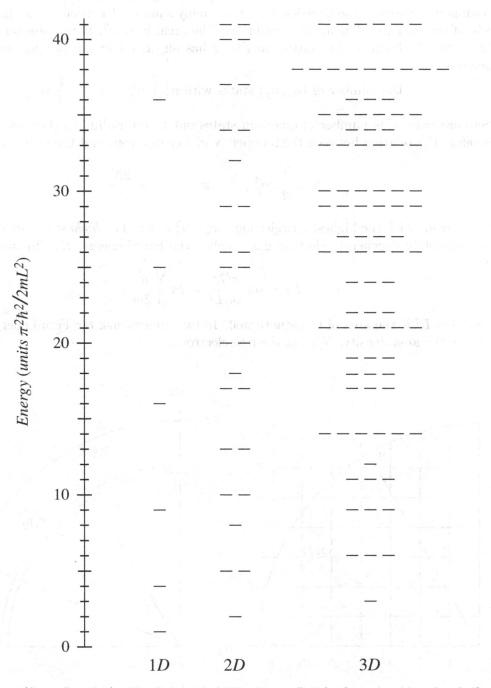

Figure 1 (from Sec. 2·7). The bottom of "Fermi seas"—the lowest energy levels for one-, two-, and three-dimensional infinite square wells of side L. No more than *two* electrons can occupy each level here.

We can associate each quantum-number pair, (n_x, n_y), either with a point or with a square on a grid, as shown in Fig. 2(a). Since the energies of the states are proportional to $n_x^2 + n_y^2$, the N free electrons will occupy the (n_x, n_y) states out to some large radius n_0, where n_0 will depend on N. How many grid points—or squares—lie within a quarter circle of radius n_0? Suppose the question was: How many squares, 1 mm on a side, lie in a quarter circle of radius 1 m? The answer would be: The area, in mm^2, of the quarter circle, which is $\pi(1000)^2/4$. Each of the squares in Fig. 2 has an area 1 in the (n_x, n_y) number space. Therefore,

the number of (n_x, n_y) states with $n_x^2 + n_y^2 \le n_0^2 \;=\; \frac{1}{4}\pi n_0^2$.

There are twice this number of quantum states out to the radius n_0 (because of the spin). Therefore the relation between the number N of free electrons and the radius n_0 is

$$N = \frac{1}{2}\pi n_0^2, \qquad \text{or} \qquad n_0^2 = \frac{2N}{\pi}.$$

The electrons with the highest energies have $n_x^2 + n_y^2 = n_0^2$. The *highest* energy of an electron in a completely degenerate electron gas is called the Fermi energy, E_F. In two dimensions, it is

$$E_F = n_0^2 \,\frac{\pi^2 \hbar^2}{2mL^2} = 2\pi\,\frac{N}{A}\,\frac{\hbar^2}{2m},$$

where $A = L^2$ is the area of the square well. In two dimensions, the Fermi energy is proportional to the area density, N/A, of the free electrons.

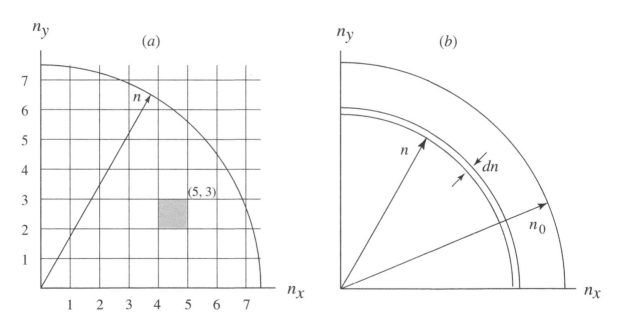

Figure 2. Number space for two dimensions. (a) The grid: No more than two electrons can be associated with any (n_x, n_y) grid point or square. (b) At temperature $T = 0$, a large number N of electrons, two to a square, occupy all the states out to a large radius n_0.

266

To find the average energy, $\overline{E}$, of a free electron in the gas, consider a thin quarter ring at a radius $n < n_0$ and with an infinitesimal width dn (infinitesimal compared to n_0); see Fig. 2(b). The number of states in the ring is $2 \times 2\pi n \, dn/4 = \pi n \, dn$. The factor 2 is for the spin states; the rest is the area of the thin quarter ring. The energy of each electron in the quarter ring is $E(n) = n^2 \pi^2 \hbar^2 / 2mA$. Thus the total energy of all the electrons in the quarter ring is

$$n^2 \frac{\pi^2 \hbar^2}{2mA} \times \pi n \, dn = \frac{\pi^3 \hbar^2}{2mA} n^3 dn .$$

The total energy of all N electrons is

$$E_{total} = \frac{\pi^3 \hbar^2}{2mA} \int_0^{n_0} n^3 \, dn = \frac{\pi^3 \hbar^2}{2mA} \frac{n_0^4}{4} = \frac{\pi^3 \hbar^2}{2mA} \frac{N^2}{\pi^2} = N\pi \frac{N}{A} \frac{\hbar^2}{2m} = N \frac{E_F}{2} .$$

The average energy of a conduction electron in the two-dimensional degenerate electron gas is $\overline{E} = E_{total}/N = E_F/2$, one-half the Fermi energy.

(b) Three dimensions—The calculation for three dimensions should now be easy, and is left for Prob. 1 (and was Prob. 2·17). Here are the answers: For a cubical box of side L, the energy levels again are $E_{n_x n_y n_z} = (n_x^2 + n_y^2 + n_z^2)(\pi^2 \hbar^2/2mL^2)$. The number of (n_x, n_y, n_z) states having $n_x^2 + n_y^2 + n_z^2 \le n_0^2$, where n_0 is a large number, is $\pi n_0^3/6$. If N electrons occupy the lowest energy states consistent with the Pauli principle, the energy at the top of the "Fermi sea" is

$$E_F = \left(3\pi^2 \frac{N}{V}\right)^{2/3} \frac{\hbar^2}{2m} ,$$

where $V = L^3$ is the volume of the box. The average energy of a conduction electron in the three-dimensional completely degenerate electron gas is

$$\overline{E} = \frac{3}{5} E_F .$$

The Fermi and average energies vary as $(N/V)^{2/3}$, where N/V is the volume density of the free electrons. (If you have calculated the potential energy of a uniform spherical distribution of charge or mass, you have seen that factor 3/5 before.)

14·2. ELECTRONS IN AN ATOM: THE PERIODIC TABLE

Around 1870, D.I. Mendeléev made a table of the known elements, ordering them by mass and "wrapping" the sequence so that elements with similar chemical properties fell into the same column. The table had 72 of the now-known 92 elements through uranium. Among the still-to-be-discovered elements were all of the inert (noble) gases, which occupy the right-most column of the modern Periodic Table of the Elements (see Table 1). The modern table is ordered by the nuclear charge Ze, and this ordering occasionally differs from one by mass. The nucleus was not discovered until 1911 (Rutherford). The charges of nuclei were not accurately determined until two years later (Henry Moseley).

The explanation of the *periodicity* of chemical properties relies on the exclusion principle and spin, and it did not come until 1926. It was in fact this triumph of the new physics that gave strongest support for the new ideas of exclusion and spin. For the complicated history, see A. Pais, *Inward Bound*, Chap. 13.

Table 1. The Periodic Table of the Elements

1 H hydrogen																	2 He helium
3 Li lithium	4 Be beryllium											5 B boron	6 C carbon	7 N nitrogen	8 O oxygen	9 F fluorine	10 Ne neon
11 Na sodium	12 Mg magnesium											13 Al aluminum	14 Si silicon	15 P phosphorus	16 S sulfur	17 Cl chlorine	18 Ar argon
19 K potassium	20 Ca calcium	21 Sc scandium	22 Ti titanium	23 V vanadium	24 Cr chromium	25 Mn manganese	26 Fe iron	27 Co cobalt	28 Ni nickel	29 Cu copper	30 Zn zinc	31 Ga gallium	32 Ge germanium	33 As arsenic	34 Se selenium	35 Br bromine	36 Kr krypton
37 Rb rubidium	38 Sr strontium	39 Y yttrium	40 Zr zirconium	41 Nb niobium	42 Mo molybdenum	43 Tc technetium	44 Ru ruthenium	45 Rh rhodium	46 Pd palladium	47 Ag silver	48 Cd cadmium	49 In indium	50 Sn tin	51 Sb antimony	52 Te tellurium	53 I iodine	54 Xe xenon
55 Cs caesium	56 Ba barium	57-71 lanthanides	72 Hf hafnium	73 Ta tantalum	74 W tungsten	75 Re rhenium	76 Os osmium	77 Ir iridium	78 Pt platinum	79 Au gold	80 Hg mercury	81 Tl thallium	82 Pb lead	83 Bi bismuth	84 Po polonium	85 At astatine	86 Rn radon
87 Fr francium	88 Ra radium	89-103 actinides	104 Rf rutherfordium	105 Db dubnium	106 Sg seaborgium	107 Bh bohrium	108 Hs hassium	109 Mt meitnerium	110 Ds darmstadtium	111 Rg roentgenium	112 Cn copernicium	113 Nh nihonium	114 Fl flerovium	115 Mc moscovium	116 Lv livermorium	117 Ts tennessine	118 Og oganesson

lanthanides

57 La lanthanum	58 Ce cerium	59 Pr praseodymium	60 Nd neodymium	61 Pm promethium	62 Sm samarium	63 Eu europium	64 Gd gadolinium	65 Tb terbium	66 Dy dysprosium	67 Ho holmium	68 Er erbium	69 Tm thulium	70 Yb ytterbium	71 Lu lutetium

actinides

89 Ac actinium	90 Th thorium	91 Pa protactinium	92 U uranium	93 Np neptunium	94 Pu plutonium	95 Am americium	96 Cm curium	97 Bk berkelium	98 Cf californium	99 Es einsteinium	100 Fm fermium	101 Md mendelevium	102 No nobelium	103 Lr lawrencium

(a) **Equivalent electrons**—The basis states of the electron in hydrogen can be specified using either of the quantum-number sets $(n, \ell, s; m_\ell, m_s)$ or $(n, \ell, s; j, m_j)$. We often omit s from the lists because it is always $1/2$ (and omitting it reduces confusion with s for $\ell = 0$). In an atom with nuclear charge Ze, each of its Z electrons is attracted by the nucleus and repelled by all the other electrons, and it is not at all clear how to specify the state of a given electron. Three- and more-body problems, whether classical or quantum, cannot be solved in closed form. There are various approximation methods and models, but they are too advanced for this text. However, if we assume, (1) that the $(n, \ell; m_\ell, m_s)$ set is good at least for *labelling* states of electrons in an atom, and (2) that for a given value of n the values of ℓ are restricted (as for hydrogen) to $0, 1, 2, \cdots, n - 1$, we can explain the periodicity of the Periodic Table. The assumptions work brilliantly.

Equivalent electrons are electrons that share the same values of n and ℓ—they belong to the same (n, ℓ) *subshell*. The values of (n, ℓ) for the first few subshells are $1s$, $2s$, $2p$, $3s$, $3p$, $\ldots$, where as usual s, p, d, $\ldots$ stand for $\ell = 0, 1, 2, \ldots$ The Pauli exclusion principle forbids two electrons in an atom to have all four of n, ℓ, m_ℓ, and m_s the same. Thus it forbids two electrons already in the same (n, ℓ) subshell to have m_ℓ and m_s both the same.

How many electrons does it take to completely fill an (n, ℓ) subshell? For $\ell = 0$, where $m_\ell = 0$, the answer is two, with $m_s = +1/2$ for one electron and $-1/2$ for the other. For $\ell = 1$, there are three choices for m_ℓ times two for m_s. Here are the "calculations" for filled s, p, and d subshells, using up and down arrows for the $m_s = +1/2$ and $-1/2$ states:

$\ell = 0$		$\ell = 1$				$\ell = 2$				
$m_\ell =$ 0		$m_\ell =$ $+1$	0	-1		$m_\ell =$ $+2$	$+1$	0	-1	-2
$\uparrow\downarrow$		$\uparrow\downarrow$	$\uparrow\downarrow$	$\uparrow\downarrow$		$\uparrow\downarrow$	$\uparrow\downarrow$	$\uparrow\downarrow$	$\uparrow\downarrow$	$\uparrow\downarrow$

Two electrons fill an s subshell, six electrons a p subshell, ten electrons a d subshell. How many electrons does it take to fill an f ($\ell = 3$) subshell? A filled subshell is *one* state, with $M_L = \sum m_\ell = 0$ and $M_S = \sum m_s = 0$. The total angular-momentum quantum numbers are $L = S = J = 0$ (using capitals for the angular momenta of the whole subshell). A completely filled subshell is a $^{2S+1}L_J = {}^1S_0$ state.

(b) **Periodicity**—The periodicity of chemical properties is a direct consequence of the sequential filling of (n, ℓ) subshells. Look at the periodic table: The first six rows have 2, 8, 8, 18, 18, and 32 elements—the first 86 elements. How can we explain these "magic numbers" in terms of the numbers of electrons, 2, 6, 10, 14, $\ldots$, in filled subshells? Like this:

(n, ℓ) subshells per row $= 1s \mid 2s\ 2p \mid 3s\ 3p \mid 4s\ 3d\ 4p \mid 5s\ 4d\ 5p \mid 6s\ 4f\ 5d\ 6p \mid$

electrons per filled subshell $=\ 2 \mid 2\ \ 6 \mid 2\ \ 6 \mid 2\ 10\ \ 6 \mid 2\ 10\ \ 6 \mid 2\ 14\ 10\ \ 6 \mid$

elements per row of the table $=\ 2 \mid\ \ 8\ \ \mid\ \ 8\ \ \mid\ \ 18\ \ \mid\ \ 18\ \ \mid\ \ \ 32\ \ \ \mid$

The subshells between the dividers fill the six rows of the periodic table in turn. How in the first line do we know, for example, that $4s$ comes before $3d$? Look at the rows of the periodic table again: 2, 2 and 6, 2 and 6, 2 and 10 and 6, $\ldots$ The structure and numerology of the table *tells* us the order that subshells are filled. Although advanced models can now calculate the filling order, we originally *got* the order from the observed periodicity of chemical properties.

The $1s$ subshell completes the first row of the periodic table. Each of the next five rows finish with six elements filling a p subshell. These six rows end with the inert gases helium, neon, argon, krypton, xenon, and radon. In their ground states, the subshells of these gases are completely filled; all six are $^{2S+1}L_J = {}^1S_0$ states. Completely filled subshells have no knobs or holes to fill or be filled by holes or knobs on other atoms. And the electrons in filled subshells are tightly bound—in each row of the periodic table, the inert gas has the largest ionization energy. The ionization energy is the minimum energy needed to take *one* electron from the ground state away to infinity. For hydrogen, we already know it is 13.6 eV. Helium has the largest ionization energy of all, 24.6 eV. The absence of holes or knobs and the tight binding is why the inert gases are chemically inert.

(c) Ground states—Table 2 gives the electron configurations, angular-momentum states, and ionization energies of the ground-states of the elements. We consider the first few elements in turn, and then introduce empirical rules that tell us how to find the values of L, S, and J for the ground states of nearly all the elements.

Hydrogen $(Z = 1)$—The ground state for the single electron has (as we already know) $n = 1$, $\ell = 0$, $m_\ell = 0$, and $m_s = \pm 1/2$. The angular-momentum state is $^{2S+1}L_J = {}^2S_{1/2}$.

Helium $(Z = 2)$—Two electrons can go into the $n = 1$, $\ell = 0$, $m_\ell = 0$ shell, but (as we already know) only if their m_s values are opposite and the total spin S is 0. The configuration is $1s^2$ (s^2 for two s electrons), and the $^{2S+1}L_J$ state is $^1S_0$. Two electrons fill the $n = 1$ shell, and the first row of the periodic table is complete. Sections 5, 6, and 7 are about helium.

Lithium $(Z = 3)$—The lowest available state for a third electron has $n = 2$ and $\ell = 0$. The electron configuration is $1s^2\,2s$. The $1s^2$ subshell has $L = S = J = 0$, and so the angular-momentum state of lithium is that of its outer electron, $^2S_{1/2}$, the same as hydrogen. Lithium starts the second row of the periodic table, directly beneath hydrogen.

The $1s^2$ inner core of lithium with charge $-2e$ partially shields the nuclear charge of $+3e$. If the shielding were complete, the $2s$ electron would only "see" a charge $+e$, and its ionization energy would be the same as that of the $n = 2$ state of hydrogen, 3.40 eV. The actual ionization energy is 5.39 eV. It is larger than 3.40 eV because an s electron, whatever its n value, has some probability of being close to the nucleus. Effectively, the $2s$ electron sees a charge of about $(5.39/3.40)^{1/2}\,e = 1.26e$.

This qualitative argument also explains why the $2s$ subshell is filled before the $2p$ subshell. A $2p$ electron is held away from the nucleus by its nonzero orbital angular momentum. Thus the effective charge it sees is closer to $+e$, and the $2p$ electron in the $1s^2\,2p$ state of lithium (an excited state), and the binding energy is 3.55 eV, close to 3.40 eV.

Beryllium $(Z = 4)$—Another electron can go into the $2s$ subshell, but only to make $S = 0$. The configuration is $1s^2\,2s^2$ and $^{2S+1}L_J = {}^1S_0$, like helium. But this closing of a subshell does not start a new row of the periodic table. In going from helium to lithium, the ionization energy drops from 24.6 eV to 5.39 eV. In going from beryllium to the next element, boron, the ionization energy only drops from 9.32 eV to 8.30 eV. New rows start when there is a large drop of ionization energy from one element to the next. In Table 2, notice the large drops right after the noble gases.

Boron $(Z = 5)$—The lowest state for a fifth electron has $n = 2$ and $\ell = 1$. The electron configuration is $1s^2\,2s^2\,2p$, so again the angular-momentum state is that of the outer electron, either $^2P_{1/2}$ or $^2P_{3/2}$. When there is a choice of angular-momentum quantum numbers, there are three *empirical* laws, called *Hund's rules*, that determine the ground state:

Table 2. The subshell structures, angular-momentum states, and ionization energies of the ground states of the first 96 elements. The noble gases, which complete the rows of the Periodic Table, are in capitals. Adapted from M. Tanabashi et al., Particle Data Group, Phys. Rev. **D98**, 030001 (2018).

	Element	Configuration	$^{2S+1}L_J$	eV		Element	Configuration	$^{2S+1}L_J$	eV
1	H Hydrogen	$1s$	$^2S_{1/2}$	13.60	49	In Indium	(Kr) $4d^{10}5s^2\,5p$	$^2P_{1/2}$	5.79
2	He HELIUM	$1s^2$	$^1S_0$	24.59	50	Sn Tin	(Kr) $4d^{10}5s^2\,5p^2$	$^3P_0$	7.34
3	Li Lithium	(He) $2s$	$^2S_{1/2}$	5.39	51	Sb Antimony	(Kr) $4d^{10}5s^2\,5p^3$	$^4S_{3/2}$	8.61
4	Be Beryllium	(He) $2s^2$	$^1S_0$	9.32	52	Te Tellurium	(Kr) $4d^{10}5s^2\,5p^4$	$^3P_2$	9.01
5	B Boron	(He) $2s^2\,2p$	$^2P_{1/2}$	8.30	53	I Iodine	(Kr) $4d^{10}5s^2\,5p^5$	$^2P_{3/2}$	10.45
6	C Carbon	(He) $2s^2\,2p^2$	$^3P_0$	11.26	54	Xe XENON	(Kr) $4d^{10}5s^2\,5p^6$	$^1S_0$	12.13
7	N Nitrogen	(He) $2s^2\,2p^3$	$^4S_{3/2}$	14.53	55	Cs Cesium	(Xe) $\qquad6s$	$^2S_{1/2}$	3.89
8	O Oxygen	(He) $2s^2\,2p^4$	$^3P_2$	13.62	56	Ba Barium	(Xe) $\qquad6s^2$	$^1S_0$	5.21
9	F Fluorine	(He) $2s^2\,2p^5$	$^2P_{3/2}$	17.42	57	La Lanthanum	(Xe) $\quad5d\,6s^2$	$^2D_{3/2}$	5.58
10	Ne NEON	(He) $2s^2\,2p^6$	$^1S_0$	21.56	58	Ce Cerium	(Xe) $4f\,\,5d\,6s^2$	$^1G_4$	5.54
11	Na Sodium	(Ne) $3s$	$^2S_{1/2}$	5.14	59	Pr Praesodymium	(Xe) $4f^3\quad6s^2$	$^4I_{9/2}$	5.47
12	Mg Magnesium	(Ne) $3s^2$	$^1S_0$	7.65	60	Nd Neodymium	(Xe) $4f^4\quad6s^2$	$^5I_4$	5.53
13	Al Aluminum	(Ne) $3s^2\,3p$	$^2P_{1/2}$	5.99	61	Pm Promethium	(Xe) $4f^5\quad6s^2$	$^6H_{5/2}$	5.58
14	Si Silicon	(Ne) $3s^2\,3p^2$	$^3P_0$	8.15	62	Sm Samarium	(Xe) $4f^6\quad6s^2$	$^7F_0$	5.64
15	P Phosphorus	(Ne) $3s^2\,3p^3$	$^4S_{3/2}$	10.49	63	Eu Europium	(Xe) $4f^7\quad6s^2$	$^8S_{7/2}$	5.67
16	S Sulfur	(Ne) $3s^2\,3p^4$	$^3P_2$	10.36	64	Gd Gadolinium	(Xe) $4f^7\,5d\,6s^2$	$^9D_2$	6.15
17	Cl Chlorine	(Ne)$3s^2\,3p^5$	$^2P_{3/2}$	12.97	65	Tb Terbium	(Xe) $4f^9\quad6s^2$	$^6H_{15/2}$	5.86
18	Ar ARGON	(Ne) $3s^2\,3p^6$	$^1S_0$	15.76	66	Dy Dysprosium	(Xe) $4f^{10}\quad6s^2$	$^5I_8$	5.94
19	K Potassium	(Ar) $\quad4s$	$^2S_{1/2}$	4.34	67	Ho Holmium	(Xe) $4f^{11}\quad6s^2$	$^4I_{15/2}$	6.02
20	Ca Calcium	(Ar) $\quad4s^2$	$^1S_0$	6.11	68	Er Erbium	(Xe) $4f^{12}\quad6s^2$	$^3H_6$	6.11
21	Sc Scandium	(Ar) $3d\,\,4s^2$	$^2D_{3/2}$	6.56	69	Tm Thulium	(Xe) $4f^{13}\quad6s^2$	$^2F_{7/2}$	6.18
22	Ti Titanium	(Ar) $3d^2\,4s^2$	$^3F_2$	6.83	70	Yb Ytterbium	(Xe) $4f^{14}\quad6s^2$	$^1S_0$	6.25
23	V Vanadium	(Ar) $3d^3\,4s^2$	$^4F_{3/2}$	6.75	71	Lu Lutetium	(Xe) $4f^{14}5d\,6s^2$	$^2D_{3/2}$	5.43
24	Cr Chromium	(Ar) $3d^5\,4s$	$^7S_3$	6.77	72	Hf Hafnium	(Xe) $4f^{14}5d^2\,6s^2$	$^3F_2$	6.83
25	Mn Manganese	(Ar) $3d^5\,4s^2$	$^6S_{5/2}$	7.43	73	Ta Tantalum	(Xe) $4f^{14}5d^3\,6s^2$	$^4F_{3/2}$	7.55
26	Fe Iron	(Ar) $3d^6\,4s^2$	$^5D_4$	7.90	74	W Tungsten	(Xe) $4f^{14}5d^4\,6s^2$	$^5D_0$	7.86
27	Co Cobalt	(Ar) $3d^7\,4s^2$	$^4F_{9/2}$	7.88	75	Re Rhenium	(Xe) $4f^{14}5d^5\,6s^2$	$^6S_{5/2}$	7.83
28	Ni Nickel	(Ar) $3d^8\,4s^2$	$^3F_4$	7.64	76	Os Osmium	(Xe) $4f^{14}5d^6\,6s^2$	$^5D_4$	8.44
29	Cu Copper	(Ar) $3d^{10}4s$	$^2S_{1/2}$	7.73	77	Ir Iridium	(Xe) $4f^{14}5d^7\,6s^2$	$^4F_{9/2}$	8.97
30	Zn Zinc	(Ar) $3d^{10}4s^2$	$^1S_0$	9.39	78	Pt Platinum	(Xe) $4f^{14}5d^9\,6s$	$^3D_3$	8.96
31	Ga Gallium	(Ar) $3d^{10}4s^2\,4p$	$^2P_{1/2}$	6.00	79	Au Gold	(Xe) $4f^{14}5d^{10}6s$	$^2S_{1/2}$	9.23
32	Ge Germanium	(Ar) $3d^{10}4s^2\,4p^2$	$^3P_0$	7.90	80	Hg Mercury	(Xe) $4f^{14}5d^{10}6s^2$	$^1S_0$	10.44
33	As Arsenic	(Ar) $3d^{10}4s^2\,4p^3$	$^4S_{3/2}$	9.79	81	Tl Thallium	(Xe) $4f^{14}5d^{10}6s^2\,6p$	$^2P_{1/2}$	6.11
34	Se Selenium	(Ar) $3d^{10}4s^2\,4p^4$	$^3P_2$	9.75	82	Pb Lead	(Xe) $4f^{14}5d^{10}6s^2\,6p^2$	$^3P_0$	7.42
35	Br Bromine	(Ar) $3d^{10}4s^2\,4p^5$	$^2P_{3/2}$	11.81	83	Bi Bismuth	(Xe) $4f^{14}5d^{10}6s^2\,6p^3$	$^4S_{3/2}$	7.29
36	Kr KRYPTON	(Ar) $3d^{10}4s^2\,4p^6$	$^1S_0$	14.00	84	Po Polonium	(Xe) $4f^{14}5d^{10}6s^2\,6p^4$	$^3P_2$	8.41
37	Rb Rubidium	(Kr) $\quad5s$	$^2S_{1/2}$	4.18	85	At Astatine	(Xe) $4f^{14}5d^{10}6s^2\,6p^5$	$^2P_{3/2}$	—
38	Sr Strontium	(Kr) $\quad5s^2$	$^1S_0$	5.69	86	Rn RADON	(Xe) $4f^{14}5d^{10}6s^2\,6p^6$	$^1S_0$	10.75
39	Y Yttrium	(Kr) $4d\,\,5s^2$	$^2D_{3/2}$	6.22	87	Fr Francium	(Rn) $\qquad7s$	$^2S_{1/2}$	4.07
40	Zr Zirconium	(Kr) $4d^2\,5s^2$	$^3F_2$	6.63	88	Ra Radium	(Rn) $\qquad7s^2$	$^1S_0$	5.28
41	Nb Niobium	(Kr) $4d^4\,5s$	$^6D_{1/2}$	6.76	89	Ac Actinium	(Rn) $\quad6d\,7s^2$	$^2D_{3/2}$	5.38
42	Mo Molybdenum	(Kr) $4d^5\,5s$	$^7S_3$	7.09	90	Th Thorium	(Rn) $\quad6d^27s^2$	$^3F_2$	6.31
43	Tc Technetium	(Kr) $4d^5\,5s^2$	$^6S_{5/2}$	7.28	91	Pa Protactinium	(Rn) $5f^2\,6d\,7s^2$	$^4K_{11/2}$	5.89
44	Ru Ruthenium	(Kr) $4d^7\,5s$	$^5F_5$	7.36	92	U Uranium	(Rn) $5f^3\,6d\,7s^2$	$^5L_6$	6.19
45	Rh Rhodium	(Kr) $4d^8\,5s$	$^4F_{9/2}$	7.46	93	Np Neptunium	(Rn) $5f^4\,6d\,7s^2$	$^6L_{11/2}$	6.27
46	Pd Palladium	(Kr) $4d^{10}$	$^1S_0$	8.34	94	Pu Plutonium	(Rn) $5f^6\quad7s^2$	$^7F_0$	6.03
47	Ag Silver	(Kr) $4d^{10}5s$	$^2S_{1/2}$	7.58	95	Am Americium	(Rn) $5f^7\quad7s^2$	$^8S_{7/2}$	5.97
48	Cd Cadmium	(Kr) $4d^{10}5s^2$	$^1S_0$	8.99	96	Cm Curium	(Rn) $5f^7\,6d\,7s^2$	$^9D_2$	5.99

(1) Choose the highest value of S.

(2) *Then* (if there is still a choice) choose the highest value of L.

(3) *Then* (if there is still a choice) choose the smallest possible value of J if the subshell is less than half filled, or the largest value if it is more than half filled.

For boron, where S is $1/2$ and L is 1, the only choice is about J. Since the p shell is less than half filled, the choice is $J = L - S = 1/2$, and the ground-state of boron is $^2P_{1/2}$.

Carbon $(Z = 6)$—The electron configuration is $1s^2\,2s^2\,2p^2$. A priori, S could be 0 or 1, and L could be 2, 1, or 0. The table below shows how to find $^{2S+1}L_J$ using Hund's rules. In each line, the first consideration is to maximize, consistent with the exclusion principle, the number of arrows pointing up (maximizing $M_S = \sum m_s$). The second consideration is to push the arrows as far to the left (maximizing $M_L = \sum m_\ell$) as possible, consistent with the exclusion principle. The p^2 line of the table puts two up-arrows as far to the left (the high-m_ℓ side) as possible. The resulting values of M_L and M_S are the values of L and S. For example, $M_S = 1$ is as large as it can be for two electrons, and $M_S = 1$ requires $S = 1$. The third rule then gives $J = |L - S| = 0$. For carbon then, $^{2S+1}L_J = {}^3P_0$. It ought to be clear how the rest of the table is made.

The ground states of one through six electrons in a p subshell.

$m_\ell =$	+1	0	−1	$M_L = \sum m_\ell$	$M_S = \sum m_s$	$^{2S+1}L_J$	$-\mu/\mu_B$
p	↑			+1	+1/2	$^2P_{1/2}$	1/3
p^2	↑	↑		+1	+1	$^3P_0$	0
p^3	↑	↑	↑	0	+3/2	$^4S_{3/2}$	3
p^4	↑↓	↑	↑	+1	+1	$^3P_2$	3
p^5	↑↓	↑↓	↑	+1	+1/2	$^2P_{3/2}$	2
p^6	↑↓	↑↓	↑↓	0	0	$^1S_0$	0

Note how (and *why*) the values of S go up in steps of $1/2$ until the subshell is half filled, and then down in steps of $1/2$ to zero. And why there is no need for Hund's third rule when a subshell is half filled. Five rows of the periodic table end with the filling of p subshells, so the table gives the $^{2S+1}L_J$ states of 30 elements, not just six.

(d) Magnetic moments and a last example—In Sec. 13·9, we found that for atoms with $J = L + S$ the magnetic moment is $\mu = -(L + 2S)\mu_B$, where μ_B is the Bohr magneton. This gives at once the moments of the p^3 through p^6 elements in the above table. In Sec. 13·9, we calculated the ratio μ/μ_B of a $^2P_{1/2}$ state to be $-gJ = -\frac{2}{3} \times \frac{1}{2} = -\frac{1}{3}$. And the moment of a state with $J = 0$ is zero. That completes the moments for ground-state p subshells.

Iron is element number 26. Its ground-state configuration is $1s^2\,2s^2\,2p^6\,3s^2\,3p^6\,4s^2\,3d^6$. The arrow table for six d electrons, constructed according to Hund's rules, is

$m_\ell =$	+2	+1	0	−1	−2	$M_L = \sum m_\ell$	$M_S = \sum m_s$	$^{2S+1}L_J$	$-\mu/\mu_B$
	↑↓	↑	↑	↑	↑	+2	+2	$^5D_4$	6

Thus $L = 2$ and $S = 2$; and since the subshell is more than half filled, $J = L + S = 4$. Then the ground state is $^{2S+1}L_J = {}^5D_4$, and the magnetic moment is $\mu = -(L + 2S)\mu_B = -6\mu_B$. All just by putting some arrows in a table.

14·3. ELECTRONS IN AN ATOM: MORE PAULI

This section (which could be skipped) derives entirely from G. Herzberg, *Atomic Spectra and Atomic Structure*, Chap. 3. We here find *all* the angular-momentum states of two electrons in a p subshell (not just the $^3P_0$ ground state as given by Hund's rules). We make a table with *all* the (m_ℓ, m_s) combinations of two arrows that satisfy the exclusion principle. There are six ways to place the first arrow, but then only five ways to add the second one— because one way is now *excluded*. And for each placement of two arrows, it matters not which arrow goes in first. So altogether there are $(6 \times 5)/2 = 15$ ways to place two arrows. Each row in the columns in the table labelled $M_L = \sum m_\ell$ and $M_S = \sum m_s$ is an $|M_L, M_S\rangle$ state consistent with the exclusion principle. The job is to figure out what L and S states there are, and then to sort them out into 15 $|J, M\rangle$ system states.

The 15 states of two electrons in a p subshell.

	$m_\ell =$ +1	0	−1	$M_L = \sum m_\ell$	$M_S = \sum m_s$		$^{2S+1}L_J$ states
1	↑↓			+2	0	△	
2	↑	↑		+1	+1	×	
3	↑	↓		+1	0	△	
4	↓	↑		+1	0	×	
5	↓	↓		+1	−1	×	
6		↑↓		0	0	△	△ → $^1D_2$
7	↑		↑	0	+1	×	
8	↑		↓	0	0	×	× → $^3P_{2,1,0}$
9	↓		↑	0	0	•	
10	↓		↓	0	−1	×	• → $^1S_0$
11		↑	↑	−1	+1	×	
12		↑	↓	−1	0	△	
13		↓	↑	−1	0	×	
14		↓	↓	−1	−1	×	
15			↑↓	−2	0	△	

The largest value of M_L in the table is $+2$, marked with a △. This calls for $L = 2$. And since $M_L = +2$ occurs only with $M_S = 0$, the entry belongs to an $L = 2$, $S = 0$ multiplet, with $J = L = 2$. But to fill it out, an $L = 2$ multiplet also needs $M_L = +1, 0, -1$, and -2 states. Four entries with these values of M_L, all with $M_S = 0$, are marked with the △. The five △ states make up a $^{2S+1}L_J = {}^1D_2$ multiplet.

Why did we choose the third entry in the table for the $M_L = +1$ state? The fourth entry is the same. In fact, since $S = 0$, the proper choice would be the $|S, M\rangle = |0, 0\rangle$ superposition of the third and fourth entries, $(\uparrow\downarrow - \downarrow\uparrow)/\sqrt{2}$. We arbitrarily chose one of the two for the $^1D_2$ multiplet; the other will be used below. Proper symmetrization can come later.

In what remains, the largest value of M_L is $+1$, and the largest value of M_S it occurs with is $+1$. This calls for $L = 1$ and $S = 1$. Three $M_L = +1, 0$, and -1 states, each with three $M_S = +1, 0$, and -1 states, nine $|M_L, M_S\rangle$ states in all, are marked with an ×. With $L = 1$ and $S = 1$, J can be 2, 1, or 0, so the nine states also sort out as five $^{2S+1}L_J = {}^3P_2$ states, three $^3P_1$ states, and one $^3P_0$ state. Of these, the ground state is $^3P_0$ (see Sec. 2(c)).

Only one entry remains, with $M_L = M_S = 0$, a $^1S_0$ state. Altogether then, there are five

^1D states, nine ^3P states, and one ^1S state. Fifteen constituent $|M_L, M_s\rangle$ states, 15 system $|J, M\rangle$ states, the same number because they are just different ways to span the same space.

If two electrons with $\ell = 1$ had *different* values of n, the exclusion principle would already be satisfied. Then L could be any of 2, 1, or 0, and S either of 1 or 0. That would be 36 states in all (count them) instead of 15.

It gets more complicated with more electrons and with higher values of ℓ, but there are simplifications. For example, the possible $^{2S+1}L_J$ states for $6-1 = 5$ electrons in a p subshell are the same as for one p electron; the possible states for $6-2$ p electrons are the same as for two p electrons; and so on. To see this, in the first line of the above table for two p electrons, remove *that* line's two arrows from the completely filled six-p table, $\uparrow\downarrow \ \uparrow\downarrow \ \uparrow\downarrow$:

$m_\ell =$	$+1$	0	-1		$m_\ell =$	$+1$	0	-1
	—	—	—	becomes		—	—	—
1		$\uparrow\downarrow$					$\uparrow\downarrow$	$\uparrow\downarrow$

The four-p line is the "negative" of the two-p line, with $M_L = -2$ and $M_S = 0$. In each of its 15 lines, the M_L and M_S values would be reversed in sign—and therefore in total would make exactly the same set of $^{2S+1}L_J$ states.

The next table, from Herzberg, p. 132, lists the ^{2S+1}L possibilities for any number of equivalent s, p, or d electrons. Superscripts in the configuration column tell the number of electrons, and s^0, p^0, and d^0 mean *no* electrons. This is to emphasize that, as far as angular-momentum quantum numbers are concerned, a filled subshell is the same as no electrons. Numbers in parentheses in the ^{2S+1}L column tell that the state can occur more than once (and how many times). The numbers at the left tell how many rows there would be in the arrow tables. (Once, in my reckless youth, I made all the arrow tables through the d^3-d^7 row.)

	Configuration	Possible ^{2S+1}L states
1	s^0, s^2	^1S
2	s	^2S
1	p^0, p^6	^1S
6	p, p^5	^2P
15	p^2, p^4	$^1S, \ ^1D, \ ^3P$
20	p^3	$^2P, \ ^2D, \ ^4S$
1	d^0, d^{10}	^1S
10	d, d^9	^2D
45	d^2, d^8	$^1S, \ ^1D, \ ^1G, \ ^3P, \ ^3F$
120	d^3, d^7	$^2P, \ ^2D(2), \ ^2F, \ ^2G, \ ^2H, \ ^4P, \ ^4F$
210	d^4, d^6	$^1S(2), \ ^1D(2), \ ^1F, \ ^1G(2), \ ^1I, \ ^3P(2), \ ^3D, \ ^3F(2), \ ^3G, \ ^3H, \ ^5D$
252	d^5	$^2S, \ ^2P, \ ^2D(3), \ ^2F(2), \ ^2G(2), \ ^2H, \ ^2I, \ ^4P, \ ^4D, \ ^4F, \ ^4G, \ ^6S$

Appending all the possible values of J on each ^{2S+1}L state would lengthen the table considerably. For example, the possibilities for the ^3F state in the (d^2, d^8) row are $^3F_4$, $^3F_3$, and $^3F_2$. The *ground* ^{2S+1}L state, as given by Hund, is the state with the highest value of S, given at the right end in each row. And then $J = L + S$ or $|L - S|$ as the shell is more or less than half filled—here, $^3F_4$ and $^3F_2$.

14·4. TWO-ELECTRON SYMMETRIES

We consider again, in a different way, two electrons in a p subshell. In Sec. 11·3(b), we wrote the *system* spin states, $|S, M\rangle$, of two electrons in terms of the *constituent* spin states, $|m_1, m_2\rangle$. The $S = 1$ states were

$$|S, M\rangle = |1, +1\rangle = |+ +\rangle, \qquad |1, 0\rangle = \frac{1}{\sqrt{2}}(|+ -\rangle + |- +\rangle), \qquad |1, -1\rangle = |- -\rangle ,$$

where $+$ and $-$ stand for $m = +1/2$ or $-1/2$. Interchanging the m_1 and m_2 values in these states leaves them unchanged. Interchanging the m_1 and m_2 values in the $S = 0$ state changes its sign,

$$|S, M\rangle = |0, 0\rangle = \frac{1}{\sqrt{2}}(|+ -\rangle - |- +\rangle) \qquad \longrightarrow \qquad \frac{1}{\sqrt{2}}(|- +\rangle - |+ -\rangle) = -|0, 0\rangle .$$

The $S = 1$ states are *symmetric* under the interchange; the $S = 0$ state is *antisymmetric*.

The total orbital angular momentum L of two p electrons can be 2, 1, or 0. Using the table of Clebsch-Gordan coefficients (Sec. 11·5) for $j_1 = j_2 = 1$, we write the nine $|L, M\rangle$ states in terms of the orbital $|m_1, m_2\rangle$ states,

$$|L, M\rangle = |2, +2\rangle = |+1, +1\rangle$$

$$|2, +1\rangle = \frac{1}{\sqrt{2}} (|+1, 0\rangle + |0, +1\rangle)$$

$$|2, 0\rangle = \frac{1}{\sqrt{6}} (|+1, -1\rangle + 2|0, 0\rangle + |-1, +1\rangle)$$

$$|2, -1\rangle = \frac{1}{\sqrt{2}} (|0, -1\rangle + |-1, 0\rangle)$$

$$|2, -2\rangle = |-1, -1\rangle$$

$$|L, M\rangle = |1, +1\rangle = \frac{1}{\sqrt{2}} (|+1, 0\rangle - |0, +1\rangle)$$

$$|1, 0\rangle = \frac{1}{\sqrt{2}} (|+1, -1\rangle - |-1, +1\rangle)$$

$$|1, -1\rangle = \frac{1}{\sqrt{2}} (|0, -1\rangle - |-1, 0\rangle)$$

$$|L, M\rangle = |0, 0\rangle = \frac{1}{\sqrt{3}} (|+1, -1\rangle - |0, 0\rangle + |-1, +1\rangle) .$$

The five $L = 2$ states and the one $L = 0$ state are *symmetric* on interchanging the m_1 and m_2 values; the three $L = 1$ states are *antisymmetric*. Note that the $|L, M\rangle = |1, 0\rangle$ state has no $|m_1, m_2\rangle = |0, 0\rangle$ term in it, because $|0, 0\rangle$ is symmetric on making the interchange.

Until now, the rule for applying the exclusion principle has been to forbid any two electrons in a system to share all the same quantum numbers. Another statement of the principle is that the state function for the electrons has to be antisymmetric overall. For *two* electrons, we can write the total state function as this product,

$$\text{state function} = \text{orbital function} \times \text{spin function} .$$

To get overall antisymmetry, we can combine symmetric spin-1 states with antisymmetric orbital $L = 1$ (P) states, or the antisymmetric spin-0 state with symmetric $L = 2$ (D) or $L = 0$ (S) states. If the Pauli principle could be ignored, the possible states would be

$$^3D,\ ^3P,\ ^3S$$
$$^1D,\ ^1P,\ ^1S\ .$$

Of these, only the ^1D, ^3P, and ^1S states are overall antisymmetric—the same states we got by putting two arrows into a table in the 15 ways that m_ℓ and m_s were not both the same.

For two d electrons, the total orbital L can be 4, 3, 2, 1, or 0. A large enough table of Clebsch-Gordan coefficients would show that the $L = 4$ (G), $L = 2$ (D), and $L = 0$ (S) states are symmetric, and that the $L = 3$ (F) and $L = 1$ (P) states are antisymmetric. Thus the antisymmetric states of two d-subshell electrons are

$$^1G,\ ^3F,\ ^1D,\ ^3P,\ \text{and }^1S\ .$$

These are the states in the (d^2, d^8) row in the table at the end of Sec. 3. And again, Hund's rules—highest S, then (if still a choice) highest L, then (if still a choice) lowest/highest J as less than/more than half filled—make $^3F_2$ and $^3F_4$ the ground states of d^2 and d^8 atoms. See Table 2 (but note there that for some of the elements the filling of d shells is not regular).

This same cascade of alternating spin and orbital (anti-)symmetries occurs for two spin-1/2 particles with $\ell = 3$, or 4, or ... For two electrons, the approach of this section is simpler than putting arrows in a table. It is, however, not as simple to write down overall antisymmetric states of three or more electrons. We come back to this later in the chapter. But helium has only two electrons, and we are now equipped to treat that element in some detail.

14·5. THE HELIUM GROUND STATE

The Schrödinger equation for a two-electron atom is

$$\hat H\psi = \left(-\frac{\hbar^2}{2m}\nabla_1^2 - \frac{kZe^2}{r_1}\right)\psi + \left(-\frac{\hbar^2}{2m}\nabla_2^2 - \frac{kZe^2}{r_2}\right)\psi + \frac{ke^2}{r_{12}}\psi = E\psi\ .$$

Here $\psi = \psi(\mathbf{r}_1, \mathbf{r}_2)$, where $\mathbf{r}_1$ and $\mathbf{r}_2$ are the position vectors of the two electrons relative to the nucleus (see Fig. 3); $r_{12} = |\mathbf{r}_1 - \mathbf{r}_2|$ is the distance between the electrons; and Ze is the nuclear charge: $Z = 2$ for helium, 3 for singly ionized lithium, and so on. And ∇_1^2 is the Laplacian whose derivatives are with respect to the coordinates (r_1, θ_1, ϕ_1) of the "first" electron, etc. There are two kinetic-energy terms (the ∇^2 terms), and three potential-energy terms (the $1/r$ terms). The Hamiltonian is symmetrical under interchange of the labels 1 and 2 on the coordinates of the two electrons because in all respects, including their interactions with the nucleus and with each other, the electrons are indistinguishable.

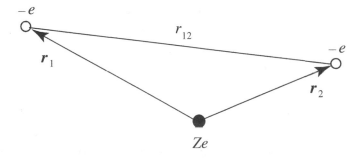

Figure 3. The coordinates for a two-electron atom.

(a) Experimental results—Figure 4 shows the experimental values of the lowest energy levels of helium when only *one* of the electrons is excited. (States with both electrons in higher states would in Fig. 4 have positive energies and are highly unstable.) The zero of energy is for one electron in the $1s$ state, and the other at rest at infinity. The ground state is at -24.6 eV; that is, the ionization energy is 24.6 eV. The figure shows the total-electron-spin $S = 0$ and $S = 1$ levels separately. In Sec. 2(c), we saw that the ground-state electron configuration is $1s^2$ and the angular-momentum state is $^{2S+1}L_J = {}^1S_0$. There is no $1s^2$ state in the $S = 1$ half of the figure because an s^2 state is spatially symmetric, and then the spin state can only be the antisymmetric $S = 0$.

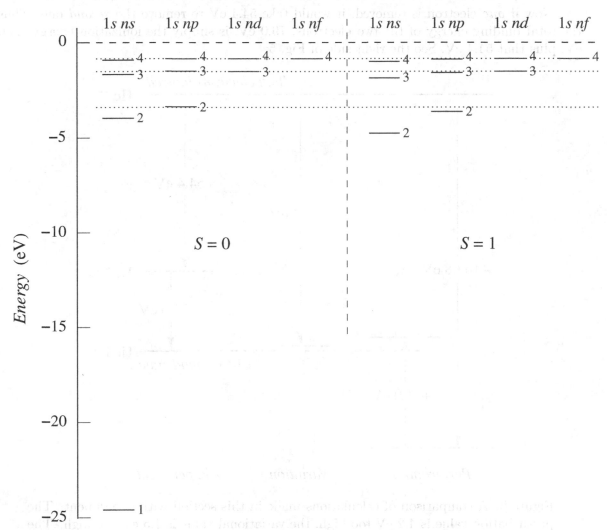

Figure 4. The ground state and first few excited states of helium when only *one* of the electrons is excited. This is indicated by the notation $1s\,n\ell$; numbers on levels tell n. The zero of energy is for one electron in the $1s$ state, and the other at rest at infinity. The $S = 0$ states are sometimes called para-helium, the $S = 1$ states ortho-helium. The dotted lines show the $n = 2$, 3, and 4 energies of hydrogen. The corresponding levels of helium are nearly the same, because the inner ($1s$) electron reduces the effective charge the outer ($n\ell$) electron sees from $+2e$ to about $+e$.

277

The bound-state energies of a single electron in the field of a nuclear charge Ze are

$$E_n = -\frac{Z^2}{n^2} \times 13.6 \text{ eV}$$

(Sec. 9·2). If the two electrons in helium did not interact, the ground-state energy of each of them would be $4 \times (-13.6) = -54.4$ eV, and it would take 108.8 eV to move both electrons to infinity. The experimental value to remove both electrons is 79.0 eV, 29.8 eV less than 108.8 eV. This is of course due to the repulsion between the electrons, which is positive in itself and pushes the electrons farther from the nucleus so that the binding is weaker.

Now if *one* electron is removed, it would take 54.4 eV to remove the *second* one. Thus the total binding energy of the two electrons, 79.0 eV, is simply the ionization energy, 24.6 eV, plus that 54.4 eV. See the right side of Fig. 5.

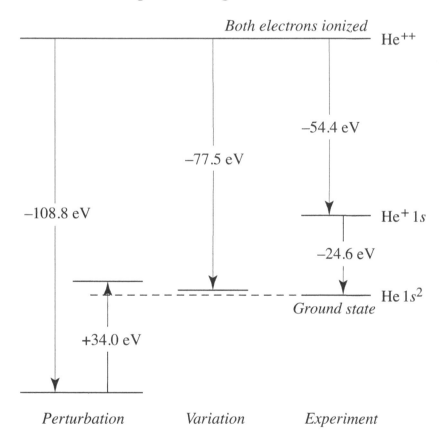

Figure 5. A comparison of calculations made in this section with experiment. The perturbation value is 4.2 eV too high, the variational value is 1.5 eV too high. The zero of energy for Fig. 4 is at the He$^+$ 1s level in this figure.

(b) **Perturbation theory**—Perturbation theory is one way to approximate the difference between -108.8 and -79.0 eV. The perturbation $\hat{H}'$ is that extra term $V(\mathbf{r}_1, \mathbf{r}_2) = ke^2/r_{12}$ in $\hat{H}$. Then the first-order correction E' to the unperturbed $E^0 = -108.8$ eV is

$$E' = \left\langle \psi_{grd}^0 \left| \frac{ke^2}{r_{12}} \right| \psi_{grd}^0 \right\rangle .$$

278

Here $\psi^0_{grd} = \psi^0_{grd}(r_1, r_2)$ is the product of two unperturbed hydrogen-like ground-state wave functions when the nuclear charge is Ze,

$$\psi^0_{grd}(r_1, r_2) = \frac{Z^3}{\pi a_0^3} e^{-Zr_1/a_0} e^{-Zr_2/a_0} .$$

This spatial state is symmetric on interchanging r_1 and r_2. (Why is the Bohr radius a_0 divided by Z everywhere here?) The E' integral is over three coordinates for each electron, and the distance r_{12} between the electrons involves all six coordinates. This makes the calculation difficult, and we leave it for the next section. The result, however, is simple:

$$E' = +\frac{5}{4} Z \times 13.6 \text{ eV} .$$

For $Z = 2$ this is 34.0 eV, which added to $E^0 = -108.8$ eV is -74.8 eV. This overshoots the actual value, -79.0 eV, by 4.2 eV (Fig. 5). But it is surprisingly good considering that the perturbation $\hat{H}'$ is not small compared to the other terms in the Hamiltonian.

(c) **The variational method**—A variational calculation with even a simple one-parameter guess at the wave function does better. (See Sec. 3·3 for the variational method.) We try a (normalized) trial wave function that has the same *form* as ψ^0_{grd} above,

$$\psi_T(r_1, r_2) = \frac{Z'^3}{\pi a_0^3} e^{-Z'r_1/a_0} e^{-Z'r_2/a_0} ,$$

where Z' will be the variational parameter. The rationale is that each electron partially screens the other from the whole nuclear charge Z. The variational method is to calculate

$$E_T(Z') = \langle \psi_T | \hat{H} | \psi_T \rangle ,$$

and then minimize E_T with respect to Z'. Because of the screening, we expect to get a value of Z' that is somewhat smaller than Z. In the calculation, $\hat{H}$ is the *whole* Hamiltonian given at the very start of this section. And the Hamiltonian itself keeps Z, not Z', in the electron-nucleus potential-energy terms. In the variational method, the *trial wave function* is a function of the variational parameter(s), the *Hamiltonian* is not.

There are five parts to $\langle \psi_T | \hat{H} | \psi_T \rangle$: (1) two kinetic-energy integrals of identical form; (2) two potential-energy integrals of identical form; and (3) the electron-electron potential-energy integral.

(1) Each of the kinetic-energy operators in $\hat{H}$ involves only one of the electrons, so the integral over the coordinates of the other particle is equal to one, from normalization. Thus

$$-\frac{\hbar^2}{2m} \langle \psi_T(r_1, r_2) | \nabla_1^2 | \psi_T(r_1, r_2) \rangle = -\frac{\hbar^2}{2m} \langle \psi_T(r_1) | \nabla_1^2 | \psi_T(r_1) \rangle ,$$

where $\psi_T(r_1) = (Z'^3/\pi a_0^3)^{1/2} e^{-Z'r_1/a_0}$. We calculated this integral (with $\psi_T \propto e^{-br}$ instead of $e^{-Z'r/a_0}$) in Sec. 3·3. A vector identity (valid only when the integration is over all space) made things easy:

$$-\frac{\hbar^2}{2m} \int \psi_T^*(r_1) \nabla_1^2 \psi_T(r_1) \, d^3\mathbf{r_1} = +\frac{\hbar^2}{2m} \int |\nabla_1 \psi_T(r_1)|^2 d^3\mathbf{r_1}$$

$$= \frac{\hbar^2}{2m} \left(\frac{Z'}{a_0}\right)^2 \int |\psi_T(r_1)|^2 d^3\mathbf{r_1} = +(Z')^2 \times 13.6 \text{ eV} ,$$

because ψ_T is normalized and $\hbar^2/2ma_0^2 = 13.6$ eV. There is one of these for each electron.

(2) The calculation for the first potential-energy term in $\hat{H}$ is

$$- \int \psi_T^*(r_1) \left(\frac{kZe^2}{r_1} \right) \psi_T(r_1)\, d^3\mathbf{r_1} = -kZe^2 \frac{Z'^3}{\pi a_0^3} 4\pi \int_0^\infty r_1\, e^{-2Z'r_1/a_0}\, dr_1$$

$$= -ZZ' \frac{ke^2}{a_0} = -2ZZ' \times 13.6 \text{ eV} .$$

The integral is the familiar $\int_0^\infty x^n e^{-bx}\, dx = n!/b^{n+1}$, and $ke^2/2a_0 = 13.6$ eV. The Z comes from $V(r)$, the Z' from ψ_T. The calculation for the second potential-energy term is identical.

(3) The electron-electron integral is, with Z replaced by Z', identical in form to the perturbation calculation, so that

$$\left\langle \psi_T(r_1, r_2) \left| \frac{ke^2}{r_{12}} \right| \psi_T(r_1, r_2) \right\rangle = +\frac{5}{4} Z' \times 13.6 \text{ eV} .$$

The sum of the two identical kinetic-energy integrals, the two identical electron-nucleus potential-energy integrals, and the electron-electron potential-energy integral is

$$E_T(Z') = \left[2(Z')^2 - 4ZZ' + 5Z'/4 \right] \times 13.6 \text{ eV} .$$

We minimize $E_T(Z')$ with respect to Z',

$$\frac{dE_T}{dZ'} = (4Z' - 4Z + 5/4) \times 13.6 \text{ eV} = 0 ,$$

and get $Z' = Z - 5/16$. For $Z = 2$, this is 1.6875, which as we expected is smaller than Z. Putting this value back into E_T, we get (do the algebra)

$$(E_T)_{\min} = -2(Z - 5/16)^2 \times 13.6 = -77.5 \text{ eV} \qquad (\text{for } Z = 2) .$$

As always with a variational calculation, this is an upper limit, and it is only 1.5 eV above the experimental value, -79.0 eV. With a computer to do the work, we could with a many-parameter trial ψ_T do much better. But this is pretty good. Figure 6 shows, for $Z = 2$, $E_T(Z')$ as a function of Z'.

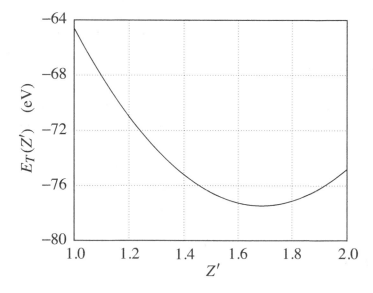

Figure 6. A plot of $E_T(Z')$ as a function of Z' when $Z = 2$ (helium). The minimum of $E_T(Z')$ is -77.5 eV at $Z' = 2 - 5/16 = 1.6875$.

14·6. THE ELECTRON-ELECTRON REPULSION INTEGRAL

It remains to calculate the integral for the interaction energy between the two electrons,

$$E' = \left\langle \psi^0_{grd}(r_1, r_2) \left| \frac{ke^2}{r_{12}} \right| \psi^0_{grd}(r_1, r_2) \right\rangle = +\frac{5}{4} Z \times 13.6 \text{ eV} .$$

Here $\psi^0_{grd}(r_1, r_2)$ is, as earlier, the product of two hydrogen-like ground-state wave functions for a nuclear charge Ze,

$$\psi^0_{grd}(r_1, r_2) = \frac{Z^3}{\pi a_0^3} e^{-Zr_1/a_0} e^{-Zr_2/a_0} .$$

The integral is over six coordinates, and there is more than one way to do it. Here we turn it into an electrostatics problem.

To make this approach clear, we first rewrite the expectation value of the potential energy of a *hydrogen* atom when the wave function depends only on r, not on θ or ϕ. Then

$$\langle V \rangle = \langle \psi | V | \psi \rangle = \int \psi^*(r) \left(\frac{-ke^2}{r} \right) \psi(r) \, d^3\mathbf{r} = \int (-e|\psi(r)|^2) \left(\frac{ke}{r} \right) d^3\mathbf{r} .$$

The rearranged integrand is the product of a negative charge density $\rho_e(r) = -e|\psi(\mathbf{r})|^2$ for the electron, and a positive potential (not potential energy) $\varphi_p(r) = ke/r$ due to the proton; see Fig. 7(a). Thus $\langle V \rangle$ may be written

$$\langle V \rangle = \int \rho_e(r) \, \varphi_p(r) \, d^3\mathbf{r} .$$

The integral over space of $\rho_e(r)$ is of course $-e$.

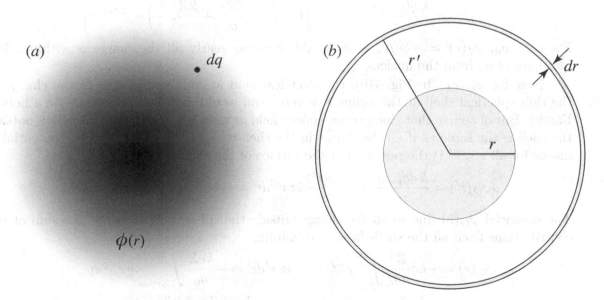

Figure 7. (a) The potential energy of an infinitesimal charge dq in a potential $\varphi(r)$ is $\varphi(r) \, dq$. (b) Construction for the calculation of the potential $\varphi(r)$ at radius r due to a charge density $\rho(r) = -e|\psi(r)|^2$.

We cast the integral for E' into this $\rho(r)\varphi(r)$ form. First we find the potential $\varphi_1(r)$ due to the charge density of one of the electrons. The density is

$$\rho_1(r) = -e\,|\psi_1(r)|^2 = -\frac{Z^3 e}{\pi a_0^3}\,e^{-2Zr/a_0} \ .$$

The hard part will be to calculate $\varphi_1(r)$ from it. When we have that, we multiply it by the identical charge density $\rho_2(r)$ of the other electron and integrate,

$$E' = \langle V \rangle = \int \rho_2(r)\,\varphi_1(r)\,d^3\mathbf{r} \ .$$

Figure 7(b) shows the volume of space inside a radius r, and (for later) a thin spherical shell with $r' > r$. The potential $\varphi_1(r)$ at a radius r is the sum of contributions from the charge inside (I) and outside (O) the sphere of radius r,

$$\varphi(r) = \varphi_I(r) + \varphi_O(r)$$

(for a moment, we omit the subscript 1). We calculate $\varphi_I(r)$ and $\varphi_O(r)$ separately.

The calculations use Gauss's law: The electrical field and potential *outside* a spherically symmetric charge distribution are the same as if all the charge were at the center of the sphere. Let $y \equiv 2Zr'/a_0$. The total charge inside the sphere of radius r is

$$q(r' < r) = -\frac{Z^3 e}{\pi a_0^3}\int_0^r e^{-2Zr'/a_0}4\pi\,r'^2 dr' = -\frac{e}{2}\int_0^{2Zr/a_0} y^2 e^{-y}dy$$

$$= +\frac{e}{2}\,(2 + 2y + y^2)e^{-y}\Big|_0^{2Zr/a_0} = -e\left[1 - \left(1 + \frac{2Zr}{a_0} + \frac{2Z^2 r^2}{a_0^2}\right)e^{-2Zr/a_0}\right] \ .$$

(The integral is a look-up; the range is not infinite. Checks: for $r = 0$, $q = 0$; for $r = \infty$, $q = -e$.) Thus the potential on the *surface* of the sphere of radius r, due to all the charge *inside* the sphere, is

$$\varphi_I(r) = \frac{k\,q(r' < r)}{r} = -\frac{ke}{r}\left[1 - \left(1 + \frac{2Zr}{a_0} + \frac{2Z^2 r^2}{a_0^2}\right)e^{-2Zr/a_0}\right] \ .$$

For $r \gg a_0$, $\varphi_I(r) \approx -ke/r$, as it should, because nearly all the charge is within a few multiples of a_0 from the nucleus.

Now for $\varphi_O(r)$. In Fig. 7(b), the electrical field at radii $r < r'$ due to the charge in the thin spherical shell at the radius r' is zero (you would just float around inside a hollow Earth). But of course that charge does make a field at radii $r > r'$. The field it makes outside the shell is the same as if all the charge in the shell were at the origin; and the potential it makes for all $r < r'$ is the potential at the surface of the shell,

$$d\varphi_O(r') = \frac{k\,dq_{shell}}{r'} = -\frac{k\rho(r')}{r'}4\pi\,r'^2 dr' = -ke\frac{Z^3}{\pi a_0^3}e^{-2Zr'/a_0}4\pi\,r'dr' \ .$$

The potential $\varphi_O(r)$ due to all the charge outside the sphere of radius r is the sum of the contributions from all the shells from r to infinity,

$$\varphi_O(r) = -ke\frac{Z^3}{\pi a_0^3}\int_r^\infty e^{-2Zr'/a_0}4\pi\,r'dr' = -\frac{keZ}{a_0}\int_{2Zr/a_0}^\infty y\,e^{-y}dy$$

$$= +\frac{kcZ}{a_0}\,(1 + y)\,e^{-y}\Big|_{2Zr/a_0}^\infty = -\frac{ke}{r}\left(\frac{Zr}{a_0} + \frac{2Z^2 r^2}{a_0^2}\right)e^{-2Zr/a_0} \ ,$$

where again $y \equiv 2Zr/a_0$. In the last step, top and bottom have been multiplied by r.

Now we add $\varphi_I(r)$ and $\varphi_O(r)$ to get $\varphi(r)$. The second term in $\varphi_O(r)$ cancels the last term in $\varphi_I(r)$, and the sum is (returning the subscript)

$$\varphi_1(r) = \varphi_I(r) + \varphi_O(r) = -\frac{ke}{r}\left[1 - \left(1 + \frac{Zr}{a_0}\right)e^{-2Zr/a_0}\right] .$$

For $r \gg a_0$, $\varphi_1(r) \approx -ke/r$, as it should.

Finally, we integrate the product of the charge density of the *second* electron with the potential due the *first* one. We already have $\rho_2(r)$ because the charge densities of the two electrons are the same,

$$\rho_2(r) = -e\,|\psi(r)|^2 = -\frac{Z^3 e}{\pi a_0^3}\,e^{-2Zr/a_0} .$$

We get

$$E' = \int_0^\infty \rho_2(r)\,\varphi_1(r)\,4\pi r^2\,dr = \frac{4Z^3 ke^2}{a_0^3}\int_0^\infty \left[e^{-2Zr/a_0} - \left(1 + \frac{Zr}{a_0}\right)e^{-4Zr/a_0}\right] r\,dr$$

$$= \frac{Zke^2}{a_0}\int_0^\infty \left(ye^{-y} - ye^{-2y} - \tfrac{1}{2}y^2 e^{-2y}\right) dy .$$

And here, because the range of integration is infinite, we can use $\int_0^\infty x^n e^{-bx}\,dx = n!/b^{n+1}$ again. The three integrals are $+1$, $-1/4$ and $-1/8$, and

$$E' = \left\langle \psi^0 \left| \frac{ke^2}{r_{12}} \right| \psi^0 \right\rangle = \frac{5}{4}Z\frac{ke^2}{2a_0} = \frac{5}{4}Z \times 13.6 \text{ eV} ,$$

as asserted earlier.

So much for the ground state of helium.

14·7. HELIUM EXCITED STATES. EXCHANGE DEGENERACY

(a) **Review**—Figure 8 shows again the lowest *excited* states of helium, all of which lie within 5 eV of $E = 0$, far above the ground state at -24.6 eV. As before, the labelling $1s\,n\ell$ of states tells that one of the electrons remains in the $1s$ state. The total spin S can now be either 0 or 1, because the electrons now have different values of n. Thus each state on one side of Fig. 8 has a (non-identical) twin on the other. The $S = 0$ states, antisymmetric in spin, are symmetric in space; the $S = 1$ states, symmetric in spin, are antisymmetric in space (Sec. 4). For example, in the approximation that ignores the interaction between the two electrons, the $S = 1$, $1s\,2p$ state function is

$$\text{space} \times \text{spin} = \left[\psi_{100}(\mathbf{r}_1)\psi_{21m_\ell}(\mathbf{r}_2) - \psi_{21m_\ell}(\mathbf{r}_1)\psi_{100}(\mathbf{r}_2)\right] \times |1, m_s\rangle.$$

The ψ's here are hydrogen-like wave functions with $Z = 2$. For the $2p$ state, m_ℓ is any of $+1$, 0, or -1; for the spin, m_s is any of $+1$, 0, or -1. Absent the electron-electron interaction, the $1s$ and $2p$ binding energies would be 54.4 and 13.6 eV, $Z^2 = 4$ times the energies of a single electron in the $n = 1$ and $n = 2$ states of hydrogen.

A rough Bohr-like model that does take into account the interaction of the electrons goes like this: Since the $1s$ electron sees nearly the full nuclear charge $2e$, its binding energy is about 54.4 eV (its Bohr radius is about $\frac{1}{2}a_0$); and being close to the nucleus, it reduces the charge that the more distant $n\ell$ electron sees to about e (in which case, for $n = 2$, its Bohr radius is about $4a_0$). Thus the energies of the outer electron are approximately those of the excited states of hydrogen, as Fig. 8 shows.

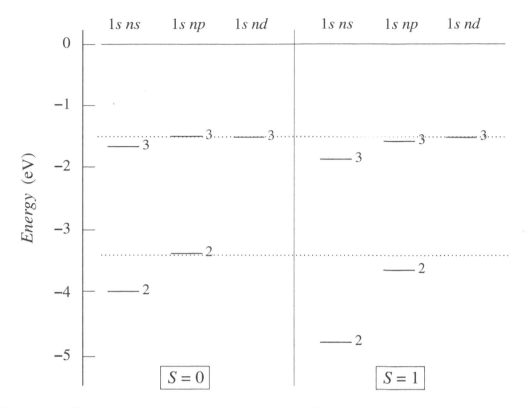

Figure 8. The lowest excited states of helium. The ground state is at -24.6 eV. Numbers on levels tell n. The $S = 0$ (para) states lie higher than the $S = 1$ (ortho) states, but the differences decrease rapidly with n and with ℓ. The dotted lines show the $n = 2$ and $n = 3$ energies of hydrogen, at -3.40 and -1.51 eV.

The differences between the energies of corresponding $S = 0$ and $S = 1$ levels in Fig. 8 might bring to mind the hyperfine splitting of the ground state of hydrogen (Sec. 11·1), the source of the 21-cm line. That splitting was due to the interaction of the intrinsic magnetic moments of the (non-identical) electron and proton, a spin-spin interaction. But here the root cause for the splitting between the spin states is the exclusion principle. This is perhaps best explained with a simple model.

(b) Identical particles and exchange degeneracy—We consider the states of two electrons in a *one*-dimensional box of width L. For the moment, we neglect the interaction between the electrons, and start with two-particle spatial wave functions that are products of one-particle wave functions,

$$\psi_{n_1 n_2}(x_1, x_2) = \psi_{n_1}(x_1)\psi_{n_2}(x_2) = \frac{2}{L}\sin\left(n_1\frac{\pi x_1}{L}\right)\sin\left(n_2\frac{\pi x_2}{L}\right).$$

Figure 9(a) shows contours for the product of the sines themselves (that is, without the normalization factor $2/L$) when $n_1 = n_2 = 1$. Then the value at the center of the contours is 1, and the contours are drawn for values 0.9, 0.5, and 0.1. For brevity in Fig. 9, $y_i \equiv \pi x_i/L$. This spatial wave function is symmetric on reflection across the $y_1 = y_2$ diagonal, or, equivalently, under the exchange of labels 1 and 2. Therefore the spin state is the antisymmetric $S = 0$. This is analogous to the ground state of helium.

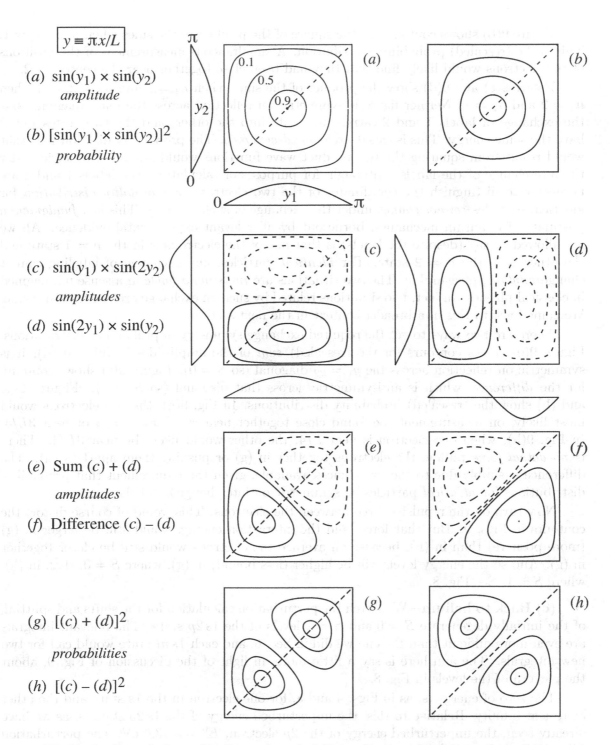

$y \equiv \pi x/L$

(a) $\sin(y_1) \times \sin(y_2)$
    amplitude

(b) $[\sin(y_1) \times \sin(y_2)]^2$
    probability

(c) $\sin(y_1) \times \sin(2y_2)$
    amplitudes

(d) $\sin(2y_1) \times \sin(y_2)$

(e) Sum (c) + (d)
    amplitudes

(f) Difference (c) − (d)

(g) $[(c) + (d)]^2$
    probabilities

(h) $[(c) − (d)]^2$

Figure 9. Rescaled amplitudes and probabilities for two non-interacting electrons in a *one*-dimensional box. The rescaling sets the value at the center of each set of contours equal to one (or, when the contours are dashed, minus one). In each panel, the contours are at 0.9, 0.5, and 0.1 times that peak value, as indicated in (a). I thank Piotr Zyla for the contours in the bottom four panels. This diagrammatic approach follows C.W. Sherwin, *Introduction to Quantum Mechanics*.

285

Figure 9(b) shows contours for the square of the product of the sines when $n_1 = n_2 = 1$. This is the (rescaled) probability distribution. A simultaneous measurement of the positions of the electrons would likely find that they had been close together near the center, $L/2$.

Figure 9(c) and 9(d) show the product of the sines when $n_1 = 1$ and $n_2 = 2$, and when $n_1 = 2$ and $n_2 = 1$. Neither figure is symmetric on reflection across the main diagonal, but the exchange of labels 1 and 2 carries one state into the other, and the two states would have the same energy. This is called *exchange degeneracy*. The probability distributions that would result from squaring the two product wave functions would not be symmetric under the interchange of the labels. Although, for purposes of calculating, the labels 1 and 2 are necessary to distinguish the coordinates of the two electrons, a *probability distribution* for identical particles *cannot change* under the exchange of such labeling. This is a *fundamental* postulate of quantum mechanics, borne out by all relevant experimental evidence. All we are allowed to say (despite the labels) is that one of the electrons is in the $n = 1$ state and the other is in the $n = 2$ state. For *identical* particles, any exchange of labeling cannot change *anything* measurable. The two situations are *indistinguishable* in a sense unimagined in classical physics. In fact, two situations related by such an exchange are one and the same. We come back to this in a broader context in the next section.

There are two ways to get the required exchange symmetry of probability distributions. Figure 9(e) shows contours for the (rescaled) *sum* of the amplitudes in (c) and (d); it is symmetric on reflection across the $y_1 = y_2$ diagonal (so $S = 0$). Figure 9(f) shows contours for the *difference*, which is antisymmetric across that diagonal (so $S = 1$). Figures 9(g) and (h) show the (rescaled) probability distributions. In Fig. 9(g), the two electrons would most likely, on measurement, be found close together near $x_1 = x_2 = L/4$ or near $3L/4$. In Fig. 9(h), when one electron is near $L/4$, the other would likely be near $3L/4$. There is *no actual force* pulling the electrons together in (g) or pushing them apart in (h). The difference is simply due to the two choices available, given the requirement that probability distributions for *identical* particles be symmetric on interchange of labels.

Now turn on the repulsive force between the electrons. This would of course distort the contours in Figs. 9. But that force, and the interaction energy, would still be larger in (g) (more positive) than in (h), because on average the electrons would still be closer together in (g). And so the energy levels will be higher (less bound) in (g), where $S = 0$, than in (h), where $S = 1$. See Fig. 8.

(c) Back to helium—We sketch the perturbation calculation for the shifts and splitting of the initially degenerate $S = 0$ and $S = 1$ levels of the $1s\,2p$ state. The relevant integrals are even more difficult than the one we did in Sec. 6, and each $1s\,n\ell$ state would call for two new integrals. The aim here is say a little more, in light of the discussion of Fig. 9, about the pattern of the levels in Fig. 8.

The zero of energy is, as in Figs. 4 and 8, for one electron in the $1s$ state and the other at rest at infinity. Relative to this, the unperturbed energy of the $1s\,2p$ state is, as we have already seen, the unperturbed energy of the $2p$ electron, $E^0 = -13.6$ eV. The perturbation $\hat{H}'$ is $V(\mathbf{r}_1, \mathbf{r}_2) = ke^2/r_{12}$, where r_{12} is the distance between the electrons. In terms of single-particle $\psi_{n\ell m_\ell}$ states, the unperturbed two-particle states are

$$|1s, 2p\rangle = \psi_{100}(\mathbf{r}_1)\,\psi_{21m_\ell}(\mathbf{r}_2) \qquad \text{and} \qquad |2p, 1s\rangle = \psi_{21m_\ell}(\mathbf{r}_1)\,\psi_{100}(\mathbf{r}_2)\,,$$

where $m_\ell = +1$, 0, or -1. We could start with the symmetric and antisymmetric superpositions of these states, but shall see them emerge as the energy eigenstates from the calculation.

With two degenerate basis states, we get a 2-by-2 matrix eigenvalue problem, familiar from Chaps. 10 and 13. The diagonal matrix elements are

$$J \equiv \langle 1s, 2p | \hat{H}' | 1s, 2p \rangle = \langle 2p, 1s | \hat{H}' | 2p, 1s \rangle .$$

The two integrals are equal because the integrands are

$$|\psi_{100}(\mathbf{r}_1)|^2 \times ke^2/r_{12} \times |\psi_{21m_\ell}(\mathbf{r}_2)|^2 \quad \text{and} \quad |\psi_{21m_\ell}(\mathbf{r}_1)|^2 \times ke^2/r_{12} \times |\psi_{100}(\mathbf{r}_2)|^2 ,$$

and the integrations for each of $\mathbf{r}_1$ and $\mathbf{r}_2$ are over all of space. J, called the direct integral, is obviously positive, and (perhaps not obviously) is the same for $m_\ell = +1$, 0, and -1. The off-diagonal matrix elements are

$$K \equiv \langle 1s, 2p | \hat{H}' | 2p, 1s \rangle = \langle 2p, 1s | \hat{H}' | 1s, 2p \rangle .$$

The two integrals are equal in magnitude because one turns into the other on exchange of the labels 1 and 2. K is called the exchange integral, and like J it is real and positive. The perturbation matrix for $\hat{H}'$ is

$$\hat{H}' = \begin{pmatrix} H'_{11} & H'_{12} \\ H'_{21} & H'_{22} \end{pmatrix} = \begin{pmatrix} J & K \\ K & J \end{pmatrix}.$$

The eigenvalue equation is

$$\begin{pmatrix} J & K \\ K & J \end{pmatrix} \begin{pmatrix} c_1 \\ c_2 \end{pmatrix} = \lambda \begin{pmatrix} c_1 \\ c_2 \end{pmatrix},$$

and the eigenvalues are $E'_\pm = J \pm K$: Remember $\lambda_\pm = a \pm b$ (Secs. 10·1 and 13·4)! The corresponding eigenstates are of course the symmetric and antisymmetric superpositions of the initial basis states,

$$\frac{1}{\sqrt{2}} \big(|1s, 2p\rangle + |2p, 1s\rangle \big) \quad \text{and} \quad \frac{1}{\sqrt{2}} \big(|1s, 2p\rangle - |2p, 1s\rangle \big) .$$

The states with the definite energies are the states with the definite symmetries. Figure 10 shows the structure of the solution. The symmetric space function gives the higher energy because the electrons are closer together. The integrals can of course be done, but more advanced methods, with integrations done by computer, would be needed to do a good job.

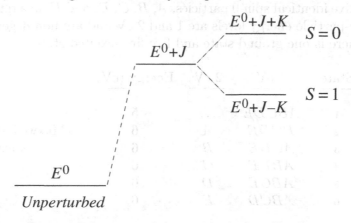

Figure 10. The pattern of the perturbation calculation for the energy levels of the $1s\,2p$ states of helium. The same pattern, with different values of E^0, J, and K, would apply to other $1s\,n\ell$ states.

14·8. FERMIONS AND BOSONS. HOW TO COUNT. SYMMETRIES

(a) Two kinds of particles—A particle, whether it be elementary or composite, with an intrinsic angular momentum that is half-integral is a *fermion*. Examples are the electron, proton, neutron, and any atom made up of an odd number of protons + neutrons + electrons. Any set of *identical* fermions in the same system obeys the Pauli exclusion principle: No two of them in the same system can share all the same quantum numbers. An equivalent statement is that the total state function (space plus spin) must be antisymmetric under the exchange of any two of the particles. Then the *probability* distribution will be symmetric under that interchange. So far, this whole chapter has been about the consequences of the exclusion principle for the electrons in an atom and for the conduction electrons in a block of metal.

A particle, whether it be elementary or composite, with an intrinsic angular momentum that is integral is a *boson*. Examples are the photon, π mesons, the deuteron, and any atom made up of an even number of protons + neutrons + electrons. The total state function for identical bosons must be *symmetric* under the exchange of any two of them, and then the probability distribution will be symmetric too. Bosons do *not* obey the exclusion principle. Just the opposite! They *like* to be in the same state.

The rules for a composite structure—those made of an odd number of spin-1/2 particles have half-integral spin, those made of an even number have integral spin—follow from the rules for the addition of angular momenta. Helium-4 (two protons, two neutrons, two electrons) is a boson; the rare isotope helium-3, with only one neutron in its nucleus, is a fermion. Composite particles are not identical unless they are in the same *internal* state (such as ^{4}He atoms all in the ground state). The properties of ^{4}He and ^{3}He in bulk at low temperatures are entirely different. The overall requirement for both fermions and bosons is that nothing *measurable* distinguishes states related by interchanges of identical particles. Such states are in fact, in a profound way, all the same *single* state.

(b) Three ways to count—Classical statistical mechanics counts states using *Maxwell-Boltzmann* (MB) statistics. Identical particles, such as the molecules of a gas of oxygen, are thought of as being distinguishable—even when one cannot imagine how one could mark them. Consider five identical spin-0 particles, A, B, C, D, and E, in a quantum system whose two lowest single-particle energy levels are 1 and 2 eV and are non-degenerate. According to MB statistics, there is one ground state and five first-excited states:

State	1 eV	2 eV	Energy (eV)	
1	$ABCDE$	...	5	
2	$BCDE$	A	6	Maxwell−Boltzmann
3	$ACDE$	B	6	statistics
4	$ABDE$	C	6	
5	$ABCE$	D	6	
6	$ABCD$	E	6	

There are five states with energy 6 eV, because there are five choices for which of the particles (assumed to be distinguishable) is in the 2-eV state.

288

Now put five identical spin-0 bosons (such as π^0s, or ^{4}He atoms) into the above system. There is only *one* state with energy 6 eV:

State	1 eV	2 eV	Energy (eV)	
1	● ● ● ● ●	...	5	Bose−Einstein
2	● ● ● ●	●	6	statistics

In quantum mechanics, when the particles are identical, the *number* of particles in each state is the end of it; the question about *which* of those particles are in which state becomes a non-question. The particles are *indistinguishable* in a way not imagined before 1900, and not fully understood until the 1920s.

Lastly, put five identical spin-1/2 particles into the same system, which also has 3- and 4-eV states. We already know that no more than two can go in each spatial state:

State	1 eV	2 eV	3 eV	4 eV	Energy (eV)	
1	↑↓	↑↓	↑	...	9	
2	↑↓	↑↓	↓	...	9	Fermi−Dirac
3	↑↓	↑↓	...	↑	10	statistics
4	↑↓	↑↓	...	↓	10	
5	↑↓	↑	↑↓	...	10	
6	↑↓	↓	↑↓	...	10	

There are two ground states and four first-excited states. All through this chapter we have being doing something that from the classical point of view is radical: We did not assume that the particles were somehow, even if only *in principle*, distinguishable; we did not count all the possible permutations of the 10^{23} or so conduction electrons between (n_x, n_y, n_z) states in a block of metal. There was just *one* state. We snuck in quantum statistics early on.

(c) Symmetries—As we have seen with helium, the spatial wave function for two electrons can be either symmetric or antisymmetric, but only because the spin state can compensate by being either antisymmetric or symmetric. However, ^{4}He atoms in the ground state and π^0s have no spin, and as bosons the spatial wave function of a system of two or more of them can only be symmetric. For example, two π^0s can be in the state (a, b) of Fig. 9, or in the state (e, g), but they cannot be in the state (f, h).

As last examples, we consider some simplest three-particle states. Just as we have often, at least to start, written two-particle states as simple products of two one-particle states, we write the three-particle states as simple products of three one-particle states. This would be the starting point, as it was for helium, for calculating energy shifts caused by secondary interactions using perturbation theory.

Bosons—Suppose that a spin-0 boson has eigenstates ψ_a, ψ_b, ψ_c, ..., with energies E_a, E_b, E_c, ... Three of these bosons can all go into any one of the eigenstates, or two can go into one state and one into another, or the three can go into three distinct states. For example,

289

State	E_a	E_b	E_c	Total E
1	$\cdots$	$\bullet\,\bullet\,\bullet$	$\cdots$	$3E_b$
2	$\bullet$	$\cdots$	$\bullet\,\bullet$	$E_a + 2E_c$
3	$\bullet$	$\bullet$	$\bullet$	$E_a + E_b + E_c$

Let $\psi_a(1)$ indicate that particle 1 is in state a, etc. And let ψ_{caa} indicate the three-particle state $\psi_c(1)\psi_a(2)\psi_a(3)$, etc. Then the symmetric linear superpositions for the three examples are

$$\psi_1 = \psi_{bbb}\,, \qquad \psi_2 = \frac{1}{\sqrt{3}}[\psi_{acc} + \psi_{cac} + \psi_{cca}]\,,$$

$$\psi_3 = \frac{1}{\sqrt{6}}[\psi_{abc} + \psi_{acb} + \psi_{cab} + \psi_{cba} + \psi_{bca} + \psi_{bac}]\,.$$

Exchange, say, the second and third subscripts in each term on the right side of ψ_2 and it is the same ψ_2. There are two other linearly independent superpositions of the three states in ψ_2, but neither of them is symmetric and neither of them exists in Nature. There are five other linearly independent superpositions of the six states in ψ_3, but none of them is symmetric and none of them exists in Nature.

Fermions—In Sec. 2(c), we found that the angular-momentum ground state of the elements with three equivalent p electrons is $^{2S+1}L_J = {}^4S_{3/2}$. There are four symmetric $S = 3/2$ spin states, with $M_S = +3/2, +1/2, -1/2,$ and $-3/2$,

$$|S, M_S\rangle = |3/2, M_S\rangle = |\uparrow\uparrow\uparrow\rangle,\ \frac{1}{\sqrt{3}}\big[|\uparrow\uparrow\downarrow\rangle + |\uparrow\downarrow\uparrow\rangle + |\downarrow\uparrow\uparrow\rangle\big],\ \frac{1}{\sqrt{3}}\big[|\downarrow\downarrow\uparrow\rangle + |\downarrow\uparrow\downarrow\rangle\rangle + |\uparrow\downarrow\downarrow\rangle\rangle\big],\ |\downarrow\downarrow\downarrow\rangle.$$

There is a single antisymmetric $L = 0$ orbital state,

$$|L, M_L\rangle = |0, 0\rangle = \frac{1}{\sqrt{6}}\big[|+0-\rangle - |+-0\rangle + |0-+\rangle - |0+-\rangle + |-+0\rangle - |-0+\rangle\big]\,,$$

where $+1$, 0, are -1 are the m_ℓ values of the individual p electrons, and stand for the individual-particle kets $|\ell, m_\ell\rangle = |1, m_\ell\rangle$. Any of the spin states pairing with this orbital state makes an antisymmetric total state. (Why is there no $|000\rangle$ state in the mix for $|L, M_L\rangle = |0, 0\rangle$?)

The table at the end of Sec. 3 lists ^2P and ^2D as the other possible states of three equivalent p electrons. But with three $\ell = 1$ electrons, why is there no $L = 3$ (F) state? Because an $L = 3$ state of three p electrons would be symmetric (why?), and it would need to pair with an antisymmetric spin state. But there *is* no antisymmetric *spin* state of three (or more) electrons, and the calculation no longer splits cleanly into space times spin. Now one has to consider the six *mixed* states, $|m_\ell, m_s\rangle$, with $m_\ell = +1$, 0, -1, and $m_s = +1/2$, $-1/2$. We won't do that here.

Besides, we already know scores of angular-momentum states of electrons that satisfy the exclusion principle, even if we cannot easily write out explicitly antisymmetric state functions for them—because the ground states of the first 96 elements (Table 2, Sec. 2), and *all* the possible states of electrons in s, p, and d subshells (given in the table at the end of Sec. 3), are such states. We can always get the states of equivalent electrons by putting arrows in tables.

(d) Particle symmetries are forever—The Hamiltonian for helium that began Sec. 5 is symmetric under the interchange of the electron labels 1 and 2. We need labels on particle coordinates (and spins) to even set up a Hamiltonian. But for identical particles nothing we can measure can depend on the ordering of the labels: Any scrambling of the labels of identical particles must leave the Hamiltonian unchanged. For two particles, this means that $\hat{H}(1,2) = \hat{H}(2,1)$, as for helium. Here a particle label signifies both its position and spin (if the particle has spin); and the Hamiltonian can include all the interactions that the identical particles have with one another or with other non-identical particles (such as electrons with a nucleus).

Let $\Psi(S)$ be a properly symmetrized state function, including time dependence, for N identical bosons. It can be a superposition of energy eigenstates, $\Psi(S) = \sum \Psi_{E_n}(S) = \sum \psi_{E_n}(S)e^{-iE_n t/\hbar}$. The time-dependent Schrödinger equation is

$$i\hbar \frac{\partial \Psi(S)}{\partial t} = \hat{H}(1, 2, \ldots, N)\Psi(S) \ .$$

The right-hand side is symmetric on interchange of any of the N particle labels because $\hat{H}$ and $\Psi(S)$ are both symmetric. Therefore the infinitesimal increment of $\Psi(S)$ in the infinitesimal time interval dt is symmetric: The symmetry is forever. Similarly, an antisymmetric state $\Psi(A)$ of identical fermions stays antisymmetric forever.

The role of the symmetries of identical particles for understanding the properties of solids, liquids, and gases—of matter in bulk—belongs to a course on quantum statistical mechanics. We will, however, use those symmetries again in Chap. 15 for black-body radiation, and in Chap. 16 for scattering angular distributions of identical bosons and of identical fermions.

PROBLEMS (The first two repeat the last two problems of Chap. 2.)

1. *The three-dimensional degenerate electron gas.*
Derive the equations for the Fermi energy, E_F, and the average energy per electron, $\overline{E}$, for the three-dimensional degenerate electron gas,

$$E_F = \left(3\pi^2 \frac{N}{V}\right)^{2/3} \frac{\hbar^2}{2m} \qquad \text{and} \qquad \overline{E} = \frac{3}{5} E_F \ .$$

2. *The free-electron gas in copper.*
The atomic number of copper is 29, the molar mass is 63.5 g/mol, the density is 8.96 g/cm^3. There is one free electron per atom. Show that the Fermi energy, E_F, is 7.0 eV, and that the corresponding "Fermi temperature," T_F, is 8.2×10^4 K, where $E_F = kT_F$ and $k = 8.617\times10^{-5}$ eV/K is the Boltzmann constant.

The 7.0-eV "depth" of the "Fermi sea" of free electrons in copper is so much larger than ordinary thermal energies ($kT = 1/40$ eV at 290 K) that nearly all the electrons lie too far below the surface to absorb those energies—all nearby states are already occupied.

The most loosely bound electrons in a block of metal are those at the top of the Fermi sea. The figure shows the relation between the depth of the Fermi sea and the minimum energy, ϕ, needed to remove a free electron from a metal; ϕ is called the work function and appears in Einstein's photoelectric equation (see Sec. 1·1).

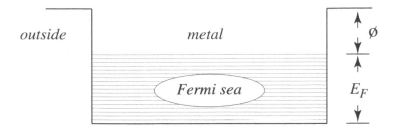

3. *Densities of filled energy eigenstates for the degenerate electron gas.*

Let us write the number dN of energy eigenstates in the infinitesimal range between E and $E + dE$ as

$$dN = \rho(E)\, dE \ ,$$

where $\rho(E)$ is the density of states per unit of energy. At $T = 0$, $\rho(E)$ is zero for $E > E_F$. Show that for $E < E_F$ at $T = 0$ the densities ρ_2 and ρ_3 in two and three dimensions are

$$\rho_2(E) = 4\pi m \left(\frac{L}{2\pi\hbar}\right)^2 \qquad \text{and} \qquad \rho_3(E) = 4\pi \left(2m\right)^{3/2} \left(\frac{L}{2\pi\hbar}\right)^3 E^{1/2} \ .$$

Dividing by $L^2 = A$ or $L^3 = V$, we get the densities of states per unit area or volume. Since ρ_2 is independent of energy, $\overline{E} = E_F/2$, as found in Sec. 1.

The figure shows $\rho_3(E)$, which varies as $E^{1/2}$ up to E_F. It also shows how the distribution changes for two nonzero values of T that are small compared to the Fermi temperature. For more on that, see a text on statistical mechanics.

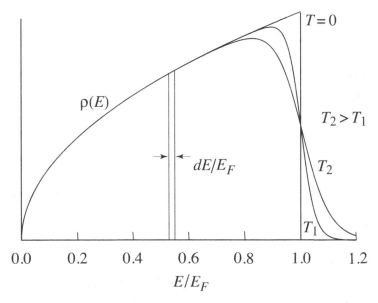

4. *The radius of a white-dwarf star.*

(Adapted from Prob. 5·35, D.J. Griffiths, *Introduction to Quantum Mechanics*, 2nd ed.) Many stars end up as white dwarfs. When fuel is exhausted, a star cools down and contracts under gravity. It is prevented from collapse only because squeezing the now highly degenerate electrons (all the electrons are "free") into a smaller volume would increase the total energy

E_e of the electrons more than it would decrease the gravitational potential energy E_g. If we model the star as a uniformly dense sphere, then E_g is

$$E_g = -\frac{3}{5}\frac{GM^2}{R},$$

where G is Newton's constant and M and R are the mass and radius of the star.

(a) Find the radius R that minimizes $E_e + E_g$. The star has about equal numbers of electrons, protons, and neutrons, so that nearly all the mass of the star comes from the nucleons. Write R in terms of the mass m of an electron, the mass μ of a proton or a neutron, and M. Is there anything surprising about the dependence of R on M?

(b) Evaluate R for $M = 2.0 \times 10^{30}$ kg, the mass of the Sun. Data: $m = 9.1 \times 10^{-31}$ kg, $\mu = 1.7 \times 10^{-27}$ kg, $G = 6.7 \times 10^{-11}$ m^3 kg^{-1} s^{-2}. Find the density of this white dwarf. This is large, but it is still much smaller than nuclear densities, which are about 2×10^{17} kg/m^3.

5. *Ground states of equivalent d electrons.*

Use the rules of Pauli and Hund to make an arrow table for one through ten equivalent d electrons, and find the $^{2S+1}L_J$ ground state for each. Check using elements in Table 2 where the d subshells fill regularly. Write down by inspection the magnetic moments of ground states for five through ten d electrons, and calculate the others.

6. *States of three equivalent p electrons.*

Make an arrow table to find *all* the ^{2S+1}L states of three equivalent p electrons. Check with the table at the end of Sec. 3.

7. *More ground states.*

(a) Write down the possible ^{2S+1}L states for two equivalent f electrons using the symmetry properties of spin and orbital angular-momentum states. Do *not* make a table.

(b) Elements 24, 41, 42, 44, and 45 are irregulars in the first half of the periodic table, in that an incompletely filled d subshell "steals" an s electron from an s^2 subshell (see Table 2). Can you make up a simple rule for getting the ground-state $^{2S+1}L_J$ configurations of these irregulars?

8. *Comparisons.*

First-order perturbation theory gave

$$E_{grd} = (2Z^2 - 5Z/4) \times 13.6 \text{ eV}$$

for the energy needed to remove *both* electrons from the ground state of helium or from a two-electron helium-like ion with a nuclear charge Z. The variational method, with a simple guess at the wave function, gave

$$E_{grd} = 2(Z - 5/16)^2 \times 13.6 \text{ eV}.$$

(a) Show that the two differ by $(25/128) \times 13.6 = 2.66$ eV, independent of Z.

(b) Make a table showing the variational values of E_{grd} for He, Li$^+$, Be^{++}, and B^{+++}, beside the experimental values -79.0, -198, -371, and -599 eV. Note that the absolute errors are independent of Z.

(c) Add another column to the table showing the energy needed to remove the *first* electron in each case. Use the experimental values of E_{grd} here.

9. *Three against two.*

The variational calculation of $E_T = \langle \psi_T | \hat{H} | \psi_T \rangle$ for helium gave

$$E_T(Z') = \left[2(Z')^2 - 4ZZ' + 5Z'/4 \right] \times 13.6 \text{ eV}$$

as a function of the variational parameter Z'. The first of the three terms here is the sum of the kinetic energies of the two electrons, the second term is the sum of the two electron-nucleus potential energies, and the last term is the electron-electron potential energy. Then minimization of $E_T(Z')$ gave $Z' = Z - 5/16$ and $E_T(Z' = Z - 5/16) = -2(Z - 5/16)^2 \times 13.6 \text{ eV}$.

(a) Evaluate the total kinetic energy at the minimum.

(b) Evaluate the total potential energy at the minimum.

(c) What is the ratio of these two values? Does this remind you of anything?]

10. *Rules, rules, rules.*

(a) *Fermi-Dirac rules*—Two spatial states are available for three identical spin-1/2 particles. So, counting spin states, there are four states for each particle. How many ways can one put the three particles into these states?

(b) *Bose-Einstein rules*—Four spatial states are available for three identical spin-0 particles. How many ways can one put the particles into these states?

(c) *Maxwell-Boltzmann rules*—How many ways can one put three identical (but assumed distinguishable) spin-0 particles into the four states of part (b)?

11. *Exchange operators.*

Let the ket $|1, 2, 3, \ldots, N\rangle$ denote a state of N identical particles. Here "2" denotes the coordinates of the particle that carries the label 2, including spin if there is spin. The *exchange operator* P_{ij} exchanges the *particles* i and j (not, after exchanges, the particles in the *positions* i and j in the sequence). Thus

$$P_{13} |1, 2, 3, \ldots, N\rangle = |3, 2, 1, \ldots, N\rangle ,$$

and

$$P_{12} P_{13} |1, 2, 3, \ldots, N\rangle = |3, 1, 2, \ldots, N\rangle ,$$

not $|2, 3, 1, \ldots, N\rangle$. The inverse of P_{31} is obviously P_{13}: P_{ij} is its own inverse.

(a) Will P_{ij} commute with the Hamiltonian for the system? Do the operators P_{12} and P_{13} commute? Do P_{12} and P_{34}? Do P_{13} and P_{24}?

(b) Show that the eigenvalues of P_{ij} are ± 1. Show that the boson states ψ_1, ψ_2, and ψ_3 and the fermion states $|S, M\rangle = |3/2, M_S\rangle$ and $|L, M_L\rangle = |0, 0\rangle$ in Sec. 8(c) are eigenstates of P_{12}, and find the eigenvalues.

(c) In general, a two-particle ket $|1, 2\rangle$ is not an eigenstate of P_{12}. Show, however, that the *projection* operators $P_{12}^{\pm} \equiv (1 \pm P_{12})/2$ operating on $|1, 2\rangle$ make eigenstates with eigenvalues $+1$ and -1; P_{12}^+ and P_{12}^- project *out* the even and odd parts of $|1, 2\rangle$. Illustrate with the two-spin kets $|\uparrow\uparrow\rangle$ and $|\uparrow\downarrow\rangle$.

12. *Symmetries with determinants.*

One way to construct antisymmetric states is with determinants. For example, the antisymmetric spin state of two spin-1/2 particles, labeled 1 and 2, is

$$\frac{1}{\sqrt{2}} \begin{vmatrix} \uparrow_1 & \uparrow_2 \\ \downarrow_1 & \downarrow_2 \end{vmatrix} = \frac{1}{\sqrt{2}} \left(\uparrow_1 \downarrow_2 - \downarrow_1 \uparrow_2 \right) = \frac{1}{\sqrt{2}} \left(\uparrow\downarrow - \downarrow\uparrow \right).$$

The possible states of particle 1 go in the first column, the possible states of particle 2 go in the second column, ... In the right-most expression, *place* tells the particle. Determinants used for this purpose are called Slater determinants, after J.C. Slater. Use a determinant to find the antisymmetric $L = 0$ state of three p electrons (see Sec. 8(c)). How would you get the symmetric $L = 0$ state of three p electrons from this?

15. TIME-DEPENDENT PERTURBATIONS. PLANCK AND EINSTEIN

15.1. *Sudden Changes*

15.2. *Time-Dependent Perturbation Theory*

15.3. *Magnetic Resonance (Again)*

15.4. *Hydrogen in an Electromagnetic Wave*

15.5. *Averaging over Polarizations and Frequencies*

15.6. *The Boltzmann Factor*

15.7. *Planck's Oscillators*

15.8. *Black-Body Radiation*

15.9. *Einstein's A and B Coefficients*

15.10. *Decay Lifetimes*

     Problems

In Chapter 13, we calculated the small shifts of energy eigenvalues when a time-independent perturbation $\hat{H}'$ is added to a Hamiltonian $\hat{H}^0$ whose eigenstates we already know. In the first half of this chapter, we calculate changes brought about when a *time-dependent* perturbation $\hat{H}'(t)$ is added to $\hat{H}^0$. And now the question is different: How do the *probabilities* of being in the eigenstates of $\hat{H}^0$ change with time? We have already, in Chapter 10, calculated the time-dependent probabilities that a spin-1/2 particle is in the spin up and down states of a static magnetic field $\mathbf{B}_0$ in the presence of a circling magnetic field $\mathbf{B}_1(t)$. The solution there was exact, but an exact solution is rarely obtainable. The first example will be to try first-order time-dependent perturbation theory on that spin-1/2 problem. The second example will be to find (approximately) how an electromagnetic wave changes the probabilities that an atom is in its energy levels. We start with a single incident wave, then obtain an average for a continuum of waves with incoherent frequencies and polarizations. The result will used later in the chapter.

In the second half of the chapter, we calculate, as did Planck in 1900, the average energy of a quantized oscillator as a function of temperature. The frequency modes of electromagnetic radiation in a box are (mathematically) very similar to those of quantized oscillators, and Planck's black-body radiation energy density follows quickly. In the physics of 1900, there was of course no such thing as a quantized oscillator. Pais again: "[His] only justification for the [steps he made] was that they gave him what he wanted," a result in accord with experimental data.

For a system in thermal equilibrium, the relative populations of constituents in energy levels are given by ratios of Boltzmann factors, and the populations are smaller at higher energies. In 1916, Einstein, considering the processes that maintain these ratios, showed that thermal radiation alone cannot explain the lower numbers at higher energies. He added spontaneous decays from the upper levels, the rates of which he was able to relate to Planck's black-body equation. The chapter ends with a calculation of the lifetime of an excited state of hydrogen.

15-1. SUDDEN CHANGES

One way things can change is—suddenly. Figure 1(a) shows a particle in the ground state of an infinite square well of width L. At $t = 0$, the width is suddenly doubled symmetrically from L to $2L$. What are the probabilities, if measured, that the particle would be found in the eigenstates of the new well? To find out, we simply expand the "old" wave function at $t = 0$, $\Psi_1^L(x,0)$, in terms of the new energy eigenstates, $\psi_n^{2L}(x)$. The *new* eigenstates are the relevant basis states in the *new* situation. For convenience, we put the origin at the center of the well. Then, for example, the probability amplitude for being in the new ground state, $\psi_1^{2L}(x)$, is

$$c_1^{2L} = \langle \psi_1^{2L}(x) | \Psi_1^L(x,0) \rangle = \int_{-L/2}^{+L/2} \sqrt{\frac{2}{2L}} \cos\left(\frac{\pi x}{2L}\right) \sqrt{\frac{2}{L}} \cos\left(\frac{\pi x}{L}\right) dx \; .$$

Note the limits of integration. The result is $c_1^{2L} = 8/3\pi$, and $|c_1^{2L}|^2 = 0.721$. (Why will the c_n^{2L} for all the odd-parity eigenstates, those with $n = 2, 4, 6, \ldots$, be zero?)

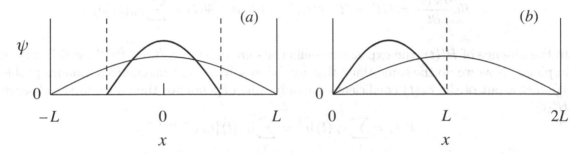

Figure 1. (a) The width of the well is suddenly doubled symmetrically. (b) The width is doubled asymmetrically. The heavy and light curves show the ground-state wave functions before and after.

Figure 1(b) shows another way in which the width of the well can be doubled. Here it is simpler to place the origin at one of the walls. Of course, the placement of the origin does not affect the result. Now

$$c_1^{2L} = \langle \psi_1^{2L}(x) | \Psi_1^L(x,0) \rangle = \int_0^L \sqrt{\frac{2}{2L}} \sin\left(\frac{\pi x}{2L}\right) \sqrt{\frac{2}{L}} \sin\left(\frac{\pi x}{L}\right) dx \; ,$$

which gives $c_1^{2L} = 4\sqrt{2}/3\pi$ and $|c_1^{2L}|^2 = 0.360$, half the previous result. It is evident in the figure that the product of the two functions is larger in (a) than in (b). And now the probabilities of being found, if measured, in the odd-parity eigenstates are not zero.

Two remarks: (1) How fast is sudden? The time dependence of a state is $e^{-iEt/\hbar}$. This repeats in a time $\tau = 2\pi\hbar/E$. Sudden means in a time very short compared to τ. (2) The expansion of an "old" state in terms of new ones requires that the new states include the full space of the old state. If we suddenly squash the box to half its width, the new states do not fully encompass the old. Some other approach would be needed.

There are more interesting problems than a suddenly widening box. An unstable hydrogen nucleus (^{3}H$^+$) decays to a helium nucleus (^{3}He^{++}), thus suddenly doubling the nuclear charge. What is the probability that an electron initially in the hydrogen ground state would be found in the singly ionized helium ground state immediately after the decay? See Prob. 1.

15·2. TIME-DEPENDENT PERTURBATION THEORY

(a) Exact relations—Let E_n and $|n\rangle$ be the eigenvalues and orthonormal eigenstates of a Hamiltonian $\hat{H}^0$, where $\hat{H}^0$ has no explicit time dependence: $\hat{H}^0|n\rangle = E_n|n\rangle$. Those eigenstates might be spatial wave functions $\psi_n(x)$ or $\psi_{n\ell m}(r, \theta, \phi)$, or spin states $|\pm z\rangle$.

Let $\hat{H}'(t)$ be an additional *time-dependent* piece of the full Hamiltonian $\hat{H}$. This $\hat{H}'(t)$ will also be a function of coordinates x, or (r, θ, ϕ), or perhaps of spin, but we suppress that dependence in the notation until we come to specific cases. We are not looking for the effect of $\hat{H}'(t)$ on the energy eigenvalues, which in the circumstances considered here would be negligible. Instead, we want to find how $\hat{H}'(t)$ brings about *transitions* between the eigenstates $|n\rangle$ of $\hat{H}^0$. We had an example of this sort in Sec. 10·5: A small magnetic field $B_1(t)$ rotating in the xy plane changed the probabilities that a spin-1/2 particle was in the energy eigenstates $|+z\rangle$ and $|-z\rangle$ of a large static field $B_0\hat{\mathbf{z}}$. (We revisit that problem in the next section.)

With $\hat{H} = \hat{H}^0 + \hat{H}'(t)$, the time-dependent Schrödinger equation is

$$i\hbar \frac{\partial \Psi(t)}{\partial t} = (\hat{H}^0 + \hat{H}')\Psi(t), \qquad \text{where} \quad \Psi(t) = \sum_n c_n(t)\,|n\rangle \;.$$

In the absence of $\hat{H}'(t)$, the expansion coefficients are $c_n(t) = c_n(0)e^{-iE_nt/\hbar}$, which vary only in phase. Now we do the same thing that we did in solving that magnetic-resonance problem: We factor out of the $c_n(t)$ coefficients the only time dependence they have in the absence of $\hat{H}'(t)$,

$$\Psi(t) = \sum_n c_n(t)\,|n\rangle = \sum_n a_n(t)\,|n\rangle\, e^{-iE_nt/\hbar} \;.$$

In the absence of $\hat{H}'(t)$, the $a_n(t)$ are simply the time-independent coefficients $c_n(0)$, and the probabilities $|c_n(t)|^2 = |c_n(0)|^2$ are independent of time. The goal is to find the time dependence, brought about by $\hat{H}'(t)$, of the factors $a_n(t)$. In the end, we can recover

$$c_n(t) = a_n(t)\, e^{-iE_nt/\hbar} \;,$$

where $|c_n(t)|^2 = |a_n(t)|^2$. If the *magnitudes* of the $a_n(t)$ are time-dependent, then so also will be the probabilities of being in the various states.

Putting the factored $\Psi(t)$ into the time-dependent Schrödinger equation and using $\hat{H}^0|n\rangle = E_n|n\rangle$, we get

$$i\hbar \sum_n \dot{a}_n(t)\,|n\rangle\, e^{-iE_nt/\hbar} + \sum_n a_n(t)\, E_n|n\rangle\, e^{-iE_nt/\hbar} =$$

$$\sum_n a_n(t)\, E_n|n\rangle\, e^{-iE_nt/\hbar} + \sum_n a_n(t)\, \hat{H}'(t)|n\rangle\, e^{-iE_nt/\hbar} \;,$$

where $\dot{a}_n(t) = da_n/dt$. The second term on the left cancels the first term on the right, which is why we factored $e^{-iE_nt/\hbar}$ out of $c_n(t)$. There remains

$$i\hbar \sum_n \dot{a}_n(t)\,|n\rangle\, e^{-iE_nt/\hbar} = \sum_n a_n(t)\, \hat{H}'(t)|n\rangle\, e^{-iE_nt/\hbar} \;.$$

So far, this is exact.

To isolate, say, da_1/dt, we take the scalar product of $\langle 1|$ with the above equation. Using the orthonormality of the eigenfunctions of $\hat{H}^0$, $\langle m|n\rangle = \delta_{mn}$, we get

$$i\hbar \frac{da_1}{dt} e^{-iE_1 t/\hbar} = \sum_n a_n(t)\langle 1|\hat{H}'(t)|n\rangle e^{-iE_n t/\hbar} .$$

The matrix elements $\langle 1|\hat{H}'(t)|n\rangle$ retain the time dependence of $\hat{H}'(t)$, because the scalar products are integrals over spatial coordinates or sums over spin components, not over time. Rearranging, we get

$$\frac{da_1}{dt} = -\frac{i}{\hbar} \sum_n H'_{1n}(t) e^{-i\omega_{n1} t} a_n(t) ,$$

where $H'_{1n}(t) \equiv \langle 1|\hat{H}'(t)|n\rangle$ and $\omega_{n1} \equiv (E_n - E_1)/\hbar$. More generally,

$$\frac{da_m}{dt} = -\frac{i}{\hbar} \sum_n H'_{mn}(t) e^{-i\omega_{nm} t} a_n(t) ,$$

where $H'_{mn}(t) \equiv \langle m|\hat{H}'(t)|n\rangle$ and $\omega_{nm} \equiv (E_n - E_m)/\hbar$. If $\hat{H}'(t) = 0$, then $da_n/dt = 0$ for all n, and the a_n are the constants $c_n(0)$.

The rate of change of $a_m(t)$ at any instant depends in general on the values of all the $a_n(t)$ at that instant. The structure of the coupled equations is easier to see in matrix form:

$$\begin{pmatrix} da_1/dt \\ da_2/dt \\ da_3/dt \\ \vdots \end{pmatrix} = -\frac{i}{\hbar} \begin{pmatrix} H'_{11} & H'_{12} e^{-i\omega_{21} t} & H'_{13} e^{-i\omega_{31} t} & \cdots \\ H'_{21} e^{+i\omega_{21} t} & H'_{22} & H'_{23} e^{-i\omega_{32} t} & \cdots \\ H'_{31} e^{+i\omega_{31} t} & H'_{32} e^{+i\omega_{32} t} & H'_{33} & \cdots \\ \vdots & \vdots & \vdots & \ddots \end{pmatrix} \begin{pmatrix} a_1(t) \\ a_2(t) \\ a_3(t) \\ \vdots \end{pmatrix} .$$

The rate of change of $a_1(t)$ is $-i/\hbar$ times the scalar product of the first row of the matrix and the column of coefficients, and so on. Again, the H'_{mn} themselves generally are time dependent: $H'_{mn} = H'_{mn}(t)$. And as was the case with time-*independent* perturbation theory, selection rules will often make many of the H'_{mn}'s—sometimes nearly all of them—zero.

This is still exact (but complicated).

(b) First-order perturbation theory—The problem we shall solve (approximately) is one in which the system is initially in one of the eigenstates of $\hat{H}^0$. For example, let $a_1(0) = 1$ and all the other $a_n(0) = 0$. First-order perturbation theory *fixes* these $t = 0$ values of the a_n on the *right* side of the above matrix equation. Then we have

$$\begin{pmatrix} da_1/dt \\ da_2/dt \\ da_3/dt \\ \vdots \end{pmatrix} \approx -\frac{i}{\hbar} \begin{pmatrix} H'_{11} & H'_{12} e^{-i\omega_{21} t} & \cdots \\ H'_{21} e^{+i\omega_{21} t} & H'_{22} & \cdots \\ H'_{31} e^{+i\omega_{31} t} & H'_{32} e^{+i\omega_{32} t} & \cdots \\ \vdots & \vdots & \ddots \end{pmatrix} \begin{pmatrix} 1 \\ 0 \\ 0 \\ \vdots \end{pmatrix} = -\frac{i}{\hbar} \begin{pmatrix} H'_{11} \\ H'_{21} e^{+i\omega_{21} t} \\ H'_{31} e^{+i\omega_{31} t} \\ \vdots \end{pmatrix} .$$

The vector on the right is the first column of the Hamiltonian array. If, instead, the system were in the state $|2\rangle$ at $t = 0$, then the vector on the right would be the second column of the array. This of course is no longer exact.

Given the elements $H'_{mn}(t)$, the first-order equations for the $a_n(t)$ can be integrated. For $a_1(0) = 1$,

$$a_1(t) \approx 1 - \frac{i}{\hbar} \int_0^t H'_{11}(t')\, dt'$$

$$a_2(t) \approx 0 - \frac{i}{\hbar} \int_0^t H'_{21}(t')\, e^{+i\omega_{21}t'}\, dt' ,$$

and so on. If, instead, we started in state 2, with $a_2(0) = 1$, then

$$a_1(t) \approx 0 - \frac{i}{\hbar} \int_0^t H'_{12}(t')\, e^{-i\omega_{21}t'}\, dt'$$

$$a_2(t) \approx 1 - \frac{i}{\hbar} \int_0^t H'_{22}(t')\, dt' .$$

Now $H'_{12}(t) = H'_{21}(t)^*$ (the matrix is Hermitian). Therefore, $a_1(t)$ starting with $a_2(0) = 1$ would equal $-a_2^*(t)$ starting with $a_1(0) = 1$. And so $|a_1(t)|^2$ when $a_2(0) = 1$ would equal $|a_2(t)|^2$ when $a_1(0) = 1$. *The perturbation acts equally to excite and de-excite.* This equality of *induced* transition rates was first discovered by Einstein in 1916, when quantum ideas were in the air but there was no quantum mechanics. We shall come to his argument in Sec. 9; it proves this reciprocity of induced rates far more generally, and it tells something *else* that can happen. And see Fig. 10·7, which shows the *exactly* calculated oscillations between the $|+z\rangle$ and $|-z\rangle$ states of a spin-1/2 particle caused by a time-dependent magnetic field. The rates up and down are the same.

In many cases (including all those below), the diagonal elements H'_{nn} are zero. If, say, $H'_{11} = 0$ and the system starts in the state $|1\rangle$, then to first order $da_1/dt = 0$, and $a_1(t) = a_1(0) = 1$. Thus first-order theory will not do well here if $\hat{H}'(t)$ also produces large values of other coefficients. The next section gives an example of when the first-order theory works well and when it does not.

Second-order theory feeds the results of the first-order theory back into the original equations. We shall not do that here.

15·3. MAGNETIC RESONANCE (AGAIN)

We try the theory on the only time-dependent problem we have already solved.

(a) Review—In Sec. 10·5, we solved exactly the problem of how the magnetic moment $\boldsymbol{\mu} = \mu\boldsymbol{\sigma}$ of a spin-1/2 particle interacts with a static magnetic field $\mathbf{B}_0 = B_0\hat{\mathbf{z}}$ plus a small field $\mathbf{B}_1(t)$ that rotates in the xy plane. The full Hamiltonian was

$$\hat{H}^0 + \hat{H}'(t) = -\boldsymbol{\mu} \cdot \mathbf{B}_0 - \boldsymbol{\mu} \cdot \mathbf{B}_1 = -\mu B_0 \begin{pmatrix} 1 & 0 \\ 0 & -1 \end{pmatrix} - \mu B_1 \begin{pmatrix} 0 & e^{+i\omega t} \\ e^{-i\omega t} & 0 \end{pmatrix},$$

where ω is the rotation angular frequency of $\mathbf{B}_1(t)$. The eigenvalues and eigenstates of $\hat{H}^0$ are $E_\pm = \mp\mu B_0$ and $|\pm z\rangle$. The general spin state is

$$|\chi(t)\rangle = \begin{pmatrix} c_+(t) \\ c_-(t) \end{pmatrix} = \begin{pmatrix} a_+(t)\, e^{-iE_+t/\hbar} \\ a_-(t)\, e^{-iE_-t/\hbar} \end{pmatrix} = \begin{pmatrix} a_+(t)\, e^{+i\omega_0 t/2} \\ a_-(t)\, e^{-i\omega_0 t/2} \end{pmatrix},$$

300

where $\hbar\omega_0 \equiv 2\mu B_0$, and $|c_+(t)|^2 + |c_-(t)|^2 = |a_+(t)|^2 + |a_-(t)|^2 = 1$. All of B_0, B_1, and ω could be varied by turning knobs. The rotating field causes the amplitudes $a_\pm(t)$ to vary with time. If at $t = 0$ the spin is up, then a measurement at some later time would find the spin to be down with a probability of

$$|a_-(t)|^2 = |c_-(t)|^2 = A(\omega) \sin^2\left(\tfrac{1}{2}\Omega t\right) ,$$

where $A(\omega) \equiv \omega_1^2/\Omega^2$, $\Omega^2 \equiv (\omega - \omega_0)^2 + \omega_1^2$, and $\hbar\omega_1 \equiv 2\mu B_1$. At resonance ($\omega = \omega_0$), $A(\omega_0) = 1$, $\Omega = |\omega_1|$, and the spin spirals around the z axis, oscillating back and forth between the $|+z\rangle$ and $|-z\rangle$ states. If instead ω is far from ω_0 (in units of ω_1), then the spin just makes small spirals out and back from the $|+z\rangle$ state.

(b) First-order perturbation theory—The (exact) coupled equations for a two-state system are, from Sec. 2(a),

$$\begin{pmatrix} da_1/dt \\ da_2/dt \end{pmatrix} = -\frac{i}{\hbar} \begin{pmatrix} H'_{11} & H'_{12}e^{-i\omega_{21}t} \\ H'_{21}e^{+i\omega_{21}t} & H'_{22} \end{pmatrix} \begin{pmatrix} a_1 \\ a_2 \end{pmatrix} .$$

For spin, the state labels $(1,2)$ become $(+,-)$, ω_{21} is $\omega_0 \equiv 2\mu B_0/\hbar$, and the elements $H'_{mn}(t)$ are those in the $\hat{H}'(t)$ matrix at the start of this section. With these replacements, we get

$$\begin{pmatrix} da_+/dt \\ da_-/dt \end{pmatrix} = +\frac{i\omega_1}{2} \begin{pmatrix} 0 & e^{+i(\omega-\omega_0)t} \\ e^{-i(\omega-\omega_0)t} & 0 \end{pmatrix} \begin{pmatrix} a_+ \\ a_- \end{pmatrix} .$$

These are exactly the exact coupled equations we solved exactly in Sec. 10·5.

We begin again with the spin up at $t = 0$: $a_+(0) = 1$ and $a_-(0) = 0$. First-order theory *fixes* these values on the *right* side of the matrix equation. In this approximation,

$$\frac{da_+}{dt} \approx 0 \quad \text{and} \quad \frac{da_-}{dt} \approx \frac{i\omega_1}{2}e^{-i(\omega-\omega_0)t} .$$

Integrating, we get $a_+(t) = 1$ (not a good approximation if $a_-(t)$ ever gets large). And

$$a_-(t) \approx \frac{i\omega_1}{2}\int_0^t e^{-i(\omega-\omega_0)t'} dt' = -\frac{i\omega_1}{2}\frac{e^{-i(\omega-\omega_0)t'}}{i(\omega-\omega_0)}\bigg|_0^t$$

$$\approx \frac{i\omega_1}{\omega-\omega_0}e^{-i(\omega-\omega_0)t/2}\sin\left[\tfrac{1}{2}(\omega-\omega_0)t\right] ,$$

where we have used $(e^\theta - 1) = e^{\theta/2}(e^{\theta/2} - e^{-\theta/2}) = 2i\,e^{\theta/2}\sin\tfrac{1}{2}\theta$. Then

$$|a_-(t)|^2 = |c_-(t)|^2 \approx \frac{\omega_1^2}{(\omega-\omega_0)^2}\sin^2\left[\tfrac{1}{2}(\omega-\omega_0)t\right] = A'(\omega)\sin^2(\tfrac{1}{2}\Omega't) .$$

We compare $A'(\omega)$ and Ω' with the exact quantities $A(\omega)$ and Ω given in (a):

$$A'(\omega) \equiv \frac{1}{(\omega-\omega_0)^2/\omega_1^2} \approx \frac{1}{1 + (\omega-\omega_0)^2/\omega_1^2} = A(\omega) \quad \text{only when} \quad (\omega-\omega_0)^2 \gg \omega_1^2$$

$$\Omega' \equiv \omega - \omega_0 \approx \left[(\omega-\omega_0)^2 + \omega_1^2\right]^{1/2} = \Omega \quad '' \quad '' \quad '' \quad '' .$$

Figure 2 compares $A'(\omega)$ and $A(\omega)$ as functions of $(\omega-\omega_0)/\omega_1$. First-order theory, as is usual with perturbation theories, works well only when the effects of $\hat{H}'(t)$ are small. Here that occurs only when ω is far off resonance.

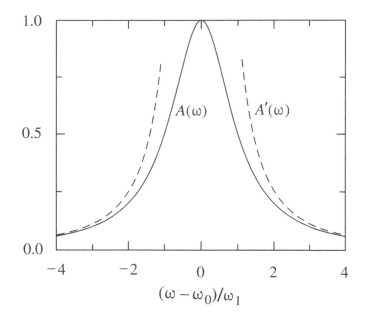

Figure 2. The first-order approximation $A'(\omega)$ (dashed) differs more and more from the exact $A(\omega)$ (solid) as ω approaches ω_0. First-order theory works well only far from resonance.

The *fractional* error made on $A(\omega)$ by first-order perturbation theory is

$$\frac{A'(\omega) - A(\omega)}{A(\omega)} = A'(\omega)$$

(show this). For $(\omega - \omega_0)/\omega_1 = \pm 2, \pm 3, \pm 5$, and ± 10, the fractional errors on $A(\omega)$ are 25, 11, 4, and 1%. What happens to the first-order approximation as ω approaches ω_0? Then

$$|c_-(t)|^2 \approx \omega_1^2 \frac{\sin^2\left[\frac{1}{2}(\omega - \omega_0)t\right]}{(\omega - \omega_0)^2} \xrightarrow{\omega \to \omega_0} \left(\frac{\omega_1 t}{2}\right)^2,$$

which quickly becomes unphysical.

15·4. HYDROGEN IN AN ELECTROMAGNETIC WAVE

(a) Perturbation theory—A plane wave of linearly polarized, monochromatic light is incident upon a hydrogen atom. The oscillating electric field is $\mathcal{E}(t) = \mathcal{E}_0 \cos \omega t$, and the polarization axis will define the z axis. The force on the electron in the atom is $F(t) = -e\mathcal{E}(t)$. The perturbation Hamiltonian is

$$\hat{H}'(z, t) = V'(z, t) = -\int_0^z F(t)\,dz = +e\mathcal{E}_0 \cos \omega t \int_0^z dz = e\mathcal{E}_0 z \cos \omega t \ .$$

If at $t = 0$ the atom is in the state i, what is the probability that at later times it is in some other state f? For example, i and f might be the $|n\ell m\rangle$ states $|100\rangle$ and $|210\rangle$ of hydrogen. The matrix element of $\hat{H}'(z, t)$ between i and f is

$$H'_{fi}(t) = \langle f|\hat{V}'(z, t)|i\rangle = e\mathcal{E}_0 \langle f|z|i\rangle \cos \omega t = V'_{fi} \cos \omega t \ ,$$

where $V'_{fi} \equiv e\mathcal{E}_0 \langle f|z|i\rangle$ and

$$\langle f|z|i\rangle = \langle n'\ell'm'|r\cos\theta|n\ell m\rangle = \langle n'\ell'|r|n\ell\rangle\langle \ell'm'|\cos\theta|\ell m\rangle \ .$$

302

The kets $|n\ell\rangle = R_{n\ell}(r)$ and $|\ell m\rangle = Y_\ell^m(\theta,\phi)$ are the radial and angular parts of the hydrogen state. The first bracket in the product is an integral over r, the second is one over θ and ϕ. In Sec. 13·5, a recurrence relation for the spherical harmonics showed that the angular integral is zero unless $\ell' = \ell \pm 1$ and $m' = m$. That relation also gave us integrals without having to integrate (as it will again below).

The calculations for the Stark effect (the effect of a *static* electric field) in hydrogen (Sec. 13·6), and for the fine structure (Sec. 13·8), involved integrals between initially *degenerate* states of hydrogen: states with the same value of n. Here the integrals will be between states with $n' \neq n$. We calculate one such integral below.

The recipe from Sec. 2(b) to get $a_f(t)$ to first-order is

$$a_f(t) \approx -\frac{i}{\hbar}\int_0^t H'_{fi}(t')\,e^{+i\omega_0 t'}\,dt' = -\frac{iV'_{fi}}{\hbar}\int_0^t \cos\omega t'\,e^{+i\omega_0 t'}\,dt' \ ,$$

where $\omega_0 = (E_f - E_i)/\hbar$ and $V'_{fi} = e\mathcal{E}_0\langle f|z|i\rangle$. Here ω is positive, but ω_0 can be positive or negative. Then

$$a_f(t) \approx -\frac{iV'_{fi}}{2\hbar}\int_0^t \left[e^{i(\omega_0+\omega)t'} + e^{i(\omega_0-\omega)t'}\right]dt'$$

$$\approx -\frac{iV'_{fi}}{2\hbar}\left[\frac{e^{i(\omega_0+\omega)t}-1}{i(\omega_0+\omega)} + \frac{e^{i(\omega_0-\omega)t}-1}{i(\omega_0-\omega)}\right].$$

The $(e^{i\alpha t}-1)$ numerators here can never be greater in magnitude than 2 (why?). Transitions to or from the lowest states of hydrogen are at frequencies of order 10^{15} Hz (see Fig. 1·4). If $E_f > E_i$, so that $\omega_0 > 0$, then $\omega_0 + \omega$ will be extremely large, and the first term in the equation will be completely negligible. But if ω is near ω_0, then the second term is not negligible. Using $(e^\theta - 1) = e^{\theta/2}(e^{\theta/2} - e^{-\theta/2}) = 2i\,e^{\theta/2}\sin\frac{1}{2}\theta$ again, we get

$$a_f(t) \approx -\frac{iV'_{fi}}{\hbar}e^{i(\omega_0-\omega)t/2}\frac{\sin[\frac{1}{2}(\omega_0-\omega)t]}{\omega_0-\omega} \ .$$

Then the probability of being in the final state f as a function of time is

$$\boxed{|c_f(t)|^2 = |a_f(t)|^2 \approx \frac{|V'_{fi}|^2}{\hbar^2}\frac{\sin^2[\frac{1}{2}(\omega_0-\omega)t]}{(\omega_0-\omega)^2} = \frac{e^2\mathcal{E}_0^2}{\hbar^2}|\langle f|z|i\rangle|^2\frac{\sin^2[\frac{1}{2}(\omega_0-\omega)t]}{(\omega_0-\omega)^2} \ .}$$

If instead $E_f < E_i$, so that $\omega_0 < 0$, it is only the *first* term (with $\omega_0 + \omega$) that matters. But otherwise everything is the same.

We can compare the form of this time-dependent probability $|c_f(t)|^2$ for being excited from the state i to the state f with the time-dependent probability $|c_-(t)|^2$ from Sec. 3 for a spin-1/2 particle to be excited from the state $|+\rangle$ to the state $|-\rangle$,

$$|c_-(t)|^2 \approx \omega_1^2\frac{\sin^2[\frac{1}{2}(\omega-\omega_0)t]}{(\omega-\omega_0)^2} = \frac{4\mu^2 B_1^2}{\hbar^2}\frac{\sin^2[\frac{1}{2}(\omega-\omega_0)t]}{(\omega-\omega_0)^2} \ .$$

An oscillating field—electric in one case, magnetic in the other—leads to an *oscillating* probability of being in an initially unoccupied state, not to a steadily increasing probability. And again, unless $|c_f(t)|^2$ is small compared to 1, the approximation will not be good.

(b) A matrix element—We are not done until we have calculated a matrix element. We calculate

$$V'_{fi} = e\mathcal{E}_0 \langle f|z|i \rangle = e\mathcal{E}_0 \langle n'\ell'm'|r\cos\theta|n\ell m \rangle = e\mathcal{E}_0 \langle n'\ell'|r|n\ell \rangle \langle \ell'm'|\cos\theta|\ell m \rangle$$

for $i = |100\rangle$ and $f = |210\rangle$. For the angular integral $\langle 10|\cos\theta|00 \rangle$, we use the recurrence relation from Sec. 13·5,

$$\cos\theta\, Y_\ell^m(\theta,\phi) = \sqrt{\frac{(\ell+1)^2 - m^2}{4(\ell+1)^2 - 1}}\, Y_{\ell+1}^m(\theta,\phi) + \sqrt{\frac{\ell^2 - m^2}{4\ell^2 - 1}}\, Y_{\ell-1}^m(\theta,\phi)\,,$$

and the orthonormality of the spherical harmonics. Here $\ell = 0$, $m = 0$, and $\ell' = \ell + 1 = 1$, so by inspection $\langle 10|\cos\theta|00 \rangle = 1/\sqrt{3}$. Tables in Secs. 9·3 and 13·6 include the $R_{n\ell}(r)$ functions for the r integration. The integral needed is the oft-used $\int_0^\infty u^n e^{-bu} du = n!/b^{n+1}$. With $u \equiv r/a_0$, where a_0 is the Bohr radius, we get

$$\langle 21|r|10 \rangle = \int_0^\infty \frac{1}{\sqrt{24a_0^3}}\frac{r}{a_0}e^{-r/2a_0}\frac{2}{\sqrt{a_0^3}}e^{-r/a_0}r^3 dr = \frac{a_0}{\sqrt{6}}\int_0^\infty u^4 e^{-3u/2}du = \frac{2^8 a_0}{3^4\sqrt{6}}\,.$$

Altogether,

$$V'_{fi} = e\mathcal{E}_0\langle 210|z|100 \rangle = e\mathcal{E}_0\langle 21|r|10 \rangle\langle 10|\cos\theta|00 \rangle = \frac{2^7\sqrt{2}}{3^5}e\mathcal{E}_0 a_0\,,$$

and then,

$$|c_f(t)|^2 = |a_f(t)|^2 \approx \frac{|V'_{fi}|^2}{\hbar^2}\frac{\sin^2[\frac{1}{2}(\omega_0 - \omega)t]}{(\omega_0 - \omega)^2} = \frac{2^{15}}{3^{10}}\left(\frac{e\mathcal{E}_0 a_0}{\hbar}\right)^2\frac{\sin^2[\frac{1}{2}(\omega_0 - \omega)t]}{(\omega_0 - \omega)^2}\,.$$

For present purposes, however, this result is just a (big) step toward getting, in the next section, something more important.

The table gives the matrix elements $\langle n'\ell'0|r\cos\theta|n\ell 0 \rangle$ between hydrogen states with $m = 0$ and $n \leq 3$. The elements are real and the array is symmetric about the main diagonal, so there is no need to fill the table out. The elements with $\Delta n \neq 0$ are used to calculate transition probabilities; those with $\Delta n = 0$ were used for the Stark effect.

Matrix elements $\langle n'\ell'0|r\cos\theta|n\ell 0 \rangle$ of $z = r\cos\theta$
between hydrogen states with $m = 0$ and $n \leq 3$.

	$|100\rangle$	$|200\rangle$	$|210\rangle$	$|300\rangle$	$|310\rangle$	$|320\rangle$
$\langle 100|$	0	0	$\frac{2^7\sqrt{2}}{3^5}a_0$	0	$\frac{3^3\sqrt{2}}{2^7}a_0$	0
$\langle 200|$		0	$-3a_0$	0	$\frac{2^{10}3^3}{5^6}a_0$	0
$\langle 210|$			0	$\frac{2^7 3^3\sqrt{6}}{5^6}a_0$	0	$\frac{2^{12}3^3\sqrt{3}}{5^7}a_0$
$\langle 300|$				0	$-3\sqrt{6}\,a_0$	0
$\langle 310|$					0	$-3\sqrt{3}\,a_0$
$\langle 320|$						0

15·5. AVERAGING OVER POLARIZATIONS AND FREQUENCIES

Now we modify the results of the previous section to find the rates of transitions between states when an atom is bathed in electromagnetic radiation coming from every which way, with random polarizations, and with a broad continuum of frequencies.

(a) **Averaging over polarizations**—In the example just completed, the relevant matrix element was $\langle 210|z|100\rangle$, because the electric field oscillated along the z axis, and then the matrix elements $\langle 21\pm1|z|100\rangle$ are zero. (Why?) A field oscillating along the x or y axis would drive transitions between the $|100\rangle$ and $|21\pm1\rangle$ states, but not between the $|100\rangle$ and $|210\rangle$ states. (Why are $\langle 210|x|100\rangle$ and $\langle 210|y|100\rangle$ both zero?) For random polarizations of the electric field, we need to do some kind of averaging, because linear oscillations of the field do not contribute equally to different relevant matrix elements

Let $\hat{\mathbf{n}}$ be a unit vector that makes angles β_x, β_y, and β_z with the three axes. Then $\hat{\mathbf{n}} = (\cos\beta_x, \cos\beta_y, \cos\beta_z)$, and

$$\hat{\mathbf{n}} \cdot \hat{\mathbf{n}} = \cos^2\beta_x + \cos^2\beta_y + \cos^2\beta_z = 1 .$$

By symmetry, the value of each $\cos^2\beta_i$ averaged over random directions is $1/3$. Thus for *randomly* polarized light, we would divide the result we got for $|c_f(t)|^2$ in the previous section (the equation in the box) by 3. In general, we might calculate the matrix elements for all three axes of polarization, and divide the sum of them by 3:

$$\frac{1}{3}\left(|\langle f|x|i\rangle|^2 + |\langle f|y|i\rangle|^2 + |\langle f|z|i\rangle|^2\right) = \frac{1}{3}\sum_{i=1}^{3}|\langle f|x_i|i\rangle|^2 .$$

(b) **Averaging over frequencies**—The electromagnetic energy density (the energy per unit volume) of the monochromatic wave $\mathcal{E}(t) = \mathcal{E}_0\cos\omega t$ considered in the previous section is

$$u = \frac{\epsilon_0\mathcal{E}^2(t)}{2} + \frac{B^2(t)}{2\mu_0} = \epsilon_0\mathcal{E}^2(t) = \epsilon_0\mathcal{E}_0^2\cos^2\omega t .$$

(The electric and magnetic fields contribute equally.) The time average of this is $\bar{u} = \epsilon_0\mathcal{E}_0^2/2$, and so $\mathcal{E}_0^2 = 2\bar{u}/\epsilon_0$. For a broad continuum of frequencies, let the energy density in the narrow frequency interval $d\omega$ between ω and $\omega + d\omega$ be $du = \rho(\omega)\,d\omega$. We consider the effect of each interval and sum over the *intensities*, which assumes that there is no phase coherence between one frequency interval and another.

We start with $|c_f(t)|^2$ for a monochromatic wave from the previous section. First, we replace $\mathcal{E}_0^2$ with $2\bar{u}/\epsilon_0$; then we replace $\bar{u}$ with $du = \rho(\omega)\,d\omega$ and integrate over all frequencies:

$$|c_f(t)|^2 = \frac{e^2\mathcal{E}_0^2}{\hbar^2}|\langle f|z|i\rangle|^2 \frac{\sin^2[\frac{1}{2}(\omega_0-\omega)t]}{(\omega_0-\omega)^2}$$

$$\to \frac{2e^2\bar{u}}{\epsilon_0\hbar^2}|\langle f|z|i\rangle|^2 \frac{\sin^2[\frac{1}{2}(\omega_0-\omega)t]}{(\omega_0-\omega)^2}$$

$$\to \frac{2e^2}{\epsilon_0\hbar^2}|\langle f|z|i\rangle|^2 \int_0^\infty \rho(\omega)\frac{\sin^2[\frac{1}{2}(\omega_0-\omega)t]}{(\omega_0-\omega)^2}\,d\omega .$$

Now look at Fig. 3. The peaking curve, $\sin^2[\frac{1}{2}(\omega_0-\omega)t]/(\omega_0-\omega)^2$, is negligibly small unless ω is near ω_0. And we are assuming that the continuum energy density $\rho(\omega)$ varies slowly across the peak. Thus the peaking curve is effectively an (unnormalized) delta function, proportional to $\delta(\omega-\omega_0)$, and we can take $\rho(\omega)$, evaluated at $\omega = \omega_0$, outside the integral.

305

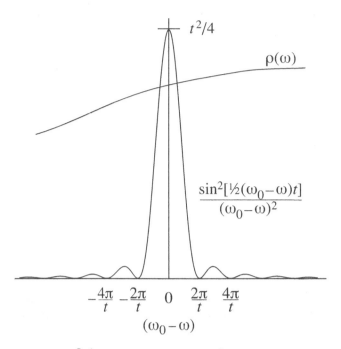

Figure 3. The function $\sin^2[\frac{1}{2}(\omega_0 - \omega)t]/(\omega_0 - \omega)^2$ for ω near ω_0. The height of the peak increases with time as t^2, and its width narrows as $1/t$. A broad continuum of energy density $\rho(\omega)$ would vary slowly across the narrowing peak, and the value of $\rho(\omega)$ at $\omega = \omega_0$ would be all that matters.

What remains of the integral is

$$\int_0^{+\infty} \frac{\sin^2[\frac{1}{2}(\omega_0 - \omega)t]}{(\omega_0 - \omega)^2} \, d\omega \; .$$

With a change of variable, $x = \frac{1}{2}(\omega_0 - \omega)t$, so that $dx = -\frac{1}{2}t \, d\omega$, we get

$$\int_{-\infty}^{+\infty} \frac{\sin^2[\frac{1}{2}(\omega_0 - \omega)t]}{(\omega_0 - \omega)^2} \, d\omega = \frac{2}{t} \frac{t^2}{4} \int_{-\infty}^{+\infty} \frac{\sin^2 x}{x^2} dx = \frac{\pi t}{2} \; .$$

The integral itself, a look-up, is π—but only because we extended the lower limit of integration from 0 to $-\infty$. This extension does not matter at all for large frequencies. (Why not?)

(c) The important results—Putting the pieces together, we get the probability of being in the final state f at time t, having been in the initial state i at $t = 0$:

$$|c_f(t)|^2 \approx \frac{2e^2}{\epsilon_0 \hbar^2} \times \frac{1}{3} \sum_{i=1}^3 |\langle f|x_i|i\rangle|^2 \times \rho(\omega_0) \times \frac{\pi t}{2} = \frac{\pi e^2}{3\epsilon_0 \hbar^2} \left(\sum_{i=1}^3 |\langle f|x_i|i\rangle|^2 \right) \rho(\omega_0) \, t \; .$$

This is a remarkable result! Now instead of oscillating sinusoidally with time, as it would for a monochromatic wave, $|c_f(t)|^2$ increases linearly with time! And therefore the rate of change of $|c_f(t)|^2$ is constant:

306

$$R(i \rightarrow f) = \frac{d |c_f(t)|^2}{dt} = \frac{\pi e^2}{3\epsilon_0 \hbar^2} \left(\sum_{i=1}^{3} |\langle f|x_i|i\rangle|^2 \right) \rho(\omega_0) = B_{if}\, \rho(\omega_0)\ ,$$

where, for use below, we write

$$B_{if} \equiv \frac{\pi e^2}{3\epsilon_0 \hbar^2} \sum_{i=1}^{3} |\langle f|x_i|i\rangle|^2\ .$$

15·6. THE BOLTZMANN FACTOR

(a) A simple example—Consider a system such as a gas of molecules in thermal equilibrium at an absolute temperature T. The probability that an atom or molecule in the system is in a particular state that has the energy E is proportional to

$$\text{the Boltzmann factor} = e^{-E/kT} = e^{-\beta E}\ .$$

Here $k = 1.381 \times 10^{-23}$ joule/kelvin$=8.617 \times 10^{-5}$ eV/kelvin is the Boltzmann constant, and kT is an energy. We sometimes use $\beta \equiv 1/kT$ to avoid writing a lot of $1/kT$'s. At $T = 290$ K ($=17\ °\text{C} = 62\ °\text{F}$), $kT = 1/40$ eV and $\beta = 40$ eV^{-1}.

The general proof of the Boltzmann factor belongs in a course on statistical mechanics, but we can show how it arises in a simple case—the fall-off of density with altitude of an ideal gas in an isothermal atmosphere. The real atmosphere is of course not isothermal—it is usually colder in the mountains.

The atmospheric pressure $p = p(z)$ at an arbitrary altitude z is just the weight per unit area of the atmosphere above that altitude. Higher up, there is less matter above and so the pressure is less. The mean translational kinetic energy K of a molecule at temperature T is $\frac{3}{2}kT$, so in an isothermal atmosphere K is independent of altitude. Energies associated with any internal degrees of freedom of a molecule will also be independent of altitude. The gravitational potential energy of a molecule is $V(z) = mgz$, where m is its mass and g is the gravitational acceleration. Consider a horizontal slice of unit area and vertical thickness dz, like a large pizza box, at altitude z. The weight of the molecules in the box is $mg \times n\, dz$, where $n = n(z)$ is the number of molecules per unit volume at altitude z. This weight is supported by a higher pressure on the lower surface of the slice.

The ideal gas law is $pV = NkT$, or $p = (N/V)kT = nkT$, where N is the total number of molecules (not moles) in the volume V. The pressure difference between the top and bottom of the box needed to support the weight $mgn\, dz$ of the molecules in the box is

$$p(z + dz) - p(z) = dp = kT\, dn = -mgn\, dz$$

(dz is positive, dp and dn are negative). Thus

$$\frac{dn}{n} = -\frac{mg}{kT}\, dz\ .$$

Integrating from $z = 0$, where the number density is n_0, to altitude z, we get

$$\ln\left(\frac{n}{n_0}\right) = -\frac{mg}{kT}\, z \quad \text{and} \quad n(z) = n_0\, e^{-mgz/kT}\ .$$

307

Thus the number density of molecules falls off exponentially with altitude, and so the probability of finding any particular molecule falls off in the same way. The Boltzmann factor is

$$e^{-\beta E} = e^{-\beta(K+V)} = e^{-\beta K} e^{-\beta V(z)} .$$

Since, as noted above, K is independent of altitude, so also is $e^{-\beta K}$. Therefore, the *ratios* of densities $n(z)$ at different altitudes z are given by ratios of either $e^{-\beta V(z)} = e^{-mgz/kT}$ or $e^{-\beta E}$.

(b) Population ratios—The Boltzmann factor is often used (as above) to get ratios of probabilities, even when the absolute probabilities themselves are unknown. What is the ratio of probabilities that a hydrogen atom is in its $n = 2$ and $n = 1$ states in thermal equilibrium at $T = 290$ K and at $T = 20 \times 290 = 5800$ K? The latter is the approximate temperature at the surface of the Sun. Here $E_1 = -13.6$ eV, $E_2 = -3.40$ eV, and $\Delta E = E_2 - E_1 = 10.2$ eV. And there are four times as many $n = 2$ states as $n = 1$ states (why?). Therefore,

$$\frac{\text{Prob}(E_2)}{\text{Prob}(E_1)} = \frac{4\,e^{-\beta E_2}}{e^{-\beta E_1}} = 4\,e^{-\beta \Delta E} .$$

At $T = 290$ K, $\beta = 40$ eV^{-1}, and the ratio is about 10^{-176}. At $T = 5800$ K, $\beta = 2$ eV^{-1}, and the ratio is about 5×10^{-9}. This is a small *fraction* of atoms, but a large *number* of atoms per *mole*. A change by a factor 20 in the exponent changed the ratio by a factor of about 10^{167}.

Now if a spark or a cosmic ray or light of high enough frequency passes through a gas, some of the atoms are excited—and this is *not* a state of thermal equilibrium. The excited atoms will fall back to lower states—but how quickly? What are the lifetimes of excited states in the absence of "supporting" thermal radiation? The basic relation was discovered by Einstein in 1916—one of Einstein's greatest hits. His discovery finds the decay rates of atoms *not* in thermal equilibrium by considering atoms *in* thermal equilibrium with black-body radiation—and then yanking the black-body radiation away. So first we need to understand black-body radiation.

15·7. PLANCK'S OSCILLATORS

(a) Quantized oscillators—To begin, we suppose that a large number of identical one-dimensional oscillators are in thermal equilibrium with the surroundings. The temperature is T, and the oscillator energy levels are $E_n = (n + \frac{1}{2})\hbar\omega$, where $n = 0, 1, 2, \ldots$ However, to avoid in what follows some irrelevant factors $e^{-\hbar\omega/2kT}$, we shift the zero of energy so that $E_n = n\hbar\omega$, $n = 0, 1, 2, \ldots$ We calculate, as a function of T, the number of oscillators in each energy level. We can if we wish add $\frac{1}{2}\hbar\omega$ back to energies at the end.

Let N_n be the number of oscillators having the energy $E_n = n\hbar\omega$. The total number of oscillators is then

$$N_{total} = N_0 + N_1 + N_2 + N_3 + \cdots$$

Since the energy levels are equally spaced, all the numbers N_n are simply related to N_0 by powers of a single Boltzmann factor,

$$\frac{N_1}{N_0} = \frac{N_2}{N_1} = \frac{N_3}{N_2} = \frac{N_4}{N_3} = \cdots = e^{-\hbar\omega/kT} .$$

Let $y \equiv e^{-\hbar\omega/kT}$. Then $N_1 = yN_0$, $N_2 = yN_1 = y^2N_0$, and so on, and

$$N_{total} = N_0\left(1 + y + y^2 + y^3 + \cdots\right) = \frac{N_0}{1-y}.$$

The series in y is the binomial expansion of $(1-y)^{-1}$. It converges for $-1 < y < +1$, which is the case here. At very low temperatures ($kT \ll \hbar\omega$), $y = e^{-\hbar\omega/kT}$ is small, and nearly all the oscillators are in the ground state. At high temperatures ($kT \gg \hbar\omega$), y is nearly 1, and nearly all the oscillators are in excited states. The table and Fig. 4 show the percentages of all the oscillators in the lowest few states for three values of $\hbar\omega/kT$:

$n =$	0	1	2	3	4	> 4	$\overline{n}$
$\hbar\omega/kT = 3$	95.0	4.73	0.24	0.012	0.0006	$\simeq 0$	0.0524
$\hbar\omega/kT = 1$	63.2	23.3	8.55	3.15	1.16	0.67	0.582
$\hbar\omega/kT = 1/3$	28.3	20.3	14.6	10.4	7.47	18.9	2.528

Assuming $\hbar\omega$ is fixed, kT increases by a factor of three from one row to the next.

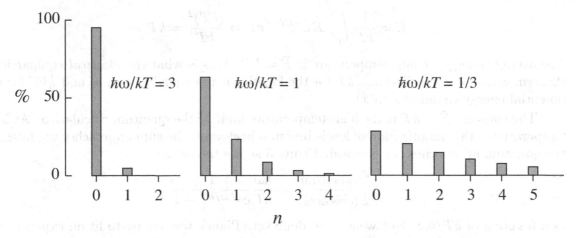

Figure 4. The percentages of oscillators having energies $E_n = n\hbar\omega$ for three values of $\hbar\omega/kT$.

The total energy in the nth level is the energy $E_n = n\hbar\omega$ of that level times the number of oscillators N_n having that energy. The *total* energy, summed over *all* the oscillators and *all* the levels, is

$$E_{total} = \hbar\omega\left(0 \times N_0 + 1 \times N_1 + 2 \times N_2 + 3 \times N_3 + \cdots\right)$$
$$= N_0\hbar\omega\left(1 \times y + 2 \times y^2 + 3 \times y^3 + \cdots\right)$$
$$= N_0\hbar\omega\frac{y}{(1-y)^2}.$$

The y-dependent factor here is the derivative of the y-dependent factor $1/(1-y)$ in N_{total}. The average energy $\overline{E}$ of an oscillator is

$$\boxed{\overline{E} = \frac{E_{total}}{N_{total}} = \frac{N_0\hbar\omega\,y}{(1-y)^2} \bigg/ \frac{N_0}{1-y} = \frac{\hbar\omega}{1/y - 1} = \frac{\hbar\omega}{e^{+\hbar\omega/kT} - 1}.}$$

(This is the energy *above* the ground-state energy $\frac{1}{2}\hbar\omega$ we subtracted off from each oscillator.) At low temperatures, $e^{+\hbar\omega/kT}$ is large, and the average energy $\overline{E}$ is close to zero. At high temperatures, $e^{+\hbar\omega/kT} \simeq 1 + \hbar\omega/kT$, and

$$\overline{E} = \frac{\hbar\omega}{e^{+\hbar\omega/kT} - 1} \approx kT .$$

The derivation so far has been for one-dimensional oscillators. But in three dimensions $V(x, y, z) = V_x(x) + V_y(y) + V_z(z)$ for the oscillator (Sec. 2·7), and there are simply three times as many independent one-dimensional oscillators.

(b) Quantum versus classical—Classically, oscillator energies are of course continuous, with $E \geq 0$. The Boltzmann factor $e^{-E/kT}$ is then a *continuous* function of E, and the probability distribution is $Ce^{-E/kT}$, where

$$C \int_0^\infty e^{-E/kT} dE = CkT ,$$

so that $C = 1/kT$ normalizes the distribution. Then

$$\overline{E} = \frac{1}{kT} \int_0^\infty E \, e^{-E/kT} dE = \frac{(kT)^2}{kT} = kT .$$

The average energy at *any* temperature is $\overline{E} = kT$. This is what the *classical* equipartition theorem gives for an oscillator: $\frac{1}{2}kT$ for the kinetic energy (quadratic in p) and $\frac{1}{2}kT$ for the potential energy (quadratic in x).

The classical $\overline{E} = kT$ is the high-temperature limit of the quantum calculation. At high temperatures, the quantization of levels becomes irrelevant, the sum approaches the integral, the quantum approaches the classical. Figure 5 shows the ratio

$$\frac{\overline{E}(\text{quantum})}{\overline{E}(\text{classical})} = \frac{\hbar\omega}{kT} \frac{1}{e^{+\hbar\omega/kT} - 1}$$

as a function of $kT/\hbar\omega$. But we are not done yet. Planck was trying to fit an experimental curve about radiation in a cavity.

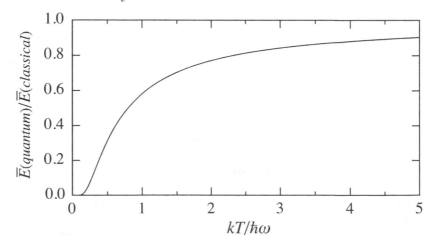

Figure 5. The ratio $\overline{E}(\text{quantum})/\overline{E}(\text{classical})$ as a function of $kT/\hbar\omega$ (not of $\hbar\omega/kT$).

310

15·8. BLACK-BODY RADIATION

Perhaps read page 2 of Chapter 1 again here.

(a) **Radiation in a box**—We consider electromagnetic radiation in a one-dimensional box with walls that are perfectly conducting. A perfect conductor cannot support an electromagnetic field, and so the fields vanish at the walls. Then the eigenstates are sine waves with an integral number $n = 1, 2, 3, \ldots$, of half-wavelengths $\lambda/2$ from wall to wall. If the width of the box is L, then

$$n\frac{\lambda}{2} = L \; .$$

So far, this is analogous to a particle in an infinite square well (Sec. 2·4). But now the relation between the energy and momentum is the relativistic linear $E = pc$, not the non-relativistic quadratic $E = p^2/2m$. And $p = h/\lambda$. Thus the energies of the oscillations are

$$E_n = p_n c = \frac{h}{\lambda_n}c = 2\pi\hbar c \times \frac{n}{2L} = n\hbar\omega \; ,$$

where $\omega \equiv \pi c/L$. The energies are *linear* in the quantum number n, just like the harmonic oscillator, and are quantized by (a different) $\hbar\omega$. Paralleling the numbers $N_0, N_1, N_2, \ldots$, of oscillators in the states with $n = 0, 1, 2, \ldots$, are numbers $N_1, N_2, N_3, \ldots$, with $n = 1, 2, 3, \ldots$, of states with energies $\hbar\omega, 2\hbar\omega, 3\hbar\omega, \ldots$ Then the math is the same as it was for the oscillators, and

$$\overline{E} = \frac{\hbar\omega}{e^{+\hbar\omega/kT} - 1} \; .$$

And again, going to a three-dimensional box with sides L just makes more independent states, with indices n_x, n_y, and n_z.

(b) **Counting states (again)**—In three dimensions, the wave functions for the electromagnetic energy eigenstates have the same form as those for particles in a cube,

$$\psi_{n_x n_y n_z}(x,y,z) = \sqrt{\frac{8}{L^3}} \, \sin\left(\frac{n_x\pi}{L}x\right) \sin\left(\frac{n_y\pi}{L}y\right) \sin\left(\frac{n_z\pi}{L}z\right) \; .$$

The wave numbers are $k_i = n_i\pi/L$, $i = x, y, z$. Then

$$k_x^2 + k_y^2 + k_z^2 = (n_x^2 + n_y^2 + n_z^2)\frac{\pi^2}{L^2}, \qquad \text{or} \qquad n^2 = \frac{L^2}{\pi^2}k^2 \; ,$$

where $k^2 = k_x^2 + k_y^2 + k_z^2$ and $n^2 = n_x^2 + n_y^2 + n_z^2$. The calculation of the number of states between n and $n + dn$ in number space was sketched in Prob. 2·17 and again in Sec. 14·1. We associate each quantum-number triple, (n_x, n_y, n_z), either with a point on a grid in number space or with the 1-by-1-by-1 cube with that point at its corner; see Fig. 14·2. Each 1-by-1-by-1 cube has a volume equal to one in the number space. The number of cubes in the positive octant out to some large radius n in the number space is $1/8 \times 4\pi n^3/3$. The volume between radii n and $n + dn$ is the derivative of this, or

$$\rho(n)\,dn = \frac{1}{8} \times 4\pi n^2\,dn = \frac{1}{2} \times \pi n^2\,dn \; .$$

What we want here, however, is $\rho(\omega)\,d\omega$, so we translate. Since $n = 2L/\lambda = Lk/\pi = L\omega/\pi c$, we get

$$\frac{1}{2} \times \pi n^2\,dn = \frac{\pi}{2} \times \frac{L^2}{\pi^2 c^2}\frac{L}{\pi c}\omega^2\,d\omega = \frac{V}{2\pi^2 c^3}\omega^2\,d\omega \; .$$

311

We multiply this by two for the two polarization states of electromagnetic waves, to get

$$\text{the number of states between } \omega \text{ and } \omega + d\omega = \rho(\omega)\,d\omega = \frac{V}{\pi^2 c^3}\,\omega^2\,d\omega\,,$$

where $V = L^3$ is the volume of the box.

(c) The black-body spectrum—The Planck distribution gives the thermal energy per unit volume of electromagnetic radiation in otherwise empty space in the angular frequency range ω to $\omega + d\omega$. It is the product of the average energy of a state with frequency ω and the number of states per unit volume in the range $d\omega$:

$$du = \rho(\omega)\,d\omega = \frac{1}{\pi^2 c^3}\,\omega^2\,d\omega \times \frac{\hbar\omega}{e^{+\hbar\omega/kT}-1} = \frac{\hbar}{\pi^2 c^3}\,\frac{\omega^3\,d\omega}{e^{\hbar\omega/kT}-1}\,.$$

This result, derived by Planck in 1900, is the first appearance of the constant h. Figure 6 shows the energy density distribution as a function of the cycle frequency ν rather than of the radian frequency ω.

It is left for a problem to show that the peak of $\rho(\omega)$ occurs at a frequency ω_{max} that is proportional to the temperature, and that the area under $\rho(\omega)$ is proportional to T^4. Thus ω_{max} at the Sun's surface is about 20 times ω_{max} at the Earth's surface, and the thermal energy density at the Sun's surface is about 160,000 times that on Earth.

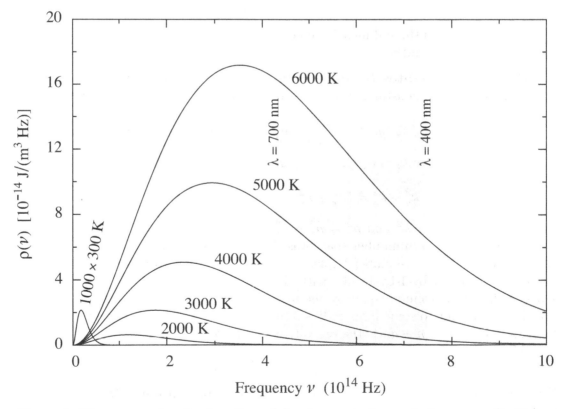

Figure 6. The energy-density function $\rho(\nu)$, where ν is the cycle frequency (in Hz), not the angular frequency ω. The nominal range over which we can see is shaded; it is less than an octave wide.

15·9. EINSTEIN'S A AND B COEFFICIENTS

(a) Transitions—A gas of atoms or molecules in a box is in thermal equilibrium with the black-body radiation in the box. The temperature is high enough that some appreciable fraction of the molecules is in excited states. Let N_i and N_f be the numbers of molecules in two arbitrary states i and f, where $E_f > E_i$. *Which* molecules are in which of these or other states is constantly changing as the molecules exchange energies with the thermal radiation and in collisions with one another. But aside from small random fluctuations, the numbers N_i and N_f remain constant, and the ratio of N_i and N_f is given by the ratio of Boltzmann factors. Figure 7 shows three processes that change the number of molecules in the upper state f. We ignore changes due to molecular collisions, but they would not spoil the argument.

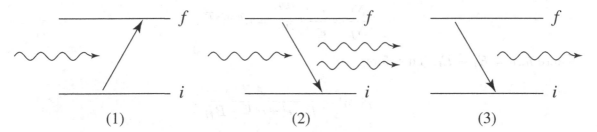

Figure 7. (1) Absorption, (2) stimulated emission, (3) spontaneous emission.

(1) A molecule in the lower state i can *absorb* a quantum of energy, a photon, from the radiation field and be excited to the upper state f. Energy is conserved. The rate of increase of N_f due to excitation from N_i is proportional to the energy density $\rho(\omega_0)$ of the black-body radiation at the transition frequency, $\omega_0 = (E_f - E_i)/\hbar$, and to the number of molecules N_i in the lower state,

$$\frac{dN_f}{dt}\bigg|_{absp} = +B_{if}\rho(\omega_0)N_i \qquad \text{(absorption)} .$$

The product $B_{if}\rho(\omega_0)$ is the rate $R(i \rightarrow f)$ of transitions, per atom or molecule, between the states i and f. The equation for B_{if} is in a box at the end of Sec. 5, but Einstein could not have known it in 1916.

(2) A molecule in the upper state f can be *de*-excited to the lower state i by an incident photon. To conserve energy, this requires the creation of a second photon. This is *stimulated emission*: The incident photon causes a *second* quantum to be emitted. Energy is conserved. The rate of decrease of N_f due to stimulated emission is

$$\frac{dN_f}{dt}\bigg|_{stim} = -B_{fi}\rho(\omega_0)N_f \qquad \text{(stimulated emission)} .$$

(3) A molecule in state f can also by itself drop down to state i with emission of a photon. This is *spontaneous emission*. The rate of spontaneous decay from f to i is

$$\frac{dN_f}{dt}\bigg|_{spon} = -AN_f \qquad \text{(spontaneous emission)} .$$

It is proportional to N_f. In Sec. 7·5, in calculating time-and-energy Fourier transforms, we showed that $A = 1/\tau$, where τ is the mean decay lifetime. But we did not know then how to calculate τ.

313

(b) Thermal equilibrium—Aside from random fluctuations, the net effect of the three processes in a state of thermal equilibrium is to make *no* change of N_f,

$$\frac{dN_f}{dt} = B_{if}\rho(\omega_0)N_i - B_{fi}\rho(\omega_0)N_f - AN_f = 0 \ .$$

Solving for $\rho(\omega_0)$, we get

$$\rho(\omega_0) = \frac{AN_f}{B_{if}N_i - B_{fi}N_f} = \frac{A}{B_{if}(N_i/N_f) - B_{fi}} \ .$$

The ratio N_i/N_f at thermal equilibrium is the ratio of Boltzmann factors,

$$\frac{N_i}{N_f} = \frac{e^{-E_i/kT}}{e^{-E_f/kT}} = e^{\hbar\omega_0/kT} \ ,$$

where $\hbar\omega_0 = E_f - E_i$. Therefore,

$$\rho(\omega_0) = \frac{A}{B_{if}\, e^{\hbar\omega_0/kT} - B_{fi}} \ .$$

Now Einstein *did* know Planck's black-body $\rho(\omega)$, given at the end of Sec. 7:

$$\rho(\omega) = \frac{\hbar}{\pi^2 c^3} \frac{\omega^3}{e^{\hbar\omega/kT} - 1} \ .$$

Comparing Einstein's equation with Planck's at the transition frequency ω_0, we get

$$\boxed{\quad B_{fi} = B_{if} \qquad \text{and} \qquad A = \frac{\hbar\omega_0^3}{\pi^2 c^3} B_{fi} \ . \quad}$$

Earlier we found that incident radiation is equally likely to drive a molecule up from i to f and down from f to i. That is what $B_{if} = B_{fi}$ means, derived by Einstein before there was quantum mechanics. And he found something that was completely new in 1916—the relation between the proportionality constants A for *stimulated* emission and B_{fi} for *spontaneous* emission. That is a lot to have learned from such a simple yet elegant argument. And now, knowing how to calculate B_{fi}, we can calculate A and the lifetime $\tau = 1/A$ for a decay.

15·10. DECAY LIFETIMES

Spontaneous decays of excited atoms dominate when the $f \to i$ transition energy $\hbar\omega_0$ is much larger than kT. Then $\rho(\omega)$ at $\omega = \omega_0$ is negligible, and there is negligible supporting thermal radiation. For example, whereas kT at room temperature is about $1/40 = 0.025$ eV, the smallest $E_f - E_i$ of all the transitions between the hydrogen levels shown in Fig. 1·4 is about 0.66 eV. Therefore, starting in such a circumstance with N_0 atoms in an excited state at $t = 0$, the number remaining at later times is

$$N(t) = N_0 e^{-At} = N_0 e^{-t/\tau} \ .$$

We calculate the mean life of an excited hydrogen level.

314

(a) Example: the mean life of a hydrogen state—The hydrogen $|n\ell m\rangle = |210\rangle$ state decays to the $|100\rangle$ state. In the last section, we found that

$$A = \frac{\hbar\omega_0^3}{\pi^2 c^3} B_{fi} ,$$

and $B_{fi} = B_{if}$. And in Sec. 5, we found that

$$B_{fi} = \frac{\pi e^2}{3\epsilon_0 \hbar^2} \sum_{i=1}^{3} |\langle f|x_i|i\rangle|^2 ,$$

where the $\langle f|x_i|i\rangle$ are the matrix elements between the initial and final states for $x_i = x$, y, and z. The elements $\langle 210|x|100\rangle$ and $\langle 210|y|100\rangle$ both equal zero (again, why?). The elements $\langle 100|z|210\rangle$ and $\langle 210|z|100\rangle$ are equal, and we calculated the latter in Sec. 4(b):

$$\langle 100|z|210\rangle = \langle 210|z|100\rangle = \frac{2^7 \sqrt{2}}{3^5} a_0 .$$

The transition frequency ω_0 is given by

$$\hbar\omega_0 = E_2 - E_1 = \frac{3}{4}|E_1| = \frac{3}{4}\frac{mk^2 e^4}{2\hbar^2} .$$

Putting all the pieces together, with $k \equiv 1/4\pi\epsilon_0$ and $a_0 = \hbar^2/mke^2$, we get

$$A = \frac{\hbar\omega_0^3}{\pi^2 c^3} \times \frac{\pi e^2}{3\epsilon_0 \hbar^2} \times \frac{2^{15}}{3^{10}} a_0^2 = \frac{3^3}{2^7}\frac{2^{15}}{3^{11}}\frac{1}{\hbar^2}\left(\frac{mk^2 e^4}{\hbar^2}\right)^3 \frac{e^2}{4\pi\epsilon_0 \hbar^2 c^3}\left(\frac{\hbar^2}{mke^2}\right)^2$$

$$= \left(\frac{2}{3}\right)^8 \left(\frac{ke^2}{\hbar c}\right)^5 \frac{mc^2}{\hbar} = \left(\frac{2}{3}\right)^8 \alpha^5 \frac{mc^2}{\hbar} ,$$

where $\alpha = ke^2/\hbar c$ is the fine-structure constant. Using $\alpha = 1/137$, $mc^2 = 5.11 \times 10^5$ eV, and $\hbar = 6.58 \times 10^{-16}$ eV s, we get the mean lifetime of the hydrogen $|210\rangle$ state:

$$A = 6.28 \times 10^8 \text{ s}^{-1} \quad \text{and} \quad \tau = 1.59 \times 10^{-9} \text{ s} .$$

The factor $\alpha^5 mc^2/\hbar$ itself equals 1.61×10^{10} s^{-1}. The calculation of any other allowed ($\Delta\ell = \pm 1$, $\Delta m = 0, \pm 1$) decay would involve this factor.

Suppose we could "pump" a large number of atoms into an excited state at which $\rho(\omega)$ at $\omega = \omega_0$ is negligible. Then one decay photon could trigger an emission, so there would be two photons; which might trigger two decays so there would be four photons; which might trigger ... A chain reaction, an avalanche of millions of photons all at the same frequency! *Light amplification by stimulated emission of radiation*. The laser stems from Einstein's 1916 paper.

PROBLEMS

1. *A sudden change:* $^3\mathrm{H}^+$ *decay to* $^3\mathrm{He}^{++}$.

The nucleus of a heavy form of hydrogen, $^3\mathrm{H}^+$, consists of one proton and two neutrons (the superscript 3 tells the total number of nucleons—protons plus neutrons). This "triton" is unstable: One of the neutrons decays to a proton, electron, and antineutrino ($n \to pe^-\bar{\nu}$). The e^- and $\bar{\nu}$ fly off, leaving a $^3\mathrm{He}^{++}$ nucleus (2 protons, 1 neutron) behind. The half-life is 12.3 yrs.

Before the decay, there is an e^- in the hydrogen ground state. The e^- from the $n \to pe^-\bar{\nu}$ decay is energetic and is long gone in a time that is short compared to the period of the motion of the atomic e^-. Use the sudden-change approximation to calculate the probability that the *atomic* electron would be found, if measured immediately after the decay, in the ground state of $^3\mathrm{He}^+$. The answer is $2^9/3^6 = 512/729 = 70.23\%$.

Note: Radioactivity was discovered by Henri Becquerel in 1896. It took some years to sort out that there were different kinds of radioactivity, involving three different kinds of "rays," labeled α, β, and γ. This was before anything was known about the structure of atoms, or even what else atoms were made of besides electrons. *Alpha rays* were charged particles that in ordinary nuclear decays would only penetrate a sheet or two of paper. Eventually they were found to be the same as $^4\mathrm{He}$ nuclei (a tightly bound state of two protons and two neutrons). *Beta rays* were charged particles that in ordinary decays would penetrate a few sheets of metal foil. They were found to be electrons or positrons. *Gamma rays* were neutral and were far more penetrating than alpha or beta rays. They were found to be high-energy photons. The easily detected particle in $n \to pe^-\bar{\nu}$ decay is the e^-, and so this and other decays that produce an electron or positron are called β decays.

2. *More sudden changes.*

(a) A spin-1/2 particle is spin-up along a magnetic field $\mathbf{B}_0$. A second field $\mathbf{B}_1$ is suddenly turned on at right angles to $\mathbf{B}_0$. Find as a function of $\mathbf{B}_1/\mathbf{B}_0$ the probability that, if measured immediately after, the spin will be up along the new total field.

(b) A particle is in the ground state of the oscillator well, $V(x) = \frac{1}{2}m\omega^2 x^2$. An electric field $\mathcal{E}_0$ is suddenly turned on. Show that the probability the particle would be found in the new ground state is $\exp(-q^2\mathcal{E}_0^2/2m\hbar\omega^3)$. See Fig. 13·1 and Prob. 13·1.

3. *Do it again.*

Solve (again) exactly the coupled two-state system of equations

$$\begin{pmatrix} da_+/dt \\ da_-/dt \end{pmatrix} = +\frac{i\omega_1}{2}\begin{pmatrix} 0 & e^{+i(\omega-\omega_0)t} \\ e^{-i(\omega-\omega_0)t} & 0 \end{pmatrix}\begin{pmatrix} a_+ \\ a_- \end{pmatrix}$$

for $a_+(t)$ and $a_-(t)$ when $a_+(0) = 0$ and $a_-(0) = 1$. Show that $|a_+(t)|^2 + |a_-(t)|^2 = 1$ at all times.

4. *A gaussian wave.*

In Sec. 13·2(a), we calculated the shifts of energy eigenvalues caused by a static electric field $\mathcal{E}_0$ acting on a charged particle in an oscillator well. The perturbation Hamiltonian there was $\hat{H}' = -q\mathcal{E}_0\hat{x}$. Changing $\mathcal{E}_0$ to $\mathcal{E}(t)$ for that same oscillator in a time-dependent electric field,

316

we get

$$\hat{H}'(t) = -q\mathcal{E}(t)\hat{x} = -q\mathcal{E}(t)\sqrt{\frac{\hbar}{2m\omega_0}}\begin{pmatrix} 0 & \sqrt{1} & 0 & 0 & \cdots \\ \sqrt{1} & 0 & \sqrt{2} & 0 & \cdots \\ 0 & \sqrt{2} & 0 & \sqrt{3} & \cdots \\ 0 & 0 & \sqrt{3} & 0 & \cdots \\ \vdots & \vdots & \vdots & \vdots & \ddots \end{pmatrix}.$$

Then from Sec. 2(a) of this chapter, the (exact) coupled equations for the expansion coefficients $a_n(t)$ are

$$\begin{pmatrix} da_0/dt \\ da_1/dt \\ da_2/dt \\ \vdots \end{pmatrix} = +\frac{iq\mathcal{E}(t)}{\sqrt{2m\hbar\omega_0}}\begin{pmatrix} 0 & \sqrt{1}\,e^{-i\omega_0 t} & 0 & \cdots \\ \sqrt{1}\,e^{+i\omega_0 t} & 0 & \sqrt{2}\,e^{-i\omega_0 t} & \cdots \\ 0 & \sqrt{2}\,e^{+i\omega_0 t} & 0 & \cdots \\ \vdots & \vdots & \vdots & \ddots \end{pmatrix}\begin{pmatrix} a_0(t) \\ a_1(t) \\ a_2(t) \\ \vdots \end{pmatrix}.$$

The values of ω_{10}, ω_{21}, ω_{32}, ... , are all equal to the oscillator frequency ω_0.

(a) Let the electric field vary from $t = -\infty$ to $t = +\infty$ as $\mathcal{E}(t) = \mathcal{E}_0 \exp(-t^2/\tau^2)$. Show that in first-order theory the probability the oscillator would be found in the first excited state at $t = +\infty$ is

$$\frac{\pi q^2 \mathcal{E}_0^2 \tau^2}{2m\hbar\omega_0}e^{\omega_0^2\tau^2/2}.$$

Ignore decays back to the ground state.

(b) What value of τ maximizes the probability in (a)?

5. *Hydrogen matrix elements for* $m' = m = +1$.

Make a table like the one at the end of Sec. 4 for the hydrogen matrix elements

$$\langle n'\ell'm'|r\cos\theta|n\ell m\rangle\,,$$

where $m = m' = +1$ and $n \leq 3$. How can you use the table in Sec. 4 to check your results?

6. *Ratios of matrix elements.*

Find these ratios of hydrogen $\langle n'\ell'm'|r\cos\theta|n\ell m\rangle$ elements:

$$\langle 540|z|430\rangle : \langle 541|z|431\rangle : \langle 542|z|432\rangle : \langle 543|z|433\rangle\,.$$

Normalize so that the first element equals one. This is a short problem. What stays the same? What changes? Maybe you can do it in your head.

7. *Paramagnetism.*

Paramagnetism is a weak form of magnetism, unlike ferromagnetism. The constituent atoms or molecules of paramagnetic substances have an intrinsic magnetic moment μ. Consider a block of N paramagnetic, spin-1/2 atoms. In an external magnetic field B, the energies $E_\pm$ of the spin-up and -down states with respect to the field direction are $-\mu B$ and $+\mu B$. At thermal equilibrium, the probabilities an atom is spin up or down are $P_\pm = C \exp(\pm\mu B/kT)$, where C is a normalization constant.

317

Find the net magnetic moment $M = N\bar{\mu}$ of the block as a function of T and B, where $\bar{\mu}$ is the mean moment along B per atom. The figure shows the result, but the axis labels were accidentally lost. Please label the axes properly and tell what function that curve is.

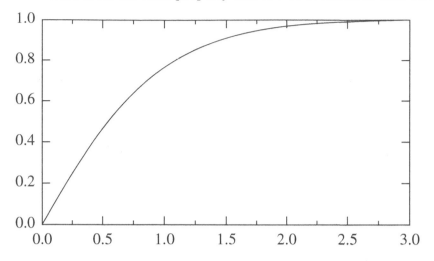

8. *Get some numbers.*

Use a calculator to calculate the numbers in the last row of the table just in front of Fig. 4.

9. *Specific heat of a crystal.*

The specific heat of a substance is the amount the energy of a mole of it increases when the temperature is increased by 1 K: $C = \partial\bar{E}/\partial T$.

10. *Properties of black-body radiation.*

(a) Figure 6 shows the Planck distribution,

$$\rho(\omega)\, d\omega = \frac{\hbar}{\pi^2 c^3} \frac{\omega^3\, d\omega}{e^{\hbar\omega/kT} - 1}\ ,$$

for several temperatures. Show that the peak of the distribution is at $\omega_{max} = 2\pi\nu_{max} = C\,(kT/\hbar)$. To find the number C, let $x \equiv \hbar\omega/kT$, and differentiate $x^3/(e^x - 1)$. Use a calculator to get C to three places using the hunt-and-peck method of Prob. 2·5. Start near $x = 3$.

(b) Show that the integral of the energy density over all frequencies is

$$u_{total} = \int_0^\infty \rho(\omega)\, d\omega = T^4 \left(\frac{k^4}{\pi^2 c^3 \hbar^3}\right) \int_0^\infty \frac{x^3}{e^x - 1}\, dx = aT^4\ .$$

Look up the integral and show that $a = \pi^2 k^4/15c^3\hbar^3$. The relation $\frac{1}{4}caT^4 = \sigma T^4$ is called the Stefan-Boltzmann law; it is the energy emitted per unit time per unit area by a black body. Put in the numbers and show that $\sigma = 5.67 \times 10^{-8}$ J s^{-1} m^{-2} K^{-4}.

(c) Calculate ν_{max} and u_{total} at 290 K and at 5800 K. We see at wavelengths in the range 400-to-700 nm. What are the peak wavelengths at 290 K and at 5800 K?

11. *When spontaneous emission dominates.* The two mechanisms for downward transitions, spontaneous emission and emission stimulated by ambient radiation (which here we take to be black-body radiation), compete. The rates are A and $B_{21}\rho(\omega_0)$, where ω_0 is the transition angular frequency.

(a) Show that near room temperature ($T = 290$ K) spontaneous emission dominates for frequencies ν well above 5×10^{12} Hz, whereas stimulated emission dominates for ν well below this value. To which realm of frequencies does visible light belong? That is why for transitions in the optical region we use $\tau = 1/A$.

(b) What are the relative contributions of the two processes for the $n = 2$ to $n = 1$ transition in hydrogen at the surface of the Sun, where $T = 5800$ K?

(c) At what temperature would the two processes contribute equally to this transition?

12. *All three $|21m\rangle$ states decay at the same rate.*

Calculate the $\langle 211|x|100\rangle$ and $\langle 211|y|100\rangle$ matrix elements for hydrogen. The radial part $\langle 21|r|10\rangle$ and the full element $\langle 210|z|100\rangle$ are in Sec. 4(b). Thus show that the lifetimes of the three $|21m\rangle$ states are the same.

13. *More lifetimes.*

(a) Calculate the spontaneous decay lifetimes for the transitions $|310\rangle \rightarrow |100\rangle$, $|300\rangle \rightarrow |210\rangle$, $|310\rangle \rightarrow |200\rangle$, and $|320\rangle \rightarrow |210\rangle$. The matrix elements are at the end of Sec. 4(b). All that is involved here is a bit of calculator work.

(b) Draw a diagram showing the six energy levels. Add arrows showing those four and the $|210\rangle \rightarrow |100\rangle$ transition. Write the lifetimes along the arrows.

16. SCATTERING

16.1. *Solid Angle*

16.2. *Classical Particle Scattering. Rutherford*

16.3. *The Scattering Amplitude*

16.4. *The Born Approximation*

16.5. *Kinematics*

16.6. *Partial-Wave Theory*

16.7. *Partial-Wave Examples*

16.8. *Scattering of Identical Particles*

Problems

The early studies of scattering of particles by targets used collimated beams of decay products from radioactive sources and low-energy beams of electrons from accelerators. Beams of electrons run into matter also produced beams of x rays. But for about 100 years now, larger and larger accelerators have produced beams of atoms, nuclei, and particles of ever greater energy and intensity.

The scattering angular distributions of collisions can tell us about the forces between beam and target. However, much more than simple $a + b \rightarrow a + b$ (elastic) scattering can occur. Experiments at the higher energies have led to the discovery of scores of unexpected massive particles ($E = mc^2$ at work), out of range at low energies. At the energies available at large accelerators, and especially at the vastly higher energies that cosmic-ray protons can have, there are sometimes showers of hundreds of particles. But to analyze even simple two-body *in*elastic reactions is more than we can do here. And even for purely elastic scattering, the energies will be nonrelativistic and, except in the last section, the particles spinless.

With these severe limitations, the questions we can tackle here are these: (1) Given a spherically symmetric potential energy $V(r)$ for the interaction between beam and target, what is the scattering angular distribution as a function of energy? (2) Given an experimental angular distribution, what are the separate scattering amplitudes for the angular momentum states with $\ell = 0, 1, 2, \ldots$, between beam and target? Sections 3 and 4 address the first of these questions, sections 6 and 7 the second.

The last section is about the scattering of identical particles. It shows, in a simple and fundamental way, the remarkable difference between fermions and bosons, due to the different requirements of exchange symmetry for the two classes of particles.

Section 1 is about solid angle, Section 2 is about cross sections and angular distributions in classical scattering. They are probably worth your time, even if you already know something about these subjects.

16·1. SOLID ANGLE

Figure 1(a) shows an arc of a circle with its center at O; the length of the arc is s and the radius of the circle is r. The angle subtended by the arc from O—the angle between the radial lines—is by definition $\theta = s/r$ radians. A complete circle about O subtends $2\pi r/r = 2\pi$ radians. The angle subtended by any object in a plane from a point O in the plane is the angle between the radial lines drawn from O to the far sides of the object; see Fig. 1(b). Symmetry is often useful: A side of a square subtends $2\pi/4$ radians from the center of the square, and $2\pi/8$ radians from an opposite corner.

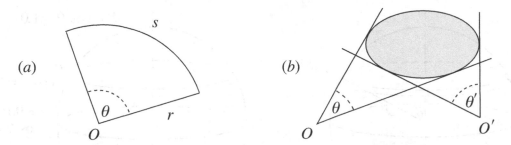

Figure 1. (a) Defining angles in radians: $\theta \equiv s/r$. (b) The angle subtended by an object depends on where it is viewed from.

Solid angle is defined in an analogous way. Figure 2 shows an area A on the surface of a sphere; the radius of the sphere is r and its center is at O. The solid angle subtended by A from O is by definition $\Omega = A/r^2$ *ster*adians (stereo = solid, in space; as in solid geometry). The whole surface of the sphere subtends $4\pi r^2/r^2 = 4\pi$ steradians from O. The solid angle subtended by an object in space from a point O is the fraction of the 4π steradians surrounding O that is blocked off by the object. Symmetry is again useful: A side of a cube subtends $4\pi/6$ steradians from the center of the cube. What solid angle does the side subtend from an opposite corner of the cube?

An infinitesimal area dA on the surface of a sphere of radius r subtends an infinitesimal solid angle $d\Omega = dA/r^2$ from the center. In spherical coordinates, the infinitesimal area on the surface of a sphere between θ and $\theta + d\theta$ and between ϕ and $\phi + d\phi$ is

$$dA = r\, d\theta \times r \sin\theta\, d\phi = r^2 \sin\theta\, d\theta\, d\phi .$$

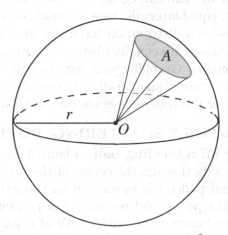

Figure 2. Defining solid angles in steradians: $\Omega \equiv A/r^2$.

321

Thus the solid angle subtended by this dA from the center is

$$d\Omega = dA/r^2 = \sin\theta \, d\theta \, d\phi \ .$$

The scattering problems considered in this chapter will be symmetric about the z axis; that is, the scattering angular distribution will not depend on ϕ. Integrating $d\Omega$ over ϕ gives the still infinitesimal $d\Omega = 2\pi \sin\theta \, d\theta$; see Fig. 3(a).

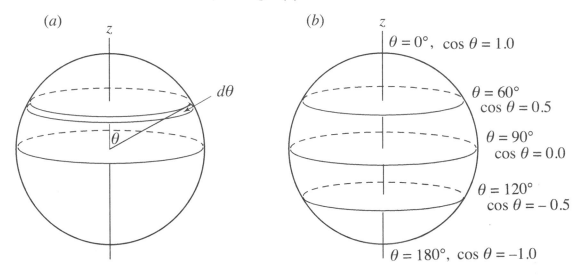

Figure 3. (a) The infinitesimal solid angle between θ and $\theta + d\theta$ is $d\Omega = 2\pi \sin\theta \, d\theta$. (b) Cutting a sphere at equal intervals of $\cos\theta$ divides the pole-to-pole diameter into equal lengths, the surface of the sphere into equal areas, and the 4π steradians about the center into equal solid angles.

Equal intervals of $d\theta$ subtend solid angles that differ by the factor $\sin\theta$, which runs from 0 at the poles to 1 at the equator. Another way to write $d\Omega = 2\pi \sin\theta \, d\theta$ is

$$\boxed{d\Omega = 2\pi \, d(\cos\theta) \ .}$$

(A minus sign is absorbed by doing integrals over $\cos\theta$ in the direction it increases, from south to north.) Therefore, equal intervals, large or small, of $\cos\theta$ subtend *equal* solid angles; see Fig. 3(b). If you measure a distribution for which all directions from an origin, or all points on the surface of a sphere, are equally likely, graphing your data in bins of equal width in θ will give a sine distribution. Graphing your data in bins of equal width in $\cos\theta$ will give a flat distribution. For this reason, angular-distribution data is usually plotted versus $\cos\theta$ instead of θ. Uniformity over the sphere deserves the simplest distribution, one that is flat.

16·2. CLASSICAL PARTICLE SCATTERING. RUTHERFORD

(a) **BB's scattering off a bowling ball**—Figure 4 shows a bowling ball, fixed at the origin, of radius R. The z axis through the center of the bowling ball is the *scattering axis*. A hail of BB's (small round pellets) is incident from the left, moving in the $+z$ direction. The flux is a uniform I BB's per second per unit area perpendicular to the scattering axis. The perpendicular distance between the initial path of a particular BB and the scattering axis is its *impact parameter b*. If $b > R$, the BB passes undeflected (the radius of a BB is negligible compared to R). The *total scattering cross section*, $\sigma(\text{total})$, is the area on a plane perpendicular to the scattering axis over which incident BB's are scattered through *any* angle. Thus here $\sigma(\text{total})$ is πR^2; this is the *cross-sectional area* of the bowling ball (not its surface area, which is four times larger). The total *number* of BB's scattered out of the beam per second is the product of the incident flux I and the total cross section:

$$\text{number of BB's scattered per second} = I\sigma(\text{total}) = I\pi R^2 .$$

What would $\sigma(\text{total})$ be if we were to take into account a small radius a of the BB's?

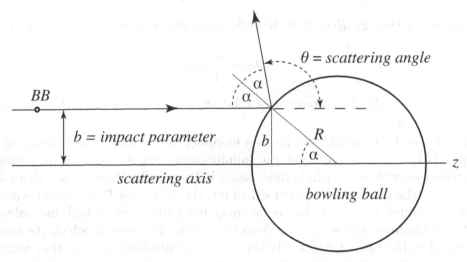

Figure 4. Scattering of a BB by a bowling ball. The spherical coordinate system is tilted onto its side.

If $b < R$, the BB bounces off the bowling ball. We shall assume elastic scattering, where the angles of incidence and reflection, α, with respect to the normal at the surface of the ball are equal. The *change of direction* of the BB—it was going this way, now it is going that way—is the *scattering angle*. This is the spherical-coordinate polar angle θ, measured from the z axis. From Fig. 4, we get

$$2\alpha + \theta = \pi \qquad \text{and} \qquad b = R\sin\alpha .$$

Thus the relation between the impact parameter b and the scattering angle θ is, for $b < R$,

$$b(\theta) = R\sin\tfrac{1}{2}(\pi - \theta) = R\cos(\tfrac{1}{2}\theta) .$$

This is independent of the energy of the BB's and of the azimuthal angle ϕ.

As the impact parameter b increases from 0 to R, the scattering angle θ decreases from $180°$ to $0°$. Thus the cross section for scattering through angles greater than some specified angle θ_0 is simply the area of the circle of radius $b(\theta_0)$,

$$\sigma(\theta > \theta_0) = \pi b^2(\theta_0) = \pi R^2 \cos^2(\tfrac{1}{2}\theta_0) ,$$

where $b \leq R$. For example,

$$\sigma(\theta > 0°) = \pi R^2 \cos^2 0° = \pi R^2 = \sigma(\text{total}) ,$$

and

$$\sigma(\theta > 90°) = \pi R^2 \cos^2 45° = \frac{1}{2}\pi R^2 = \frac{1}{2}\sigma(\text{total}) .$$

Half of all the BB's that are scattered are scattered through more than $90°$.

The *differential* cross section $d\sigma$ for impact parameters between b and $b + db$, where $b < R$, is the area of the ring of radius b and width db, $d\sigma = 2\pi b\, db$. BB's incident over this area will be scattered into angles between θ and $\theta + d\theta$, where $b(\theta) = R\cos(\frac{1}{2}\theta)$. Using $\sin\theta = 2\sin(\frac{1}{2}\theta)\cos(\frac{1}{2}\theta)$, we get

$$d\sigma = 2\pi b\, db = -\pi R^2 \cos(\tfrac{1}{2}\theta)\,\sin(\tfrac{1}{2}\theta)\, d\theta$$
$$= -\frac{R^2}{4}\, 2\pi \sin\theta\, d\theta = \frac{R^2}{4}\, 2\pi\, d(\cos\theta)\ .$$

We saw in Sec. 1 that $2\pi\, d(\cos\theta)$ is the solid angle $d\Omega$ between θ and $\theta + d\theta$. Thus

$$d\sigma = \frac{R^2}{4}\, d\Omega\ .$$

Integrating $d\sigma$ over the 4π steradians of solid angle into which the BB's are scattered gives $\sigma(\text{total}) = \pi R^2$ again.

The relation between $d\Omega$ and $d\sigma$ is a mapping between an infinitesimal solid angle $d\Omega$ into which BB's are scattered and the infinitesimal area $d\sigma$ on a plane perpendicular to the scattering axis through which those same BB's were incident. Since there is no θ or ϕ dependence in the coefficient relating $d\Omega$ to $d\sigma$, the scattered BB's scatter equally in every direction. Consider a counter that is far away from the bowling ball and subtends a solid angle $d\Omega$ (of whatever shape) as seen from the origin. The rate at which the counter counts BB's is equal to the rate at which BB's cross the infinitesimal area $d\sigma$ that maps into $d\Omega$:

$$\text{number of BB's scattered into the counter per second} = I\, d\sigma = I\frac{R^2}{4}\, d\Omega\ .$$

This is proportional to $d\Omega$ and is independent of the direction to the counter. Counters subtending equal solid angles, no matter in what direction, will count BB's at the same rate. (The angular dependence of Rutherford scattering will be very different.) Figure 5 shows $d\sigma/d(\cos\theta)$ versus $\cos\theta$. Equal intervals of $\cos\theta$ subtend equal solid angles (see Fig. 3), and the distribution is flat.

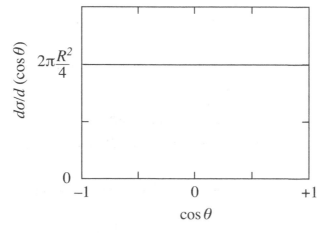

Figure 5. The angular distribution for BB's scattering from a bowling ball.

Imagine that the bowling ball is silvered, and that a plane wave of light is reflected geometrically from it (angle of incidence = angle of reflection). Now walk in a circular path, centered on and distant from the ball, along which θ increases. Your pupil (the counter) subtends a constant infinitesimal solid angle from the ball. The spot on the ball the light is reflected from changes as you walk, but the level of brightness you see does not.

[A note: "Differential cross section" often means either $d\sigma/d\Omega$ or $d\sigma/d(\cos\theta)$ instead of $d\sigma$. The units are still those of area, because Ω and $\cos\theta$ are dimensionless.]

(b) Rutherford scattering—Rutherford scattering is about the scattering of a parallel beam of particles, each having charge q, mass m, and energy E, by a charge Q. The force on q is the Coulomb force,

$$\mathbf{F} = \frac{kqQ}{r^2}\hat{\mathbf{r}} \ .$$

Here (again) $k \equiv 1/4\pi\epsilon_0$, and $\hat{\mathbf{r}}$ is the unit vector from Q toward q. The inverse-square law leads to hyperbolic orbits, bent away from Q for repulsive scattering and toward Q for attractive scattering. (We take Q to be fixed in position, perhaps as part of a crystal lattice.) With the change of a few symbols, our results for the attractive case would apply to the scattering of a comet swarm by a star. Figure 6 shows a trajectory for like charges.

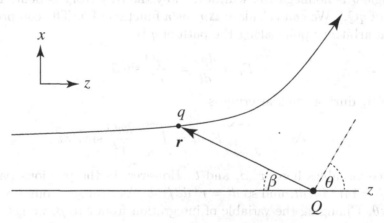

Figure 6. A hyperbolic trajectory for Rutherford scattering of like charges. We take Q to be fixed in position.

Since the Coulomb force is of infinite range, any particle, no matter how large its impact parameter, will be scattered through *some* angle. Thus the area perpendicular to the scattering axis over which incident particles are scattered—the total cross section—is infinite. A comet that passes light years distant from the Sun will still be deflected very, very slightly.

The total energy of an incident particle is the sum of its kinetic and potential energies,

$$E = \frac{1}{2}mv_0^2 = \frac{1}{2}mv^2 + \frac{kqQ}{r} = \frac{1}{2}mv_0'^2 \ .$$

Here v_0 and v_0' are the speeds long before and long after the particle passes Q, and v is the speed at some intermediate time when the particle is at a distance r from Q. And since Q is fixed in position and energy is conserved, v_0 and v_0' are equal. If q and Q are of like sign, then a particle that is heading directly at Q ($b=0$) is running straight up a potential-energy hill

325

and will stop and reverse direction when the initial kinetic energy $\frac{1}{2}mv_0^2$ is turned entirely into potential energy. Thus

$$E = \frac{1}{2}mv_0^2 = \frac{kqQ}{D} \ ,$$

where

$$D \equiv \frac{kqQ}{\frac{1}{2}mv_0^2} = \frac{kqQ}{E}$$

is the distance of closest approach for head-on repulsive scattering. The parameter $D \equiv |kqQ|/E$ will be useful for the attractive as well as for the repulsive case.

Another preliminary: In Fig. 6, the component of velocity of q perpendicular to the position vector $\mathbf{r}$ is $v_\perp = r\,d\beta/dt$, where β is the angle between $\mathbf{r}$ and the negative z axis. The magnitude of the angular momentum $\mathbf{L}$ about Q is $|\mathbf{r} \times \mathbf{p}| = rp_\perp = rmv_\perp = mr^2 d\beta/dt$. The angular momentum is initially $mv_0 b$, and it is conserved. Thus $L = mv_0 b = mr^2 d\beta/dt$, a relation we shall need in a moment.

A particle scattered through an angle θ acquires a component of momentum perpendicular to the z axis of

$$\Delta p_x = \pm mv_0 \sin\theta \ .$$

(The polar angle θ is nonnegative, whichever way the trajectory is bent. In Fig. 6, Δp_x will have the sign of qQ.) We can calculate Δp_x as a function of b. The component of $\mathbf{F}$ in the x direction at an arbitrary point along the path of q is

$$F_x = \frac{dp_x}{dt} = \frac{kqQ}{r^2}\sin\beta \ .$$

The change of p_x during the scattering is

$$\Delta p_x = \int_{-\infty}^{+\infty} F_x\,dt = \int_{-\infty}^{+\infty} \frac{kqQ}{r^2}\sin\beta\,dt \ .$$

There are three variables here, r, β, and t. However, in the previous paragraph we found that $L = mv_0 b = mr^2 d\beta/dt$, and so $dt = r^2\,d\beta/v_0 b$. As t ranges from $-\infty$ to $+\infty$, β sweeps from 0 to $\pi - \theta$. Changing the variable of integration from t to β, we get

$$\Delta p_x = \int_0^{\pi-\theta} \frac{kqQ}{r^2}\sin\beta\,\frac{r^2}{v_0 b}\,d\beta = \frac{kqQ}{v_0 b}\int_0^{\pi-\theta}\sin\beta\,d\beta$$

$$= \frac{kqQ}{v_0 b}\big[\cos 0 - \cos(\pi - \theta)\big] = \frac{kqQ}{v_0 b}(1 + \cos\theta) \ .$$

This nice cancellation of factors of r^2 occurs only for the inverse-square-law force.

Equating the expressions for Δp_x at the start and the end of the preceding paragraph and solving for b, we get

$$b(\theta) = \frac{|kqQ|}{mv_0^2}\frac{(1 + \cos\theta)}{\sin\theta} = \frac{D}{2}\frac{\cos^2(\frac{1}{2}\theta)}{\sin(\frac{1}{2}\theta)\cos(\frac{1}{2}\theta)} = \frac{D}{2}\cot(\tfrac{1}{2}\theta) \ .$$

Here we have used half-angle relations and $D \equiv |kqQ|/E$. This $b(\theta) = \frac{1}{2}D\cot(\frac{1}{2}\theta)$ may look much like the BB-bowling ball $b(\theta) = R\cos(\frac{1}{2}\theta)$, but $\cot(\frac{1}{2}\theta)$ and $\cos(\frac{1}{2}\theta)$ are very different functions.

326

As b gets larger, θ gets smaller. Thus the cross section for scattering through angles greater than θ_0 is

$$\sigma(\theta > \theta_0) = \pi b^2(\theta_0) = \frac{\pi D^2}{4}\cot^2(\tfrac{1}{2}\theta_0) \ .$$

For example, $\sigma(\theta > 90°) = \pi D^2/4$. The differential cross section $d\sigma$ is obtained in exactly the same way as for the BB-bowling ball example: We evaluate $d\sigma = 2\pi b\,db$ using the relation between b and θ. From $d(u/v) = (v\,du - u\,dv)/v^2$, we get $d\cot(\tfrac{1}{2}\theta) = -\tfrac{1}{2}d\theta/\sin^2(\tfrac{1}{2}\theta)$. Then

$$d\sigma = 2\pi b\,db = -2\pi\frac{D^2}{8}\cot(\tfrac{1}{2}\theta)\frac{d\theta}{\sin^2(\tfrac{1}{2}\theta)} = -2\pi\frac{D^2}{8}\frac{\cos(\tfrac{1}{2}\theta)}{\sin^3(\tfrac{1}{2}\theta)}\,d\theta \ .$$

Multiplying top and bottom by $\sin(\tfrac{1}{2}\theta)$ and using $\sin\theta = 2\sin(\tfrac{1}{2}\theta)\cos(\tfrac{1}{2}\theta)$ and $2\pi\sin\theta\,d\theta = -2\pi d(\cos\theta)$, we get

$$\boxed{d\sigma = \frac{D^2}{16}\frac{d\Omega}{\sin^4(\tfrac{1}{2}\theta)} = \left(\frac{kqQ}{4E}\right)^2\frac{d\Omega}{\sin^4(\tfrac{1}{2}\theta)} \ .}$$

Figure 7 is a plot of $1/\sin^4(\tfrac{1}{2}\theta)$ versus $\cos\theta$. Multiplied by $2\pi D^2/16$, the curve would be $d\sigma/d(\cos\theta)$. The BB-bowling ball angular distribution was flat and independent of energy. Rutherford scattering is very sharply peaked in the forward direction and is energy dependent, since D^2 is proportional to $1/E^2$.

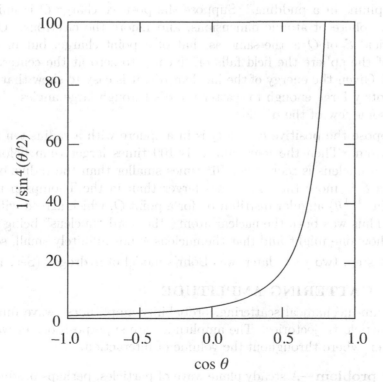

Figure 7. A plot of $1/\sin^4(\tfrac{1}{2}\theta)$ versus $\cos\theta$. The curve reaches 400 at $\cos\theta = 0.9$ and 4×10^4 at $\cos\theta = 0.99$. Multiplied by $2\pi D^2/16$, the curve would be $d\sigma/d(\cos\theta)$.

The main results of this section on *classical* scattering are

		BB — bowling ball	Rutherford
$b(\theta)$	$=$	$R\cos(\frac{1}{2}\theta)$	$\frac{1}{2}D\cot(\frac{1}{2}\theta)$
$\sigma(\text{total})$	$=$	πR^2	∞
$\sigma(\theta > \theta_0)$	$=$	$\pi R^2\cos^2(\frac{1}{2}\theta_0)$	$\frac{1}{4}\pi D^2\cot^2(\frac{1}{2}\theta_0)$
$d\sigma$	$=$	$\frac{1}{4}R^2\,d\Omega$	$\frac{1}{16}D^2\sin^{-4}(\frac{1}{2}\theta)\,d\Omega$

Here $b = b(\theta)$ is the impact parameter, R is the radius of the bowling ball, and $D = |kqQ|/E$ is the distance of closest approach for head-on repulsive point-charge scattering.

(c) The nuclear atom—In 1909, Hans Geiger (who later invented the Geiger counter) and Ernest Marsden (an undergraduate), working in Ernest Rutherford's laboratory at Manchester University, were studying the scattering of a collimated beam of α particles by a thin gold foil. (An α particle is a tightly bound state of two protons and two neutrons; a ^{4}He nucleus is an α particle.) The α particles came from the decay of radium nuclei. Geiger and Marsden were surprised—in fact, astonished—to find that a small fraction of the α's were scattered through large angles. Rutherford: "It was almost as incredible as if you fired a 15-inch shell at a piece of tissue paper and it came back and hit you."

Here is why they were astonished. A few years earlier, J.J. Thompson, the discoverer of the electron, had proposed a provisional model of the atom in which the positive charge and nearly all the mass of the atom fill its volume, and the electrons are somehow embedded in this "like plums in a pudding." Suppose the positive charge Q is uniformly distributed throughout a sphere of atomic dimensions, and ignore the electrons. Outside the sphere, the electric field $\mathcal{E}$ of Q is the same as that of a point charge, but in going inward from the surface of the sphere the field falls off linearly to zero at the center. (Show this using Gauss's law.) Given the energy of the incident α's, it is easy to show that the electric field is nowhere remotely large enough to scatter the α's through large angles—but scatter through large angles some few of the α's do.

Now suppose the positive charge Q is in a sphere with a radius ten times smaller than that of the atom. Then the maximum $\mathcal{E}$ is 100 times larger than before (why?). In fact the radius of a nucleus is more than 10^4 times smaller than the radius of its atom, so that the maximum $\mathcal{E}$ is more than 10^8 times larger than in the Thompson model. Rutherford derived the $\sin^{-4}(\frac{1}{2}\theta)$ angular distribution for a point Q, which agreed with the results of the experiment. Thus was born the nuclear atom—the word "nucleus" being borrowed from cell biology. For how one might find that the nucleus is not infinitely small, see Prob. 4.

The next step, two years later, was Bohr's model of hydrogen (Sec. 1·4).

16·3. THE SCATTERING AMPLITUDE

In quantum-mechanical scattering, probability amplitudes (wave functions) replace deterministic particle trajectories. The amplitudes are superpositions of waves scattered outward from everywhere throughout the volume of interaction.

(a) The problem—A steady plane wave of particles, perhaps produced by an accelerator, is streaming in the $+z$ direction. The number of particles per second crossing a unit area perpendicular to the scattering axis is I. The momentum of each particle is $\mathbf{p}_i = \hbar\mathbf{k}_i = \hbar k\hat{\mathbf{z}}$, where $k = 2\pi/\lambda$ is the wave number. The particle energies are $E = p^2/2m = \hbar^2 k^2/2m$. The incident wave is partially scattered on interacting with a spherically symmetric potential energy $V(r)$, which (for the present) is fixed at the origin. With $\mathbf{r}$ the position vector with respect to the scattering center,

$$\mathbf{k}_i \cdot \mathbf{r} = k\hat{\mathbf{z}} \cdot (x\hat{\mathbf{x}} + y\hat{\mathbf{y}} + z\hat{\mathbf{z}}) = kz = kr\cos\theta \ .$$

The wave function for the incident beam, normalized to one particle per unit volume, is

$$\psi_i(\mathbf{r}) = e^{i\mathbf{k}_i \cdot \mathbf{r}} = e^{ikz} = e^{ikr\cos\theta}$$

(Sec. 4·1). In the region of interaction, the wave function will be complicated, but very far from the origin, there is just the incident wave plus a radial wave of the scattered particles, as in Fig. 8. At large r, the full wave function, incident plus scattered, is

$$\boxed{\psi(\mathbf{r}) = e^{ikz} + f(\theta, k)\frac{e^{ikr}}{r}}$$

(see Prob. 5). The e^{ikr}/r part is an outgoing radial wave with its amplitude falling off as $1/r$ and its intensity as $1/r^2$. Given $V(r)$, the problem is to find the θ- and k-dependent *scattering amplitude* $f(\theta, k)$. There is no azimuth ϕ in f because we are taking $V(r)$ to be spherically symmetric. The differential cross section $d\sigma$ is equal to the absolute square of the scattered wave function times the frontal area $dA = r^2 d\Omega$ of a counter at a distance r,

$$\boxed{d\sigma = \frac{|f(\theta, k)\, e^{ikr}|^2}{r^2}\, r^2 d\Omega = |f(\theta, k)|^2 d\Omega \ ,}$$

where $d\Omega$ is the solid angle the counter subtends from the origin. The rate the counter will count particles is $I \times d\sigma$, where I is the incident flux. The whole purpose of this section is to get an equation for $f(\theta, k)$ in terms of $V(r)$.

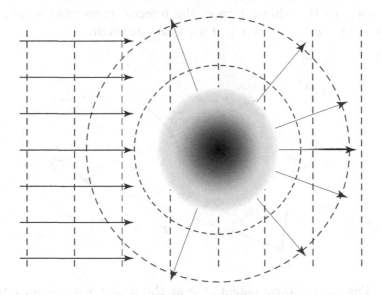

Figure 8. A plane wave incident upon a scattering center produces at a distance radially outgoing waves.

329

The Schrödinger equation for $\psi(\mathrm{r})$ is

$$-\frac{\hbar^2}{2m}\nabla^2\psi + V(r)\psi = E\psi, \qquad \text{or} \qquad \left(\nabla^2 + k^2\right)\psi = \frac{2m}{\hbar^2}V(r)\,\psi\ ,$$

where $k^2 = 2mE/\hbar^2$. The incident plane wave e^{ikz} is a solution of the homogeneous equation: $\left(\nabla^2 + k^2\right)e^{ikz} = 0$ (show this). The scattered wave only exists because of $V(r)$. To calculate $f(\theta, k)$ from $V(r)$, we will need this *integral* equation for $\psi(\mathrm{r})$,

$$\psi(\mathbf{r}) = e^{ikz} - \frac{2m}{\hbar^2}\int \frac{e^{ik|\mathbf{r}-\mathbf{r}'|}}{4\pi|\mathbf{r}-\mathbf{r}'|}V(r')\psi(\mathbf{r}')\,d^3\mathbf{r}'\ .$$

Here r is the vector to a distant particle detector, whereas r′ is a vector that ranges over the volume of space in which $V(r')$ is not zero. The integration is over $\mathbf{r}'$, with r held fixed. To understand how this integral form of the Schrödinger equation comes from the differential form, we begin with a simpler and more familiar problem.

(b) The electrostatic potential and Green functions—In electrostatics, there is both an integral and a differential equation for the relation between a *potential* $\varphi(\mathbf{r})$ (not the potential energy) and the charge density $\rho(\mathbf{r}')$ that is the source of $\varphi(\mathbf{r})$. Let r be a *field* point—a point at which we want $\varphi(\mathbf{r})$; and let r′ be a *source* point—a point at which ρ is not zero (see Fig. 9). The *integral* equation is

$$\varphi(\mathbf{r}) = \frac{1}{4\pi\epsilon_0}\int\limits_{source} \frac{\rho(\mathbf{r}')}{|\mathbf{r}-\mathbf{r}'|}\,d^3\mathbf{r}'\ .$$

Here r is fixed while r′ roams over the space in which $\rho(\mathbf{r}')$ is not zero. For a point charge q at r′, $\rho(\mathbf{r}') = q\,\delta(\mathbf{r}')$, where $\delta(\mathbf{r}')$ is a three-dimensional Dirac delta function, zero everywhere except at r′, and infinite there in just such a way that the integral over any volume that includes r′ is one. In this simplest case, the integral gives $\varphi(\mathbf{r}) = q/4\pi\epsilon_0|\mathbf{r}-\mathbf{r}'|$ for any $\mathbf{r} \neq \mathbf{r}'$—or just $\varphi(r) = q/4\pi\epsilon_0 r$ for $r > 0$ if q is at the origin.

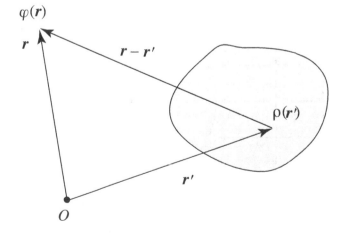

Figure 9. The electrostatic potential φ at the *field* point r is an integral over the *source* points r′ where the charge distribution ρ is not zero. The field point can be anywhere, including within the region in which ρ is not zero.

The *differential* equation relating $\varphi(\mathbf{r})$ and $\rho(\mathbf{r})$ is

$$\nabla^2 \varphi(\mathbf{r}) = -\frac{\rho(\mathbf{r})}{\epsilon_0} \ ,$$

which applies at every point $\mathbf{r}$ in space. There is no $\mathbf{r}'$ here; the ∇^2 operation is with respect to $\mathbf{r}$. Example: In Sec. 13·2(b), we modeled the proton as a charge $+e$ spread uniformly throughout a sphere of radius R. The potential was

$$\varphi(r < R) = \frac{1}{4\pi\epsilon_0} \frac{e}{2R} \left(3 - \frac{r^2}{R^2} \right), \qquad \text{and} \qquad \varphi(r \geq R) = \frac{1}{4\pi\epsilon_0} \frac{e}{r} \ .$$

For the charge-filled $r < R$ space, and φ only a function of r, we get

$$\nabla^2 \varphi(r) = \frac{1}{r} \frac{d^2(r\varphi)}{dr^2} = -\frac{1}{4\pi\epsilon_0 r} \frac{e}{2R} \frac{6r}{R^2} = -\frac{1}{\epsilon_0} \frac{e}{4\pi R^3/3} = -\frac{\rho(r)}{\epsilon_0} \ ,$$

as required. For the charge-free $r > R$ space, $r\varphi = e/4\pi\epsilon_0$ and $\nabla^2 \varphi(r) = 0$, as required.

For clues on how to relate the differential Schrödinger equation to an integral equation, we look at the relation between the two electrostatic equations. The right side of the differential equation, $\nabla^2 \varphi(\mathbf{r}) = -\rho(\mathbf{r})/\epsilon_0$, appears in the integral equation for $\varphi(\mathbf{r})$, primed and changed in sign, and with $4\pi|\mathbf{r} - \mathbf{r}'|$ in the denominator:

$$\varphi(\mathbf{r}) = \frac{1}{4\pi\epsilon_0} \int\limits_{source} \frac{\rho(\mathbf{r}')}{|\mathbf{r} - \mathbf{r}'|} \, d^3\mathbf{r}' = \int\limits_{source} \frac{\rho(\mathbf{r}')}{\epsilon_0} \times \frac{1}{4\pi|\mathbf{r} - \mathbf{r}'|} \, d^3\mathbf{r}' \ .$$

The second part of the integrand,

$$G(\mathbf{r}, \mathbf{r}') = \frac{1}{4\pi|\mathbf{r} - \mathbf{r}'|} \ ,$$

is called a Green function. It is well defined for all $\mathbf{r} \neq \mathbf{r}'$, but at $\mathbf{r} = \mathbf{r}'$ it blows up in a special way. To see this, we operate on $\varphi(\mathbf{r})$ with $\nabla_{\mathbf{r}}^2$, where the subscript $\mathbf{r}$ is a reminder that the operation is with respect to the *field* coordinates:

$$\nabla_{\mathbf{r}}^2 \varphi(\mathbf{r}) = \int\limits_{source} \frac{\rho(\mathbf{r}')}{\epsilon_0} \times \nabla_{\mathbf{r}}^2 \left(\frac{1}{4\pi|\mathbf{r} - \mathbf{r}'|} \right) d^3\mathbf{r}' \ .$$

(Once again, because the $\mathbf{r}$ and $\mathbf{r}'$ coordinates are independent, we can move $\nabla_{\mathbf{r}}^2$ into the integral, where it operates only on $\mathbf{r}$-dependent functions.) To leave just $-\rho(\mathbf{r})/\epsilon_0$ on the right, the integral has to pick out the value of ρ at *only* the point $\mathbf{r}' = \mathbf{r}$. Therefore, it must be so that

$$\nabla_{\mathbf{r}}^2 G(\mathbf{r}, \mathbf{r}') = -\delta(\mathbf{r} - \mathbf{r}') \ ,$$

where $\delta(\mathbf{r} - \mathbf{r}')$ is a Dirac delta function at $\mathbf{r}'$. The proof of this is left for a problem.

(c) Integral form of the Schrödinger equation—We asserted in (a) that the integral form of the Schrödinger equation (including the homogeneous e^{ikz} solution) is

$$\psi(\mathbf{r}) = e^{ikz} - \frac{2m}{\hbar^2} \int \frac{e^{ik|\mathbf{r}-\mathbf{r}'|}}{4\pi|\mathbf{r} - \mathbf{r}'|} V(r') \, \psi(\mathbf{r}') \, d^3\mathbf{r}' \ .$$

Following the example of the electrostatic equations, we operate on this with $\nabla_{\mathbf{r}}^2 + k^2$. We can only recover the differential Schrödinger equation, $(\nabla^2 + k^2)\psi = (2m/\hbar^2)V(r)\psi$, if

$$(\nabla_{\mathbf{r}}^2 + k^2)\, G(\mathbf{r}, \mathbf{r}') = -\delta(\mathbf{r} - \mathbf{r}'), \qquad \text{where} \quad G(\mathbf{r}, \mathbf{r}') = \frac{e^{ik|\mathbf{r} - \mathbf{r}'|}}{4\pi |\mathbf{r} - \mathbf{r}'|}.$$

It is sufficient to test this at $\mathbf{r}' = 0$ (or to change to $\mathbf{s} \equiv \mathbf{r} - \mathbf{r}'$). Then, for all $r \neq 0$,

$$(\nabla^2 + k^2)\left(\frac{e^{ikr}}{4\pi r}\right) = \frac{1}{r}\frac{d^2}{dr^2}\left(r\frac{e^{ikr}}{4\pi r}\right) + k^2\left(\frac{e^{ikr}}{4\pi r}\right) = (-k^2 + k^2)\left(\frac{e^{ikr}}{4\pi r}\right) = 0.$$

But at $r = 0$, r/r is indeterminate.

To get at the behavior at the origin, we use three basic relations from vector calculus,

$$\nabla^2 f = \nabla \cdot \nabla f, \qquad \nabla(fg) = f\,\nabla g + g\,\nabla f, \qquad \nabla \cdot (f\,\nabla g) = \nabla f \cdot \nabla g + f\,\nabla^2 g.$$

The first one says that the Laplacian is the divergence of the gradient; the second and third tell how to calculate the gradient or the divergence of a product. (The gradient makes a vector from a scalar, the divergence makes a scalar from a vector.) For f only a function of r, $\nabla f(r) = (df/dr)\,\hat{\mathbf{r}}$. Using the three relations in the order given, we get

$$\nabla^2\left(\frac{e^{ikr}}{r}\right) = \nabla \cdot \nabla\left(\frac{e^{ikr}}{r}\right) = \nabla \cdot \left[\frac{1}{r}\,\nabla\left(e^{ikr}\right) + e^{ikr}\,\nabla\left(\frac{1}{r}\right)\right]$$
$$= \frac{1}{r}\,\nabla^2\left(e^{ikr}\right) + 2\,\nabla\left(e^{ikr}\right) \cdot \nabla\left(\frac{1}{r}\right) + e^{ikr}\,\nabla^2\left(\frac{1}{r}\right).$$

The terms left-to-right go ∇^2, ∇^1, ∇^0 on e^{ikr}, and ∇^0, ∇^1, ∇^2 on $1/r$, two ∇'s per term. The first term is

$$\frac{1}{r}\,\nabla^2\left(e^{ikr}\right) = \frac{1}{r^2}\frac{d^2}{dr^2}\left(re^{ikr}\right) = \frac{2ik}{r^2}e^{ikr} - k^2\left(\frac{e^{ikr}}{r}\right).$$

The second term is

$$2\,\nabla\left(e^{ikr}\right) \cdot \nabla\left(\frac{1}{r}\right) = -\frac{2ik}{r^2}e^{ikr},$$

which cancels part of the first term. The third term can stay as is. Adding the three terms and rearranging, we get

$$(\nabla^2 + k^2)\left(\frac{e^{ikr}}{4\pi r}\right) = e^{ikr}\,\nabla^2\left(\frac{1}{4\pi r}\right) = -\,e^{ikr}\delta(\mathbf{r}) = -\delta(\mathbf{r}).$$

We know that $\nabla^2\left(1/4\pi r\right) = -\delta(\mathbf{r})$ from the previous subsection (and Prob. 6); and $e^{ikr} = 1$ at $r = 0$. Therefore, we have indeed shown that operating on

$$\psi(\mathbf{r}) = e^{ikz} - \frac{2m}{\hbar^2}\int \frac{e^{ik|\mathbf{r} - \mathbf{r}'|}}{4\pi |\mathbf{r} - \mathbf{r}'|} V(r')\,\psi(\mathbf{r}')\,d^3\mathbf{r}'$$

with $(\nabla^2 + k^2)$ gives the differential Schrödinger equation, $(\nabla^2 + k^2)\,\psi = (2m/\hbar^2)V(r)\psi$.

(d) First (very good) approximations—In applications of the integral Schrödinger equation, the scattering centers will be atoms or nuclei, whereas in comparison the detectors of the scattered particles will be at astronomical distances. Thus $r \gg r'$, and we can in the denominator of the integrand of the integral equation set $|\mathbf{r} - \mathbf{r}'| = r$. (For purposes of space travel, α Centauri is equidistant from anywhere in the Solar System.) However, we cannot set $|\mathbf{r} - \mathbf{r}'| = r$ in $e^{ik|\mathbf{r}-\mathbf{r}'|}$ because, as Fig. 10 shows, a change in $|\mathbf{r} - \mathbf{r}'|$ by half a wavelength $\lambda = 2\pi/k$ can change the factor $e^{ik|\mathbf{r}-\mathbf{r}'|}$ from $+1$ to -1, or vice versa. The *magnitude* of the contribution of an infinitesimal volume to $\psi(\mathbf{r})$ at a counter depends on the value of $V(r')$ in that volume. But how the contributions of all those volumes add up to make $\psi(\mathbf{r})$ depends on differences in *phase*. And this, as we shall see, is a function of the scattering angle θ. (Changes of relative *phase* as a function of forward angle are what cause the fringes in the double-slit pattern of Sec. 1·2.)

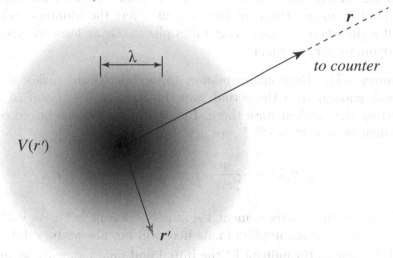

Figure 10. An example in which, as $\mathbf{r}'$ ranges over the space in which $V(r')$ is nonzero, the distance $|\mathbf{r} - \mathbf{r}'|$ to a distant counter would vary by wavelengths. The wavelength will be large at low energies and small at high energies.

We can, however, simplify things by expanding $|\mathbf{r} - \mathbf{r}'|$ in powers of the very small quantity r'/r and dropping an $(r'/r)^2$ term. Using $(1+\epsilon)^\alpha \approx 1 + \alpha\epsilon$ when ϵ is small, we get

$$|\mathbf{r} - \mathbf{r}'| = (r^2 - 2\mathbf{r} \cdot \mathbf{r}' + r'^2)^{1/2} = r\left(1 - 2\frac{\mathbf{r} \cdot \mathbf{r}'}{r^2} + \frac{r'^2}{r^2}\right)^{1/2} \approx r\left(1 - \frac{\mathbf{r} \cdot \mathbf{r}'}{r^2}\right).$$

Since $\mathbf{r}/r$ is the unit vector $\hat{\mathbf{r}}$ in the direction of $\mathbf{r}$, and $k\hat{\mathbf{r}}$ is the vector $\mathbf{k}_f$ in that direction, we get

$$k|\mathbf{r} - \mathbf{r}'| \approx kr - k\frac{\mathbf{r} \cdot \mathbf{r}'}{r} = kr - \mathbf{k}_f \cdot \mathbf{r}' \;,$$

and

$$e^{ik|\mathbf{r}-\mathbf{r}'|} \approx e^{ikr} \times e^{-i\mathbf{k}_f \cdot \mathbf{r}'}.$$

Taking the r-dependent factors $1/r$ and e^{ikr} outside the scattering integral, we get

$$\psi(\mathbf{r}) = e^{ikz} - \frac{e^{ikr}}{r} \times \frac{m}{2\pi\hbar^2} \int e^{-i\mathbf{k}_f \cdot \mathbf{r}'} V(r')\psi(\mathbf{r}') \, d^3r' \;.$$

What multiples e^{ikr}/r here is just the sought-after $f(\theta, k)$,

$$f(\theta, k) = -\frac{m}{2\pi\hbar^2} \int e^{-i\mathbf{k}_f \cdot \mathbf{r}'} V(r')\, \psi(\mathbf{r}')\, d^3\mathbf{r}' \ .$$

We finally have an equation for the amplitude for (elastic) scattering from a spherically symmetric potential energy. It still does not look simple.

16·4. THE BORN APPROXIMATION

There is a big difference between the electrostatic and the quantum equations in Sec. 3. The former cleanly separate φ and ρ—the potential φ is on the left side of the equations, the source ρ is on the right. The quantum equations have ψ on both sides. To integrate the integral equation for $f(\theta, k)$, we evidently need to know $\psi(\mathbf{r})$ over the volume in which $V(r)$ is not zero. That is, to get $f(\theta, k)$ at large r (out where the counters are), we need to know $\psi(\mathbf{r})$ at small r (in where the scattering takes place). Or at least we need some idea about how to approximate $\psi(\mathbf{r})$ at small r.

(a) Theory—The *Born approximation* (after Max Born) makes the assumption that if $V(r)$ is weak enough then the actual wave function in the region of interaction can be approximated by the incident wave there. Replacing $\psi(\mathbf{r}')$ in the boxed equation for $f(\theta, k)$ with the incident wave $e^{ikz'} = e^{i\mathbf{k}_i \cdot \mathbf{r}'}$, we get

$$f(\theta, k) = -\frac{m}{2\pi\hbar^2} \int e^{-i(\mathbf{k}_f - \mathbf{k}_i)\cdot\mathbf{r}'} V(r')\, d^3\mathbf{r}' \ .$$

The integral itself is the matrix element $V_{fi} \equiv \langle e^{i\mathbf{k}_f \cdot \mathbf{r}} | V(r) | e^{i\mathbf{k}_i \cdot \mathbf{r}} \rangle$. As with other methods of approximation, the basic assumption made here will not always be valid.

Figure 11(a) shows (in units of $\hbar$) the initial and final momenta $\mathbf{k}_i$ and $\mathbf{k}_f$ of a particle scattered through an angle θ. The *momentum transfer* is the change of momentum, $\mathbf{q} \equiv \mathbf{k}_f - \mathbf{k}_i$. For elastic scattering from a fixed scattering center, $|\mathbf{k}_i| = |\mathbf{k}_f| = k$. Then as Fig. 11(b) shows,

$$|\mathbf{q}| = q = 2k\sin(\tfrac{1}{2}\theta) \ ,$$

and

$$f(\theta, k) = -\frac{m}{2\pi\hbar^2} \int e^{-i\mathbf{q}\cdot\mathbf{r}'} V(r')\, d^3\mathbf{r}' \ .$$

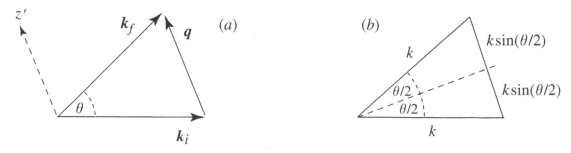

Figure 11. (a) The definition of $\mathbf{q}$, and the polar axis z' for the integration to follow. (b) The geometry for $q = 2k\sin(\tfrac{1}{2}\theta)$.

We can go part way with the integral without specifying $V(r)$. This is done most easily if we reorient the coordinates for the integration, taking the polar axis $\hat{z}'$ to lie parallel to $\mathbf{q}$. Then $\mathbf{q} \cdot \mathbf{r}'$ is $qz' = qr' \cos\theta'$, and

$$f(\theta, k) = -\frac{m}{2\pi\hbar^2} \int e^{-iqr'\cos\theta'} V(r')\, r'^2 \sin\theta'\, dr'd\theta'd\phi'$$

$$= -\frac{m}{\hbar^2} \int_0^\infty r'^2 V(r') \left(\int_0^\pi e^{-iqr'\cos\theta'}\, d(-\cos\theta') \right) dr'$$

With $u \equiv \cos\theta'$, the θ' integral is

$$\int_{-1}^{+1} e^{-iqr'u}\, du = \frac{1}{(-iqr')}e^{-iqr'u}\Big|_{-1}^{+1} = \frac{2}{qr'}\left(\frac{e^{+iqr'} - e^{-iqr'}}{2i}\right) = \frac{2}{qr'}\sin(qr')\ .$$

Then

$$\boxed{f(\theta, k) = -\frac{2m}{q\hbar^2}\int_0^\infty r'V(r')\sin(qr')\, dr'\ .}$$

The dependence of $f(\theta, k)$ on θ and k is in $q = 2k\sin(\frac{1}{2}\theta)$, which occurs both in the integrand and out front. Given $V(r)$, we can calculate $f(\theta, k)$—in the Born approximation.

(b) Yukawa scattering—The Yukawa potential energy is

$$V(r) = \frac{g}{r}e^{-r/R}\ .$$

This has the $1/r$ dependence of the Coulomb potential energy, but $V(r)$ is killed off at large r by the exponential; it might model the scattering of a particle with charge e (say a proton) by a nucleus with charge Ze surrounded (screened) by atomic electrons. Then g would be $Ze^2/4\pi\epsilon_0$.

Letting $b \equiv 1/R$, we get

$$f(\theta, k) = -\frac{2mg}{q\hbar^2}\int_0^\infty e^{-br'}\sin(qr')\, dr'$$

$$= -\frac{2mg}{q\hbar^2}\frac{1}{2i}\int_0^\infty \left(e^{(-b+iq)r'} - e^{(-b-iq)r'}\right) dr'$$

$$= -\frac{2mg}{q\hbar^2}\frac{1}{2i}\left[\frac{e^{(-b+iq)r'}}{-b+iq} + \frac{e^{(-b-iq)r'}}{b+iq}\right]_0^\infty$$

$$= +\frac{2mg}{q\hbar^2}\frac{1}{2i}\left(\frac{1}{-b+iq} + \frac{1}{b+iq}\right)$$

$$= -\frac{2mg}{\hbar^2(b^2+q^2)}\ .$$

Without the screening factor $e^{-br'}$, the integrand would be $\sin(qr')$, and the integral would be proportional to the difference of $\cos(qr')$ evaluated at $r' = 0$ and infinity. What is $\cos(\infty)$? By effectively killing off the infinite range of the Coulomb potential energy, the screening removes this problem.

Going back to $b = 1/R$ and $q = 2k\sin(\frac{1}{2}\theta)$, we get

$$\frac{d\sigma}{d\Omega} = |f(\theta, k)|^2 = \left(\frac{2mg}{\hbar^2}\right)^2 \frac{1}{(b^2 + q^2)^2} = \left(\frac{2mgR^2}{\hbar^2}\right)^2 \frac{1}{\left[1 + 4k^2R^2\sin^2(\frac{1}{2}\theta)\right]^2} \; .$$

At $\theta = 0$, this is

$$\frac{d\sigma}{d\Omega}(0) = |f(0, k)|^2 = \left(\frac{2mgR^2}{\hbar^2}\right)^2 ,$$

independent of k. Elsewhere, the angular distribution depends on the product $kR = 2\pi R/\lambda$ of the wave number and the screening distance. At very low energies, where $kR \ll 1$, the $\sin^2(\frac{1}{2}\theta)$ term scarcely comes into play, and the angular distribution is nearly flat. At high energies, where $kR \gg 1$, we get

$$\frac{d\sigma}{d\Omega} \simeq \left(\frac{2mg}{4\hbar^2 k^2}\right)^2 \frac{1}{\sin^4(\frac{1}{2}\theta)} \; ,$$

independent of R, and the angular distribution is sharply peaked. In this limit, the Yukawa scattering angular distribution approaches the Rutherford distribution, except in the very forward direction. See Fig. 12.

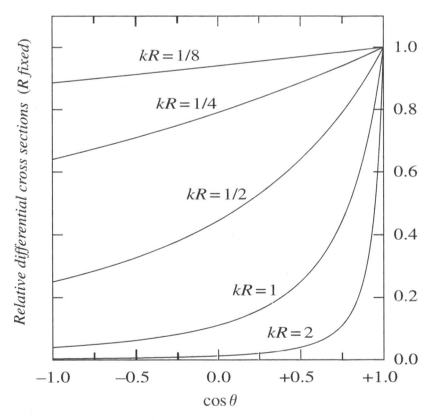

Figure 12. Angular distributions in the Born approximation for the Yukawa potential energy for several values of $kR = 2\pi R/\lambda$. For R fixed, the distributions become more sharply peaked as k (and thus the incident energy) increases.

16·5. KINEMATICS

The *kinematics* of scattering is about the constraints of conservation of energy, momentum, and angular momentum. The *dynamics* is about how the forces determine the angular distributions. In Sec. 2, we analyzed classical elastic scattering on scattering centers (a bowling ball, a charge Q) held fixed in position. The kinematics was then trivial—the elastic interaction changed neither the kinetic energy nor the magnitude of the momentum of a scattered particle; the calculations in Sec. 2 were about dynamics. We now consider the kinematics of angular distributions in the inertial frames in which experiments are actually done, and see how the scattering angle in one such frame is related to that in the "fixed-target" frame. The constraints from the conservation laws apply equally to classical and quantum collisions.

(a) Laboratory and center-of-mass frames—Figure 13(a) shows an elastic scatter of particles 1 and 2 as seen in the inertial frame in which particle 2 is initially at rest. The sum of the final-state momenta, $\mathbf{P}'_1 + \mathbf{P}'_2$, is equal to the incident momentum $\mathbf{P}_1$. This frame is called the *lab* frame, because nearly all early experiments were done with the target particles initially at rest in the actual laboratory. Until major advances were made in beam intensity and control, beam-on-beam experiments would have given negligibly small interaction rates—the beams would have passed through one another like sparse comet swarms.

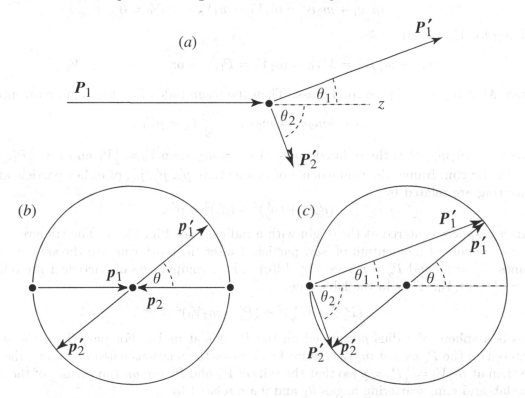

Figure 13. Elastic-scattering kinematics in the plane of the scatter: (a) in the laboratory frame; (b) in the center-of-mass frame; (c) lab and c.m. kinematics superimposed. Parts (a) and (c) are drawn for $m_1 = m_2$. Then the lab angle between the two scattered particles, $\theta_1 + \theta_2$, is 90°. The figure in Prob. 10 shows scattering for $m_2 = 4m_1$.

Figure 13(b) shows the same elastic scatter as seen in the *center-of-mass* (c.m.) frame. In this frame, the momenta of the initial-state particles are equal and opposite, $\mathbf{p}_1 = -\mathbf{p}_2$, and the total momentum is zero. Since the total momentum is conserved, the final-state c.m. momenta are also equal and opposite, $\mathbf{p}'_1 = -\mathbf{p}'_2$. Furthermore, the magnitudes of all four of the c.m. momenta are the same, p (why?). Many experiments are now done with *colliding beams* of equal momenta—the actual laboratory is the c.m. frame.

The analysis of angular distributions is simplest in the c.m. frame, where there is just a single scattering angle θ, not a θ_1 and θ_2. Even when experiments are done in the "lab" frame, the events are nearly always transformed (by computer) to the c.m. frame before the angular distributions are analyzed. So how do we transform scattering events from one inertial frame to another? And what about the fixed-target frame?

(b) The non-relativistic (Galilean) transform—Let the particle masses be m_1 and m_2. In the lab frame, initially $P_1 = m_1 V_1$ and $P_2 = 0$. In the c.m. frame, initially $p_1 = m_1 v_1$, $p_2 = m_2 v_2$, and $p_1 + p_2 = 0$ (v_2 is negative). In the lab frame, the center-of-mass frame is moving forward with velocity V_0. The initial velocities in the two frames are related by

$$v_1 = V_1 - V_0 \qquad \text{and} \qquad v_2 = 0 - V_0 \ .$$

To get V_0, we multiply v_1 by m_1, v_2 by m_2, and add,

$$m_1 v_1 + m_2 v_2 = m_1 V_1 - m_1 V_0 - m_2 V_0 = 0 \ .$$

Solving for V_0, we get

$$(m_1 + m_2)V_0 = MV_0 = m_1 V_1 = P_1, \qquad \text{or} \qquad V_0 = \frac{m_1}{M}V_1 \ ,$$

where $M \equiv m_1 + m_2$ is the total mass. Then the magnitude of all four c.m. momenta is

$$p = -m_2 v_2 = m_2 V_0 = \frac{m_2}{M}P_1 = \mu V_1 \ ,$$

where $\mu \equiv m_1 m_2 / M$ is the reduced mass. If $m_1 = m_2$, then $V_0 = \frac{1}{2}V_1$ and $p = \frac{1}{2}P_1$.

In the c.m. frame, the components of momentum, p'_x, p'_y, p'_z, of either particle after the scattering are related by

$$(p'_x)^2 + (p'_y)^2 + (p'_z)^2 = p^2 \ .$$

This is a sphere centered at the origin with a radius p; see Fig. 13(b). The transverse (x and y) components of momentum of, say, particle 1 after the scattering are the same in the two frames, $P'_x = p'_x$ and $P'_y = p'_y$; see Fig. 13(c). The z components of particle 1 are related by $P'_z = p'_z + m_1 V_0$. Thus in the lab frame,

$$(P'_x)^2 + (P'_y)^2 + (P'_z - m_1 V_0)^2 = p^2 \ .$$

This is a sphere of radius p centered on the P_z axis at $m_1 V_0$. For particle 2, the sphere is centered on the P_z axis at $m_2 V_0$. Figure 13(c) shows the relations when $m_1 = m_2$; the centers are then at $m_1 V_0 = \frac{1}{2}P_1 = p$ (so that the tails of $\mathbf{P}'_1$ and $\mathbf{P}'_2$ are *on* the surface of the sphere). The lab and c.m. scattering angles θ_1 and θ are related by

$$\tan \theta_1 = \frac{\sin \theta}{\cos \theta + m_1/m_2}$$

(see Prob. 9). If $m_1 = m_2$, then $\theta_1 = \frac{1}{2}\theta$ and $\theta_1 + \theta_2 = 90°$. (For an example with $m_1 \neq m_2$, see the figure in Prob. 10.)

The radius p of the sphere increases with energy. The kinetic energy in the c.m. frame is

$$K_{cm} = \frac{p_1^2}{2m_1} + \frac{p_2^2}{2m_2} = \frac{p^2}{2}\left(\frac{1}{m_1} + \frac{1}{m_2}\right) = \frac{p^2}{2\mu},$$

where μ is the reduced mass. Thus $p = (2\mu K_{cm})^{1/2}$, and it is easy to show that $K_{cm} = (m_2/M)K_{lab}$, where $K_{lab} = P_1^2/2m_1$. If $m_1 = m_2$, then $K_{cm} = \frac{1}{2}K_{lab}$.

(c) The fixed-target scattering angle—Figure 14(a) shows, at an arbitrary instant after an elastic scatter, the position vectors $\mathbf{r}_1'$ and $\mathbf{r}_2'$ of particles 1 and 2 in the c.m. frame. This is the only inertial frame in which those vectors are antiparallel. Figure 14(b) shows at the same instant the position vector $\mathbf{r}$ of particle 1 relative to particle 2. This is the position of particle 1 in the fixed-target frame (which is only an inertial frame if m_2 is infinite). Only in the c.m. frame are $\mathbf{r}_1'$ and $\mathbf{r}$ parallel, and so the c.m. scattering angle θ is equal to the scattering angle in the fixed-target frame.

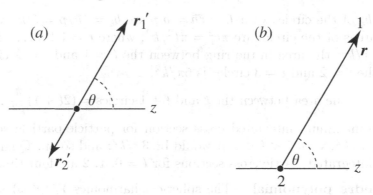

Figure 14. (a) The positions $\mathbf{r}_1'$ and $\mathbf{r}_2'$ of particles 1 and 2 in the c.m. frame at some instant after an elastic scatter. (b) The position $\mathbf{r}$ of particle 1 relative to particle 2 in the frame in which particle 2 is fixed in position. Only in the c.m. frame is $\mathbf{r}$ parallel to $\mathbf{r}_1'$, and only then is θ the same in (a) and (b).

16·6. PARTIAL-WAVE THEORY

The analysis of the rest of this chapter is about elastic scattering in the c.m. inertial frame. We break down the scattering amplitude $f(\theta, k)$ into *partial waves*, one for each value of the orbital angular-momentum quantum number ℓ. Each partial wave is treated as an independent channel, with no regard for what might be going on in the other ℓ channels. Partial-wave expansions are mainly used when the incident momentum is low enough that only a few waves, say up to $\ell = 5$ or 6, are needed to describe the angular distributions.

(a) A rough picture—Classically, the angular momentum of a particle incident with impact parameter b and momentum p is $L = bp$. In Fig. 15, we look in along the scattering axis at a scattering center, its range indicated by shading. In Fig. 15(a), the circle is drawn for an impact parameter b such that $L = \ell\hbar$, where $\ell = 1$. (Of course, classically ℓ is continuous; and quantum mechanically, there is an ℓ-dependent wave function, not an impact parameter.) Particles with values of $\ell > 1$ will pass too far from the scattering center to be much affected by it. In Fig 15(b), the momentum p is four times larger than in (a), and so the impact parameter b for $\ell = 1$ is one-fourth what it is in (a). The circles in (b) show that particles with ℓ up to 3 or 4 will now be affected too.

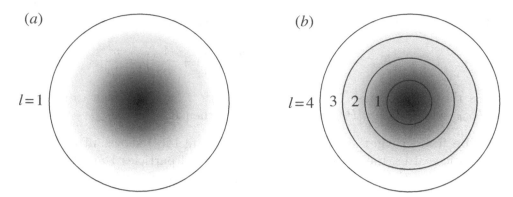

Figure 15. Looking in along the scattering axis at a scattering center, its range indicated by shading. The rings are at impact parameters at which $L = \ell\hbar$. The incident momentum in (b) is four times larger than in (a).

The radii b_ℓ of the circles with $L = \ell\hbar = b_\ell p$ are $b_\ell = \ell\hbar/p = \ell/k$, where k is the wave number. The areas of the circles are $\pi b_\ell^2 = \pi\ell^2/k^2$, where $\ell = 1, 2, 3, \ldots$ The area inside the $\ell = 1$ circle is π/k^2; the area in the ring between the $\ell = 1$ and $\ell = 2$ circles is $3\pi/k^2$; the area between the $\ell = 2$ and $\ell = 3$ circles is $5\pi/k^2$,

$$\text{the area between the } \ell \text{ and } \ell + 1 \text{ circles} = (2\ell + 1)\frac{\pi}{k^2} \ .$$

Classically, the maximum integrated cross section for particle-particle scattering with $0 < \ell < 1$ would be π/k^2; with $1 < \ell < 2$, it would be $3\pi/k^2$; and so on. Quantum mechanically, the maximum integrated elastic cross sections for $\ell = 0, 1, 2$ are four times these values.

(b) Legendre polynomials—The spherical harmonics $Y_\ell^m(\theta, \phi)$ are a complete, orthonormal set for expansions of angular distributions in three dimensions. However, the azimuthal symmetry about the scattering axis means that the only harmonics needed here are those with no ϕ dependence—those with $m = 0$. It is usual then to switch to the *ordinary Legendre polynomials* $P_\ell(\theta)$, which are proportional to the harmonics with $m = 0$,

$$P_\ell(\theta) = \sqrt{\frac{4\pi}{2\ell + 1}}\ Y_\ell^0(\theta)\ .$$

The factor multiplying $Y_\ell^0(\theta)$ makes $P_\ell(\theta=0) = 1$ for all ℓ, the standard convention. The ordinary Legendre polynomials are polynomials in $\cos\theta$. The first few are

$P_0(\theta) = 1,$	$P_3(\theta) = \frac{5}{2}\cos^3\theta - \frac{3}{2}\cos\theta,$
$P_1(\theta) = \cos\theta,$	$P_4(\theta) = \frac{1}{8}(35\cos^4\theta - 30\cos^2\theta + 3),$
$P_2(\theta) = \frac{3}{2}\cos^2\theta - \frac{1}{2},$	$P_5(\theta) = \frac{1}{8}(63\cos^5\theta - 70\cos^3\theta + 15\cos\theta)\ .$

Figure 16 shows them. They are a complete set for expansions of angular distributions with no ϕ dependence. They are orthogonal (though not orthonormal) because they are proportion to spherical harmonics. The even-ℓ polynomials are symmetric about $\theta = 90°$ (or $\cos\theta = 0$), the odd-ℓ polynomials are antisymmetric. From the relation between $P_\ell(\theta)$ and $Y_\ell^0(\theta)$, we get

$$\int_{\theta,\phi} |P_\ell(\theta)|^2\, d\Omega = \frac{4\pi}{2\ell + 1}\int_{\theta,\phi} |Y_\ell^0(\theta)|^2\, d\Omega = \frac{4\pi}{2\ell + 1}\ .$$

340

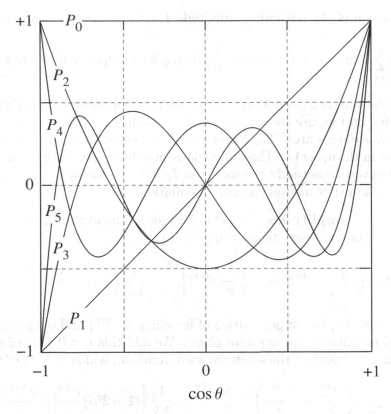

Figure 16. The first six ordinary Legendre polynomials.

(c) The partial-wave expansion—In Sec. 3(a), we found that far from the scattering center $(r \gg 0)$ the *form* of the full wave function, incident plus scattered, is

$$\psi(\mathbf{r}) = e^{ikr\cos\theta} + f(\theta, k)\frac{e^{ikr}}{r} \ .$$

The expansion of the incident plane wave in terms of the Legendre polynomials is

$$e^{ikr\cos\theta} = \sum_{\ell=0}^{\infty} a_\ell(kr)\, P_\ell(\theta) = a_0(kr)P_0(\theta) + a_1(kr)P_1(\theta) + a_2(kr)P_2(\theta) + \cdots$$

Each term in the sum is a product of kr- and θ-dependent factors. Multiplying through by $P_\ell(\theta)$ and using the orthogonality of the Legendre polynomials, we get

$$\int_{\theta,\phi} P_\ell(\theta)\, e^{ikr\cos\theta}\, d\Omega = 2\pi \int_\theta P_\ell(\theta)\, e^{ikr\cos\theta} d(\cos\theta) = \frac{4\pi}{2\ell+1}\, a_\ell(kr) \ ,$$

where $d\Omega = 2\pi\, d(\cos\theta)$. Setting $u \equiv \cos\theta$, we get

$$a_\ell(kr) = \frac{1}{2}\,(2\ell+1) \int_{-1}^{+1} P_\ell(u)\, e^{ikru}\, du \ .$$

The integrands are just polynomials times an exponential.

341

The expansion of the scattering amplitude $f(\theta, k)$ in terms of Legendre polynomials is

$$f(\theta, k) = \frac{1}{k} \sum_{\ell=0}^{\infty} (2\ell + 1)\, T_\ell(k) P_\ell(\theta) = \frac{1}{k} \left[S(k) P_0(\theta) + 3P(k) P_1(\theta) + 5D(k) P_2(\theta) + \cdots \right].$$

Each term in the sum is a product of k- and θ-dependent factors. The $T_\ell(k)$ are the *partial-wave amplitudes*, and we use the same symbols for them, S, P, D, $\ldots$, we used to specify the $L = 0, 1, 2, \ldots$, atomic states. The factor $(2\ell + 1)$ taken out of each term will match up with the same factor in the $a_\ell(kr)$'s. This will make it possible to *parametrize* all the $T_\ell(k)$'s in the same way. However, to actually *calculate* the $T_\ell(k)$'s, we would need a scattering potential energy $V(r)$, or an experimental angular distribution.

(d) The $\ell = 0$ amplitudes—To get $a_0(kr)$ for the incident plane wave $e^{ikr\cos\theta}$, we put $\ell = 0$ and $P_0 = 1$ into the equation for $a_\ell(kr)$ in (c),

$$a_0(kr) = \frac{1}{2} \int_{-1}^{+1} e^{ikru}\, du = \frac{1}{2ikr} e^{ikru} \Big|_{-1}^{+1} = \frac{1}{2ik} \left(\frac{e^{ikr}}{r} - \frac{e^{-ikr}}{r} \right) = \frac{\sin(kr)}{kr}.$$

This shows that $a_0(kr)$ is a superposition of incoming (e^{-ikr}) and outgoing (e^{+ikr}) spherical waves of equal amplitude and opposite phase. We add this $\ell = 0$ part of the incident wave function to the $\ell = 0$ part of the scattered wave function, which is $S(k)\, e^{ikr}/kr$:

$$\frac{1}{2ik} \left(\frac{e^{ikr}}{r} - \frac{e^{-ikr}}{r} \right) + \frac{S}{k} \frac{e^{ikr}}{r} = \frac{1}{2ik} \left[(1 + 2iS) \frac{e^{ikr}}{r} - \frac{e^{-ikr}}{r} \right].$$

Absent in both amplitudes is $P_0(\theta)$, but $P_0(\theta) = 1$, so this is everything for $\ell = 0$.

We limit the formalism to interactions in which there is *only* elastic scattering. Examples are proton-proton $(pp \to pp)$ and π^+-proton $(\pi^+ p \to \pi^+ p)$ scattering at energies too low for "inelastic" reactions such as $pp \to pp\pi^0$ or $\pi^+ p \to \pi^+ p\pi^0$ to occur. (Even then, we are ignoring complications that occur when the particles, such as the proton, have spin.) Conservation of particles and of angular momentum then requires that the fluxes in and out be equal—and not just overall, but for each value of ℓ. That is, the factors multiplying e^{-ikr}/r and e^{ikr}/r in the previous equation have to be equal in magnitude. But they don't have to be equal in *phase*; scattering can shift the *phase* of an amplitude. The requirement is that

$$|1 + 2iS| = 1, \qquad \text{or} \qquad 1 + 2iS = e^{2i\delta_0},$$

where $\delta_0 = \delta_0(k)$ is the $\ell = 0$ *phase shift* (2δ here instead of δ is a convention). Solving for $S = S(k)$, we get

$$\boxed{S = \frac{e^{2i\delta_0} - 1}{2i} = e^{i\delta_0} \sin \delta_0.}$$

Figure 17(a) is a plot of $S = e^{i\delta_0} \sin \delta_0$ as a function of δ_0. The amplitude is a vector in the complex plane from the origin to a point on the unit circle centered at $i/2$. The phase δ_0 is the angle between the amplitude and the real axis; and then the angle subtended by the arrow from the center of the circle is $2\delta_0$. The length of S is $\sin \delta_0$. S is small and mainly real when δ_0 is near $0°$ or $180°$, and large and mainly imaginary when δ_0 is near $90°$.

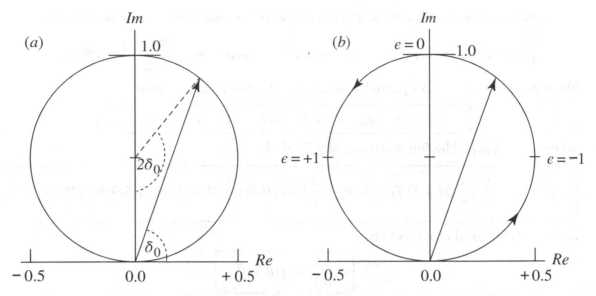

Figure 17. (a) The amplitude $S = e^{i\delta_0} \sin \delta_0$ is a vector in the complex plane from the origin to a point *on* the unit circle centered at $i/2$. (If there is any inelastic scattering, the arrow for *elastic* scattering ends somewhere within the circle.) The P, D, F, ..., elastic amplitudes are parametrized in the same way, each with its own independent phase $\delta_\ell = \delta_\ell(k)$. (b) Figure 7·4 again, showing the Breit-Wigner resonance amplitude $(i - \epsilon)/(1 + \epsilon^2)$ tracing a counterclockwise circle as the energy E increases. Here $\epsilon \equiv (E - E_0)/\frac{1}{2}\Gamma$, where E_0 is the energy when $\delta = 90°$ and Γ is the difference in energies between the $\delta = 135°$ and $45°$ points on the circle.

Now Fig. 17(a) looks like the amplitude $(i - \epsilon)/(1 + \epsilon^2)$ of a Breit-Wigner resonance, shown again in Fig. 17(b). However, the energy dependence of a resonant partial wave is a special case. In general, the partial-wave amplitudes as functions of energy follow no fixed paths. The next section has examples.

(e) The $\ell > 0$ amplitudes—Putting $\ell = 1$ and $P_1(u) = u$ into the integral equation for $a_\ell(kr)$, we get

$$a_\ell(kr) = \frac{1}{2}(2\ell + 1) \int_{-1}^{+1} P_\ell(u) e^{ikru} \, du \;.$$

Looking up the integral, we get

$$a_1(kr) = \frac{3}{2} \int_{-1}^{+1} u \, e^{ikru} \, du = \frac{3}{2} \left(u - \frac{1}{ikr} \right) \frac{e^{ikru}}{ikr} \bigg|_{-1}^{+1} \;.$$

Our interest is in amplitudes at values of r that on atomic or nuclear scales are astronomical, so we neglect the $1/r$ term in the parentheses. Then we have

$$a_1(kr) = \frac{3}{2ik} \left(\frac{e^{ikr}}{r} + \frac{e^{-ikr}}{r} \right) = 3 \frac{\cos(kr)}{ikr} \;.$$

This is again a superposition of incoming and outgoing spherical waves. Adding the factor that multiplies $P_1(\theta)$ in the expansion of $f(\theta, k)$, we get

$$\frac{3}{2ik} \left(\frac{e^{ikr}}{r} + \frac{e^{-ikr}}{r} \right) + \frac{3P}{k} \frac{e^{ikr}}{r} = \frac{3}{2ik} \left[(1 + 2iP) \frac{e^{+ikr}}{r} + \frac{e^{-ikr}}{r} \right] \;.$$

343

The constraint on the amplitude $P(k)$ has exactly the same form as the one on $S(k)$:

$$|1 + 2iP| = 1, \quad \text{so that} \quad 1 + 2iP = e^{2i\delta_0}, \quad \text{and} \quad P = \frac{e^{2i\delta_1} - 1}{2i} = e^{i\delta_1} \sin \delta_1 \ .$$

We now assert that *every* partial-wave amplitudes has this same form,

$$\boxed{S = e^{i\delta_0} \sin \delta_0, \quad P = e^{i\delta_1} \sin \delta_1, \quad D = e^{i\delta_2} \sin \delta_2, \ \ldots,}$$

where $\delta_\ell = \delta_\ell(k)$. The full scattering amplitude is

$$\boxed{f(\theta, k) = \frac{1}{k} \sum_{\ell=0}^{\infty} (2\ell + 1) T_\ell(k) P_\ell(\theta) = \frac{1}{k} \big[S(k) P_0(\theta) + 3P(k) P_1(\theta) + 5D(k) P_2(\theta) + \cdots \big],}$$

and the differential cross section is

$$\boxed{\frac{d\sigma}{d\Omega} = |f(\theta, k)|^2 \ .}$$

The actual values of the phase shifts and the partial-wave amplitudes as functions of k are usually found by measuring angular distributions and doing a *partial-wave analysis*.

16·7. PARTIAL-WAVE EXAMPLES

(a) **The angular distribution**—The scattering amplitude $f(\theta, k)$ is a series linear in the Legendre polynomials. Squaring $f(\theta, k)$ to get $d\sigma/d\Omega$ gives a series quadratic in those polynomials, with factors like $P_2^2(\theta)$ and $P_2(\theta) P_3(\theta)$. Experiments measure $d\sigma/d\Omega$, and then expand those angular distributions in series that are *linear* in the Legendre polynomials ("the Legendre polynomials are a complete set ..."). So we need to write the quadratic series of Legendre polynomials in $|f(\theta, k)|^2$ as a series linear in those polynomials.

An example will show the way. Suppose that only the $\ell = 0$ (S) and $\ell = 1$ (P) partial-wave amplitudes are nonzero. Then

$$f(\theta, k) = \frac{1}{k} \big[S P_0(\theta) + 3P P_1(\theta) \big] = \frac{1}{k} (S + 3P \cos \theta) \ ,$$

and

$$\frac{d\sigma}{d\Omega} = |f(\theta, k)|^2 = \frac{1}{k^2} \big[|S|^2 + 3(S^* P + S P^*) \cos \theta + 9|P|^2 \cos^2 \theta \big] \ .$$

We solve $P_2(\theta) = \frac{3}{2} \cos^2 \theta - \frac{1}{2}$ for $\cos^2 \theta$ to write it in terms of Legendre polynomials,

$$\cos^2 \theta = \tfrac{2}{3} P_2(\theta) + \tfrac{1}{3} = \tfrac{2}{3} P_2(\theta) + \tfrac{1}{3} P_0(\theta) \ ,$$

since $1 = P_0(\theta)$. Then with $S^* P + S P^* = 2Re(S^* P)$ and $\cos \theta = P_1(\theta)$, we get

$$\frac{d\sigma}{d\Omega} = \frac{1}{k^2} \big[(|S|^2 + 3|P|^2) P_0(\theta) + 6\, Re(S^* P)\, P_1(\theta) + 6|P|^2 P_2(\theta) \big] \ ,$$

which is linear in the Legendre polynomials. In general, we write

$$\boxed{\frac{d\sigma}{d\Omega} = \frac{1}{k^2} \sum_{n=0}^{n_{max}} B_n P_n(\theta) \ ,}$$

344

where $n_{max} = 2\ell_{max}$. The table below gives the elements in the expansion of $d\sigma/d\Omega$ when D as well as S and P waves contribute. The B_0 column gives the contributions to $P_0(\theta)$, the B_1 column gives the contributions to $P_1(\theta)$, and so on. If you cross out the rows with a D amplitude, what is left are the numbers 1, 3, 6, and 6 that are in the expansion when only S and P amplitudes contribute. Getting the rest of the table is left as a problem.

Contributions to the expansion coefficients B_n
when only the S, P, and D amplitudes are nonzero

	B_0	B_1	B_2	B_3	B_4
$|S|^2$	1				
$|P|^2$	3		6		
$|D|^2$	5		50/7		90/7
$Re(S^*P)$		6			
$Re(S^*D)$			10		
$Re(P^*D)$		12		18	

We can of course write out the whole angular distribution,

$$k^2 \frac{d\sigma}{d\Omega} = \sum_{n=0}^{4} B_n P_n(\theta) = P_0(\theta) \times \left(|S|^2 + 3|P|^2 + 5|D|^2\right)$$
$$+ P_1(\theta) \times \left[6\,Re(S^*P) + 12\,Re(P^*D)\right]$$
$$+ P_2(\theta) \times \left[6|P|^2 + \tfrac{50}{7}|D|^2 + 10\,Re(S^*D)\right]$$
$$+ P_3(\theta) \times 18\,Re(P^*D)$$
$$+ P_4(\theta) \times \tfrac{90}{7}|D|^2 .$$

But the table is a simpler way to present the numerical expansion coefficients.

(b) **Three angular distributions**—Figure 18 shows the angular distributions, all to the same scale, for an S-wave alone, for S and P waves, and for S, P, and D waves, when all the wave amplitudes are the same ($\delta_0 = \delta_1 = \delta_2$, $S = P = D$). Ignored here, however, is the factor $1/k^2$ in $d\sigma/d\Omega$. Figure 15 showed that as the momentum $p = \hbar k$ increases, waves of larger angular momenta are brought within range of the scattering $V(r)$. Thus, for example, the angular distributions in Fig. 18 might be for three cases with k_1, $k_2 = 2k_1$, and $k_3 = 3k_1$. Then the $1/k^2$ factor would scale down the magnitudes of the $S + P$ and $S + P + D$ angular distributions by factors 4 and 9 relative to the S distribution.

Note the build-up of a sharp peak in the forward direction as more partial waves are added. Partial-wave amplitudes that are more or less equal in phase make for sums of Legendre polynomials all multiplied by coefficients having the same (positive) sign. And since, as Fig. 16 shows, the Legendre polynomials themselves peak in the forward direction, this produces a forward peak. At high enough energies, elastic-scattering angular distributions are always strongly peaked in the forward direction.

345

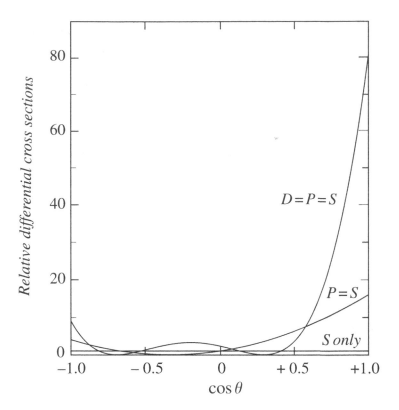

Figure 18. Angular distributions, all to the same scale, for an S-wave alone, for S and P waves, and for S, P, and D waves, when the amplitudes are equal.

(c) The total cross section—Integrating the differential cross section is trivial, because the integral of every Legendre polynomial except $P_0(\theta)$ over the range of θ is zero (why?):

$$\sigma(\text{total}) = \int_{\theta,\phi} \frac{d\sigma}{d\Omega}\, d\Omega = \frac{1}{k^2} \int_{\theta,\phi} \left(\sum_n B_n P_n(\theta)\right) d\Omega$$

$$= \frac{4\pi}{k^2} B_0 = \frac{4\pi}{k^2} \left(|S|^2 + 3|P|^2 + 5|D|^2 + \cdots\right)$$

And since $|S|^2 = |e^{i\delta_0} \sin \delta_0|^2 = \sin^2 \delta_0$, and so on, we get

$$\sigma(\text{total}) = \frac{4\pi}{k^2} \left(\sin^2 \delta_0 + 3\sin^2 \delta_1 + 5\sin^2 \delta_2 + \cdots\right)$$

There is another way to write this, using $T_\ell = e^{i\delta_\ell} \sin \delta_\ell = \cos \delta_\ell \sin \delta_\ell + i \sin^2 \delta_\ell$. From this we get $Im\, T_\ell = \sin^2 \delta_\ell = |T_\ell|^2$, and then

$$\sigma(\text{total}) = \frac{4\pi}{k^2} \left(Im\, S + 3\, Im\, P + 5\, Im\, D + \cdots\right)$$

But at the end of Sec. 6, we had

$$f(\theta, k) = \frac{1}{k} \left[S(k)P_0(\theta) + 3P(k)P_1(\theta) + 5D(k)P_2(\theta) + \cdots\right],$$

so that at $\theta = 0°$, where all the Legendre polynomials are equal to one,

$$\frac{4\pi}{k} Im f(0, k) = \frac{4\pi}{k^2} Im\, (S + 3P + 5D + \cdots) = \sigma(\text{total}) .$$

346

This is the *optical theorem*: The total cross section is directly proportional to the imaginary part of the scattering amplitude at $0°$. So finally, we can write

$$\sigma(\text{total}) = \sum_{\ell=0}^{\ell_{max}} \sigma_\ell, \quad \text{where} \quad \sigma_\ell \equiv \frac{4\pi}{k^2}(2\ell+1)\sin^2\delta_\ell = \frac{4\pi}{k^2}(2\ell+1)\,Im\,T_\ell.$$

This relation between the *elastic* scattering amplitude at $0°$ and the *total* cross section holds even at high energies, where inelastic reactions account for most of the total cross section.

(d) A resonance with background—Figure 19(a) shows an S wave with δ_0 fixed at $135°$ and a resonating P wave at five values of δ_1. Figure 19(b) shows $|S|^2+3|P|^2$, which aside from a factor $4\pi/k^2$ is the total cross section. Figure 19(c) shows the angular distributions $|S+3P\cos\theta|^2$, which aside from a factor $1/k^2$ is the differential cross section. For a narrow resonance (Γ small), k would not vary much over the resonance region.

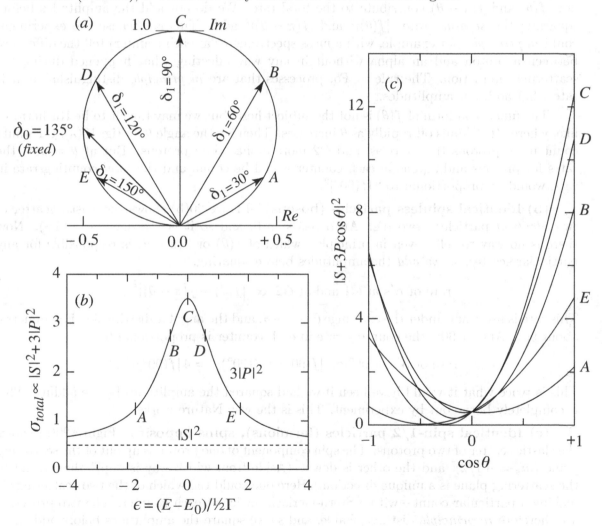

Figure 19. (a) A fixed S wave and a resonating P wave that traces the unit circle. (b) The total cross section (the factor $4\pi/k^2$ not included) as a function of energy across the resonance. (c) The angular distributions at the five marked energies.

347

16·8. SCATTERING OF IDENTICAL PARTICLES

Section 14·8 was about how the amplitudes for identical fermions or identical bosons combine when something can happen in two indistinguishable ways. For identical fermions the amplitudes subtract, for bosons they add. Figure 20 shows some elastic scatterings of particles in the c.m. frame, where the momenta before and again after the scattering are equal and opposite. Here I follow the *The Feynman Lectures*, Vol. III, Secs. 3·4 and 4·1.

(a) Non-identical particles—Figure 20(a) shows the scattering of two *non*-identical particles, a proton and an α. There are two different scattering angles, θ and $\pi - \theta$, that lead to counts in the counters $C1$ and $C2$. The *total* counting rate in each of $C1$ and $C2$ is

$$\text{total rate of particles} \ = \text{rate of protons} + \text{rate of alphas} \ \propto \ |f(\theta)|^2 + |f(\pi - \theta)|^2 \ ,$$

where $f(\theta)$ is the scattering amplitude. The rate would depend on the c.m. momentum, the incoming fluxes, and the solid angles subtended by the counters, but here we only care about how $f(\theta)$ and $f(\pi - \theta)$ contribute to the total rate. We do not add the amplitudes before squaring; the *separate* rates $|f(\theta)|^2$ and $|f(\pi - \theta)|^2$ add. This is because the experiment could *in principle* (for example, with a mass spectrometer at each counter) tell the difference between a proton and an alpha without in any way affecting what happened during the scattering interaction. The rule is: For processes that are *in principle* distinguishable, it is rates that add, not amplitudes.

The functional form of $f(\theta)$ is not the subject here, but we may take it to be Rutherford-like, where $|f(\theta)|^2$ falls off rapidly as θ increases. Then for the angle θ in Fig. 19(a), $C1$ would count more protons than alphas, and $C2$ more alphas than protons. But at $\theta = 90°$, the rates for protons and alphas in each counter would be equal, and the *total* counting rate in each would be proportional to $2|f(90°|^2$.

(b) Identical spinless particles (bosons)—Figure 20(b) shows the elastic scatter of two *identical* particles—two α's. An α has spin 0, and so it is a boson (Sec. 14·8). Now there is no way to tell—even in principle—which of $f(\theta)$ or $f(\pi - \theta)$ is responsible for any particular scatter, so we *add* the amplitudes before squaring,

$$\text{rate of } \alpha\text{'s at } C1 \text{ and at } C2 \ \propto \ \left| f(\theta) + f(\pi - \theta) \right|^2 \ .$$

This rate is invariant under the exchange $\theta \leftrightarrow \pi - \theta$, and the angular distribution is symmetric about 90°. And at 90°, the counting rate in each counter is proportional to

$$\text{rate of } \alpha\text{'s at } 90° \ \propto \ |f(90°) + f(90°)|^2 = 4|f(90°)|^2 \ .$$

This is twice what it would have been if we had squared the amplitudes before adding. This is completely borne out by experiment. This is the way Nature works.

(c) Identical spin-1/2 particles (fermions), spins opposite—Figure 20(c) shows the elastic scatter of two protons. The spin component of one proton is up out of the scattering plane ($m_s = +1/2$), and the other is down. (Aside from which way is "up," the normal to the scattering plane is a unique direction.) Here one could tell which of the two scatterings is making a particular count—with a Stern-Gerlach device at each counter. The two processes are therefore *in principle* distinguishable, and so we square the amplitudes before adding,

$$\text{rate of protons at } C1 \text{ and at } C2 \ \propto \ |f(\theta)|^2 + |f(\pi - \theta)|^2 \ .$$

At 90° the rate is proportional to $2|f(90°)|^2$.

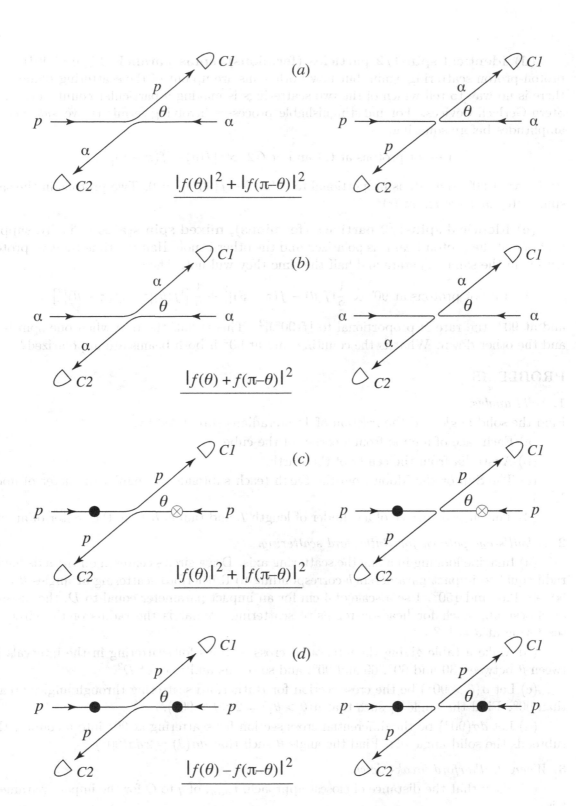

Figure 20. Elastic scattering in the c.m. system: (a) $p\alpha \to p\alpha$; (b) $\alpha\alpha \to \alpha\alpha$; (c) $pp \to pp$ with one spin up ($\bullet$) out of the plane, one down ($\otimes$); (d) $pp \to pp$ with both spins up.

349

(d) Identical spin-1/2 particles (fermions), spins parallel—Figure 20(d) shows proton-proton scattering again, but now both spins are up out of the scattering plane. Now there is no way to tell which of the two scatterings is making a particular count—even with Stern-Gerlach devices. For indistinguishable processes involving fermions, we *subtract* the amplitudes before squaring,

$$\text{rate of protons at } C1 \text{ and at } C2 \propto \left| f(\theta) - f(\pi - \theta) \right|^2 .$$

And now at $90°$, the rate is proportional to $\left| f(90°) - f(90°) \right|^2 = 0$. Two protons in the same spin state *can't scatter* at $90°$.

(e) Identical spin-1/2 particles (fermions), mixed spin states—Finally, suppose that one of the proton beams is polarized and the other is not. Half the time the two protons will be in the same m_s state and half the time they will not. Then

$$\text{rate of protons at } 90° \propto \frac{1}{2} \left| f(\theta) - f(\pi - \theta) \right|^2 + \frac{1}{2} \left[|f(\theta)|^2 + |f(\pi - \theta)|^2 \right] ,$$

and at $90°$, the rate is proportional to $|f(90°)|^2$. This is half the rate when one spin is up and the other down. What is the counting rate at $90°$ if both beams are unpolarized?

PROBLEMS

1. *Solid angles.*

Find the solid angle and the fraction of 4π steradians subtended by:

(a) Each face of a cube from a corner of the cube.

(b) Australia from the center of the Earth.

(c) The Sun or the Moon from the Earth (each subtends an angular diameter of about $1/2°$).

(d) The curved surface of a cylinder of length L and radius R from the center of an end.

2. *A bull's-eye pattern for Rutherford scattering.*

(a) Imagine looking in along the scattering axis. Draw circles centered on the axis having radii equal to impact parameters b corresponding to Rutherford scattering at angles $\theta = 30$, 60, 90, 120, and $150°$. Use a scale of 4 cm for an impact parameter equal to D, the distance of closest approach for head-on repulsive scattering. What is the radius of the circle for scattering at $\theta = 1°$?

(b) Make a table giving the integrated cross sections for scattering in the intervals between θ between 30 and $60°$, 60 and $90°$, and so on, as multiples of D^2.

(c) Let $\sigma(\theta > 90°)$ be the cross section for Rutherford scattering through angles greater than $90°$. Find the angle θ_1 such that $\sigma(\theta > \theta_1) = 2\sigma(\theta > 90°)$.

(d) Let $d\sigma(90°)$ be the differential cross section for scattering at $90°$ into a counter that subtends the solid angle $d\Omega$. Find the angle θ such that $d\sigma(\theta) = 2\, d\sigma(90°)$.

3. *Where Rutherford breaks down.*

(a) Show that the distance of closest approach, r_{min}, of q to Q for the impact parameter b is

$$r_{min} = \left(\frac{D^2}{4} + b^2 \right)^{1/2} \pm \frac{D}{2} ,$$

where $+$ and $-$ are for charges of like or unlike sign. As in Sec. 2, Q is fixed in position.

(b) For scattering of like charges at an energy E, the angular distribution ceases to vary as $\sin^{-4}(\frac{1}{2}\theta)$ for $\theta > 150°$. Find the radius R of the nucleus in terms of D and of E.

(c) A swarm of comets, all having the same velocity $\mathbf{v}_0$, approaches a star having a radius R. Find the "capture cross section," the area of the swarm front (while it is still distant) over which comets will strike the star.

4. *More Rutherford scattering.*

Show that the construction below illustrates the classical Rutherford-scattering relation $b = \frac{1}{2}D\cot(\frac{1}{2}\theta)$. The vertical lines are at a distance $\frac{1}{2}D$ from the scattering center Q—in front or in back of Q as qQ is positive or negative. The dashed lines from Q bisect the angles between the incoming and outgoing asymptotes of the (hyperbolic) orbits.

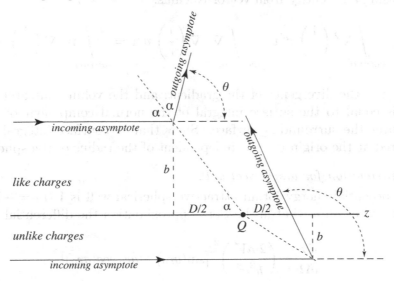

5. *Particle beam fluxes.*

(a) Use the time-dependent Schrödinger equation to show that to satisfy

$$\frac{\partial \rho}{\partial t} = \frac{\partial}{\partial t}(\Psi^*\Psi) = -\nabla \cdot \mathbf{J} \ ,$$

where $\Psi = \Psi(\mathbf{r}, t)$, the *probability current density* $\mathbf{J}(\mathbf{r}, t)$ is

$$\mathbf{J}(\mathbf{r}, t) \equiv \frac{i\hbar}{2m}\left[(\nabla\Psi^*)\Psi - \Psi^*(\nabla\Psi)\right] \ .$$

This is the three-dimensional generalization of $J(x, t)$ in Sec. 4·1. As there, when the particles all have the same energy, the factors $e^{\mp iEt/\hbar}$ in $\rho = \Psi^*\Psi$ and in $\mathbf{J}$ multiply to one, neither ρ nor $\mathbf{J}$ has any time dependence, and calculations involve only the spatial functions $\psi(\mathbf{r})$.

(b) The current density for a plane wave, $\psi(z) = Ae^{ikz}$, is $J = (\hbar k/m)|A|^2$, density times velocity (Sec. 4·1(d)). Show for a spherical wave, $\psi(r, \theta) = f(\theta, k)e^{ikr}/r$, that for $r \gg \lambda$,

$$\mathbf{J} \simeq \frac{\hbar k}{m}\frac{|f(\theta, k)|^2}{r^2}\hat{\mathbf{r}} \ .$$

6. *The Green function for electrostatics.*

To show that

$$\nabla_\mathbf{r}^2 G(\mathbf{r},\mathbf{r}') = \nabla_\mathbf{r}^2 \left(\frac{1}{4\pi|\mathbf{r}-\mathbf{r}'|}\right) = -\delta(\mathbf{r}-\mathbf{r}') \ ,$$

it spoils nothing to take $\mathbf{r}' = 0$ (or to change to $\mathbf{s} \equiv \mathbf{r} - \mathbf{r}'$). Then $G = 1/4\pi r$. Obviously,

$$\nabla^2 \left(\frac{1}{4\pi r}\right) = \frac{1}{r}\frac{\partial^2}{\partial r^2}\left(r \times \frac{1}{4\pi r}\right) = 0$$

for all $r \neq 0$. At $r = 0$, r/r is indeterminate. To catch the delta function at $\mathbf{r} = 0$, we surround it using an identity from vector calculus,

$$\int_{volume} \nabla^2\left(\frac{1}{r}\right) d^3\mathbf{r} = \int_{volume} \nabla\cdot\nabla\left(\frac{1}{r}\right) d^3\mathbf{r} = \int_{surface} \mathbf{\hat{n}}\cdot\nabla\left(\frac{1}{r}\right) dS \ .$$

The Laplacian is the divergence of the gradient; and the volume integral of the divergence of a vector is equal to the surface integral of the normal component of that vector (here a gradient) over the surrounding surface. Show that the surface integral of $\nabla(1/r)$ over a sphere centered at the origin is -4π, independent of the radius of the sphere.

7. *Born approximation for a spherical well.*

(a) The potential energy for an attractive spherical well is $V(r) = -V_0$ for $r < r_0$ and $V(r) = 0$ otherwise. Show that in the Born approximation the differential cross section is

$$\frac{d\sigma}{d\Omega} = \left(\frac{2mV_0}{\hbar^2 q^3}\right)^2 \left[\sin(qr_0) - qr_0\cos(qr_0)\right]^2 \ .$$

(b) Show that as $kr_0 \to 0$ the scattering becomes isotropic, and that then

$$\sigma \approx \pi\left(\frac{4mV_0 r_0^3}{3\hbar^2}\right)^2 \ .$$

8. *Born approximation for a Gaussian $V(r)$.*

(a) The Gaussian potential energy is $V(r) = V_0\exp(-r^2/R^2)$. Show that in the Born approximation the scattering amplitude is

$$f(\theta,k) = -\frac{\pi^{1/2}mV_0 R^3}{2\hbar^2}\exp\left[-k^2 R^2\sin^2(\tfrac{1}{2}\theta)\right] \ .$$

You may want to use $\sin(qr) = (e^{iqr} - e^{-iqr})/2i$ and complete squares to get an easier integral.

(b) Sketch the angular distribution for $kR \ll 1$ and for $kR \gg 1$. Show that the total cross section is

$$\sigma = 2\left(\frac{\pi mV_0 R^2}{2k\hbar^2}\right)^2 \left[1 - \exp(-2k^2 R^2)\right] \ .$$

352

9. *Lab versus c.m. scattering angles.*

Show that the lab scattering angle θ_1 of particle 1 in Fig. 13 and the c.m. scattering angle θ are related by

$$\tan\theta_1 = \frac{\sin\theta}{\cos\theta + m_1/m_2} \, .$$

What is $\tan\theta_2$? Show that if $m_1 = m_2$, then $\theta_1 = \frac{1}{2}\theta$ and $\theta_1 + \theta_2 = 90°$. (This is the non-relativistic limit. Relativistically, as the energy increases, the sum of the angles for equal-mass scattering slowly decreases from $90°$. This was an early test of relativistic kinematics.)

10. *Kinematics when $m_1 < m_2$.*

The figure shows the lab-frame kinematics for elastic scattering of protons on α particles, in which case $m_1/m_2 = 1/4$. The angles around the circle are c.m. angles θ. Adding, say, the two vectors marked $45°$ gives the incident momentum $\mathbf{P}_1$ of the proton; the tips of those vectors lie at opposite ends of a diameter of the circle. The figure scales linearly with P_1 (as long as relativity can be ignored).

Draw the diagram when protons are incident on deuterons, in which case $m_1/m_2 = 1/2$.

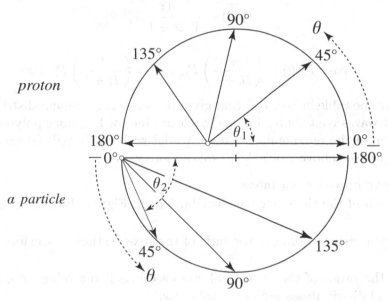

11. *Kinematics when $m_1 > m_2$.*

Draw the kinematics diagram when α-particles are incident on protons. Find the largest lab angle, θ_1, that the α can scatter through.

12. *s-wave hard-sphere scattering.*

(a) In Sec. 6(d), we found that the full $\ell = 0$ wave function is the left side of

$$\frac{1}{2ik}\left[(1+2iS)\frac{e^{ikr}}{r} - \frac{e^{-ikr}}{r}\right] = \frac{1}{kr}\, e^{i\delta_0}\sin(kr + \delta_0)\, ,$$

where $S = S(k)$ is the s-wave scattering amplitude. Show that this is equal to the right side, where $\delta_0 = \delta_0(k)$ is the s-wave phase shift.

353

(b) A plane wave scatters off an impenetrable (hard) sphere of radius R. The energy is low enough that the scattering is entirely s-wave. Continuity of the wave function requires that it vanish at $r = R$. Show then that $\delta_0 = -kR$, the total cross section is $\sigma_0 = (4\pi/k^2)\sin^2\delta_0$, and $\sigma_0 = 4\pi R^2$ in the $k \to 0$ limit.

13. *The D-wave scattering amplitude.*

(a) Find the $\ell = 2$ amplitude factor $a_2(kr)$ of the incident plane wave $e^{ikr\cos\theta}$ in the limit that r is very large.

(b) Show that the D-wave elastic scattering amplitude can be parametrized in terms of a phase shift δ_2 in the same way the S and P amplitudes were in terms of δ_0 and δ_1.

14. *Partial-wave angular distributions.*

(a) Use

$$\cos\theta \, Y_\ell^m(\theta,\phi) = \sqrt{\frac{(\ell+1)^2 - m^2}{4(\ell+1)^2 - 1}} \, Y_{\ell+1}^m(\theta,\phi) + \sqrt{\frac{\ell^2 - m^2}{4\ell^2 - 1}} \, Y_{\ell-1}^m(\theta,\phi)$$

from Sec. 13·5 and

$$P_\ell(\theta) = \sqrt{\frac{4\pi}{2\ell+1}} \, Y_\ell^0(\theta)$$

to show that

$$\cos\theta \, P_\ell(\theta) = \left(\frac{\ell+1}{2\ell+1}\right) P_{\ell+1}(\theta) + \left(\frac{\ell}{2\ell+1}\right) P_{\ell-1}(\theta) \, .$$

(b) Derive the table in Sec. 7(a) that gives the scattering angular distribution when S, P, and D partial waves contribute. Expand each product of Legendre polynomials in $|f(\theta,k)|^2$ as a *linear* sum of the polynomials. This is trivial with $P_0(\theta)P_\ell(\theta)$ (since $P_0 = 1$), and easy with $P_1(\theta)P_\ell(\theta)$ (see above), but $P_2^2(\theta)$ takes more work.

15. *Fore-aft and cross-section ratios.*

(a) For each of the three angular distributions in Fig. 18, find the ratio of the values at $0°$ and $180°$.

(b) For each distribution, find the ratio of the cross sections in the forward and backward hemispheres.

(c) Find the ratios of the three total cross sections if the values of k for the S, $S = P$, and $S = P = D$ distributions are in the ratio 1:2:3.

16. *An angular distribution from phase shifts.*

(a) The S, P, and D phase shifts are $150°$, $30°$, and $60°$. Draw the partial-wave amplitudes on the complex plane (all to the same circle).

(b) Calculate the coefficients B_0 through B_4 from these phase shifts.

(c) Draw the angular distribution.

17. *Phase shifts from an angular distribution.*
The expansion coefficients

$$B_0 = 3.617, \quad B_1 = -0.309, \quad B_2 = 7.970, \quad B_3 = 0.000, \quad B_4 = 3.214$$

fit the angular distribution shown below.

(a) Find S, P, and D phase shifts, δ_0, δ_1, and δ_2, from these coefficients. Draw the partial-wave amplitudes on the complex plane, all to the same circle.

(b) Find, by looking at your diagram, another set of phase shifts that would give the same coefficients and angular distribution.

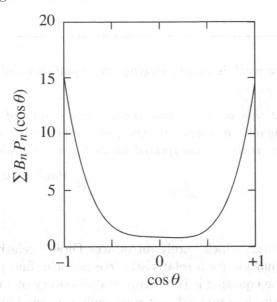

18. *Deuteron-deuteron scattering.*

Deuterons have spin 1, and as such are bosons. Consider elastic scattering of deuterons by deuterons in the c.m. system, and assume that the components m_s of spins perpendicular to the scattering plane are not affected by the interactions (the spins don't "flip"). In terms of $|f(90°)|^2$:

(a) What is $d\sigma/d\Omega$ at 90° if the spin component m_s is +1 in both beams?

(b) What is $d\sigma/d\Omega$ at 90° if m_s is +1 in one beam sand 0 in the other?

(c) What is $d\sigma/d\Omega$ at 90° if m_s is +1 in one beam and the other beam is unpolarized?

(d) What is $d\sigma/d\Omega$ at 90° if both beams are unpolarized?

(e) Show that in (c) the counting rates at 90° for deuterons with $m_s = +1$, 0, and -1 are in the ratio 6:1:1.

17. THE DIRAC EQUATION

17.1. *Dirac at Play*

17.2. *Spin!*

      Problems

A great deal of my work is simply playing with equations and seeing what they give.

&

It was found that this equation gave the particle a spin of half a quantum. And also gave it a magnetic moment. It gave just the properties that one needed for an electron. That was really an unexpected bonus for me, completely unexpected.

Paul Adrien Maurice Dirac

A very short chapter, which barely introduces Dirac's relativistic quantum mechanics. We develop Dirac's equation for a relativistic free particle, find plane-wave solutions of the equation, show how the equation is the theoretical discovery of spin, and show that only the total angular momentum is conserved, not separately the spin and orbital angular momenta. The relativistic solution of the hydrogen atom (and much more) is for your next course on quantum mechanics.

17·1. DIRAC AT PLAY

The Schrödinger equation we have spent so much time with,

$$i\hbar\frac{\partial\Psi}{\partial t} = \hat{H}\Psi = \frac{\hat{p}^2}{2m}\Psi + \hat{V}\Psi = -\frac{\hbar^2}{2m}\nabla^2\Psi + \hat{V}\Psi \,,$$

is explicitly nonrelativistic because it uses the nonrelativistic relation $K = p^2/2m$ between the kinetic energy K and momentum p of a particle. Part of the perturbation calculation of the fine structure of hydrogen in Chap. 13 involved the lowest-order relativistic correction, proportional to a p^4 term.

The relativistic energy E and momentum p of a *free* particle are

$$E = \gamma mc^2 \quad\text{and}\quad p = \gamma mv = \gamma\beta mc \,,$$

where $\beta \equiv v/c$ is the ratio of the speed of the particle to that of light, and $\gamma \equiv 1/\sqrt{1-\beta^2}$. The relation between E and p is then (show this)

$$E^2 = p^2c^2 + m^2c^4 \,.$$

Here E includes not only the kinetic energy K but also the rest-mass energy mc^2, which E reduces to when p is zero. The relativistic *kinetic* energy is $(\gamma - 1)\,mc^2$.

If we try the usual operator substitutions

$$\hat{H} \to i\hbar\frac{\partial}{\partial t}, \qquad \hat{p}_x \to -i\hbar\frac{\partial}{\partial x}, \qquad \text{etc.}\,,$$

in $E^2 = p^2c^2 + m^2c^4$, and operate on Ψ, we get

$$-\hbar^2\frac{\partial^2\Psi}{\partial t^2} = -\hbar^2c^2\,\nabla^2\Psi + m^2c^4\Psi \,.$$

This is called the Klein-Gordon equation. It is indeed relativistic, and it has its uses, but it does not describe spin-1/2 particles. We certainly need an equation for spin-1/2 particles, because the elementary constituents of matter are spin-1/2 particles. The problem is *not* with those derivative operators we equate to $\hat{H}$ and $\hat{p}$. They prove to be the properly relativistic, as the symmetry in them between x (and y and z) and t suggests.

(a) Playing with equations—Dirac began with $E = (p^2c^2 + m^2c^4)^{1/2}$. With the operator substitutions for E and p, this becomes

$$i\hbar\frac{\partial\Psi}{\partial t} = \sqrt{-\hbar^2c^2\,\nabla^2 + m^2c^4}\,\Psi \,,$$

and it is not clear what to do with the right-hand side. Far preferable on the right would be operations that are *linear* in the components of the momentum—that is, *linear* in the spatial derivatives, as is the time derivative on the left. It is the essence of special relativity that space and time (and that other pairing, momentum and energy) enter symmetrically.

So Dirac—playing with equations—looked for a way to make the equation linear by trying to write $p^2c^2 + m^2c^4$ as a perfect square. He wrote

$$p_x^2 + p_y^2 + p_z^2 + m^2c^2 = (\alpha_x p_x + \alpha_y p_y + \alpha_z p_z + \beta mc)(\alpha_x p_x + \alpha_y p_y + \alpha_z p_z + \beta mc) \,,$$

and asked what conditions had to be placed on α_x, α_y, α_z, and β to make the equation true. (*This* β will become a matrix—it is not v/c.) We write the 16 terms of the product out, keeping the order of α_x, α_y, α_z, and β, in case they are operators that do not commute with one another. We assume they commute with the components of momentum.

We get

$$p_x^2 + p_y^2 + p_z^2 + m^2c^2 = \alpha_x^2 p_x^2 + \alpha_y^2 p_y^2 + \alpha_z^2 p_z^2 + \beta^2 m^2 c^2$$

$$+ (\alpha_x \alpha_y + \alpha_y \alpha_x) p_x p_y + (\alpha_x \alpha_z + \alpha_z \alpha_x) p_x p_z + (\alpha_y \alpha_z + \alpha_z \alpha_y) p_y p_z$$

$$+ (\alpha_x \beta + \beta \alpha_x) p_x mc + (\alpha_y \beta + \beta \alpha_y) p_y mc + (\alpha_z \beta + \beta \alpha_z) p_z mc .$$

For the equation to be satisfied, all ten of the following relations have to be true:

$$\alpha_x^2 = \alpha_y^2 = \alpha_z^2 = \beta^2 = 1$$

$$\alpha_x \alpha_y + \alpha_y \alpha_x = \alpha_x \alpha_z + \alpha_z \alpha_x = \alpha_y \alpha_z + \alpha_z \alpha_y = 0$$

$$\alpha_x \beta + \beta \alpha_x = \alpha_y \beta + \beta \alpha_y = \alpha_z \beta + \beta \alpha_z = 0 .$$

Clearly, we cannot satisfy, say, both $\alpha_x^2 = \alpha_y^2 = 1$ and $\alpha_x \alpha_y + \alpha_y \alpha_x = 0$ with ordinary numbers, real or complex.

We can, however, satisfy the requirements with *matrices*, provided that we enlarge the meaning of "1" to include the identity matrix $\hat{I}$. The Pauli matrices give the clue, for in fact,

$$\hat{\sigma}_x^2 = \begin{pmatrix} 0 & 1 \\ 1 & 0 \end{pmatrix} \begin{pmatrix} 0 & 1 \\ 1 & 0 \end{pmatrix} = \begin{pmatrix} 1 & 0 \\ 0 & 1 \end{pmatrix} = \hat{I} ;$$

and, likewise, $\hat{\sigma}_y^2 = \hat{\sigma}_z^2 = \hat{I}$. Furthermore,

$$\hat{\sigma}_x \hat{\sigma}_y + \hat{\sigma}_y \hat{\sigma}_x = \begin{pmatrix} 0 & 1 \\ 1 & 0 \end{pmatrix} \begin{pmatrix} 0 & -i \\ i & 0 \end{pmatrix} + \begin{pmatrix} 0 & -i \\ i & 0 \end{pmatrix} \begin{pmatrix} 0 & 1 \\ 1 & 0 \end{pmatrix} = \begin{pmatrix} 0 & 0 \\ 0 & 0 \end{pmatrix} ;$$

and likewise for the other pairings of $\hat{\sigma}_x$, $\hat{\sigma}_y$, and $\hat{\sigma}_z$. The Pauli matrices are said to *anti-commute*, and the *anti*commutator of operators $\hat{A}$ and $\hat{B}$ is written (as in Sec. 7·3) $[\hat{A}, \hat{B}]_+$.

However, there is no fourth 2-by-2 matrix that will anticommute with the three Pauli matrices and play the role of β (see Prob. 2). Nor is there any set of 3-by-3 matrices that will do the job. But there *is* a set of 4-by-4 matrices, and one choice for them is

$$\hat{\alpha}_x = \begin{pmatrix} 0 & \hat{\sigma}_x \\ \hat{\sigma}_x & 0 \end{pmatrix}, \quad \hat{\alpha}_y = \begin{pmatrix} 0 & \hat{\sigma}_y \\ \hat{\sigma}_y & 0 \end{pmatrix}, \quad \hat{\alpha}_z = \begin{pmatrix} 0 & \hat{\sigma}_z \\ \hat{\sigma}_z & 0 \end{pmatrix}, \quad \hat{\beta} = \begin{pmatrix} \hat{I} & 0 \\ 0 & -\hat{I} \end{pmatrix}.$$

Here each of the elements in these matrices is itself a 2-by-2 matrix. In detail,

$$\hat{\alpha}_x = \begin{pmatrix} 0 & 0 & 0 & 1 \\ 0 & 0 & 1 & 0 \\ 0 & 1 & 0 & 0 \\ 1 & 0 & 0 & 0 \end{pmatrix}, \quad \hat{\alpha}_y = \begin{pmatrix} 0 & 0 & 0 & -i \\ 0 & 0 & i & 0 \\ 0 & -i & 0 & 0 \\ i & 0 & 0 & 0 \end{pmatrix}, \quad \hat{\alpha}_z = \begin{pmatrix} 0 & 0 & 1 & 0 \\ 0 & 0 & 0 & -1 \\ 1 & 0 & 0 & 0 \\ 0 & -1 & 0 & 0 \end{pmatrix},$$

$$\hat{\beta} = \begin{pmatrix} 1 & 0 & 0 & 0 \\ 0 & 1 & 0 & 0 \\ 0 & 0 & -1 & 0 \\ 0 & 0 & 0 & -1 \end{pmatrix}.$$

358

So now, provisionally, we have a way to write a fully relativistic Schrödinger equation for a free particle,

$$i\hbar \frac{\partial \Psi}{\partial t} = \hat{H}\Psi = (p^2c^2 + m^2c^4)^{1/2}\Psi$$

$$= (\hat{\alpha}_x\hat{p}_x + \hat{\alpha}_y\hat{p}_y + \hat{\alpha}_z\hat{p}_z + \hat{\beta}mc)c\Psi$$

$$= (\hat{\boldsymbol{\alpha}} \cdot \hat{\mathbf{p}}c + \hat{\beta}mc^2)\Psi$$

$$= -i\hbar c\hat{\boldsymbol{\alpha}} \cdot \nabla\Psi + \hat{\beta}mc^2\Psi \,,$$

where $\hat{\mathbf{p}} = -i\hbar\nabla$. Because the matrices are 4-by-4, the wave function Ψ will now have to have four components, $\Psi_i(x, y, z, t)$, $i = 1, 2, 3, 4$. Writing things out, we have

$$i\hbar \begin{pmatrix} \partial\Psi_1/\partial t \\ \partial\Psi_2/\partial t \\ \partial\Psi_3/\partial t \\ \partial\Psi_4/\partial t \end{pmatrix} = \begin{pmatrix} mc^2 & 0 & c\hat{p}_z & c(\hat{p}_x - i\hat{p}_y) \\ 0 & mc^2 & c(\hat{p}_x + i\hat{p}_y) & -c\hat{p}_z \\ c\hat{p}_z & c(\hat{p}_x - i\hat{p}_y) & -mc^2 & 0 \\ c(\hat{p}_x + i\hat{p}_y) & -c\hat{p}_z & 0 & -mc^2 \end{pmatrix} \begin{pmatrix} \Psi_1 \\ \Psi_2 \\ \Psi_3 \\ \Psi_4 \end{pmatrix}.$$

The matrix is of course Hermitian. The expectation value of the energy would be the row vector $\Psi^\dagger \equiv (\Psi_1^*, \Psi_2^*, \Psi_3^*, \Psi_4^*)$ times the right side of this equation, $\langle \Psi | \hat{H} | \Psi \rangle$.

(b) A plane-wave solution—To get a feel for this equation, we look for a four-component plane-wave solution traveling in the z direction with momentum $p_z = p$:

$$\Psi(z, t) = \begin{pmatrix} A_1 \\ A_2 \\ A_3 \\ A_4 \end{pmatrix} e^{i(pz - Et)/\hbar} \,.$$

The factor $(pz - Et)$ in the exponent is completely relativistic, both in form and, given the relativistic relation between E and p, in substance. We look for a solution in which the components A_i are independent of position and time. If we put Ψ into the Schrödinger equation, the operator $i\hbar\,\partial/\partial t$ on the left simply multiplies each of the four components of Ψ by E. The spatial derivatives on the right, implied by the $\hat{p}_i$ operations in the matrix, simply replace p_x and p_y with 0 and p_z with p. Cancelling the common exponential factor, we get

$$\begin{pmatrix} EA_1 \\ EA_2 \\ EA_3 \\ EA_4 \end{pmatrix} = \begin{pmatrix} mc^2 & 0 & cp & 0 \\ 0 & mc^2 & 0 & -cp \\ cp & 0 & -mc^2 & 0 \\ 0 & -cp & 0 & -mc^2 \end{pmatrix} \begin{pmatrix} A_1 \\ A_2 \\ A_3 \\ A_4 \end{pmatrix},$$

which are these four equations,

$$EA_1 = mc^2A_1 + pcA_3 \qquad EA_3 = pcA_1 - mc^2A_3$$

$$EA_2 = mc^2A_2 - pcA_4 \qquad EA_4 = -pcA_2 - mc^2A_4 \,.$$

They relate A_1 and A_3, and A_2 and A_4. Had we chosen the momentum to be along the x or y axis, the pairings would have been A_1A_4 and A_2A_3.

Solving for ratios, we get

$$\frac{A_3}{A_1} = \frac{E - mc^2}{pc} = \frac{pc}{E + mc^2} = -\frac{A_4}{A_2} \,.$$

Cross multiplying at the middle equality, we get $E^2 = p^2c^2 + m^2c^4$; the relativistic relation we began with is built into the equations. And only two of the four A_i's, say A_1 and A_2, are independent. In terms of the relativistic parameters γ and β, the ratios are

$$\frac{A_3}{A_1} = \frac{\gamma - 1}{\gamma\beta} = \frac{\gamma\beta}{\gamma + 1} = -\frac{A_4}{A_2} \ .$$

We can write

$$\chi_+ = \begin{pmatrix} 1 \\ 0 \\ (E - mc^2)/pc \\ 0 \end{pmatrix} \quad \text{and} \quad \chi_- = \begin{pmatrix} 0 \\ 1 \\ 0 \\ -(E - mc^2)/pc \end{pmatrix}$$

(which are not normalized). Then

$$\Psi(z, t) = (c_+\chi_+ + c_-\chi_-)\, e^{i(pz - Et)/\hbar} \ .$$

is the general state of a free particle moving in the z direction. Or in fact of a free *spin-1/2* particle, as we find next.

17·2. SPIN!

Because the state function Ψ now has *components*, we might suspect the components are related to spin—although it is perhaps surprising to find four components instead of two. But let us see what happens in the nonrelativistic limit, where the kinetic energy $K = E - mc^2$ reduces to $p^2/2m$ and p to mv. Then

$$\frac{A_3}{A_1} = -\frac{A_4}{A_2} = \frac{E - mc^2}{pc} \simeq \frac{p^2/2m}{pc} = \frac{p}{2mc} \simeq \frac{v}{2c} \simeq 0 \ .$$

In that limit, the A_3 and A_4 components of Ψ are negligible, and the plane-wave solution is

$$\Psi(z, t) = \begin{pmatrix} c_+ \\ c_- \end{pmatrix} e^{i(pz - Et)/\hbar} \ .$$

We have a two-component state, and there is no reason not to interpret the two-component-ness as *SPIN*! Or, actually, spin in the nonrelativistic limit.

(a) Four-by-four spin operators—We can put this theoretical discovery (or explanation) of spin—which arose *naturally* from the search for a relativistic quantum mechanics—on firmer ground. We define four-dimensional spin operators in terms of the 2-by-2 Pauli matrices,

$$\hat{S}_x = \frac{\hbar}{2}\begin{pmatrix} \hat{\sigma}_x & 0 \\ 0 & \hat{\sigma}_x \end{pmatrix}, \quad \hat{S}_y = \frac{\hbar}{2}\begin{pmatrix} \hat{\sigma}_y & 0 \\ 0 & \hat{\sigma}_y \end{pmatrix}, \quad \hat{S}_z = \frac{\hbar}{2}\begin{pmatrix} \hat{\sigma}_z & 0 \\ 0 & \hat{\sigma}_z \end{pmatrix}.$$

This is reasonable, because in the nonrelativistic limit, where the bottom two components of Ψ are negligibly small, we are effectively back to the 2-by-2 spin operators. The full $\hat{S}_z$ operator is

$$\hat{S}_z = \frac{\hbar}{2}\begin{pmatrix} 1 & 0 & 0 & 0 \\ 0 & -1 & 0 & 0 \\ 0 & 0 & 1 & 0 \\ 0 & 0 & 0 & -1 \end{pmatrix}.$$

360

The relativistic plane-wave solutions $\chi_\pm$ for the momentum in the z direction are eigenstates of the $\hat{S}_z$ operator. For example,

$$\hat{S}_z\chi_+ = \frac{\hbar}{2}\begin{pmatrix} 1 & 0 & 0 & 0 \\ 0 & -1 & 0 & 0 \\ 0 & 0 & 1 & 0 \\ 0 & 0 & 0 & -1 \end{pmatrix}\begin{pmatrix} 1 \\ 0 \\ (E-mc^2)/pc \\ 0 \end{pmatrix} = +\frac{\hbar}{2}\chi_+ \; .$$

And $\hat{S}_z\chi_- = -\frac{1}{2}\hbar\,\chi_-$.

(b) Conservation of angular momentum—The components of the orbital angular momentum operator $\hat{\mathbf{L}}$ commute with the *non*relativistic Hamiltonian $\hat{H}$ for a free particle, and with the *non*relativistic $\hat{H}$ for a particle in a spherically symmetric potential energy $V(r)$. And since those two Hamiltonians are independent of spin, the components of the spin operator $\hat{\mathbf{S}}$ commute with them too. That was why we were able to specify the states of the hydrogen atom with n, ℓ, m_ℓ, s, and m_s. When an operator $\hat{A}$ commutes with the Hamiltonian, the expectation value $\langle A \rangle$ of the physical property the operator represents is conserved, whether the particle is in an eigenstate of that operator or not. From Sec. 5·7(b)),

$$\frac{d\langle A \rangle}{dt} = \frac{i}{\hbar}\langle[\hat{H}, \hat{A}]\rangle = 0 \; .$$

Thus for those nonrelativistic cases, the orbital and spin angular momenta are *separately* conserved. We now show that neither $\hat{\mathbf{L}}$ nor $\hat{\mathbf{S}}$ commutes with even Dirac's free-particle Hamiltonian $\hat{H} = \hat{\boldsymbol{\alpha}} \cdot \hat{\mathbf{p}}c + \hat{\beta}mc^2$, but that the sum, $\hat{\mathbf{J}} = \hat{\mathbf{L}} + \hat{\mathbf{S}}$, *does* commute with it, and therefore the *total* angular momentum is still conserved.

We again need the basic rules that reduce commutators (Sec. 5·4),

$$[\hat{A} + \hat{B}, \hat{C}] = [\hat{A}, \hat{C}] + [\hat{B}, \hat{C}] \qquad \text{and} \qquad [\hat{A}\hat{B}, \hat{C}] = \hat{A}[\hat{B}, \hat{C}] + [\hat{A}, \hat{C}]\hat{B} \; ,$$

as well as the commutators of the components of position and momentum,

$$[\hat{x}_i, \hat{x}_j] = 0, \quad [\hat{p}_i, \hat{p}_j] = 0, \quad \text{and} \quad [\hat{x}_i, \hat{p}_j] = i\hbar\delta_{ij} \; ,$$

where x_i, $i = 1, 2, 3$, are x, y, z, etc. The x component of $\mathbf{L} = \mathbf{r} \times \mathbf{p}$ is $L_x = yp_z - zp_y$, and in a step or two one can find that

$$[\hat{L}_x, \hat{p}_x] = 0, \quad [\hat{L}_x, \hat{p}_y] = i\hbar\hat{p}_z, \quad \text{and} \quad [\hat{L}_x, \hat{p}_z] = -i\hbar\hat{p}_y \; .$$

Then, with Dirac's free-particle $\hat{H} = \hat{\boldsymbol{\alpha}} \cdot \hat{\mathbf{p}}c + \hat{\beta}mc^2$, we get

$$[\hat{L}_x, \hat{H}] = c\hat{\alpha}_x[\hat{L}_x, \hat{p}_x] + c\hat{\alpha}_y[\hat{L}_x, \hat{p}_y] + c\hat{\alpha}_z[\hat{L}_x, \hat{p}_z] + \hat{\beta}[\hat{L}_x, mc^2]$$
$$= i\hbar c(\hat{\alpha}_y\hat{p}_z - \hat{\alpha}_z\hat{p}_y) = i\hbar c\,(\hat{\boldsymbol{\alpha}} \times \hat{\mathbf{p}})_x \; .$$

(We assume here, as we already did in Sec. 1 for the $\hat{p}_i$, that the $\hat{x}_i$ commute with the $\hat{\alpha}_i$ and $\hat{\beta}$ operators.) By cyclic symmetry, we also get $[\hat{L}_y, \hat{H}] = i\hbar c(\hat{\boldsymbol{\alpha}} \times \hat{\mathbf{p}})_y$ and $[\hat{L}_z, \hat{H}] = i\hbar c(\hat{\boldsymbol{\alpha}} \times \hat{\mathbf{p}})_z$. Formally, we may summarize these results as

$$[\hat{\mathbf{L}}, \hat{H}] = i\hbar c\,\hat{\boldsymbol{\alpha}} \times \hat{\mathbf{p}} \; .$$

361

To get $[\hat{\mathbf{S}}, \hat{H}]$, we need the commutators of the $\hat{\mathbf{S}}$ operators with the $\hat{\alpha}_i$ and $\hat{\beta}$ operators. First, we get

$$\begin{pmatrix} \hat{\sigma}_x & 0 \\ 0 & \hat{\sigma}_x \end{pmatrix} \begin{pmatrix} 0 & \hat{\sigma}_x \\ \hat{\sigma}_x & 0 \end{pmatrix} - \begin{pmatrix} 0 & \hat{\sigma}_x \\ \hat{\sigma}_x & 0 \end{pmatrix} \begin{pmatrix} \hat{\sigma}_x & 0 \\ 0 & \hat{\sigma}_x \end{pmatrix} = \begin{pmatrix} 0 & \hat{\sigma}_x^2 \\ \hat{\sigma}_x^2 & 0 \end{pmatrix} - \begin{pmatrix} 0 & \hat{\sigma}_x^2 \\ \hat{\sigma}_x^2 & 0 \end{pmatrix} = 0 \,,$$

so that $[\hat{S}_x, \hat{\alpha}_x] = 0$. And since the Pauli matrices satisfy $[\hat{\sigma}_x, \hat{\sigma}_y] = 2i\hat{\sigma}_z$ (from $[\hat{s}_x, \hat{s}_y] = i\hbar\hat{s}_z$), we get

$$\begin{pmatrix} \hat{\sigma}_x & 0 \\ 0 & \hat{\sigma}_x \end{pmatrix} \begin{pmatrix} 0 & \hat{\sigma}_y \\ \hat{\sigma}_y & 0 \end{pmatrix} - \begin{pmatrix} 0 & \hat{\sigma}_y \\ \hat{\sigma}_y & 0 \end{pmatrix} \begin{pmatrix} \hat{\sigma}_x & 0 \\ 0 & \hat{\sigma}_x \end{pmatrix} = \begin{pmatrix} 0 & \hat{\sigma}_x\hat{\sigma}_y - \hat{\sigma}_y\hat{\sigma}_x \\ \hat{\sigma}_x\hat{\sigma}_y - \hat{\sigma}_y\hat{\sigma}_x & 0 \end{pmatrix}$$

$$= 2i \begin{pmatrix} 0 & \hat{\sigma}_z \\ \hat{\sigma}_z & 0 \end{pmatrix} = 2i\,\hat{\alpha}_z \,,$$

so that $[\hat{S}_x, \hat{\alpha}_y] = i\hbar\hat{\alpha}_z$. Similarily, $[\hat{S}_x, \hat{\alpha}_z] = -i\hbar\hat{\alpha}_y$ and $[\hat{S}_x, \hat{\beta}] = 0$. Then

$$[\hat{S}_x, \hat{H}] = c[\hat{S}_x, \hat{\alpha}_x]\hat{p}_x + c[\hat{S}_x, \hat{\alpha}_y]\hat{p}_y + c[\hat{S}_x, \hat{\alpha}_z]\hat{p}_z + [\hat{S}_x, \hat{\beta}]mc^2$$
$$= i\hbar c(\hat{\alpha}_z\hat{p}_y - \hat{\alpha}_y\hat{p}_z) = -i\hbar c(\hat{\boldsymbol{\alpha}} \times \hat{\mathbf{p}})_x \,.$$

Finally then,

$$[\hat{\mathbf{S}}, \hat{H}] = -i\hbar c\,\hat{\boldsymbol{\alpha}} \times \hat{\mathbf{p}} = -[\hat{\mathbf{L}}, \hat{H}] \,,$$

and

$$[\hat{\mathbf{L}} + \hat{\mathbf{S}}, \hat{H}] = [\hat{\mathbf{J}}, \hat{H}] = 0 \,.$$

The *total* angular momentum $\mathbf{J} = \mathbf{L} + \mathbf{S}$ is conserved, even though the orbital and spin angular momenta are not.

PROBLEMS

1. *Relativistic relations.*
The relativistic energy and momentum of a free particle are $E = \gamma mc^2$ and $p = \gamma\beta mc$, where $\gamma = 1/\sqrt{1 - \beta^2}$ and $\beta = v/c$ is the ratio of the speed of the particle to that of light.

    (a) Show that $E^2 = p^2c^2 + m^2c^4$.

    (b) Show that $E - mc^2 \simeq p^2/2m$ and that $p \simeq mv$ when β is very small.

2. *Operator anticommutators.*

    (a) Show that the Pauli matrices anticommute with one another.

    (b) Show that all four elements of a 2-by-2 matrix that anticommutes with all three of the Pauli matrices are zero.

    (c) Show that the Dirac matrices $\hat{\alpha}_x$, $\hat{\alpha}_y$, $\hat{\alpha}_z$, and $\hat{\beta}$ anticommute with one another and that the square of each is the 4-by-4 identity matrix. When it is convenient to do so, use 2-by-2 block matrix multiplication, as in Sec. 2(b).

3. *A particle moving in the x direction.*

Find the free-particle Dirac state function for a particle moving in the $+x$ direction with momentum p. That is, write out the four equations given by the Dirac equation, assume a solution of the form $A\exp[i(px - Et)/\hbar]$, where A is a 4-component spinor, and find the relations between the components of A. Is it still the bottom two components that become negligible in the nonrelativistic limit? (Assume that $E > 0$.)

4. *A continuity equation.*

(a) Use $i\hbar\,\partial\Psi/\partial t = \hat{H}\Psi$, where $\hat{H} = c\hat{\boldsymbol{\alpha}}\cdot\hat{\mathbf{p}} + \hat{\beta}mc^2 = -i\hbar c\hat{\boldsymbol{\alpha}}\cdot\nabla + \hat{\beta}mc^2$ is the free-particle Hamiltonian at the end of Sec. 1(a) to show that in the continuity equation

$$\frac{\partial}{\partial t}(\Psi^\dagger\Psi) = -\nabla\cdot\mathbf{J},$$

the probability current density $\mathbf{J}$ is now $\mathbf{J} = c\Psi^\dagger\hat{\boldsymbol{\alpha}}\Psi$. (The components of $\Psi^\dagger$ are the complex conjugates of the components of Ψ, and $\Psi^\dagger$ is a row vector.)

(b) Find $\rho \equiv \Psi^\dagger\Psi$ and the components of $\mathbf{J}$ for the free-particle plane wave moving in the $+z$ direction. Show that $J_z = v\rho$, where v is the speed of the particle.

5. *Another conserved quantity.*

(a) Show that the 4-by-4 operator $\hat{S}_z$ for the z component of spin commutes with the Hamiltonian $\hat{H}$ for the plane-wave solution of Sec. 1(b), where $p_x = p_y = 0$. Thus the spin component *along* the direction of motion is conserved (Sec. 5·7). This makes it the axis of choice for spin.

(b) Calculate the commutator of $\hat{S}_x$ with that Hamiltonian, and thus show that a component of the spin perpendicular to the line of motion is not conserved.

6. *Dirac γ matrices.*

There are other sets of matrices besides α_x, α_y, α_z, and β that satisfy the required commutation relations. Nearly all theoretical work uses the following γ matrices (or a set like them), defined by

$$\gamma_0 = \beta, \quad \gamma_1 = \beta\alpha_x, \quad \gamma_2 = \beta\alpha_y, \quad \gamma_3 = \beta\gamma_z.$$

(And the γ and β here are not those of $p = \gamma\beta mc$.)

(a) Show that

$$\gamma_0 = \begin{pmatrix} \hat{I} & 0 \\ 0 & -\hat{I} \end{pmatrix}, \quad \gamma_1 = \begin{pmatrix} 0 & \hat{\sigma}_x \\ -\hat{\sigma}_x & 0 \end{pmatrix}, \quad \gamma_2 = \begin{pmatrix} 0 & \hat{\sigma}_y \\ -\hat{\sigma}_y & 0 \end{pmatrix}, \quad \gamma_3 = \begin{pmatrix} 0 & \hat{\sigma}_z \\ -\hat{\sigma}_z & 0 \end{pmatrix}.$$

(b) Show that then the Dirac equation for a free particle,

$$i\hbar\frac{\partial\Psi}{\partial t} = (\hat{\alpha}_x\hat{p}_x + \hat{\alpha}_y\hat{p}_y + \hat{\alpha}_z\hat{p}_z + \hat{\beta}mc)\,c\Psi\,,$$

now takes on the more symmetrical form

$$\gamma_0\frac{\partial\Psi}{c\,\partial t} + \gamma_1\frac{\partial\Psi}{\partial x} + \gamma_2\frac{\partial\Psi}{\partial y} + \gamma_3\frac{\partial\Psi}{\partial z} + i\frac{mc}{\hbar}\Psi = 0\,,$$

or

$$\sum_{j=0}^{3} \gamma_j \frac{\partial \Psi}{\partial x_j} + i\frac{mc}{\hbar}\Psi = 0 \ ,$$

where $x_j = c\,t, x, y, z$. There are other notational conventions, such as using both upper and lower indices, and making summations over repeated indices (here j) implicit, thereby discarding summation signs. For these, see a more advanced text.

SOURCES OF QUOTES

References from books include the page of the quote, those from journal articles give only the article reference. The bracketed numbers are the pages on which the quotes occur in this text. Several of the quotations are taken from the histories by Jammer and Pais: (see [2] and [3] below and the first page of Chapter 1. When they in turn are quoting others, the original source is given as well.

[v] William James, *The Varieties of Religious Experience*, Library of America Paperback Classics, New York (2010), page 201.

[v] Richard W. Hamming, *Introduction to Applied Numerical Analysis*, McGraw-Hill Book Company, New York and other cities (1971), page 31.

[2] Abraham Pais, *Subtle is the Lord*, Oxford University Press, Oxford (1982), page 371.

[3] Max Jammer, *The Conceptual Development of Quantum Mechanics*, McGraw-Hill Book Company, New York and other cities (1966), page 23.

[3] A. Einstein, Annalen der Physik (Leipzig), "Zur Theorie der Lichterzeugung und Lichtabsorption," **20**,199 (1906). The quote in English is in A. Pais, *Subtle is the Lord*, page 378.

[3-4] G. Kirsten and H. Körber, *Physiker über Physiker*, Akademie Verlag, Berlin (1975), page 201. The quote is in English in A. Pais, *Subtle is the Lord*, page 382.

[5-6] R. Gähler and A. Zeilinger, "Wave-optical experiments with very cold neutrons," American Journal of Physics **59** (1991) 316.

[7] Richard F. Feynman, Robert B. Leighton, and Matthew Sands, *The Feynman Lectures on Physics*, Vol. III, Addison-Wesley, Reading, Massachusetts (1965), page 1-1.

[9] Yogi Berra, *When You Come to a Fork in the Road, Take It!*, Yogi Berra with Dave Kaplan, Hyperion, New York (2001), book title.

[9] A.C. Elitzur and L. Vaidman, "Quantum Mechanical Interaction-Free Measurements," Foundations of Physics **23**, No. 7, (1993) 987.

[14] Joe Kane, *Running the Amazon*, Alfred A. Knopf, New York (1989), page 153.

[14] Woody Allen, *Getting Even*, Random House, New York (1966), page 58.

[17] A. Einstein, preface to Max Planck, *Where is Science Going?*, W.W. Norton & Co, New York (1933).

[17] Max Born, "Quantum Mechanics of Collision Phenomena," Zeitschrift für Physik **38**, 803, 1926. The quote is in A. Pais, *Inward Bound*, Clarendon Press, page 258.

[36] Ronald N. Bracewell, *The Fourier Transform and Its Applications*, 3rd ed., McGraw-Hill Book Company (2000), page 74.

[46] A.P. French, Edwin F. Taylor, *An Introduction to Quantum Physics*, W.W. Norton & Company (1978), page 149.

[48] Richard Feynman, *Surely You're Joking, Mr. Feynman*, W.W. Norton & Company, New York London (2018), page 102.

[93] Eugene Wigner, "The Unreasonable Effectiveness of Mathematics in the Natural Sciences," Communications in Pure and Applied Mathematics, **13**, No. 1 (1960).

[111] George E.P. Box (coauthor Norman Draper), *Empirical Model-Building and Response Surfaces*, John Wiley & Sons (1987), New York and other cities, page 424.

[130] W. Heisenberg, "Uber den anschaulichen Inhalt der quantentheoretischen Kinematik und Mechanik," Zeitschrift für Physik 33, 172 (1927). The quote in English is in M. Jammer, *The Conceptual Development of Quantum Mechanics*, page 330.

[147] Harry J. Lipkin, *Lie Groups for Pedestrians*, 2nd ed., North-Holland Publishing Company (1966), page 9.

[227] A. Einstein, B. Podolsky, and N. Rosen, "Can [the] Quantum-Mechanical Description of Reality be Considered Complete?", Physical Review **47**, 777 (1935).

[233] Augustus de Morgan, *A Budget of Paradoxes*, Longmans, Green, and Co., London (1872), page 377.

[320] Ernest Rutherford, quoted in E.N. da C. Andrade, *Rutherford and the Nature of the Atom*, Doubleday (1964), page 111.

[356] The two quotes are in A. Pais, *Inward Bound*, Clarendon Press, pages 334 and 286. The original sources are P.A.M. Dirac, interview with T. Kuhn, May 7, 1963, transcript in Niels Bohr Library, American Institute of Physics, New York, and P.A.M. Dirac, report KFKI-1977-62, Hungarian Academy of Science.

INDEX

The Contents, with 17 chapter titles and 110 section titles, might be a better place to search for major subjects.

A and *B* coefficients, Einstein's, 313-314

Allen, Woody, 14

angular momentum, 147-164, 183-203, 204-222

    addition of two, 204-222

      Clebsch-Gordan coefficients, 211, 214-215

    commutation relations for, 150-152

    eigenvalues of, 153-154

    harmonic oscillator, comparison with, 163

    isospin and, 216-218, 221-222

    operators as matrices, 157-159

    operator solution, 152-155

    orbital, 148-151, 159-162

      differential operators for, 159-160

    spherical harmonics, 161-162

      parities of, 162, 164

      a recurrence relation, 243-244

    spin 1/2 (*see also* spin 1/2), 156-157, 183-203

    states as vectors, 157-158

    uncertainty relations for, 156-157

atoms (*see* elements)

Bell's inequality, 227-230

Bell, John S., 227, 231

Berra, Yogi, 9

binding energy (ionization energy)

    of the elements, table, 271

    of hydrogen, 12

black-body radiation, 2, 3, 311-312

Bohr model of hydrogen, 11-14

    classical limit of, 15

Bohr, Niels, 1

Bohr radius (a_0), 12

Boltzmann factor, 307-308

Born approximation, 334-336, 352

    and Yukawa potential energy, 335-336

Born, Max, 17

    and probability interpretation, 21

Bose, Satendra Nath, 263

bosons, 288-291

bound states (*see* eigenvalues)

Box, George E.P., 111

Bracewell, R.N., 36

bra(c)ket (Dirac) notation, 94-98

Breit-Wigner amplitude, 139-141

and exponential decay, 138-141

and resonant scattering, 343, 347

Brillouin, L., 62

Buckingham, E., 57

central forces, 148-149

Clebsch-Gordan coefficients, 211, 214-215

commutators, defined, 103-104

commuting operators, 102, 142-143

and simultaneous eigenstates, 143

non-commuting operators, 103

and uncertainty relations (*see also* uncertainty), 133-135

Compton, Arthur Holly, 4

Compton scattering, 4, 15

conduction electrons, 52-53, 264-267

conservation of charge, 71-72

charge density, 71

continuity equation, 72

current density, 71

conservation of particles, 72-73

continuity equation, 73

particle current density, 73

simple examples, 73

particle density, 72

cryptography, quantum, 224-226

de Broglie, Louis, 4

de Broglie wavelength, 4, 6, 11, 15

and WKB method, 58-62

decay lifetimes, 138-139, 314-315

delta function wells

double, 47-48

multiple, 47

single, 34-36

d'Espagnat, Bernard, 231

dimensions

    of parameters, for estimates, 55-57

        and hydrogen, 56

        and the oscillator, 55

    of Planck constant, 3

    and quantum numbers, 57

    and scaling, 56

Dirac, Paul Adrien Maurice, 356

Dirac bra(c)ket notation, 94-98

Dirac equation

    for free particle, 353-362

        and conservation of angular momentum, 361-362

        Dirac matrices, 358, 363

        plane-wave solution, 359-360

        and spin 1/2, 360-362

    for hydrogen levels, 260

double-slit interference

    Fourier transform for, 136-137

    of light, 6

    of neutrons, 4-7

    on water, 14

effective potential energy, 166-167

    for hydrogen, 166-168

    for isotropic oscillator, 166, 174

Ehrenfest relations, 107

eigenvalues, eigenfunctions of

    angular momenta, 155

        orbital, 161-162

        spin-1/2, 157-158, 184-187

    bound states, qualitative properties of, 44

    delta function wells, 34-36, 47-48

    finite square wells, 28-34, 45-46

    helium, 276-287

        excited states, 283-287

        ground state, 276-280

    hydrogen

        energy eigenfunctions, table of, 173

        energy eigenvalues, 12-13, 170

        ground state, 43-44

        series solution, 167-173

hyperbolic-secant wells, 49-50

infinite square wells

one-dimensional, 26-28

three-dimensional, 38-39

oscillator, one dimensional

all states, 117-118, 121-122

ground state, 42-43

oscillator, three-dimensional, 125, 171, 174-179

Einstein, Albert, 3-4, 17, 227, 313

Einstein A and B coefficients, 313-314

Einstein-Podolsky-Rosen (EPR) argument, 226-227

electrons

in an atom, 267-274

in a box, 52-53, 264-267

in a molecule, 51

electron volt (eV), 2-3

elements, ground-state properties, 267-272

angular momenta and Hund's rules, 270-272

angular momenta (table), 271

electron configurations (table), 271

ionization energies (table), 271

elements, periodic table of, 268

Elitzer, A.C., 9

energy levels (*see* eigenvalues)

English, Damon, 90

entanglement, 223-232

Bell's inequality, 227-231

the EPR argument, 226-227

quantum cryptography, 224-226

equivalent electrons, 269

exchange degeneracy, 284-287

expectation values, 105-106

exponential integrals, 67

Fermi, Enrico, 263

Fermi energy, sea, 52-53, 266-267, 290-291

fermions, 288-291

Feynman, Richard, 7, 19, 48, 204, 348

fine-structure constant (α), 11-12

fine-structure, hydrogen

experiment, 248-250

theory, 250-254

finite square wells, 28-34, 45-46

fitting wavelengths in a well (WKB), 58-62

    for bouncing ball, 60-61

    for harmonic oscillator, 59-60

    for infinite square well, 59

Fourier, Joseph, 57

Fourier transforms, 135-141

    position-momentum, 135-137

    time-energy, 138-141

functions, even and odd, 45, 275-276, 284-286

Gähler, R., 5-6

Galilei, Galileo, 57

Greenstein, George, 231

guessing ground states (variational method), 62-67

    and helium, 279-280

    and hydrogen, 65-67

    and infinite square well, 63

    theory, 62-65

half wells, 45

Hamiltonian operator, 19, 21, 24

harmonic oscillator, one-dimensional, 111-129

    angular momentum, comparison with, 163

    as an approximation, 113

    classical, 112-113, 125

    commutation relations, 119-120

    and a continuous family, 128

    coupled, 129

    eigenfunctions, 118, 122

    eigenvalues, 117, 120-121

    in an electric field, 126

    ground state, 42-43

    Hermite polynomials, 118

    ladder operators, 119, 124

    matrix elements, 123-125

    Morse potential, 126-127

    operators as matrices, 123-124

    operator solution, 119-122

    perturbed, 126, 236

    series solution, 113-119

states as vectors, 123

harmonic oscillator, isotropic, 125, 171, 174-179

    degeneracies, 176

    energy eigenfunctions, table of 178

    energy eigenvalues, 176

    and hydrogen, 177-179

    series solution, 174-178

Heaviside step function, 47

Heisenberg uncertainty relations, 130-146

    and commutators, 131

    examples, 132, 134-135, 156-157

    general proof of, 133-134

    time-energy relation, 138-141

      and decay lifetimes, 138-141

      examples, 140

    uncertainties, definitions, 131

Heisenberg, Werner, 18, 130

helium, 276-287

    excited states, 283-287

      and exchange degeneracy, 284-287

    ground state, 276-280

    spectrum, 277, 284

Hermite polynomials, 118

Hilbert space, 99

hydrogen

    Bohr model, 11-14

    bound states, 11-14, 166-173

    classical instabilities, 9-11

    degeneracies, 170

    energy eigenfunctions, table of, 173, 246

    energy eigenvalues, 12-14, 170

    fine structure, 248-254

    ground state, 43-44

    hyperfine splitting, ground state, 205-208

      of deuterium, 219

      and 21-cm line, 208-209

      with Zeeman splitting, 219

    isotropic oscillator, comparison with, 177-179

    radial wave functions, 170-173

    radiation spectrum, 12-14

a related family, 181-182

    series solution, 167-173

    Stark splitting, 245-247

    Zeeman splitting, 257-258

hyperbolic-secant wells, 49-50

identical particles, 263-295

    and conduction electrons, 52-53, 264-267

    and exchange degeneracy, 284-287

    and fermions and bosons, 288-291

    and Pauli exclusion principle, 263-274

    and periodic table, 267-272

    and scattering, 348-350

infinite square wells

    one-dimensional, 26-28

    three-dimensional, 38-39

integrals, exponential, 67

interaction-free measurements, 9

interference

    of neutrons, 4-7

    of photons, 7-9

    of water waves, 14

ionization energy

    of the elements (table), 271

    of hydrogen, 12

isospin

    particle branching fractions, 216-218, 221-222

    particle multiplets, 216

Jammer, Max, 1, 3

Jeffries, H., 62

Kane, Joe, 14

Killingbeck, J.P., 128, 181

Kramers, Hendrik Anton, 62

Legendre polynomials, 340-341, 344-345

    and spherical harmonics, 340

light

    interference of, 6, 7-9

    polarization of, 199-201

    thin films and, 86-87

linear vector spaces, 94-98

Lipkin, Harry J., 146

Mach-Zehnder interferometer, 7-9

magnetic moments

    for spin 1/2, 189-190

    for atoms, 255-258

magnetic resonance, 192-196, 300-302

matrices

    for crossing boundaries, 77-78

    for crossing regions, 78-86

    for observables (as operators), 104-105

        angular momentum, 157-159

        harmonic oscillator, 123-125

        picket-fence recipe, 104

        spin, 184-185

Mermin, N. David, 231

momentum space, 135-137

mystery element Fb, 16

neutrons

    interference of, 4-7

operators

    for observables, 21, 100-105

        and commutators, 102-104

        as matrices, 104-105

        properties of, 100-102

orbital angular momentum, 148-151, 159-162

    differential operators for, 159-160

oscillator (*see* harmonic oscillator)

Pais, Abraham, 1, 2

Pedrotti, F.L., L.S., and L.M., 87

particle scattering (*see* scattering)

periodic table of the elements, 268

perturbation theory, time-dependent, 296-319

    hydrogen in an electromagnetic wave, 302-304

        averaging over waves, 305-307

    magnetic resonance, 300-302

    sudden changes, 297, 316

    theory, first-order, 298-300

perturbation theory, time-independent, 233-262

    degenerate theory, 241-258

        examples, 242-258

            fine structure in hydrogen, 247-254

hyperfine splitting in hydrogen, 205-208, 242

and 21-cm line, 208-209

Stark effect in hydrogen, 245-247

two-fold, 243

Zeeman splitting in hydrogen, 219, 257-258

matrix recipe, 241-243

nondegenerate theory, 234-241

derivations, 239-241

examples, 235-239

proton finite size, 237

and relativity, 238-239

Stark effect and oscillator, 236

recipes, 234-235

photoelectric effect, 3-4

photons

and Compton scattering, 4

interference of, 6, 7-9

picket-fence recipe for matrices, 104

Planck, Max, 2-4

Planck constant $(h, \hbar)$ 2-3

Planck oscillators, 2, 308-310

plane waves, 19-20, 71-73

densities and currents of, 72-73

exponential form, 19

scattering of (see scattering)

Podolsky, B., 227

quantum cryptography, 224-226

quantum mechanics, 1-364

provisional rules, 23-25

revised rules, 108

quantum seeing in the dark, 9

Rayleigh, Lord, 57

recurrence relations, 243-244, 260

reduced mass, 14, 15-16

relativity

Dirac's relativistic theory, 356-364

first-order corrections, 238-239

and hydrogen fine structure, 252-253

rescaling of radial equations

for hydrogen, 167

for the isotropic oscillator, 174

resonance

Breit-Wigner amplitude, 141

and exponential decay, 138-141

and scattering, 343, 347

magnetic resonance, 192-196

Rosen, N., 227

Rumsfeld, Donald Henry, 223

Rutherford, Ernest, 1

and nuclear atom, 9-11, 328

Rutherford scattering, 325-328, 336, 350-351

scalar products, 94-98

Dirac bra(c)ket notation for, 95

of functions, 97

of vectors, 94-96

scaling and dimensions, 55-57

scattering of particles in one dimension, 70-92

conservation of particles, 72-73

by a delta-function barrier, 91

by two delta-function barriers, 85-86

densities and currents, 72-73

designing with barriers, 82-86

field emission of electrons, 89-90

by a rectangular barrier, 76-82

general, 76-79

simple, 80-82

rectangular barriers in series, 82-86

by a step, 74-76

thin films and light, 86-87

weak tunneling, 87-90

scattering of particles in three dimensions, 320-355

amplitudes, 328-336, 341-350

angular distributions, 324, 327, 336, 344-345, 355

BB-bowling ball scattering, 322-325

Born approximation, 334-336, 352

classical scattering, 322-328

differential cross sections, 324, 327, 336, 344-345, 355

Green functions, 330-332, 352

identical-particle scattering, 348-350, 355

kinematics of, 337-339, 353

376

optical theorem, 346-347

partial waves, 339-347, 353-355

resonant scattering, 343, 347

Rutherford scattering

    classical, 325-328

      and nuclear atom, 38

      quantum, 336

    solid angle, 321-322. 350

    spherical waves, 328-330, 341-344

    total cross section, 323-324, 325, 346-347

    and Yukawa potential energy, 335-336

Schrödinger, Erwin, 18, 21

Schrödinger equations

    time-dependent, 20-21

    time-independent, 23-24

Schwarz inequality, 132-133

separation of variables

    for central potential energies, 148-150

    for rectangular coordinates, 37-38

    time from space, 23-24

solid angle, 321-322. 350

spectrum diagrams

    for hydrogen, 13, 171, 208, 249, 258

spherical coordinates, 39-41

spherical harmonics

    and central forces, 161-162

    and Legendre polynomials, 337

    parities of, 164

    a recurrence relation, 243-244

    selection rules and, 243-244

    table of, 162

spin 1/2, 183-203, 360-362

    eigenstates, 185-187

    Larmor precession, 191-192

    and magnetic moments, 189-190

    magnetic resonance, 192-196

    Pauli spin matrices, 158, 185

    polarization vector, 187-188

    and relativity (Dirac), 360-362

    spinors, 184

Stern-Gerlach experiments, 196-198, 203

    uncertainty relations, 156-157

    Zeeman splitting, 190

Stark effect

    and harmonic oscillator, 126, 236

    and hydrogen, 244-247

superposition

    of energy eigenstates, 25

    and exchange degeneracy, 284-286

    and magnetic resonance, 192-196

    and spatial oscillations, 45

symmetries

    of energy eigenstates, 28, 30-31, 44

    of identical particles, 288-291

      in scattering, 348-350, 355

    of spherical harmonics, 164

    of two electrons, 275-276

time dependence

    of decay rates and lifetimes, 138-141, 314-315

      and a pseudo-uncertainty relation, 138-139

    of energy eigenstates, 24-25

    of expectation values, 106-107

21-cm line, 208-209

uncertainty relations, 130-146

    and angular momentum, 156-157

    general derivation, 133-134

    and harmonic oscillator, 126

    position-momentum, 132

    time-energy, 138-140

    uncertainties defined, 131

Vaidman, L., 9

variational method for ground states, 62-67

    and helium, 279-280

    and hydrogen, 65-67

    and infinite-square-well, 63

    theory, 63-65

vectors, states as, 98-99, 123, 157-158

vector space, linear, 94-98

    Hilbert space, 99

virial theorem, 107-108

and helium, 294

wave equations

    classical, 18-19

    quantum (Schrödinger)

        time-dependent, 20-21

        time-independent, 23-24

wave function, 21-23

    normalization, 23

    probabilistic interpretation, 21-23

    qualitative properties, 44

Wentzel, Gregor, 62

Wigner, Eugene P., 228, 231

WKB (Wentzel-Kramers-Brillouin) method, 58-62

    and bouncing ball, 60-61

    and harmonic oscillator, 59-60

    and infinite square well, 59

Young, Thomas, 7

Zajonc, Arthur G., 231

Zeeman splitting

    and hydrogen, 257-258

    and hyperfine splitting, 219

    and spin 1/2, 189-190

Zeilinger, A., 5-6

9781032256023